Reviews for Other Books by Yoshihito Isogawa

"Yoshihito's books are among the most useful I own."

—BRICKSET

THE LEGO MINDSTORMS EV3 IDEA BOOK:

"A fresh approach to Lego Mindstorms building. . . . I could go on and on about all the inspiring designs, but I want to leave something for you to check out on your own."

—GEEKMOM

THE LEGO TECHNIC SERIES:

"These are an invaluable set of books to have as a reference to build mechanisms."

—JOE MENO, *BrickJournal*

"I can emphatically state that no self-respecting LEGO fan should exclude this series from their library."

—BRICKS IN MY POCKET

"These are excellent books showing a lot of great ideas for LEGO mechanisms. Even if you're an experienced builder, there are surely some ideas in here you've never seen."

—BILL WARD, Brickpile

"For anyone who loves LEGO, prototypes in LEGO, or loves mechanical assemblies, these books are definitely required viewing, and we're not sure how we lived without them for so long."

—LENORE EDMAN, Evil Mad Scientist Laboratories

THE LEGO® MINDSTORMS®
Robot Inventor
I D E A   B O O K

THE LEGO® MINDSTORMS®
Robot Inventor
IDEA BOOK

128 Simple Machines and Clever Contraptions

YOSHIHITO ISOGAWA

**no starch
press**

San Francisco

First printing

25 24 23 22 21   1 2 3 4 5 6 7 8 9

ISBN-13: 978-1-7185-0177-5 (print)
ISBN-13: 978-1-7185-0178-2 (ebook)

Publisher: William Pollock
Production Manager: Rachel Monaghan
Production Editor: Paula Williamson
Developmental Editor: Nathan Heidelberger
Cover Design: Monica Kamsvaag
Photographer: Yoshihito Isogawa
Author Photo: Sumiko Hirano
Technical Reviewer: Sumiko Hirano
Compositor: Maureen Forys, Happenstance Type-O-Rama
Proofreader: Lisa Devoto Farrell

For information on book distributors or translations, please contact No Starch Press, Inc. directly:
No Starch Press, Inc.
245 8th Street, San Francisco, CA 94103
phone: 1.415.863.9900; info@nostarch.com
www.nostarch.com

Library of Congress Control Number: 2021941878

Contents

PART 1 Basic Mechanisms

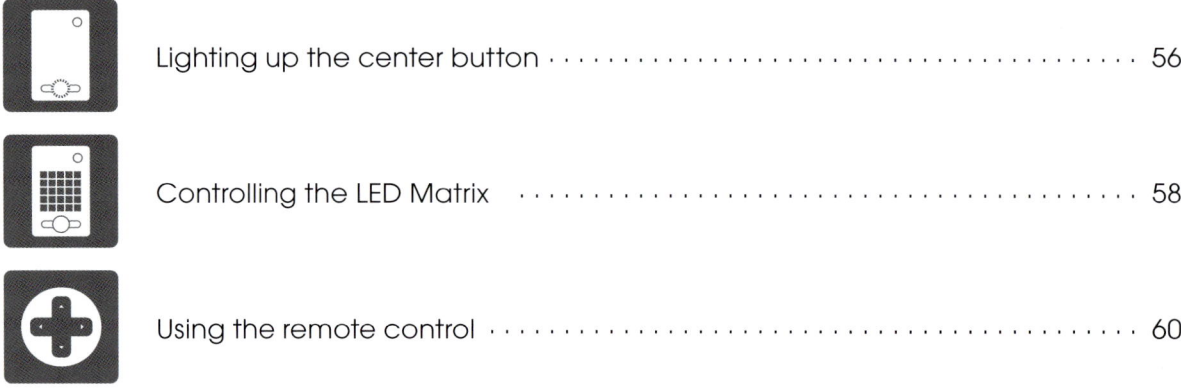
PART 2   Moving Mechanisms

PART 3 Practical Mechanisms

PART 4 Using Sensors

PART 5   Other Enjoyable Mechanisms

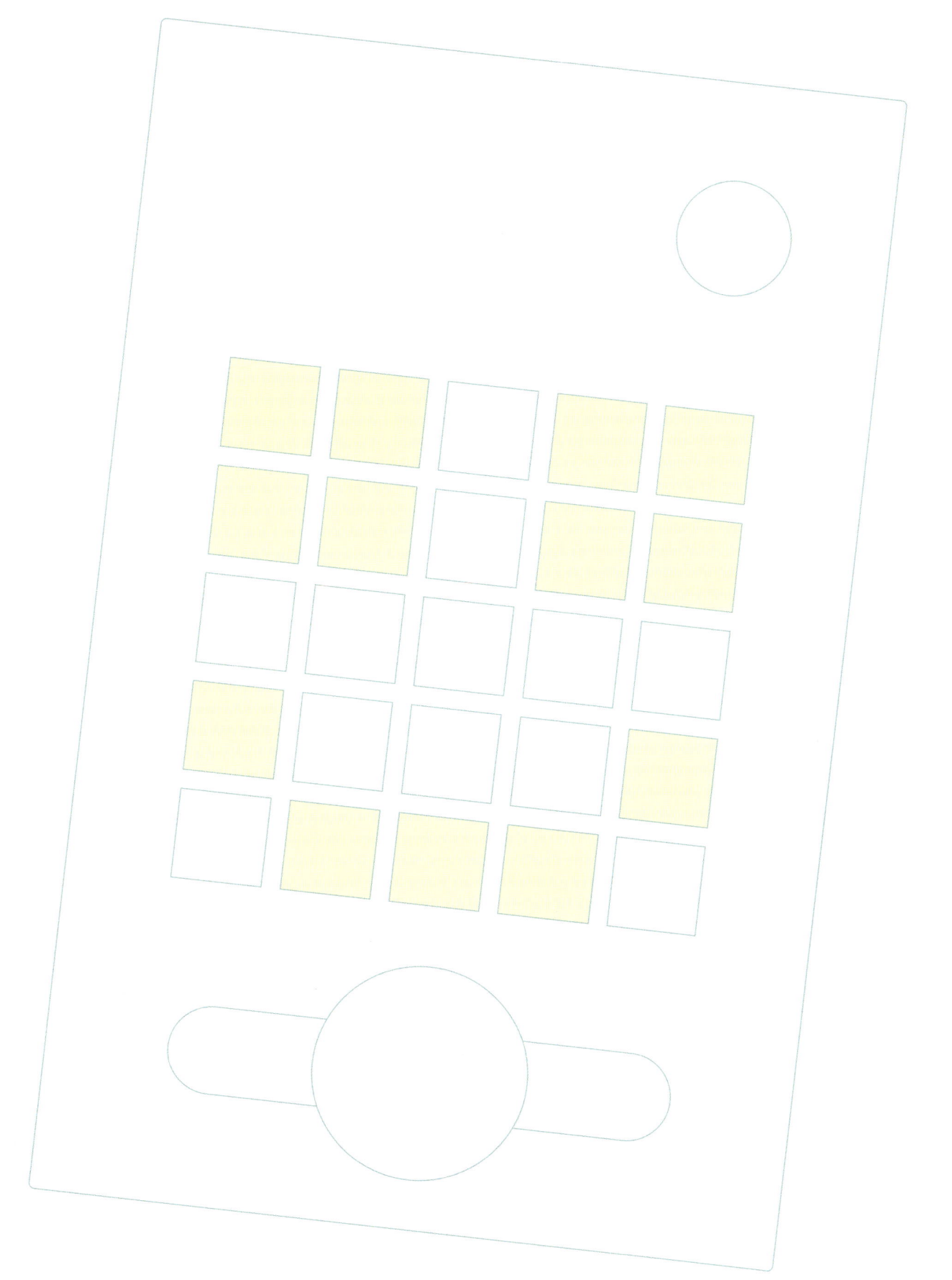

Introduction

With the LEGO MINDSTORMS Robot Inventor set and its app, you can enjoy building great robots and learn a lot. If you've already tried the models that came with it, you might be wondering how to have even more fun with the LEGO MINDSTORMS Robot Inventor set. This idea book will help you do that. The experience you get through this book will greatly enhance your imagination and creativity.

To build the models in this book, all you need is the LEGO MINDSTORMS Robot Inventor set (#51515) and a device that can run the LEGO MINDSTORMS app.

How to Use This Book

This book doesn't include step-by-step building instructions. Instead, you'll find photographs of each model taken from various angles, plus a list of the parts you'll need for the model. Look at the photographs closely and try to reproduce the models. Building in this way is like putting together a puzzle. If you aren't familiar with this method of building, start with the warm-up on the next page to get some practice.

You don't have to build this book's models in order. Flip through the pages and try making the ones you find most interesting. The book contains a variety of models, with basic models in the first half and more complex models in the second half.

The models in this book are designed to be as simple as possible so you can easily build them and understand how they work. But they're only a starting point. I invite you to use your own ideas to make the models better and cooler. When you find a model a little fragile, strengthen its weak points. If you see a different way to use a mechanism, try it out. These experiences will help you grow as a builder.

As you build more models, keep modifying them, and try combining different mechanisms into more sophisticated robots. If you already have other LEGO sets, it will be fun to combine them with the Robot Inventor set, too. The final product will be your own original model, the only one of its kind in the world. As an author, I hope that each of you will show your originality and create unique models, and that they will entertain people throughout the world.

Points to Note

- The motors in the LEGO MINDSTORMS Robot Inventor set are very powerful. Be careful not to touch the rotating gears, or you might hurt your fingers.

- Some models have parts that spin at high speeds. Be very careful not to let these parts hit your eyes.

- There are some models for drawing with a marker in the book. If possible, use a water-based marker. An oil-based marker can stain the table or floor.

- The programs in the book were created using the latest version of the app available at the time of writing. Your screen may look different depending on the version of the app (or the model of the tablet or smartphone you use).

Recommended Book

For a beginner's guide to the LEGO MINDSTORMS Robot Inventor set, check out *The LEGO MINDSTORMS Robot Inventor Activity Book* by Daniele Benedettelli (No Starch Press, 2021).

Acknowledgments

LDraw data and the LPub application were used to create the illustrations in this book. I would like to thank those involved in the development of those programs.

Warm-Up

You won't find step-by-step building instructions in this book. Instead, you'll use photographs taken from various angles to try to reproduce the model shown. Building in this way is like putting together a puzzle. You'll soon get the hang of this process and learn to enjoy it! Let's practice first.

 # 1

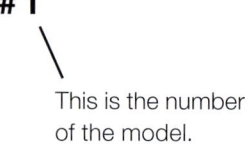

This is the number of the model.

All the parts you need for this model are shown below. Find them in your MINDSTORMS Robot Inventor set and start building!

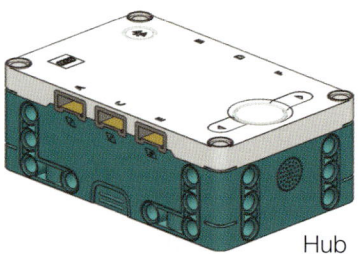

Hub

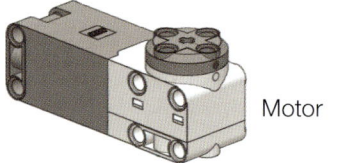

Motor

5

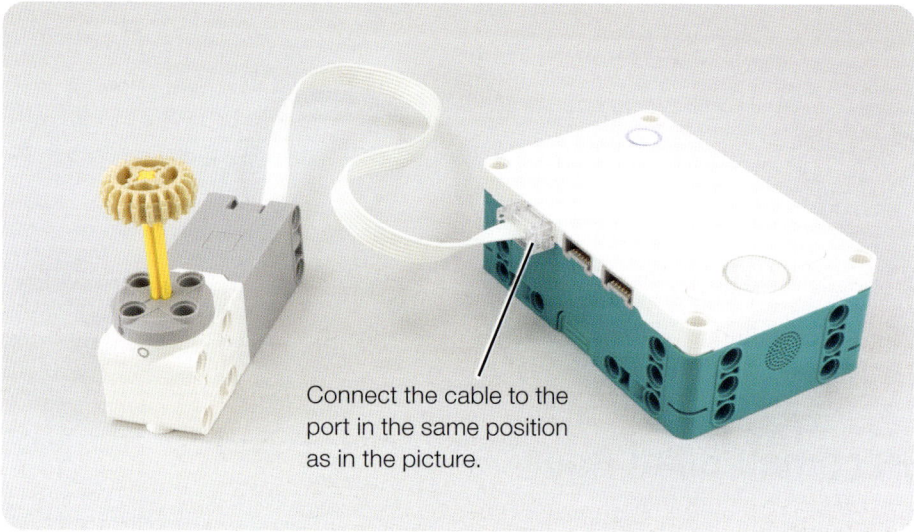

Connect the cable to the port in the same position as in the picture.

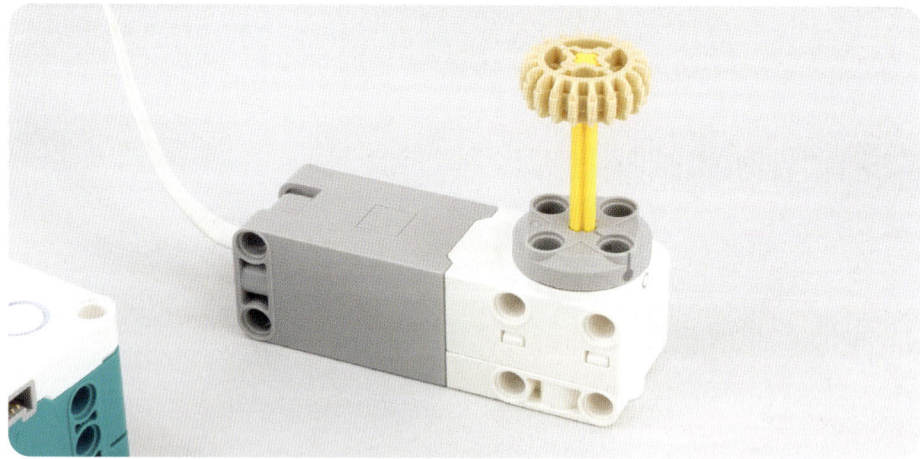

After gathering your parts, try building the model using the photos on this page and the next. To work faster, put your model in the same position as the one in the photos, and keep comparing them as you build.

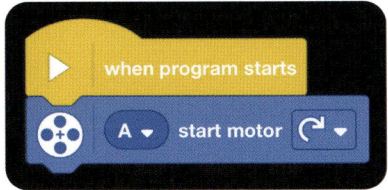

This is a program to move the model. When you create a new program, follow the steps below.

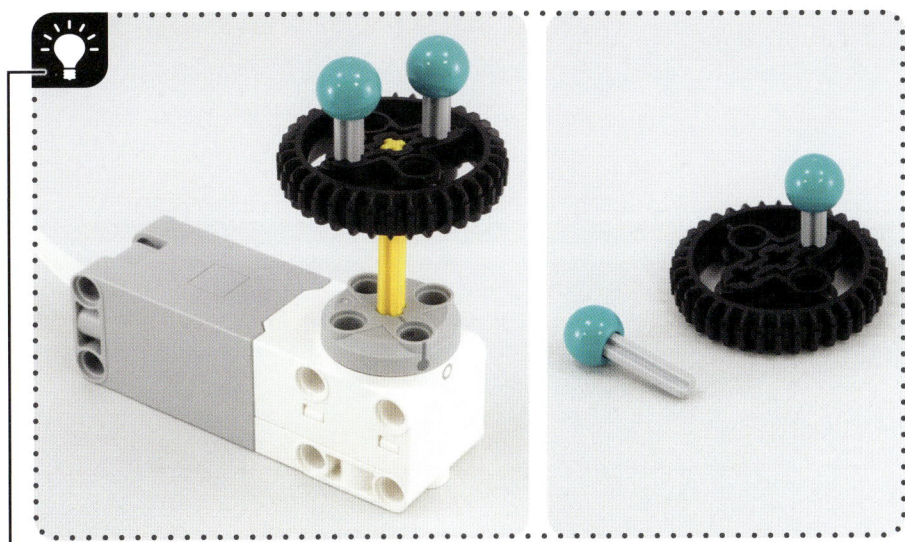

This is the hint icon, which suggests other ways of building and programming. Try to create your own unique and fun models using these tips. Please note that the parts used in the hint are not included in the parts list for each project.

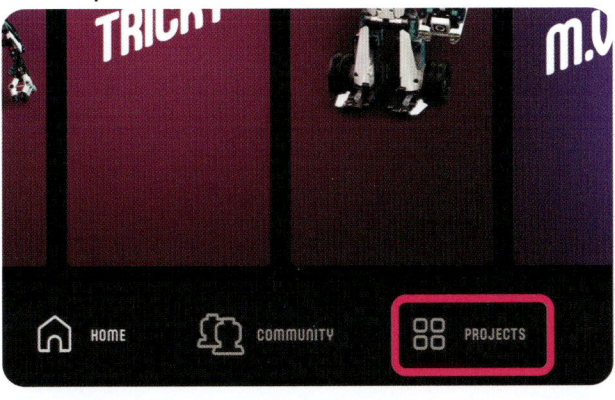

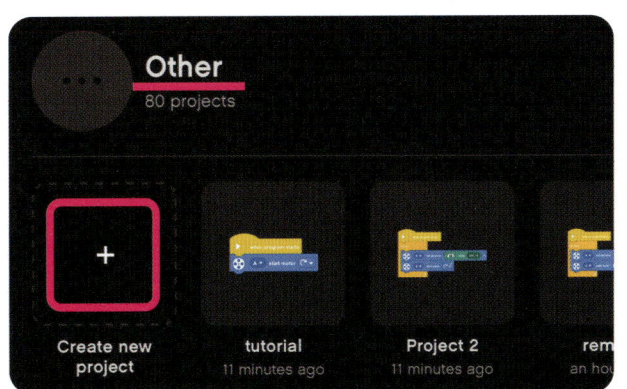

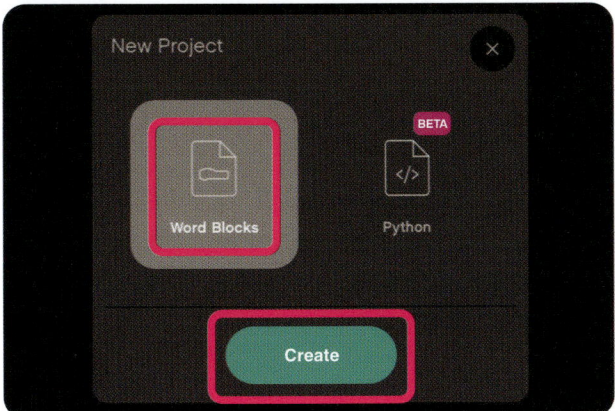

Please note that the screen may change depending on the application version.

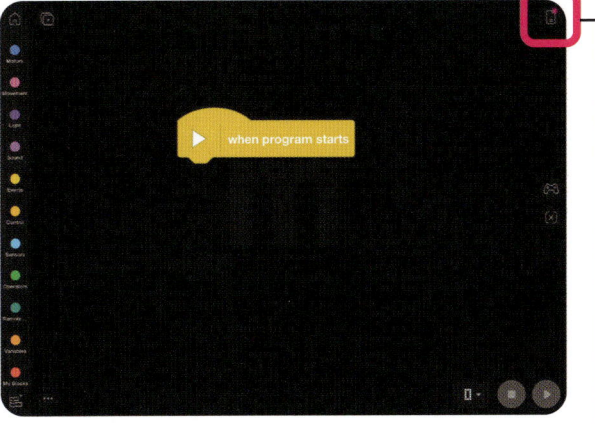

If the Hub and your tablet/smartphone are not connected, make the connection from here.

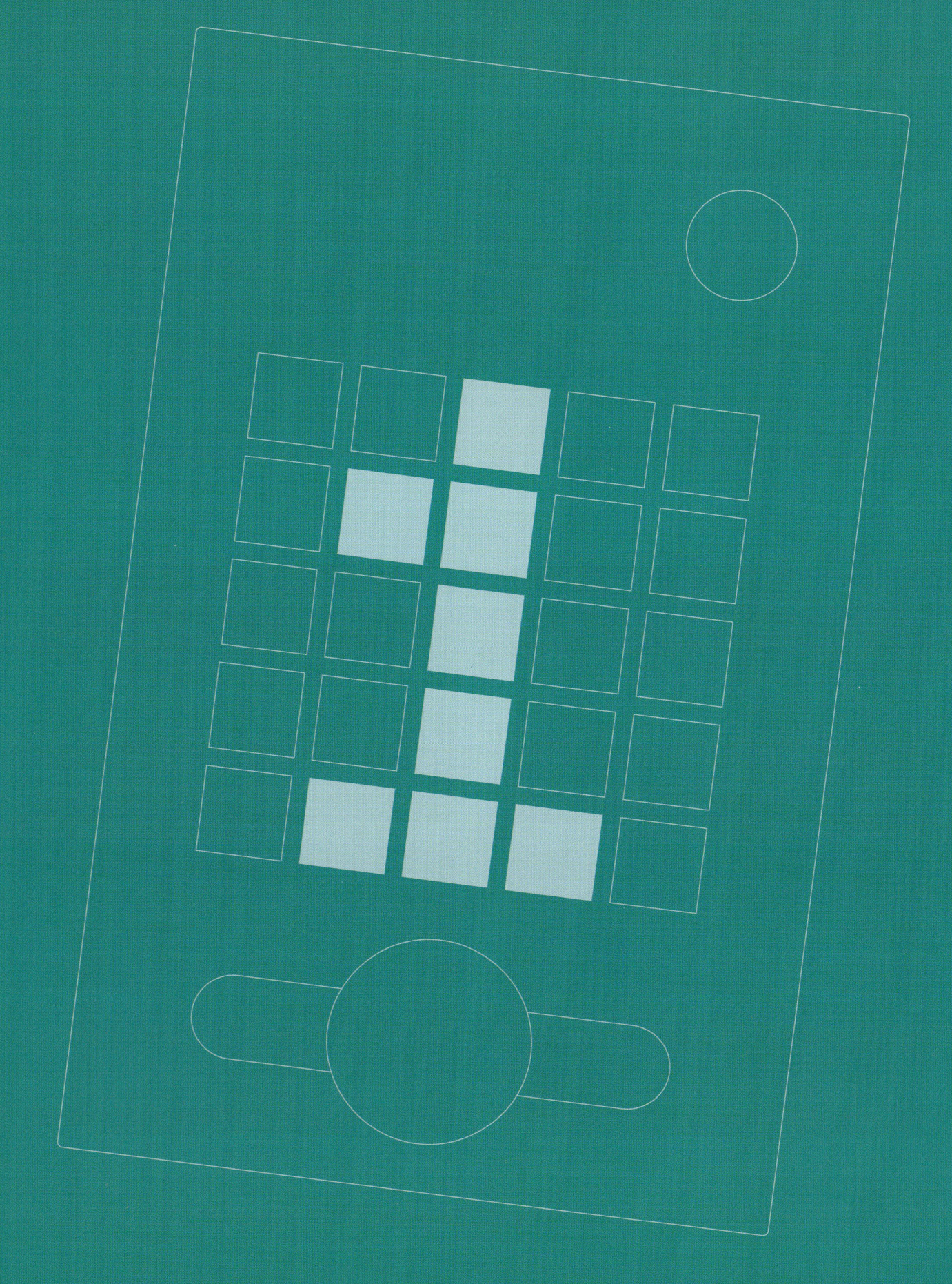

PART 1
Basic
Mechanisms

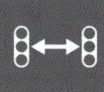

Rotating motors

1

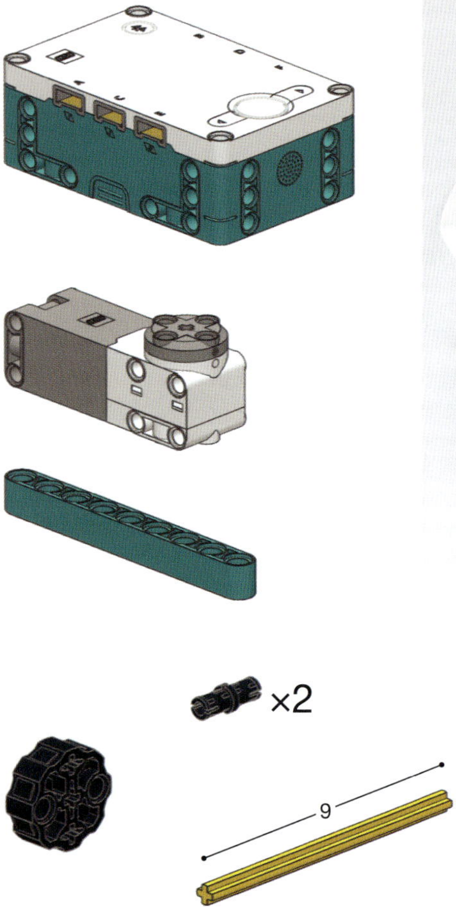

×2

9

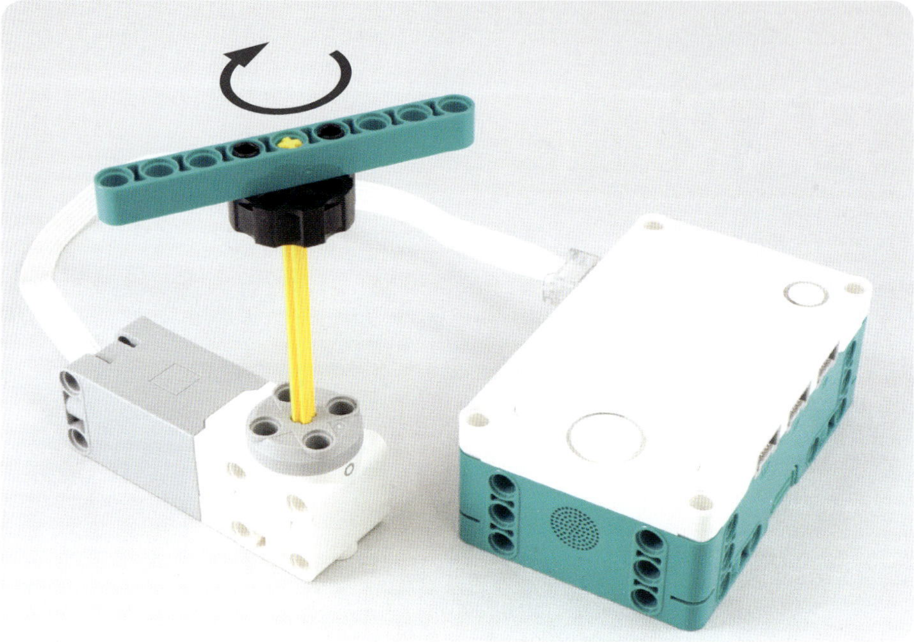

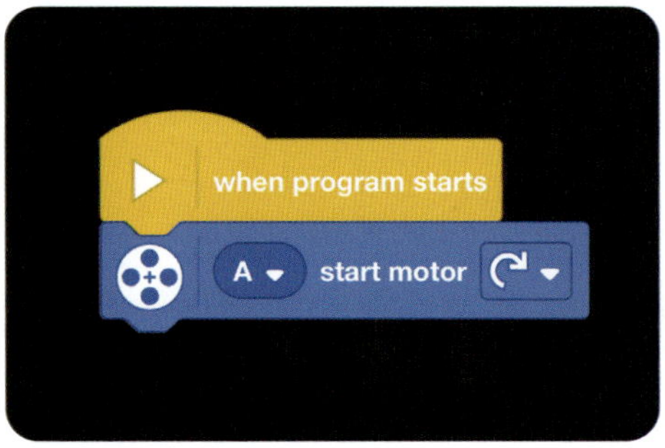

```
when program starts

A ▾  start motor  ↻ ▾
```

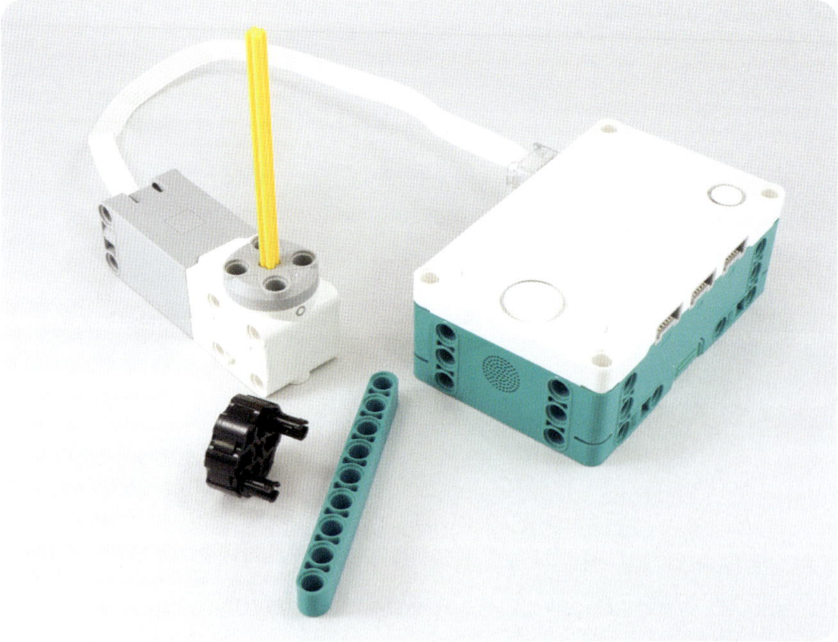

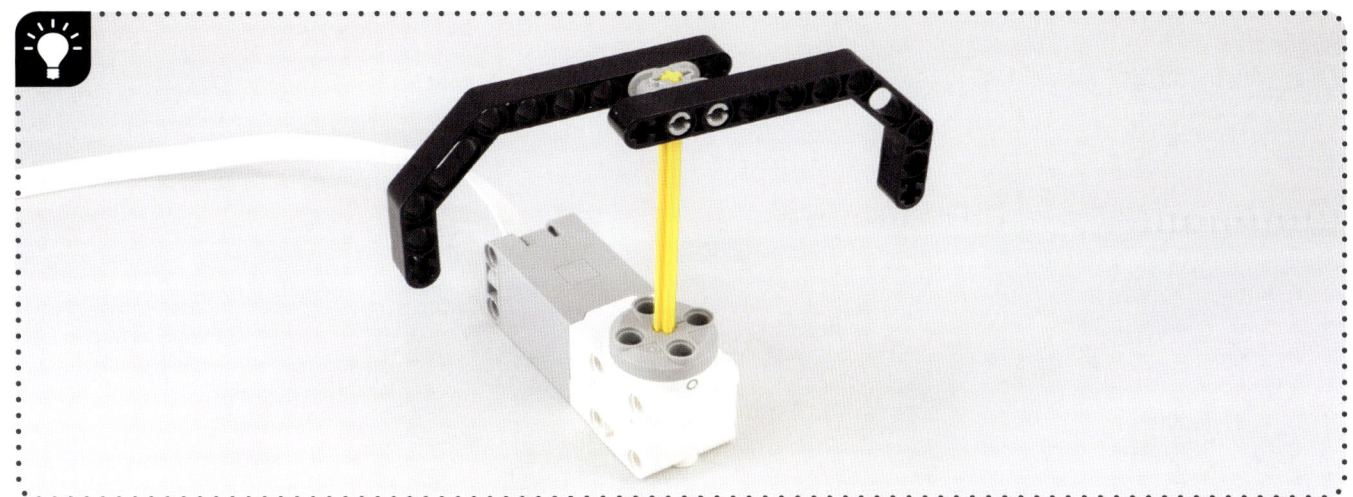

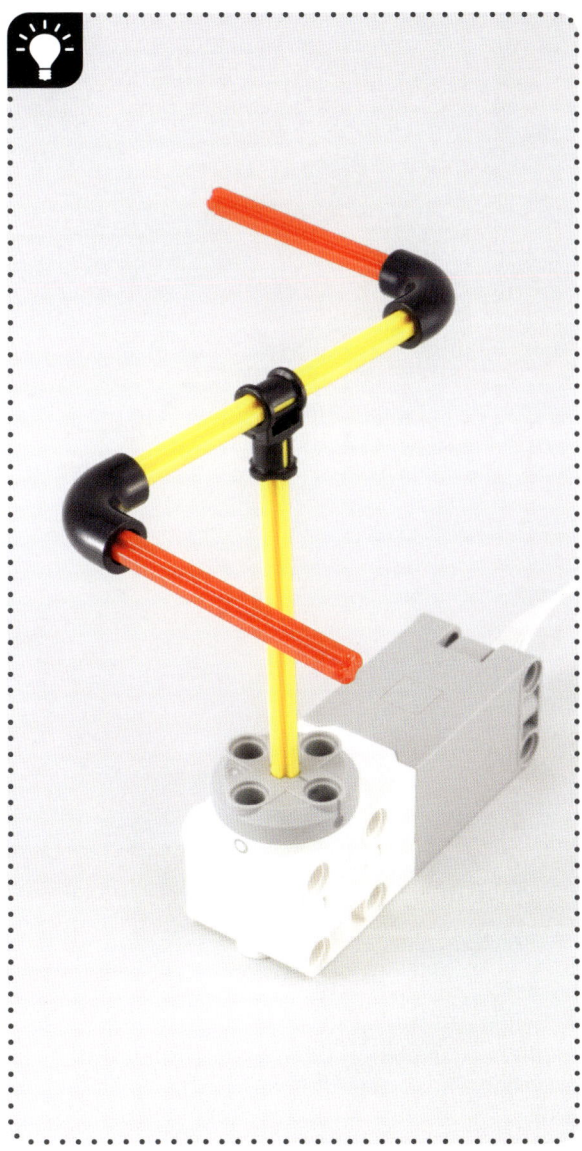

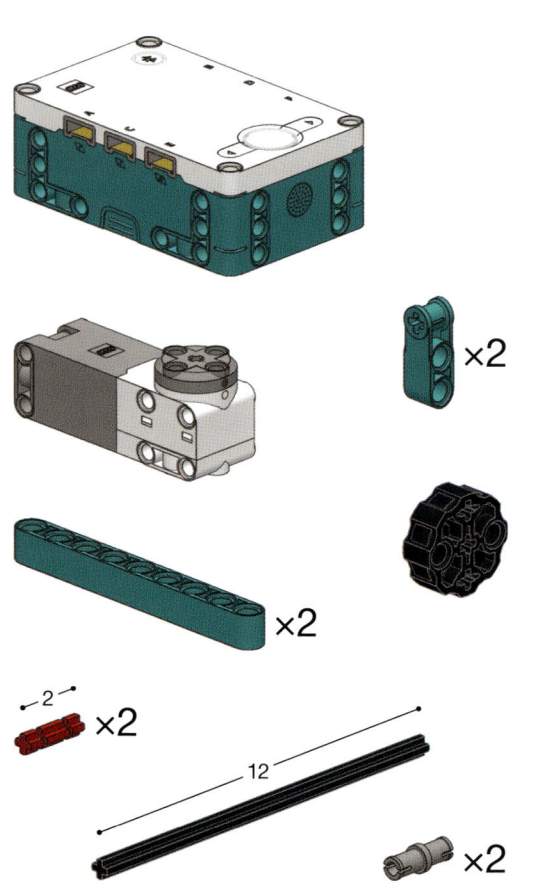

×2

×2

←2→ ×2

12

×2

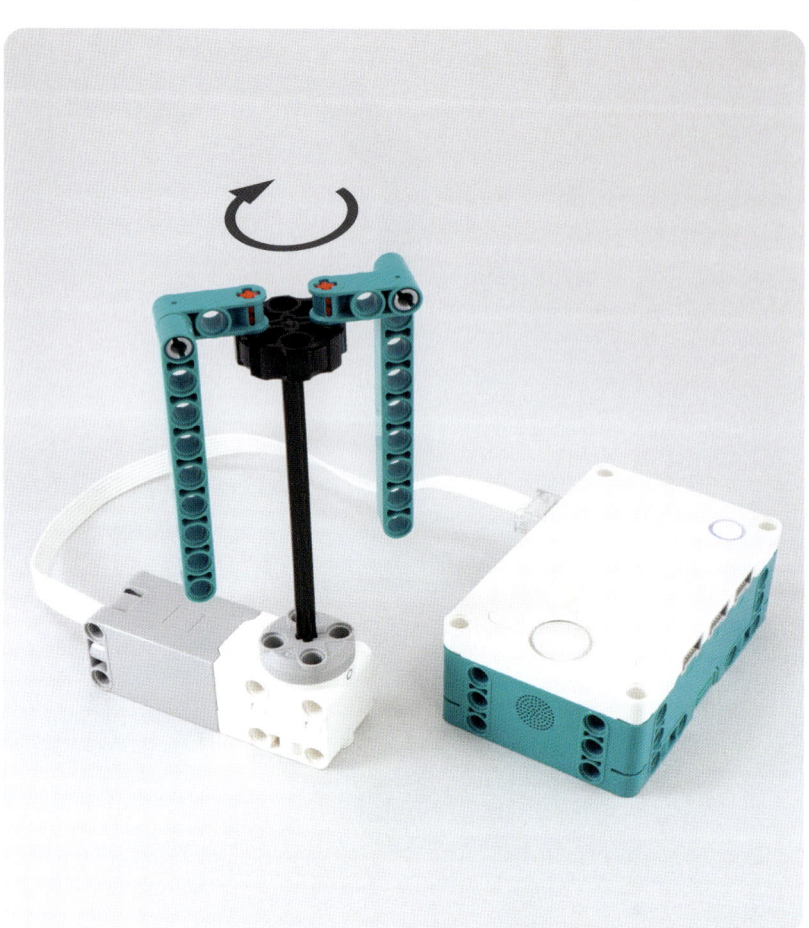

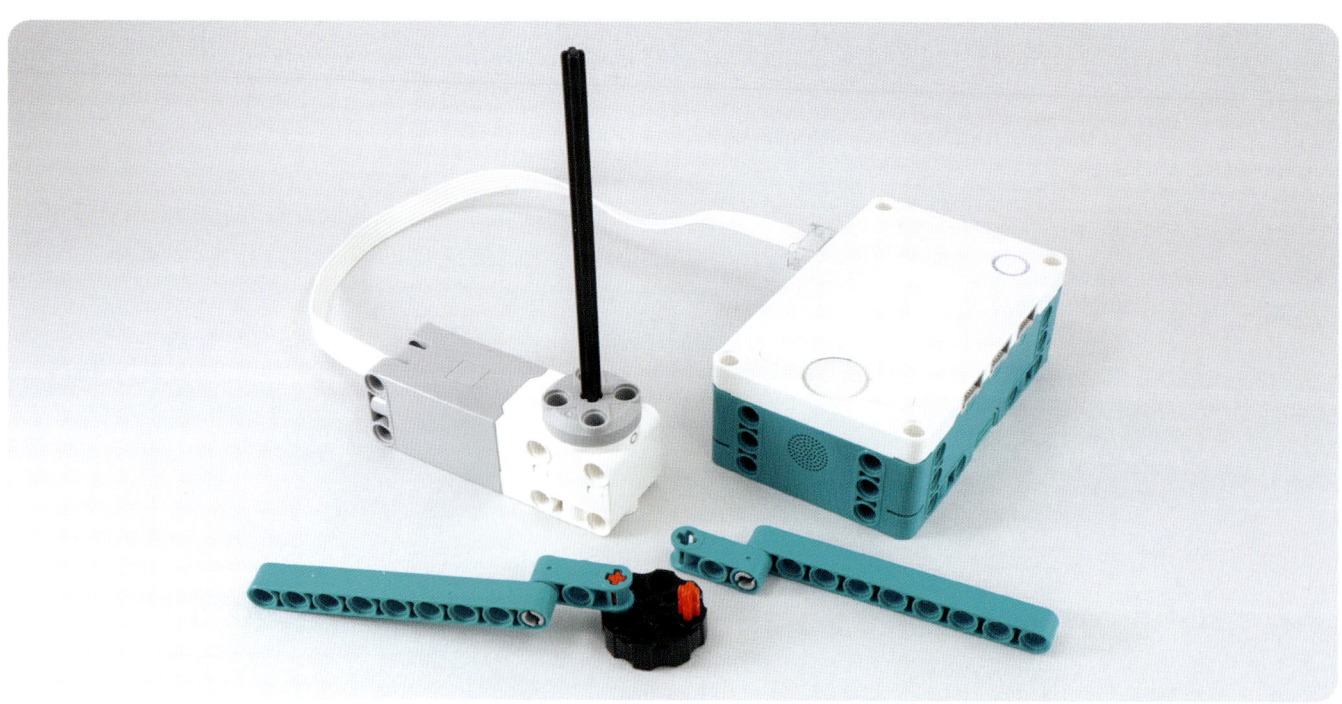

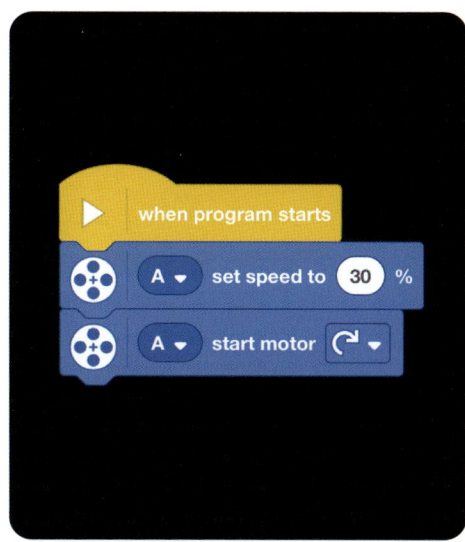

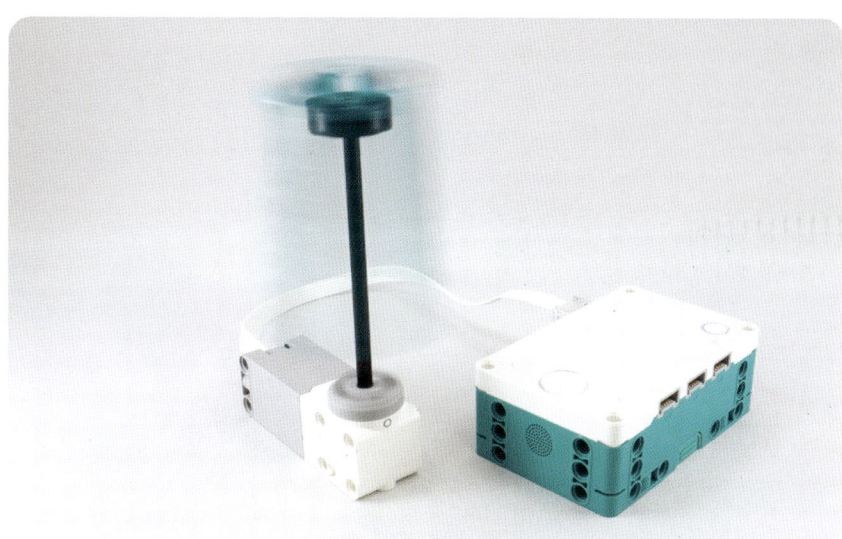

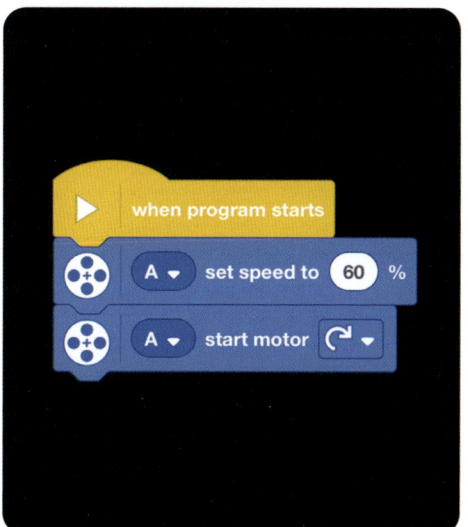

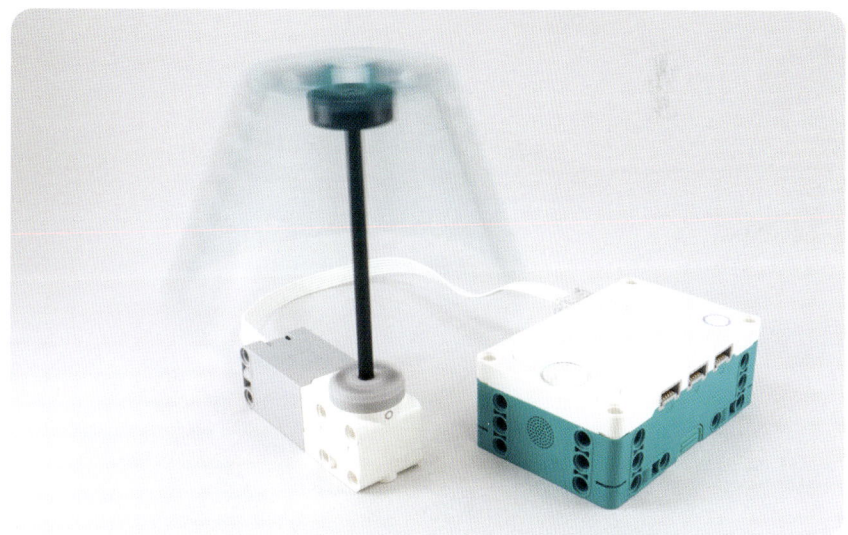

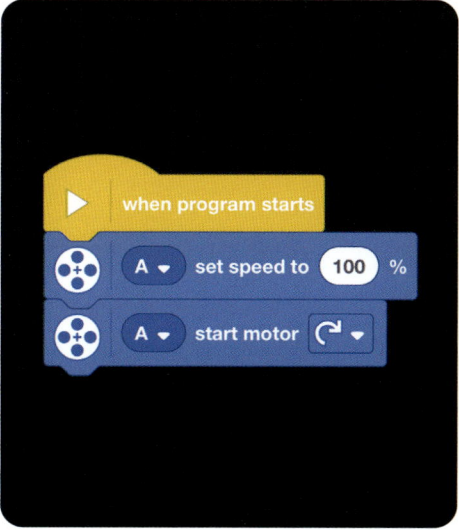

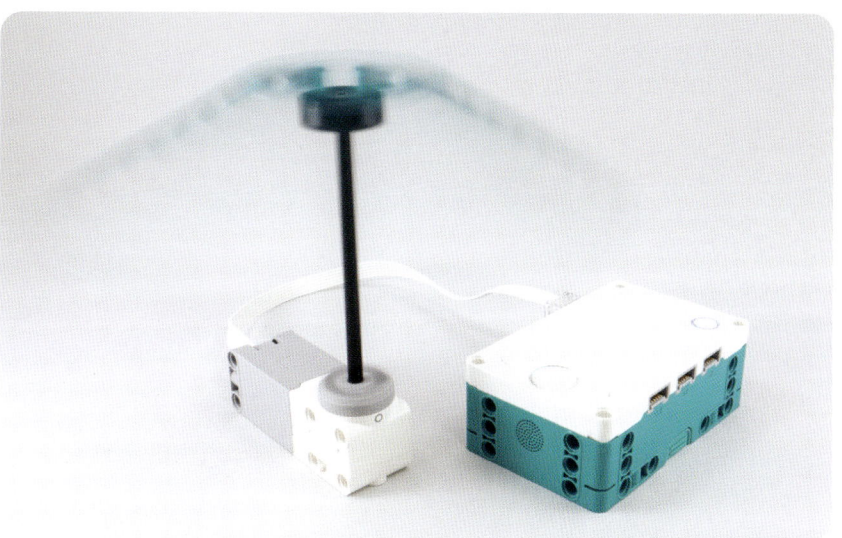

#3

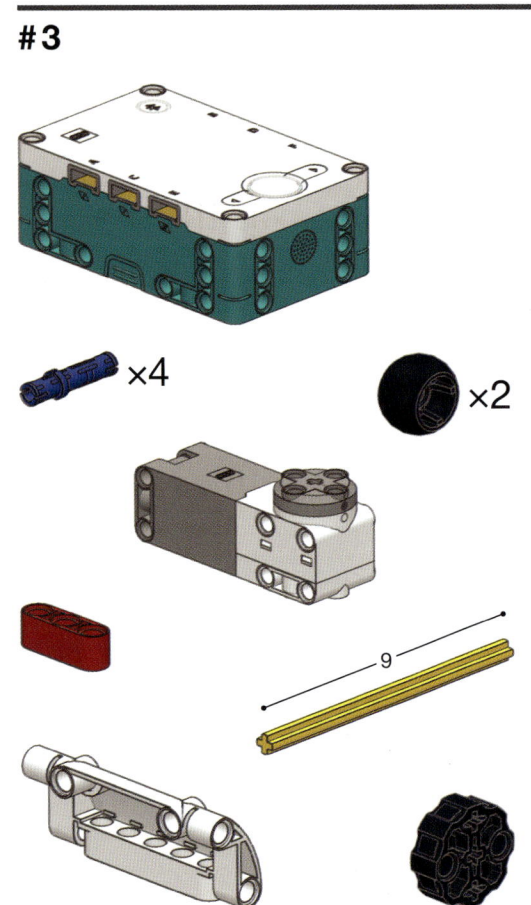

×4 ×2

9

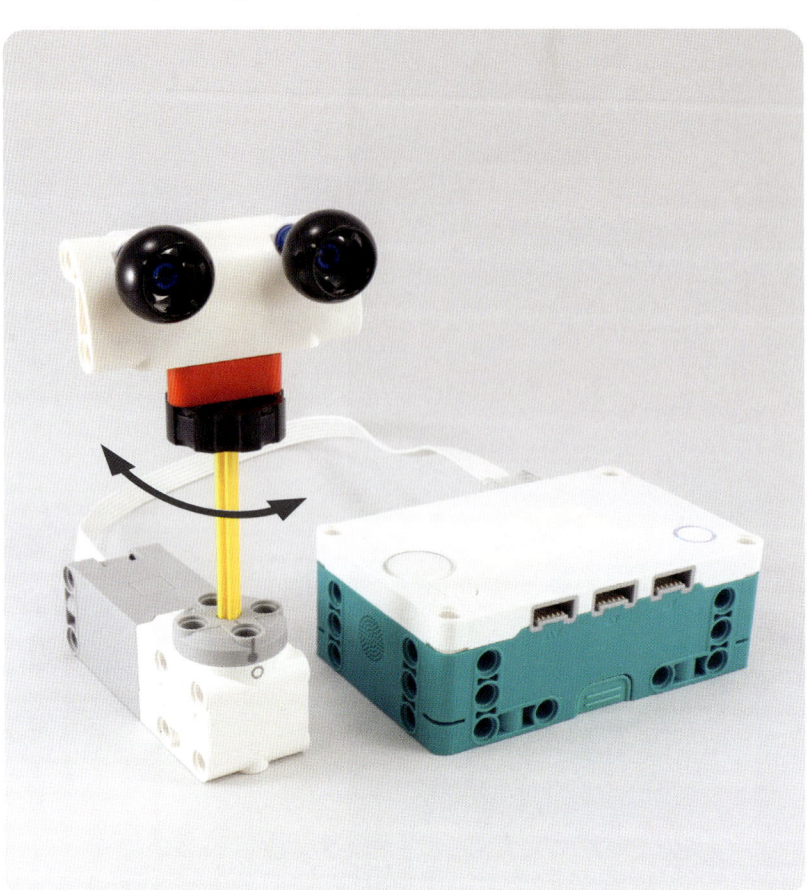

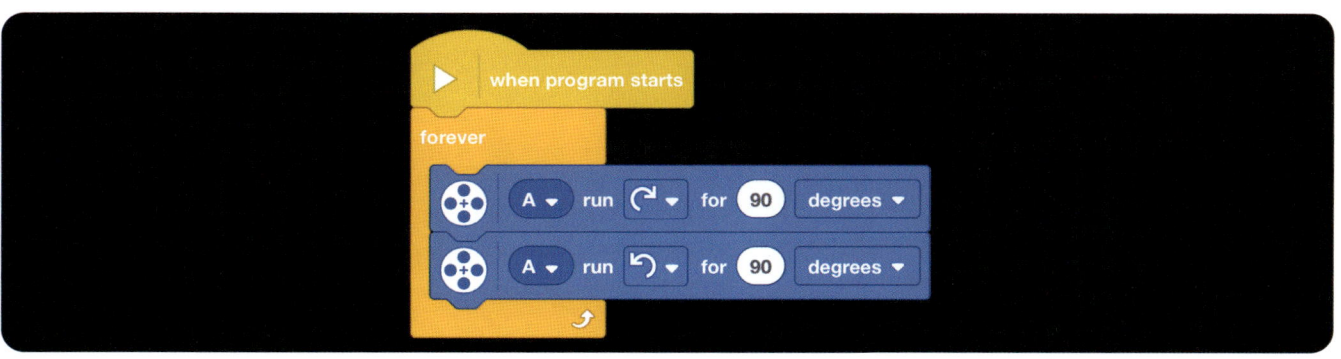

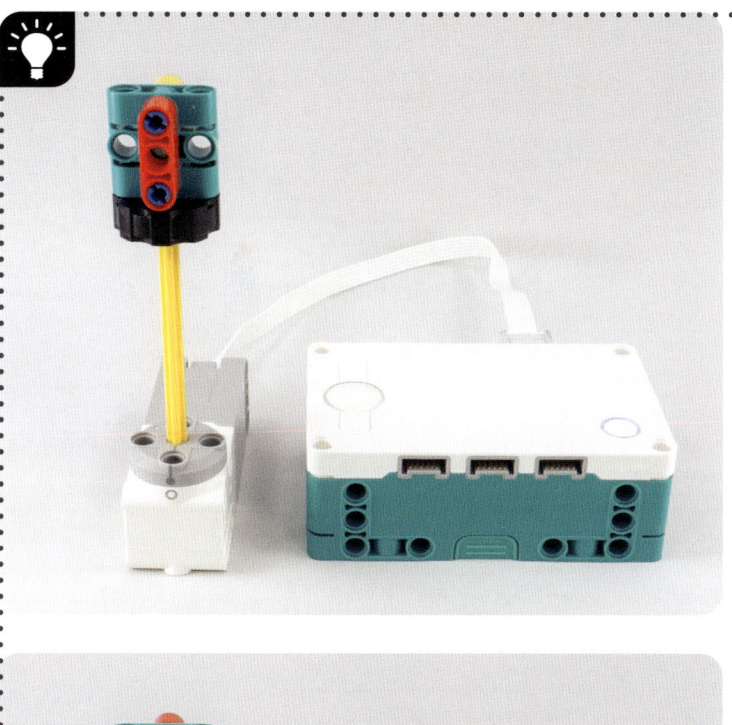

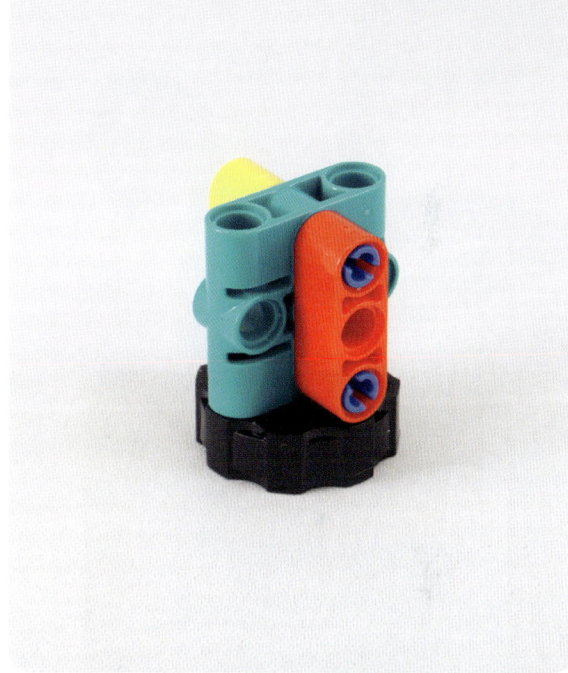

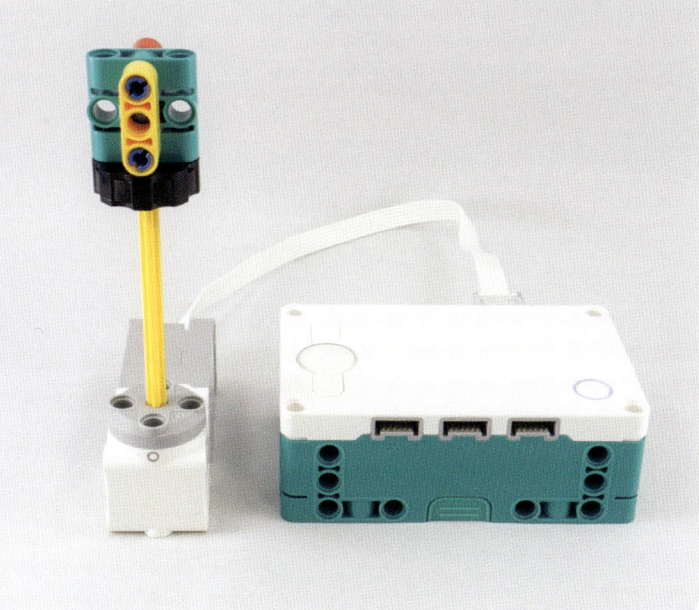

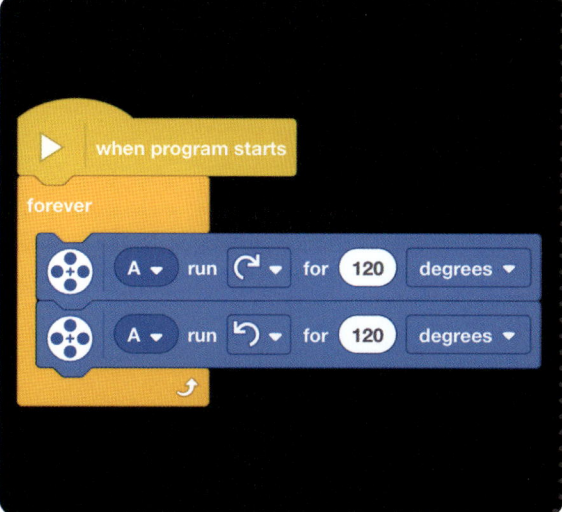

#4

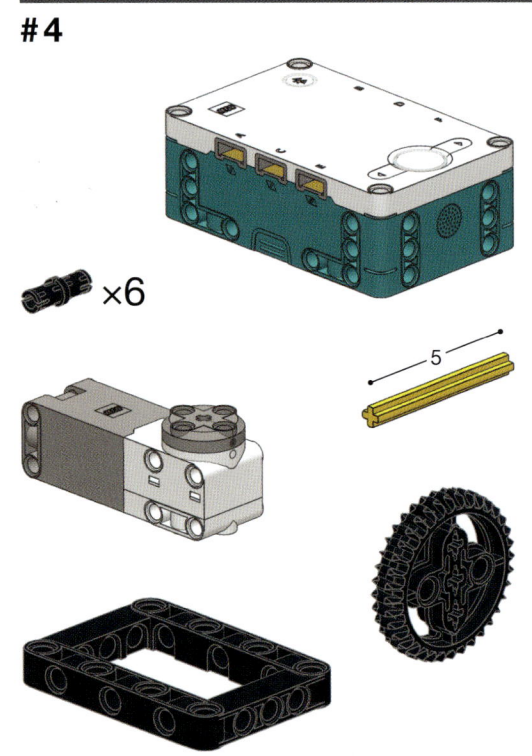

×6

5

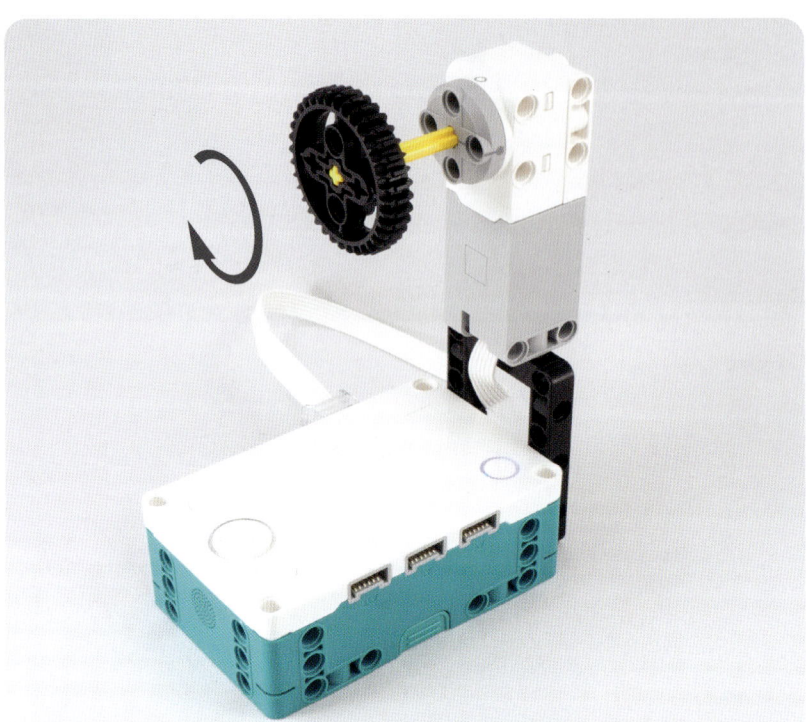

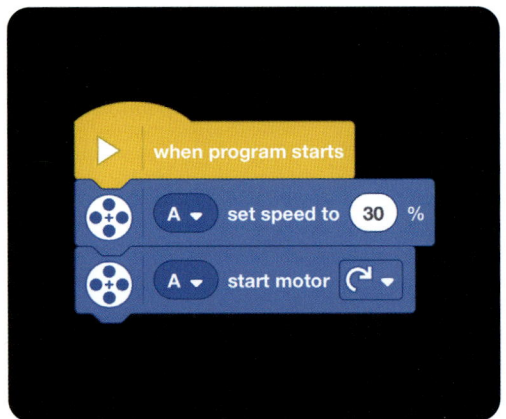

when program starts

A ▾ set speed to 30 %

A ▾ start motor ↻ ▾

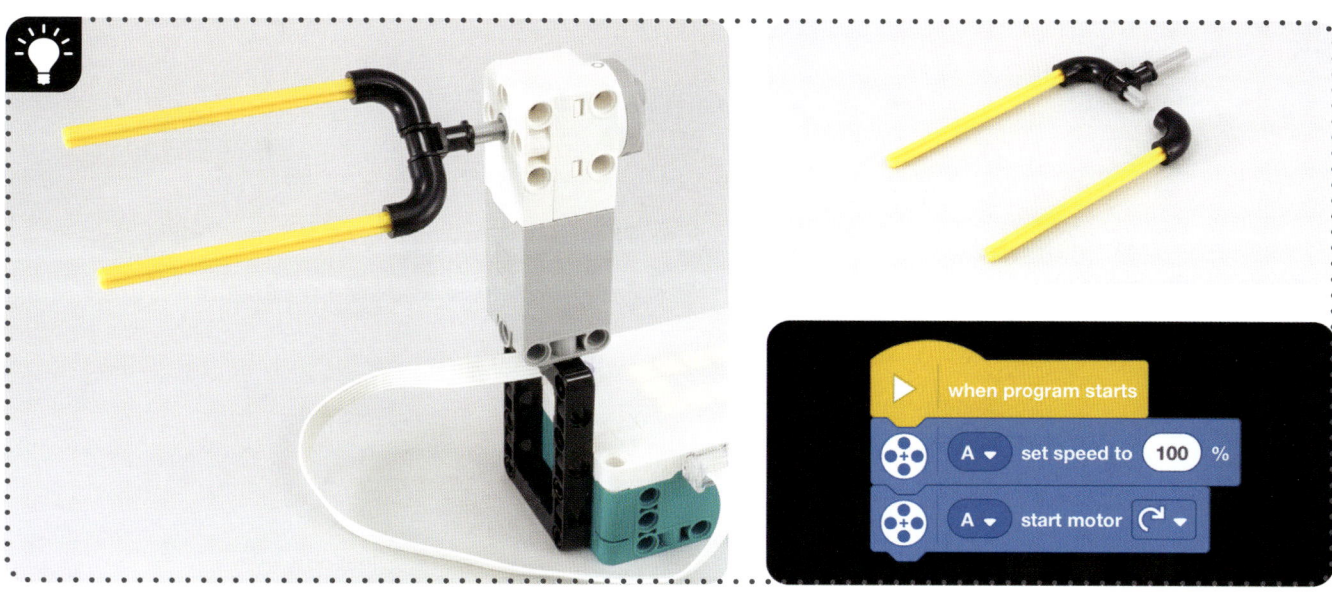

```
when program starts
  A ▾   set speed to   100  %
  A ▾   start motor  ↻ ▾
```

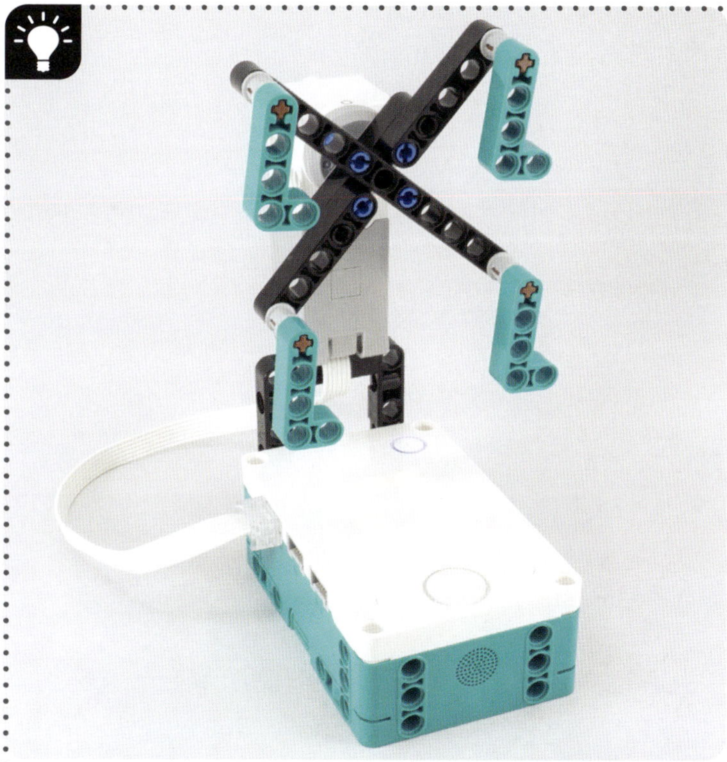

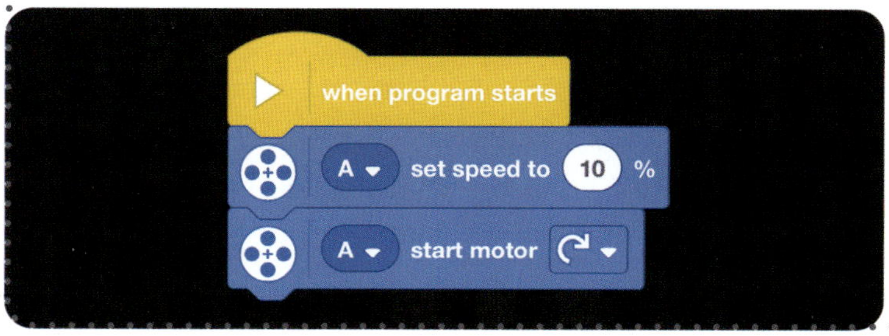

```
when program starts
  A ▾   set speed to   10  %
  A ▾   start motor  ↻ ▾
```

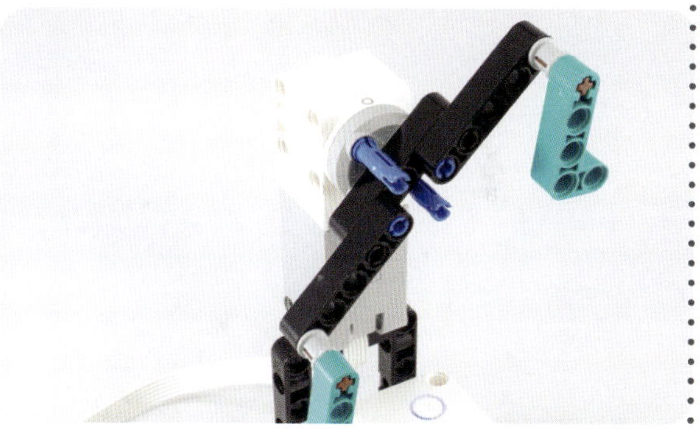

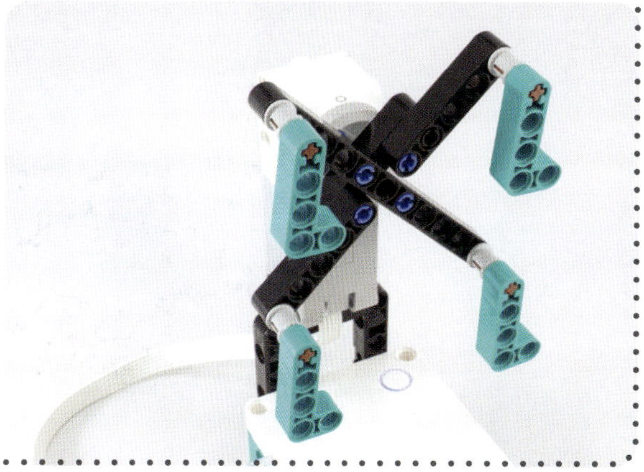

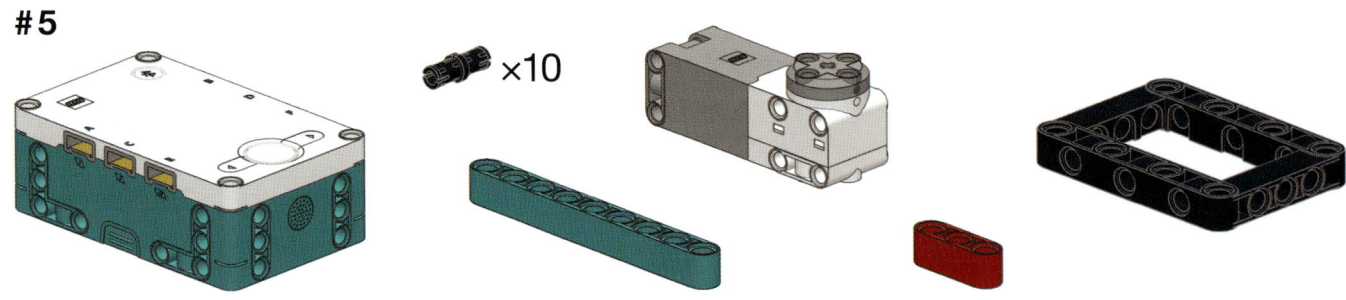

×10

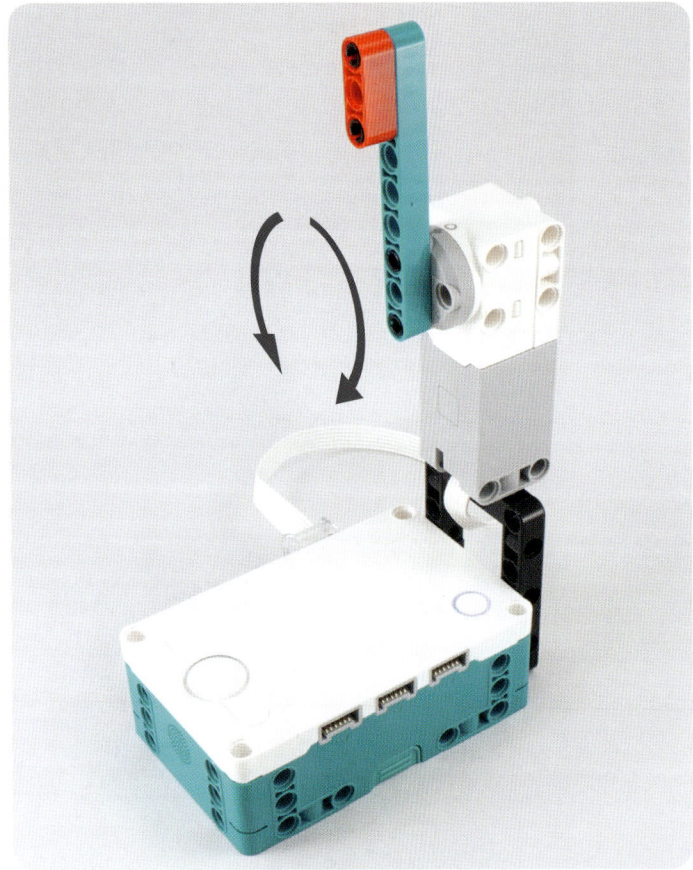

0°

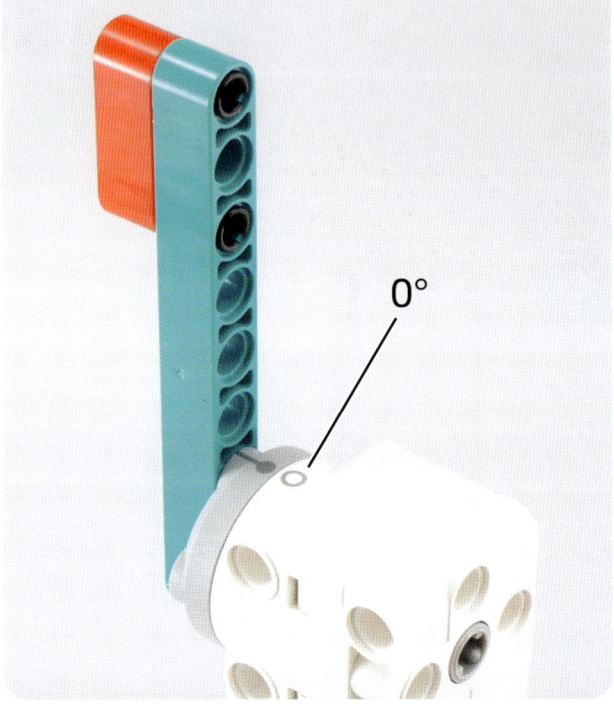

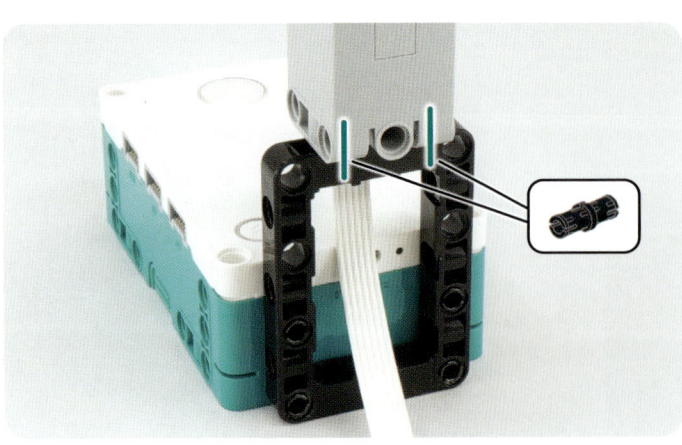

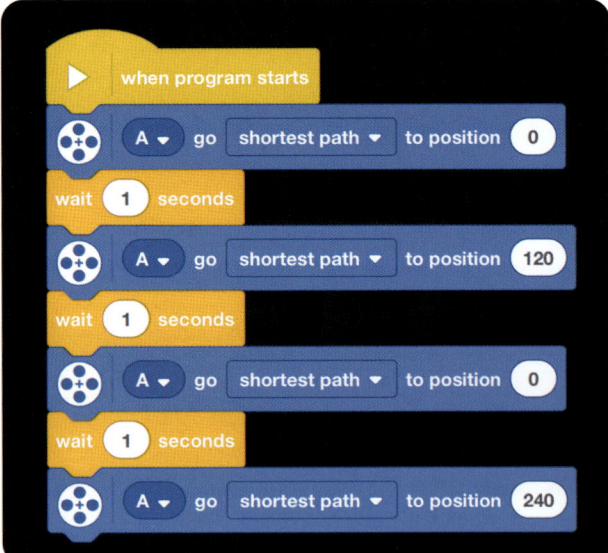

when program starts

A ▾ go shortest path ▾ to position 0

wait 1 seconds

A ▾ go shortest path ▾ to position 120

wait 1 seconds

A ▾ go shortest path ▾ to position 0

wait 1 seconds

A ▾ go shortest path ▾ to position 240

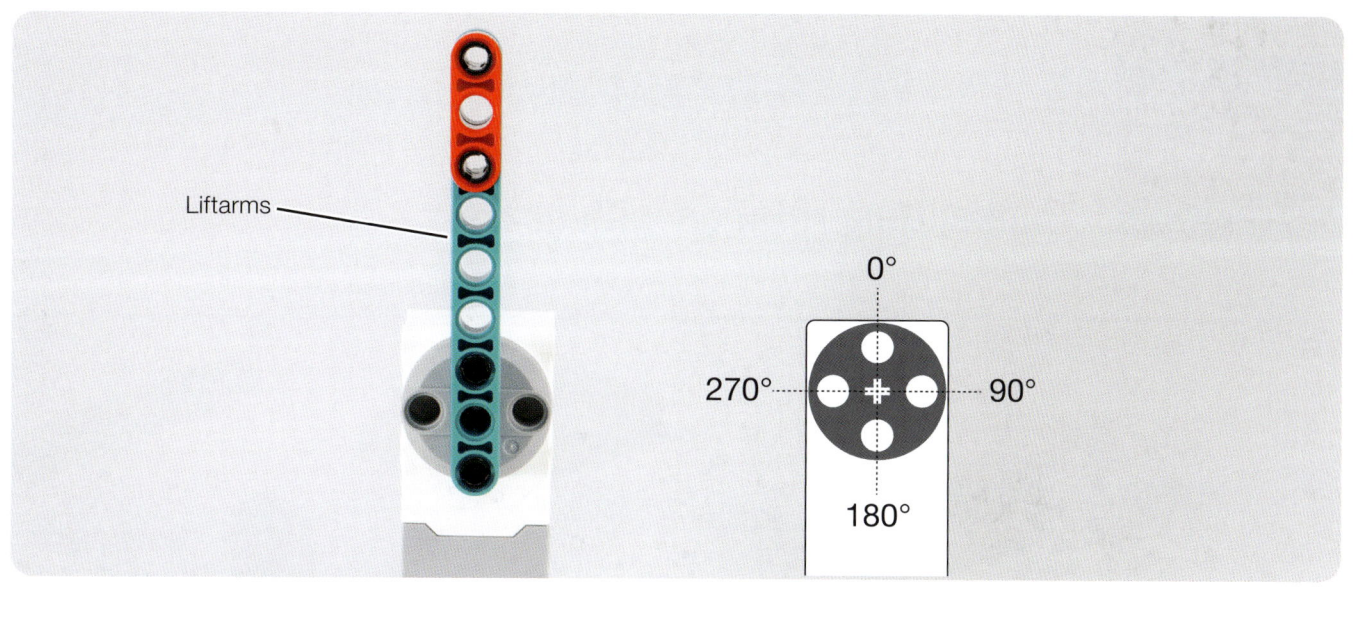

Liftarms

0°

270° 90°

180°

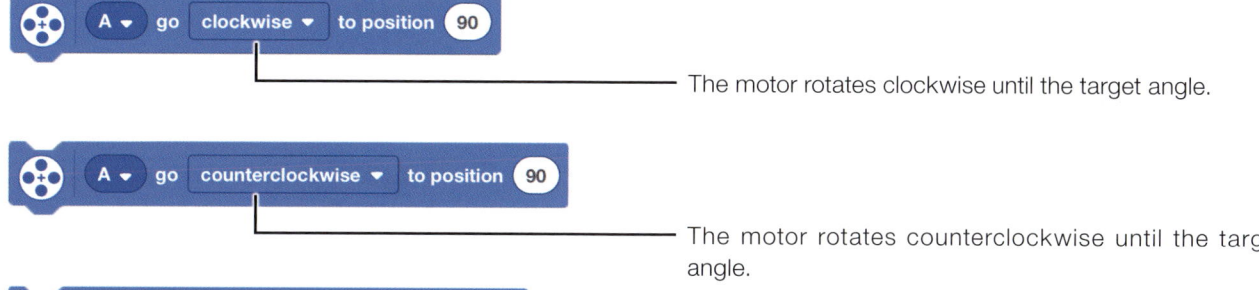

A ▾ go clockwise ▾ to position 90

———— The motor rotates clockwise until the target angle.

A ▾ go counterclockwise ▾ to position 90

———— The motor rotates counterclockwise until the target angle.

A ▾ go shortest path ▾ to position 90

———— The motor rotates in the direction that reaches the target angle faster.

💡 The motor makes the liftarms move randomly.

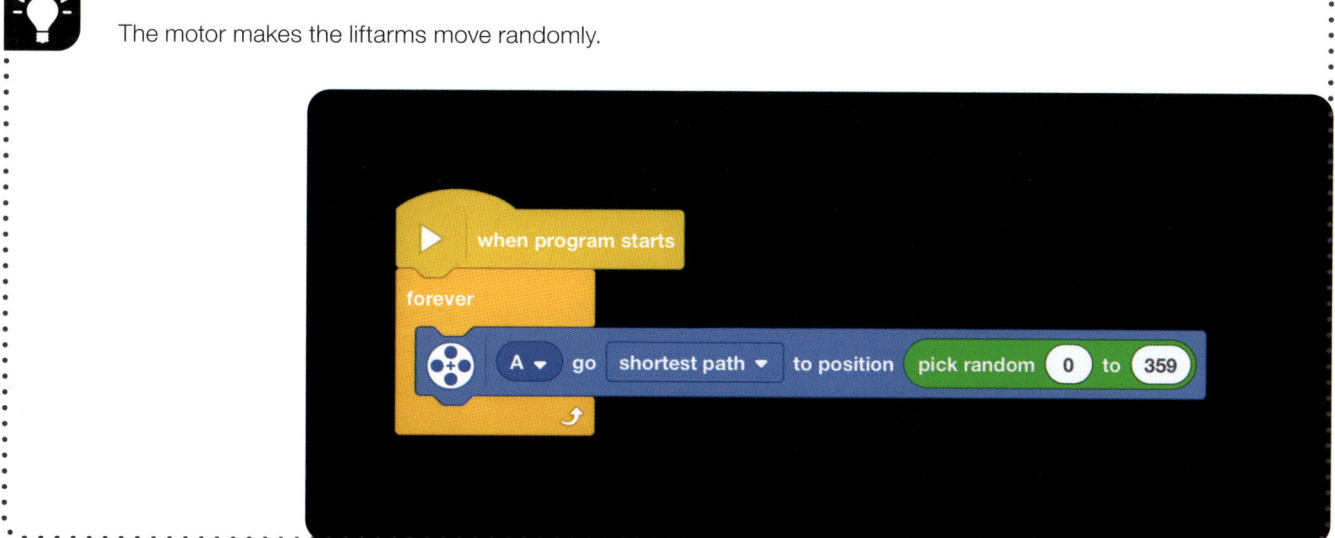

when program starts
forever
    A ▾ go shortest path ▾ to position pick random 0 to 359

Transmitting rotation with gears

#6

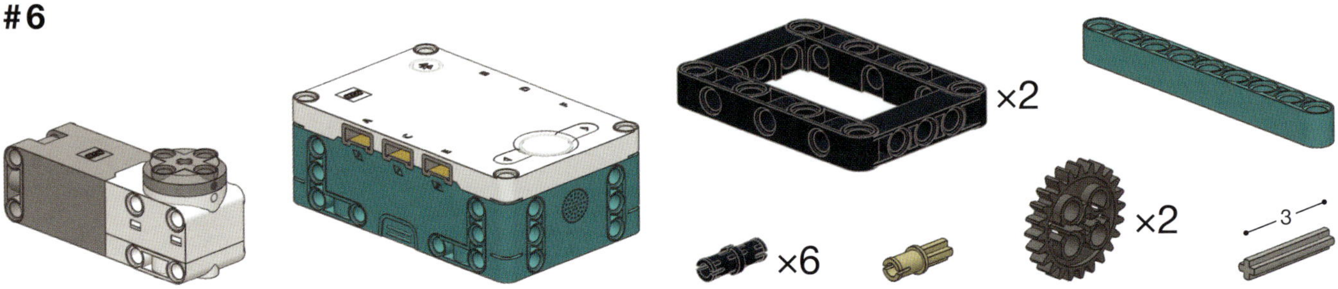

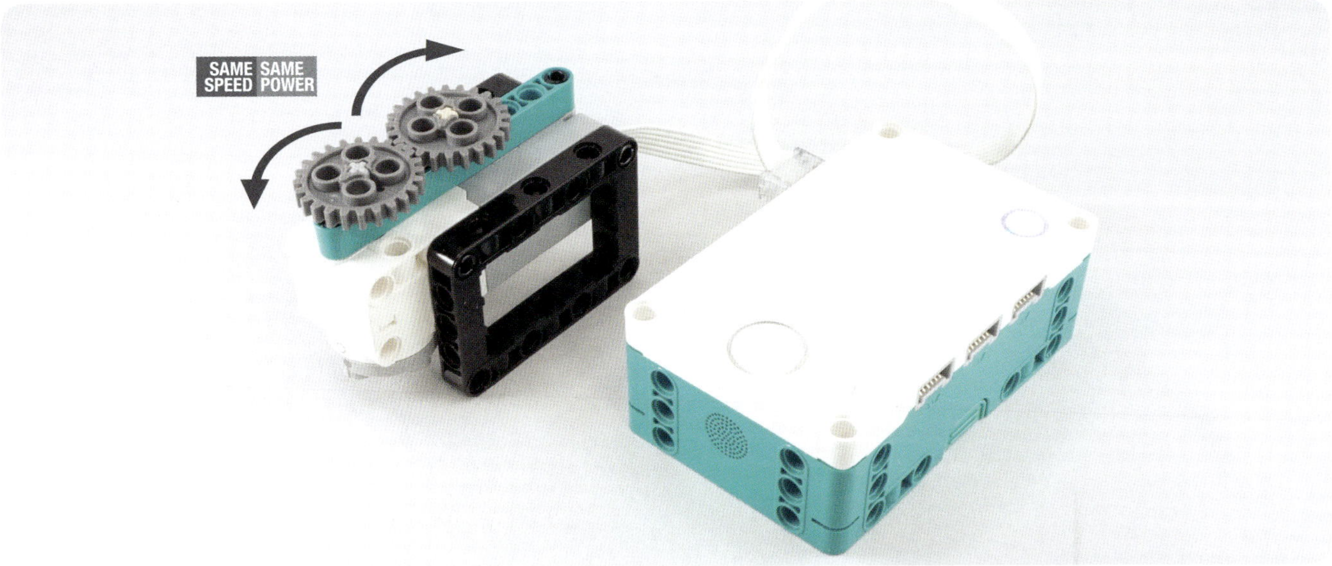

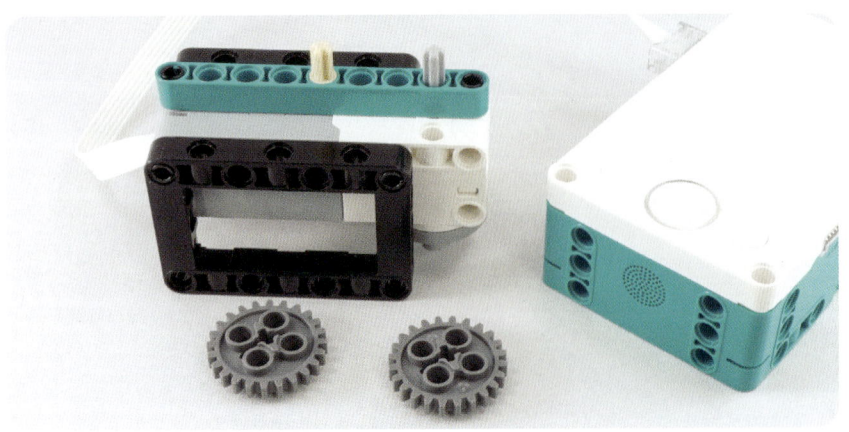

SAME SPEED | SAME POWER

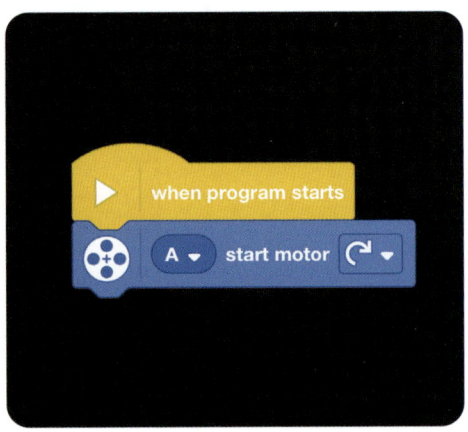

when program starts

A ▾ start motor ↻ ▾

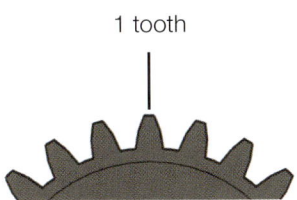
1 tooth

8 teeth

12 teeth

20 teeth

24 teeth

36 teeth

 24 ▶ 24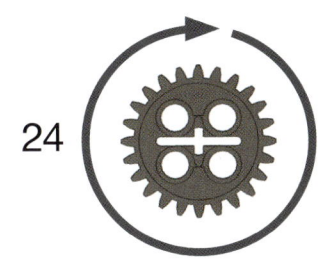

When the transmitting gear rotates once, 24 teeth are advanced.

When 24 teeth of the receiving gear are advanced, it makes exactly one rotation. The direction of rotation is reversed.

SAME SPEED **SAME POWER** When there is no change in the rotation speed, the power of the receiving gear is the same as that of the transmitting gear.

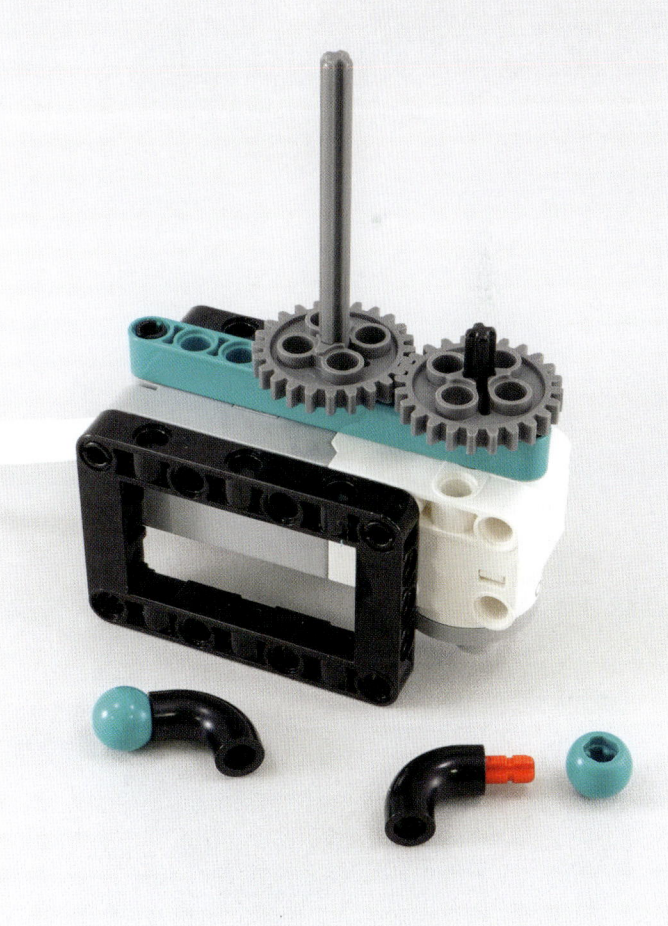

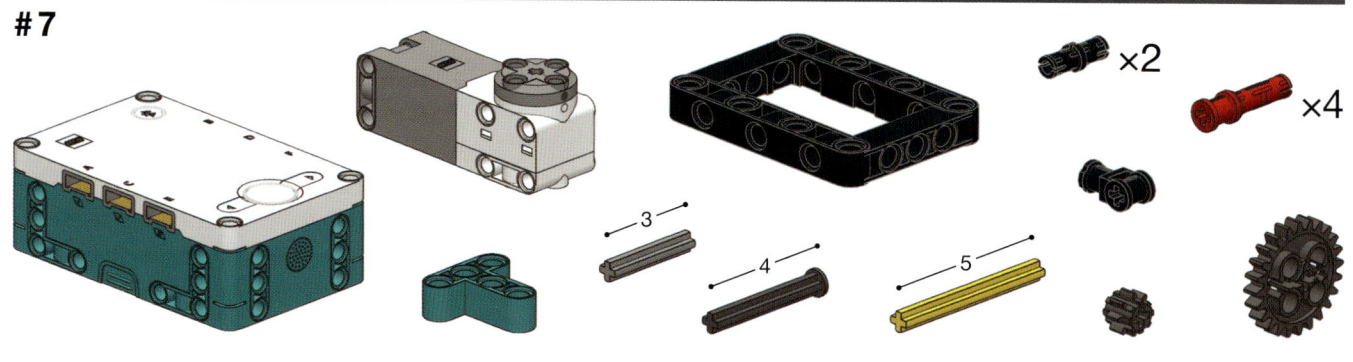

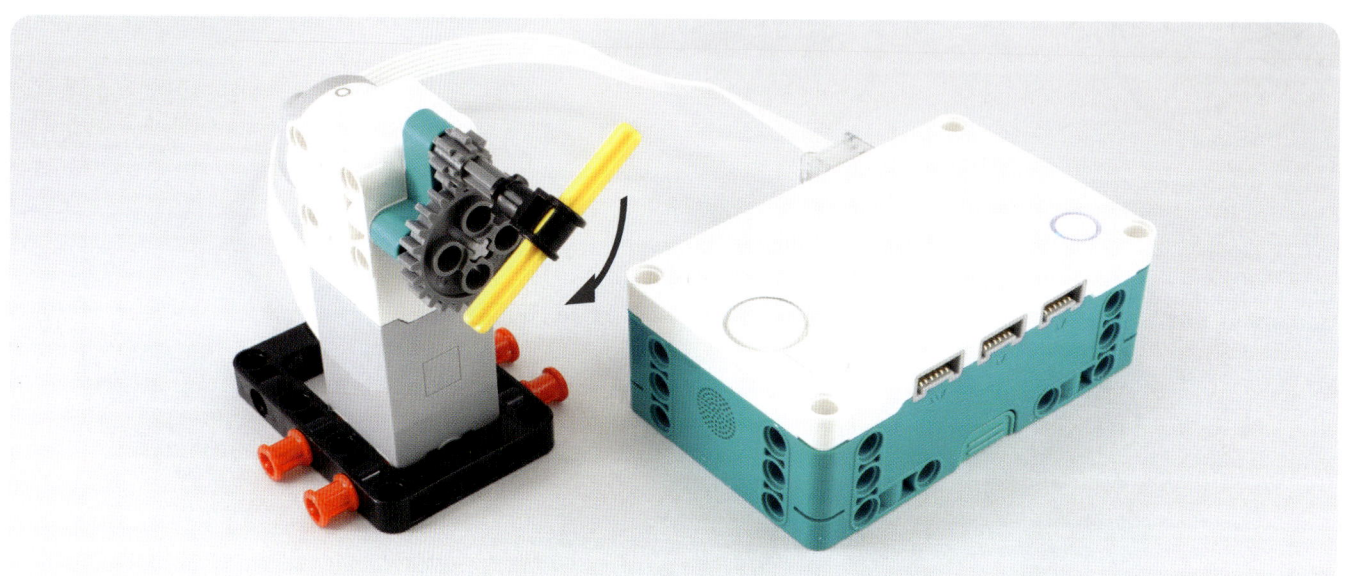

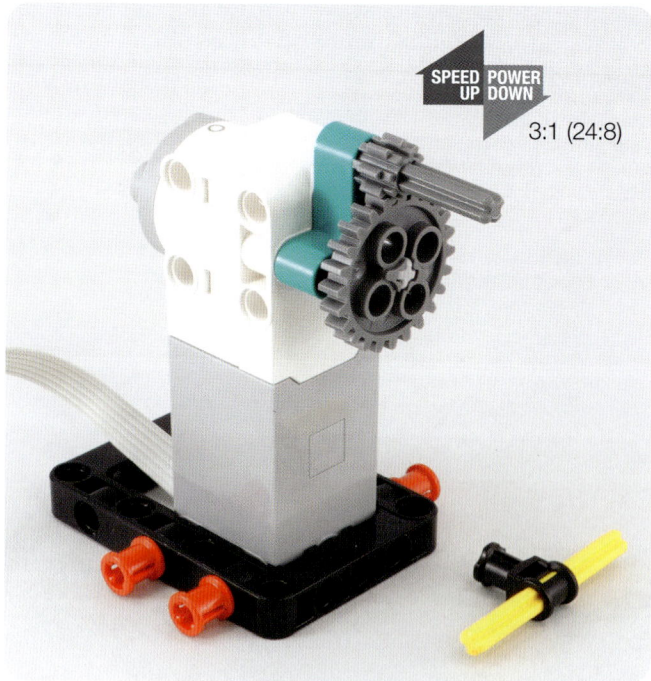

7

SPEED POWER
UP DOWN

3:1 (24:8)

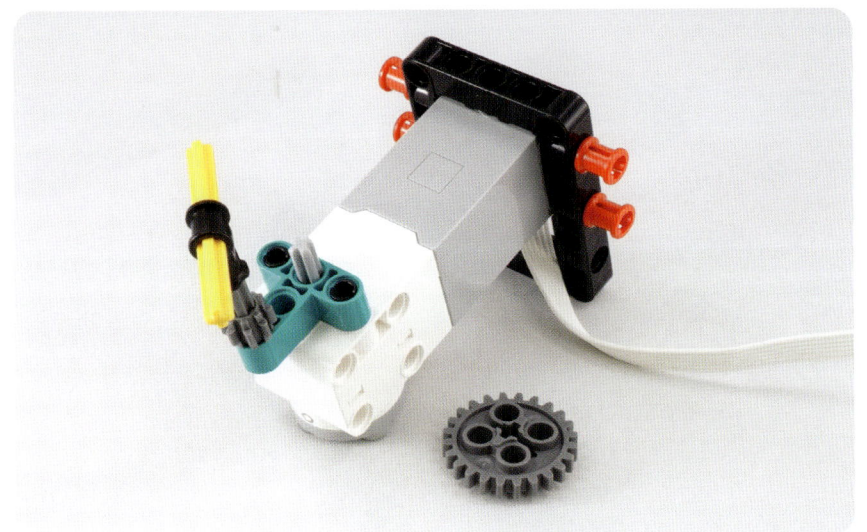

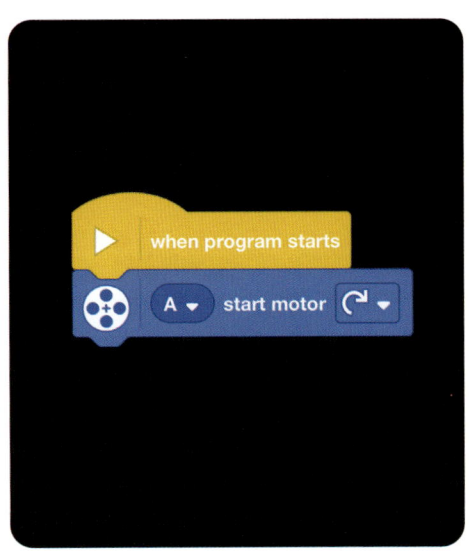

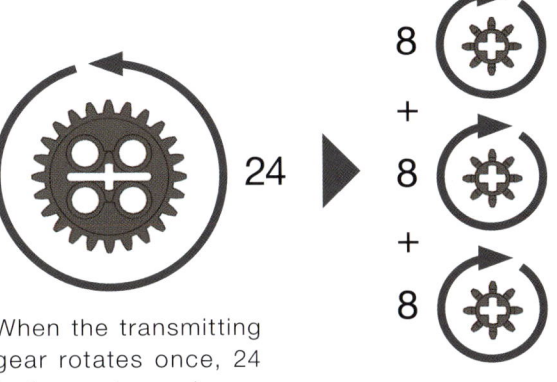

When the transmitting gear rotates once, 24 teeth are advanced.

8 + 8 + 8

The receiving gear makes one rotation when 8 teeth are advanced. Thus, it rotates 3 times when 24 teeth are advanced.

SPEED UP POWER DOWN

When the rotation speed increases, the power of the receiving gear decreases compared to that of the transmitting gear. In this model, the rotation speed is increased threefold, and the power is reduced to one-third.

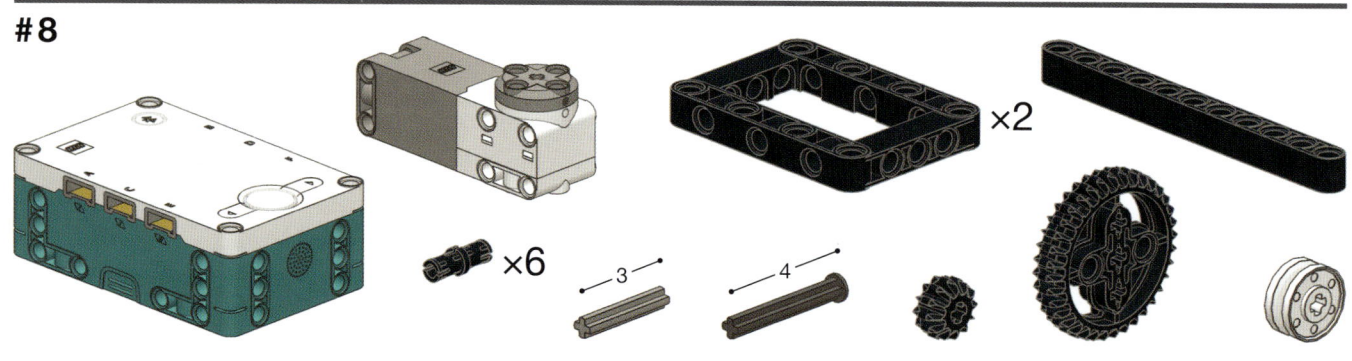

8

×2
×6
3
4

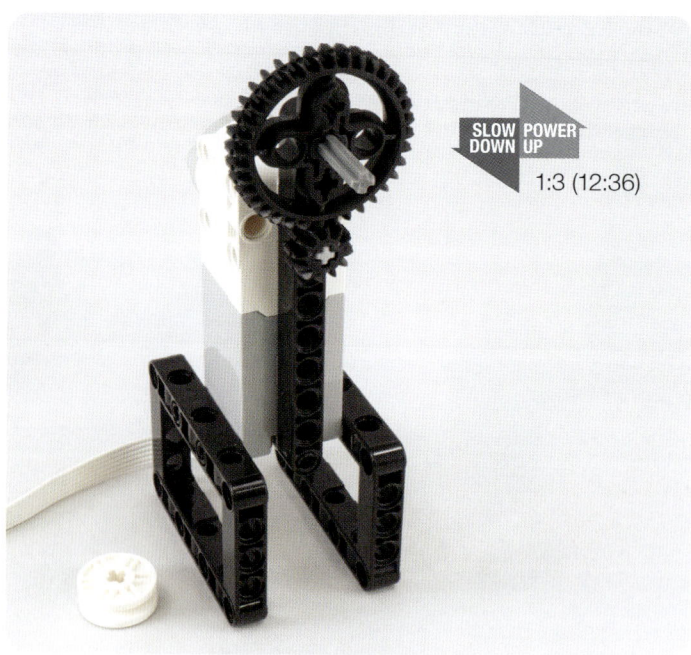

SLOW POWER
DOWN UP
1:3 (12:36)

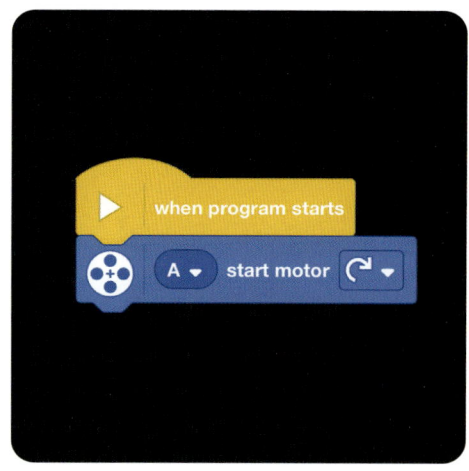

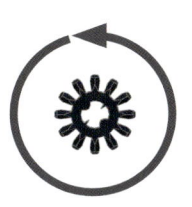 12 ▶ 12

When the transmitting gear rotates once, 12 teeth are advanced.

As the receiving gear has 36 teeth, it rotates one-third when 12 teeth are advanced.

SLOW DOWN **POWER UP**

When the rotation speed decreases, the power of the receiving gear increases compared to that of the transmitting gear. In this model, the rotation speed is reduced to one-third, and the power is increased threefold.

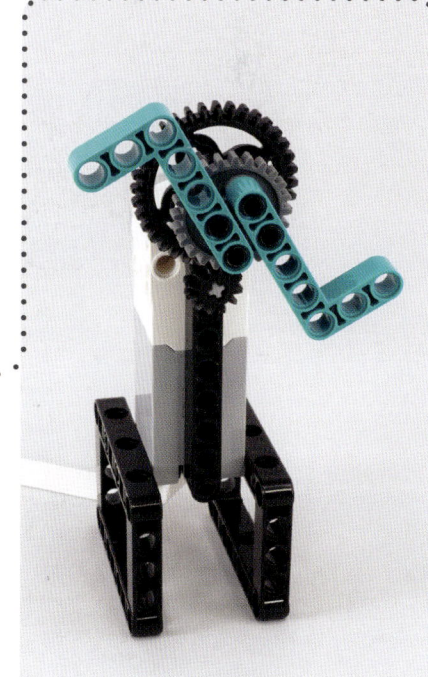

#9

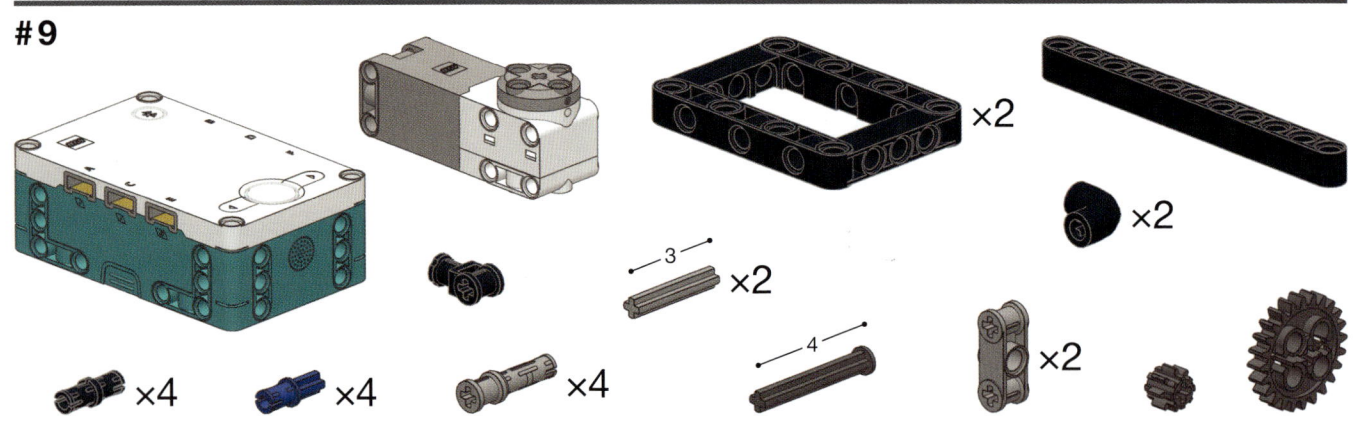

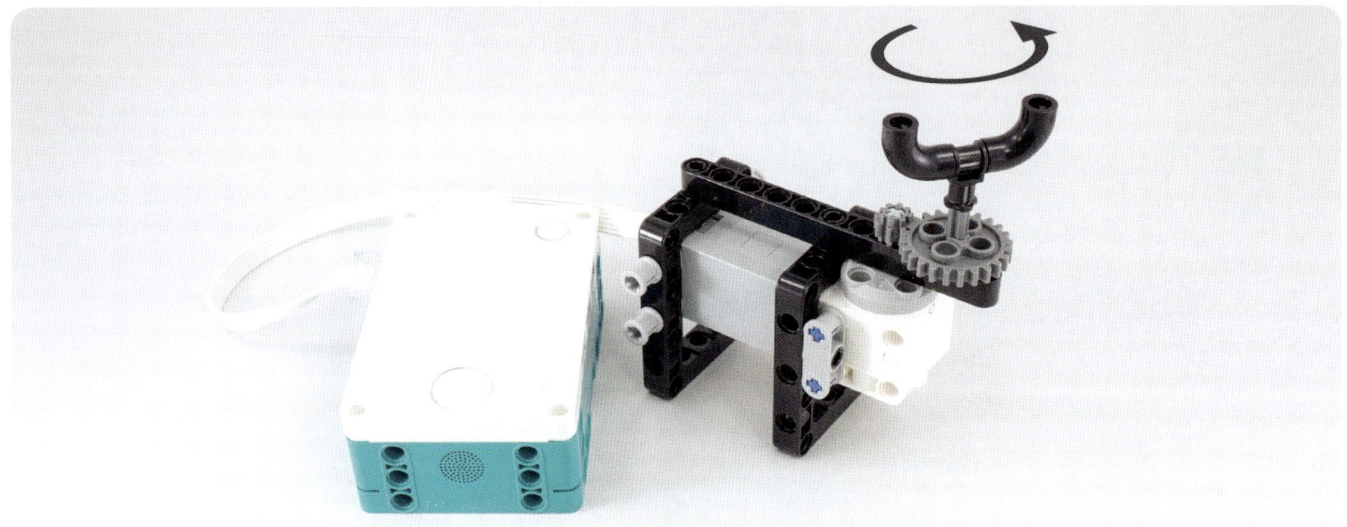

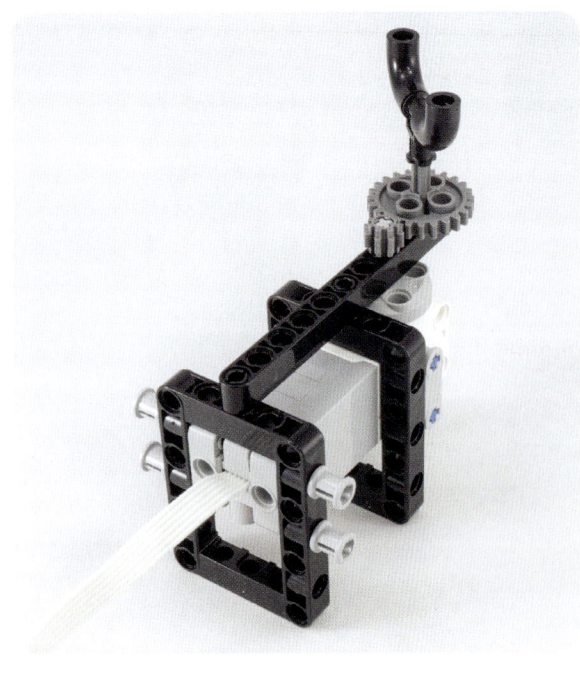

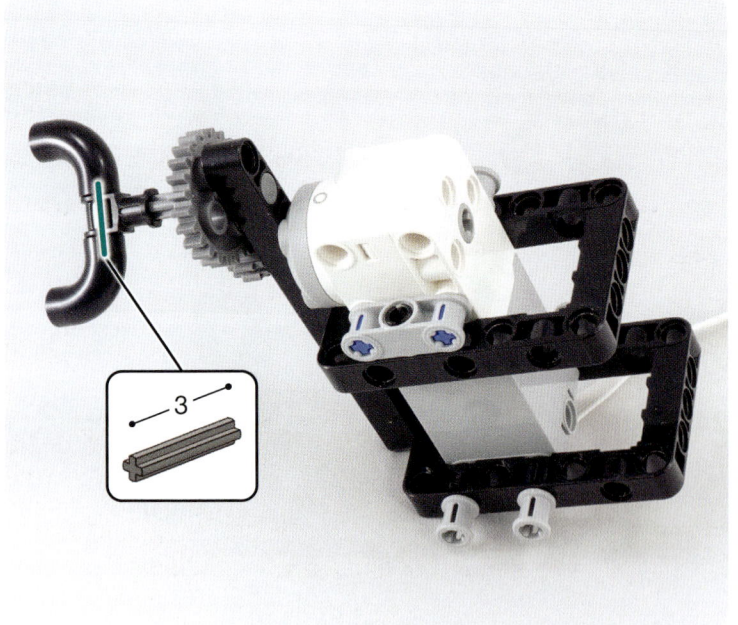

SLOW DOWN / POWER UP

1:3 (8:24)

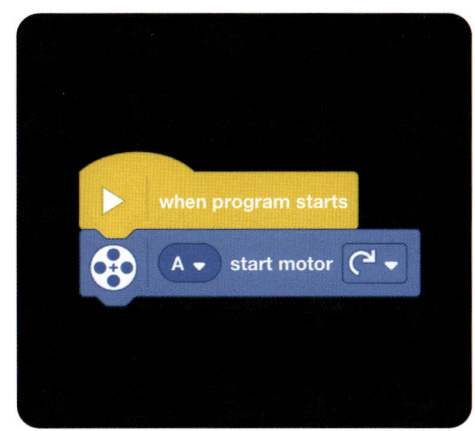

SLOW DOWN / POWER UP

3:5 (12:20)

SPEED UP / POWER DOWN

5:3 (20:12)

SPEED UP / POWER DOWN

3:1 (36:12)

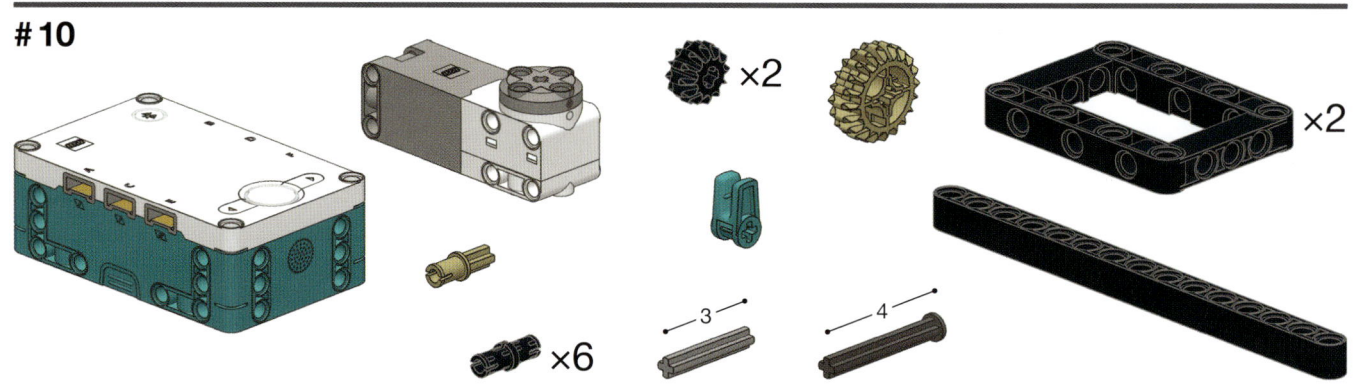

×2 ×2

×6 3 4

SAME SAME
SPEED POWER

1:1 (12:~~20~~:12)

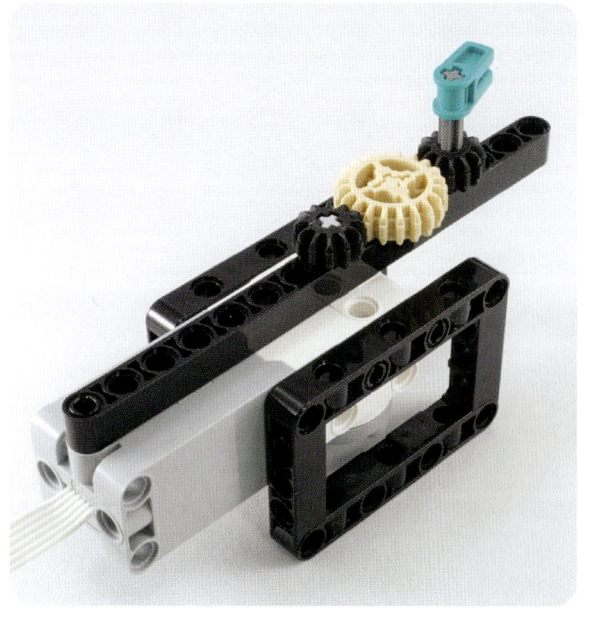

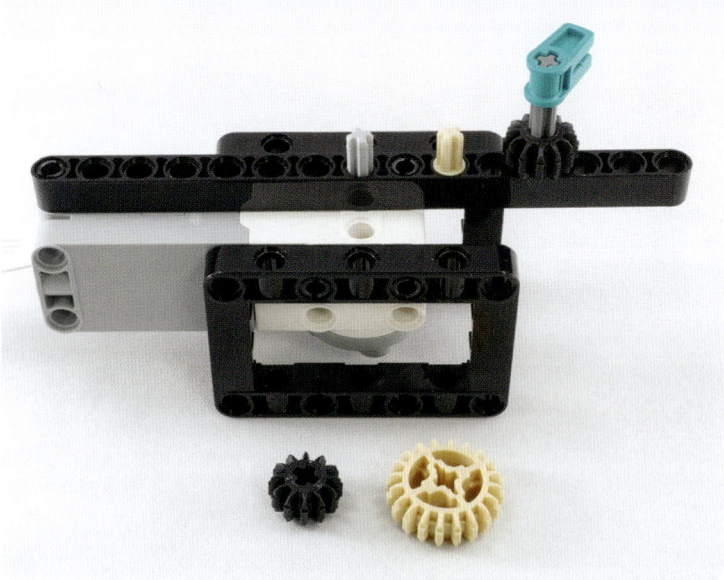

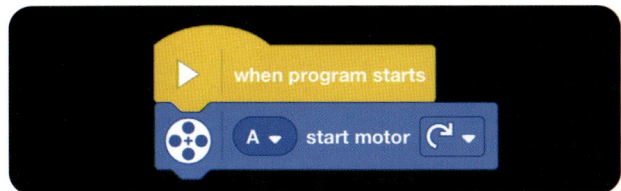

As the gear between the two gears only transmits the rotation of the same number of teeth into the next gear, there is no change in the rotation speed or power. However, the direction of rotation is reversed.

SLOW DOWN / POWER UP
5:9 (20:~~12~~:36)

SLOW DOWN / POWER UP
1:3 (8:~~24~~:24)

#11

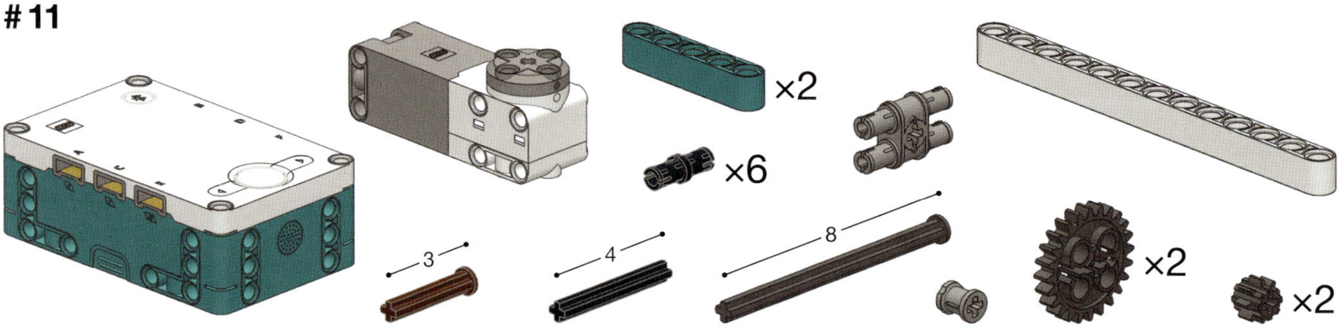

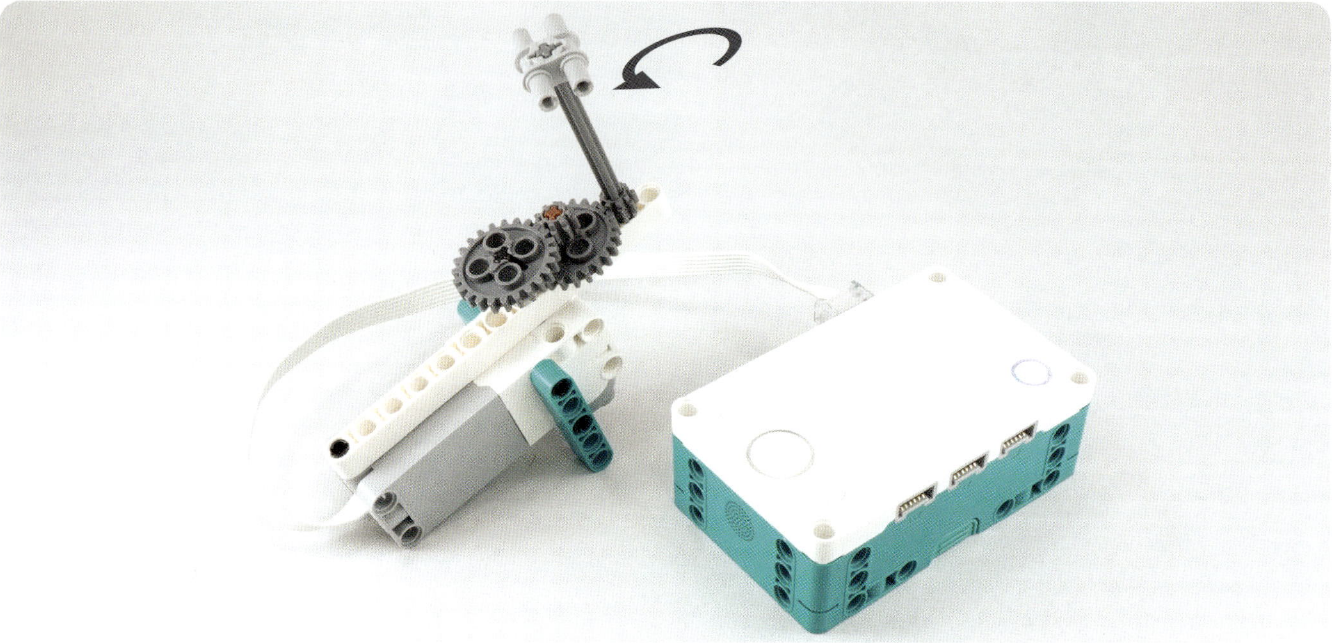

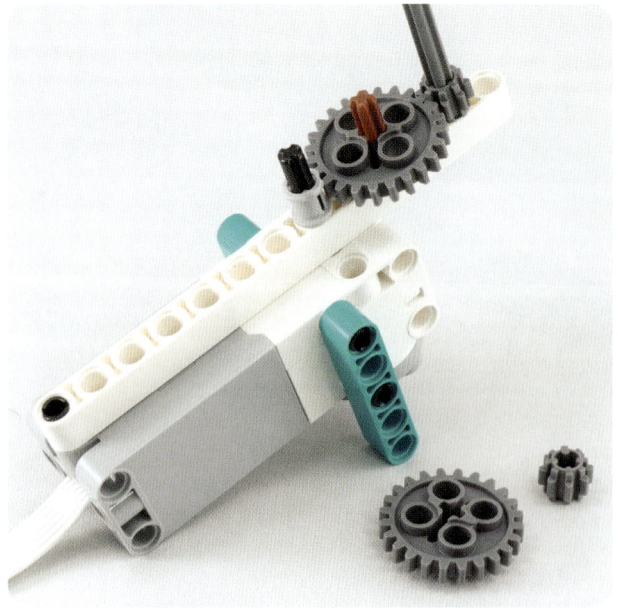

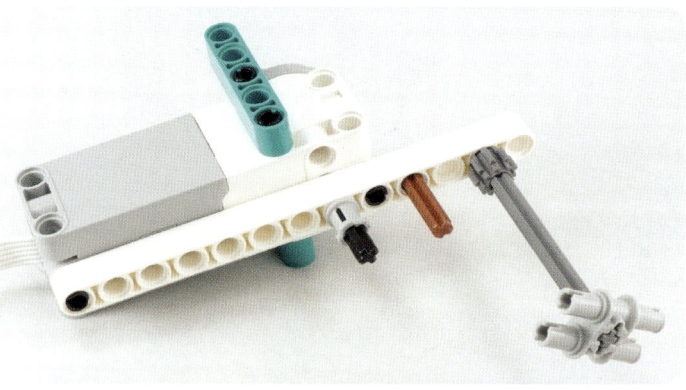

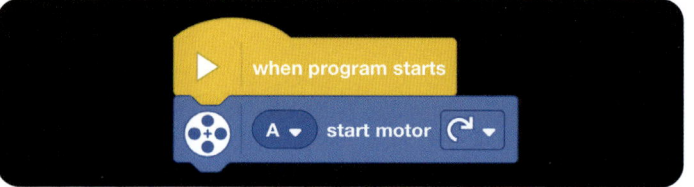

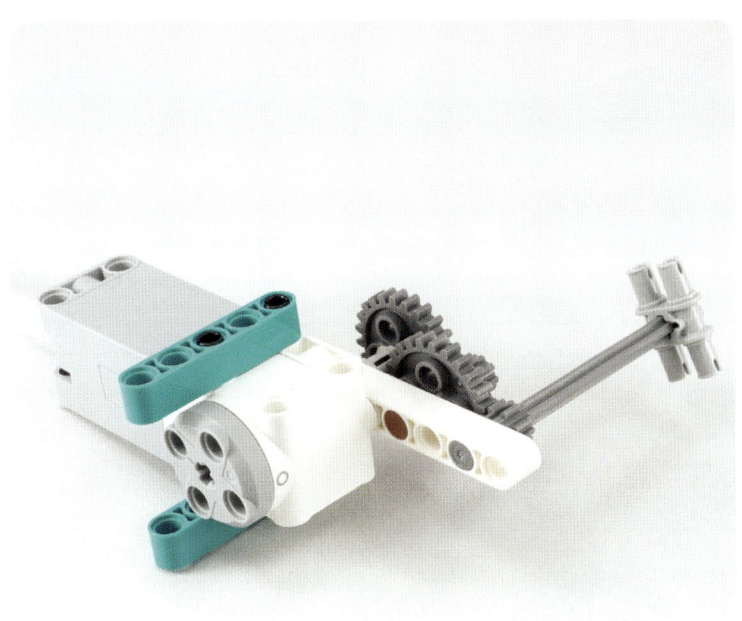

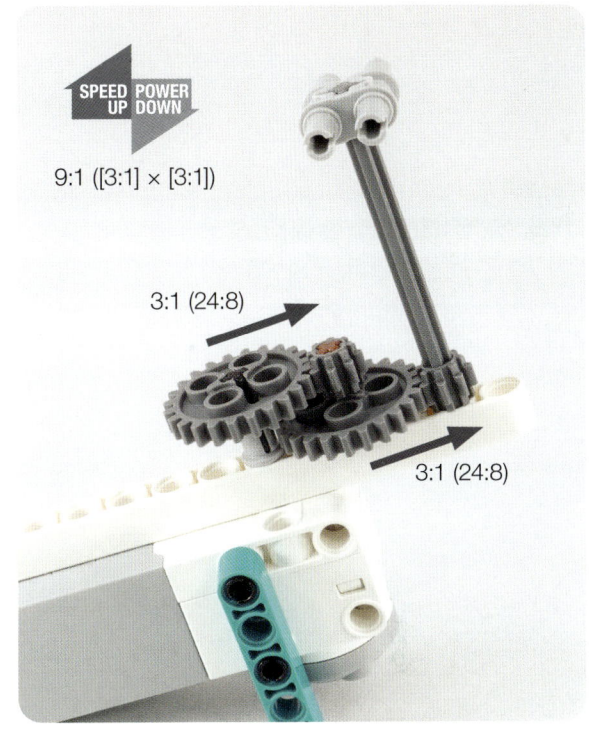

9:1 ([3:1] × [3:1])

3:1 (24:8)

3:1 (24:8)

SPEED UP **POWER** DOWN

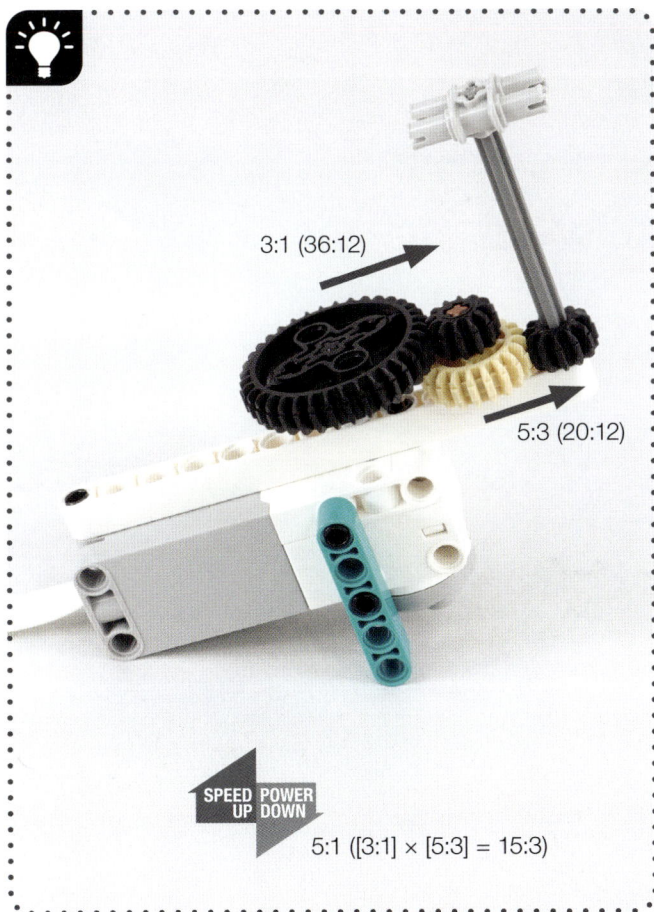

3:1 (36:12)

5:3 (20:12)

SPEED UP **POWER** DOWN

5:1 ([3:1] × [5:3] = 15:3)

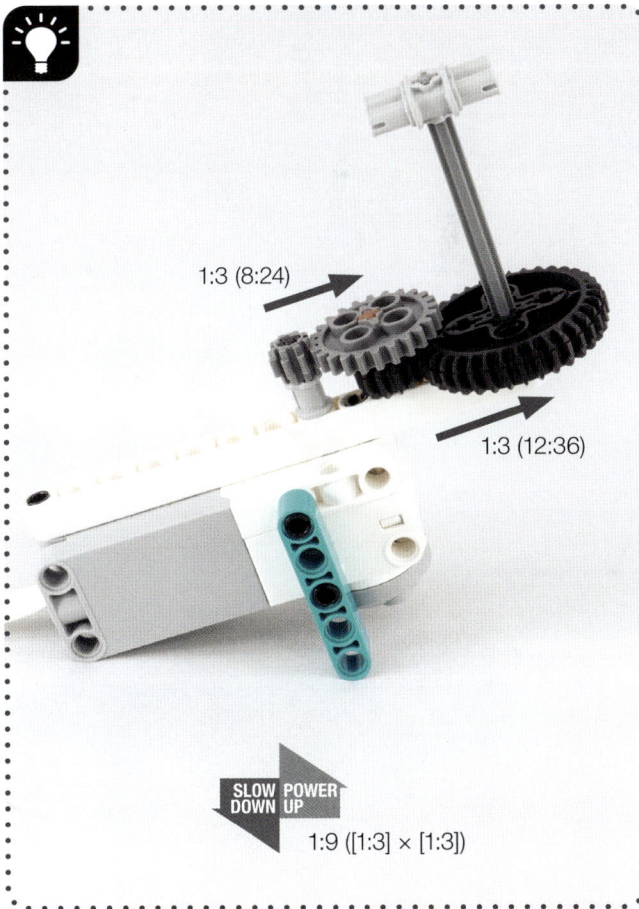

1:3 (8:24)

1:3 (12:36)

SLOW DOWN **POWER** UP

1:9 ([1:3] × [1:3])

23

Changing the angle of rotation by 90°

#12

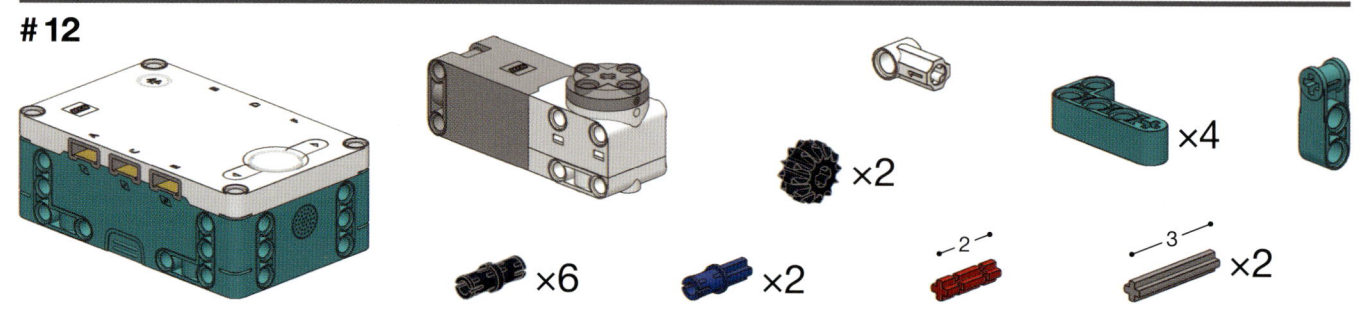

×2

×4

×6   ×2   —2—   —3— ×2

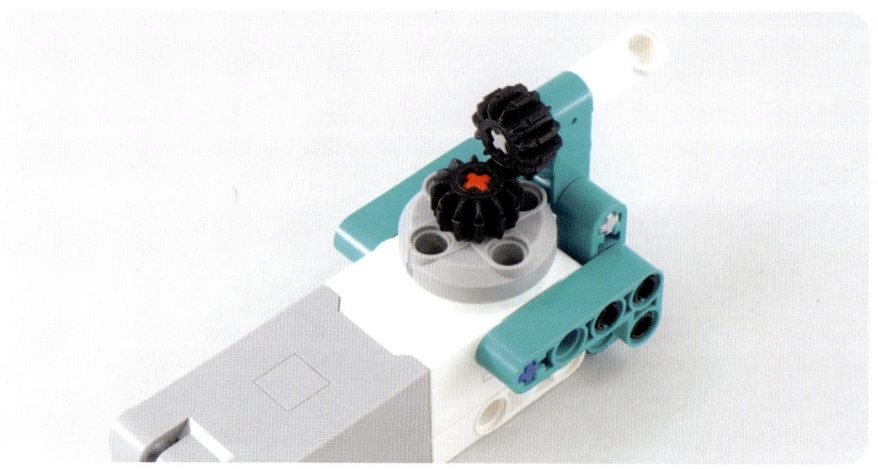

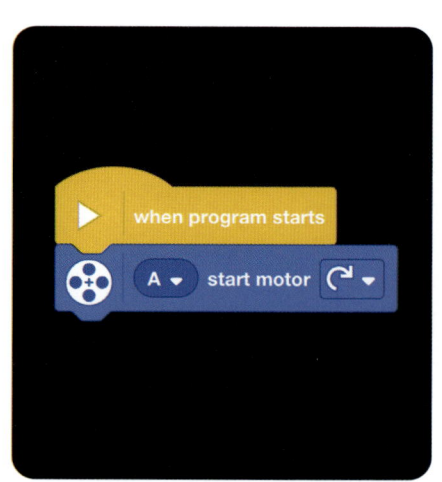

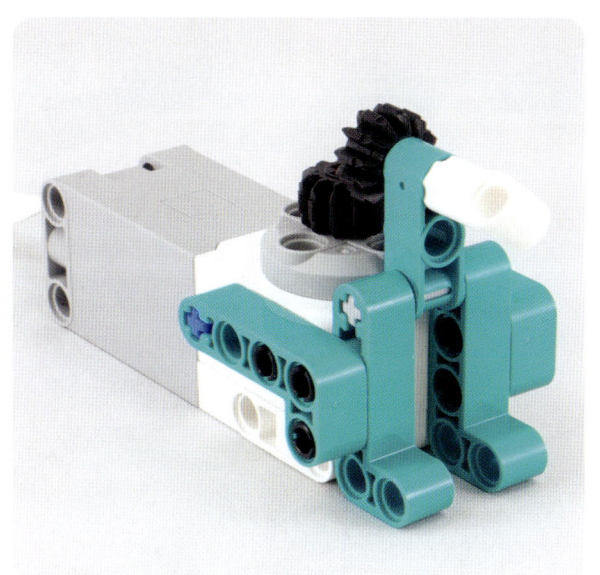

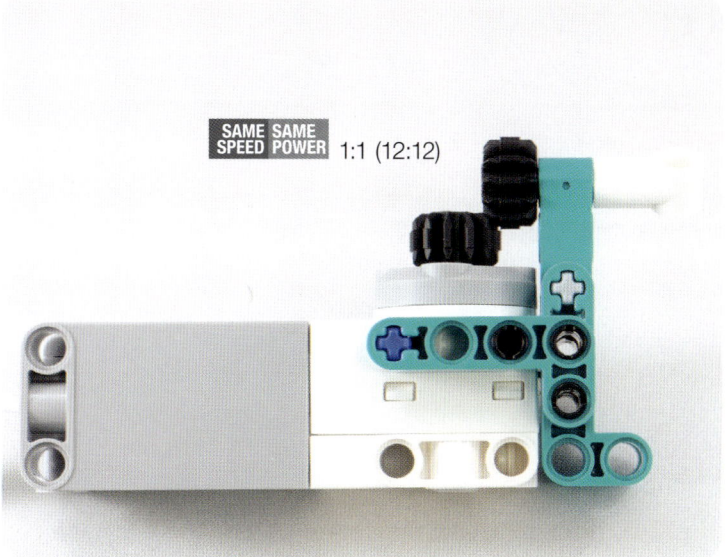

SAME SAME
SPEED POWER 1:1 (12:12)

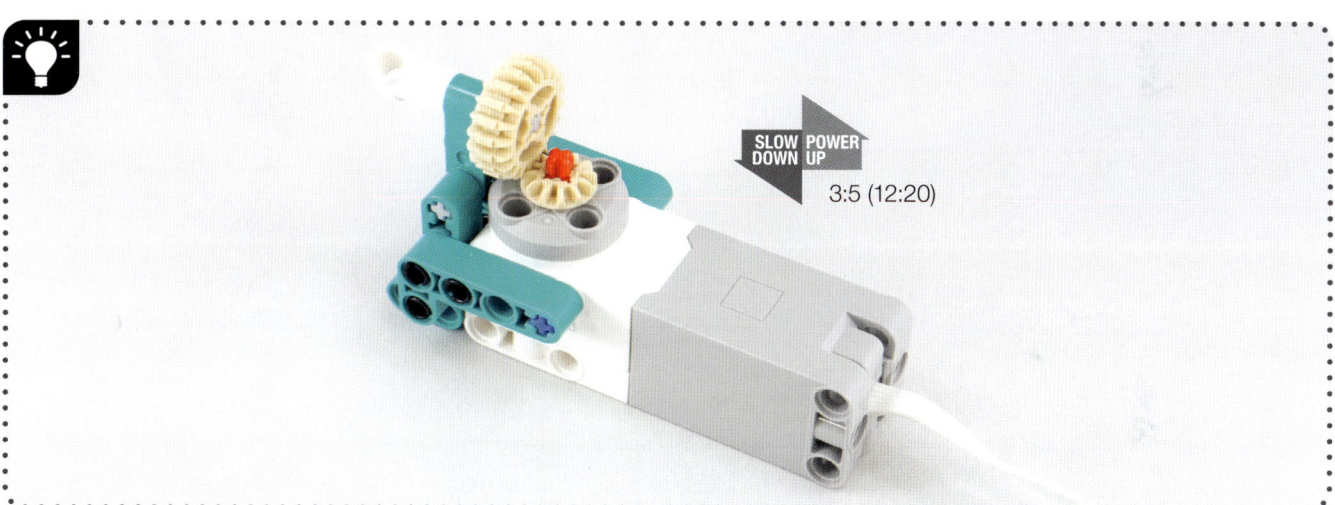

SLOW POWER
DOWN UP
3:5 (12:20)

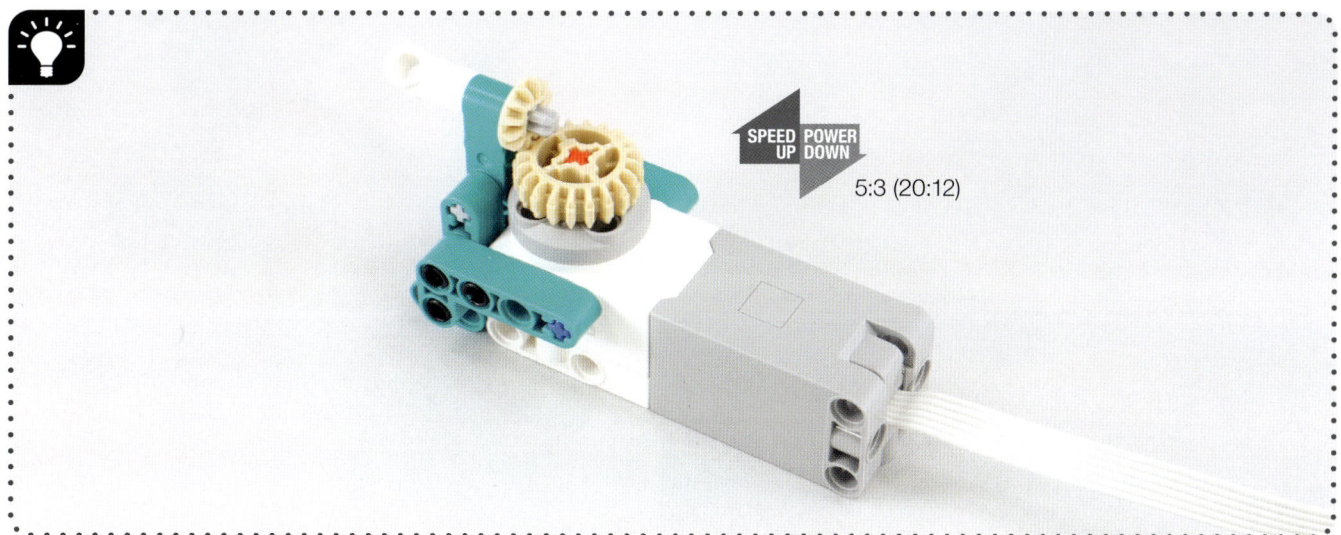

SPEED POWER
UP DOWN
5:3 (20:12)

13

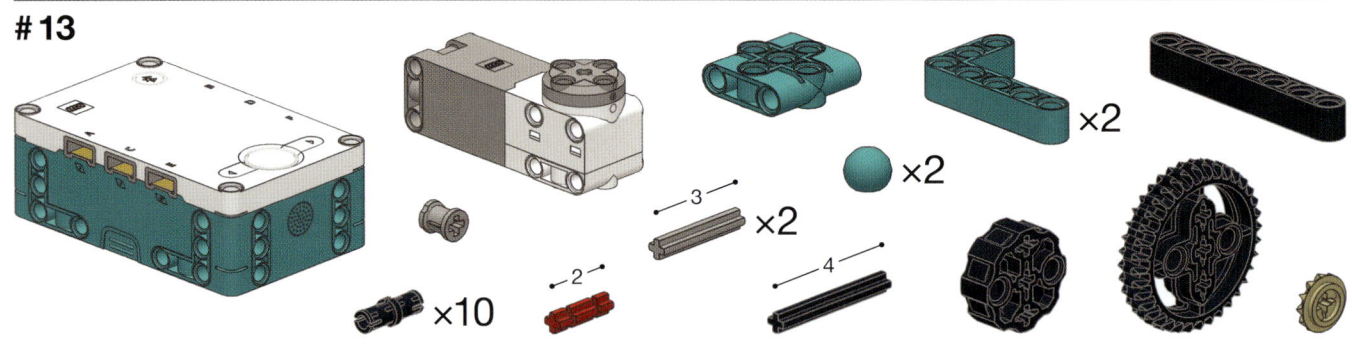

1:3 (12:36)

SLOW POWER
DOWN UP

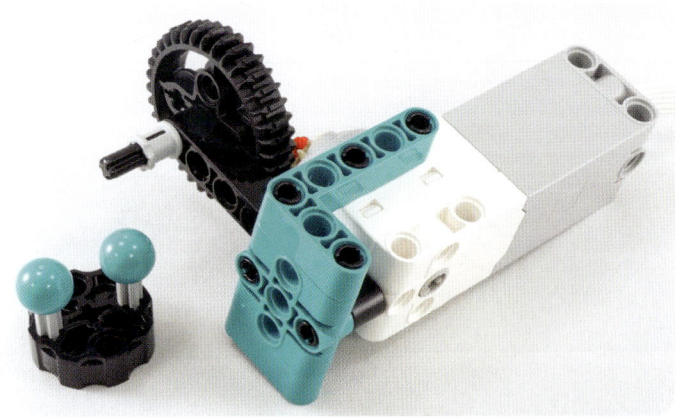

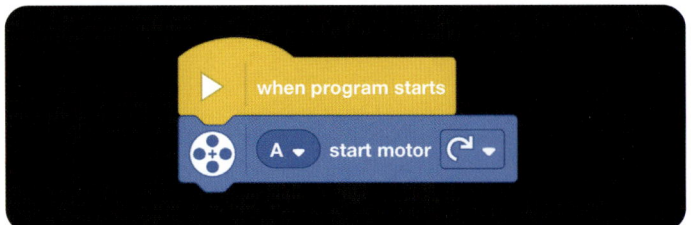

when program starts

A ▾ start motor ↻ ▾

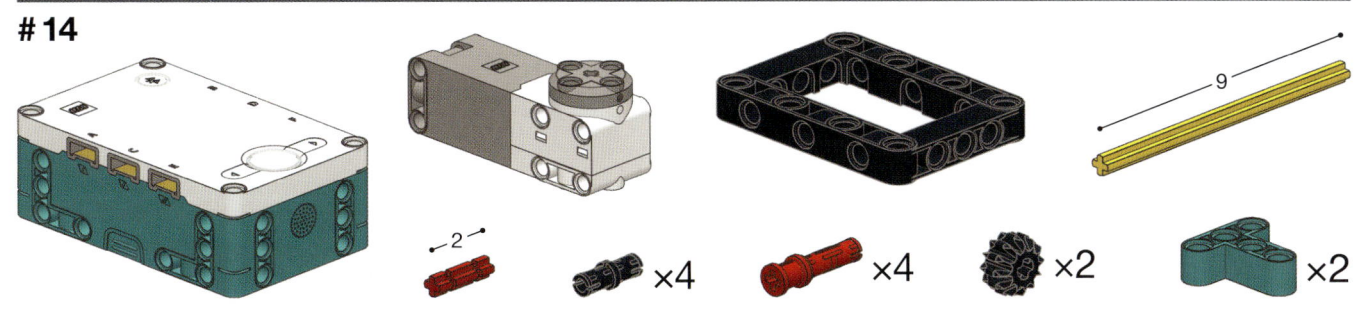

9

×2

×4

×4

×2

×2

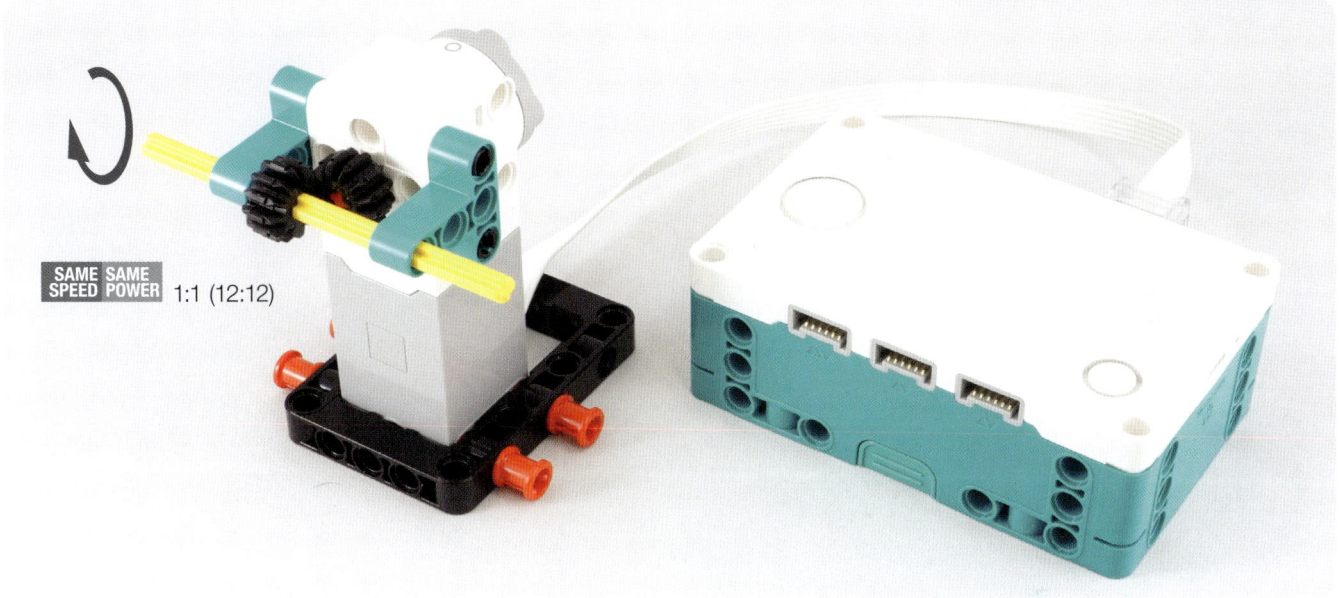

SAME SAME
SPEED POWER 1:1 (12:12)

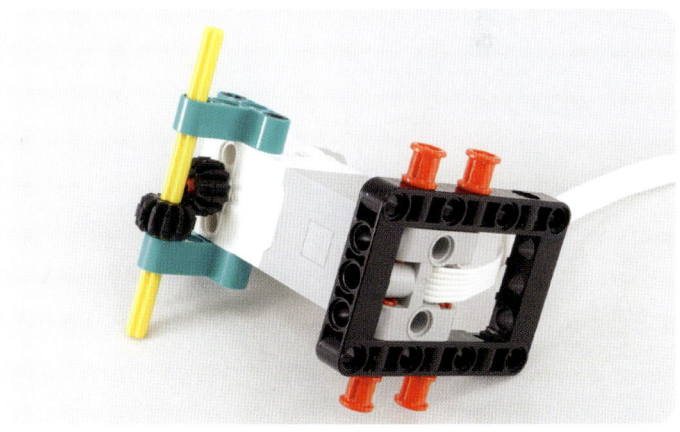

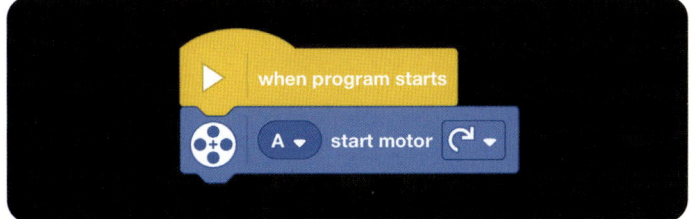

when program starts

A ▾ start motor ↻ ▾

Oscillating mechanisms

×2   ×3

when program starts

A ▾ start motor ↻ ▾

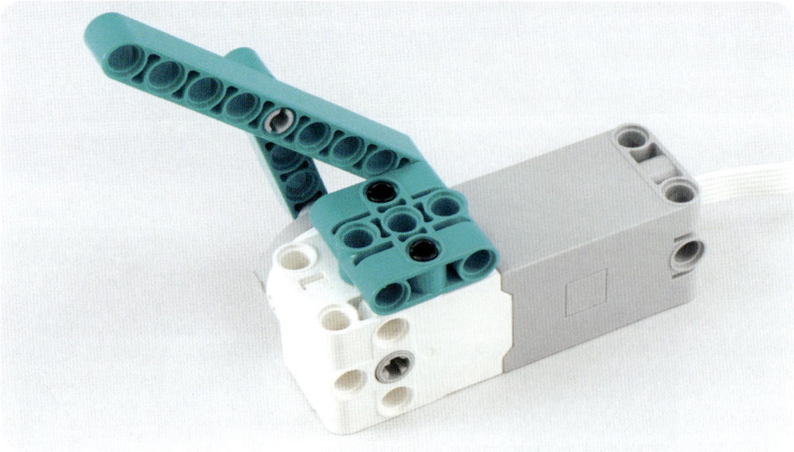

28

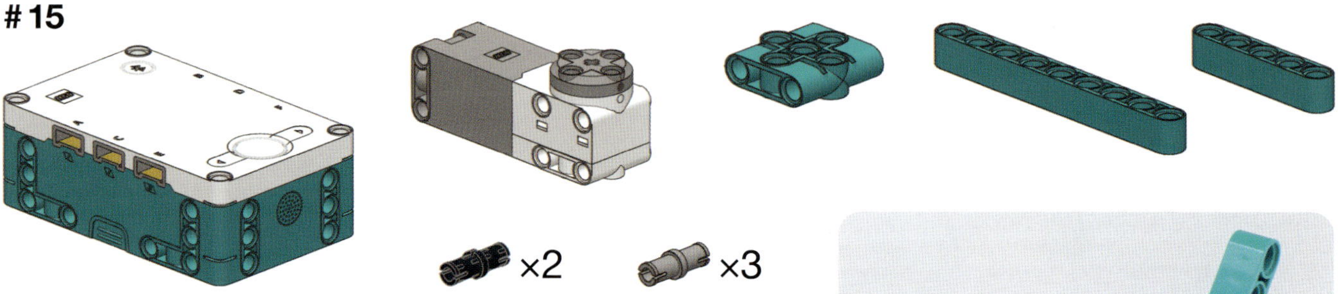

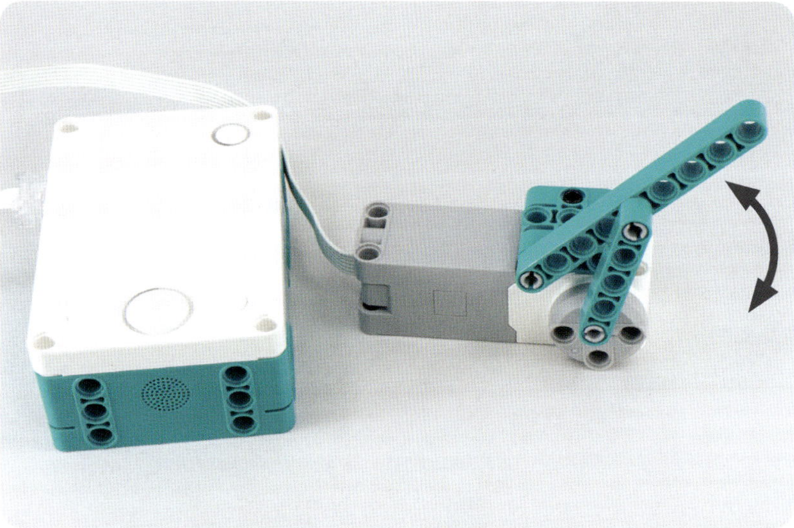

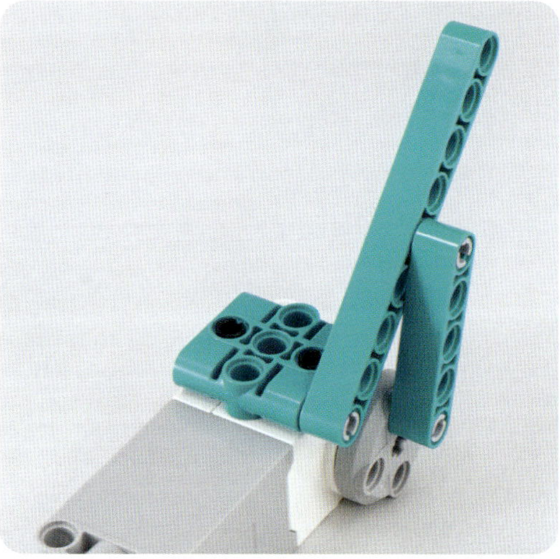

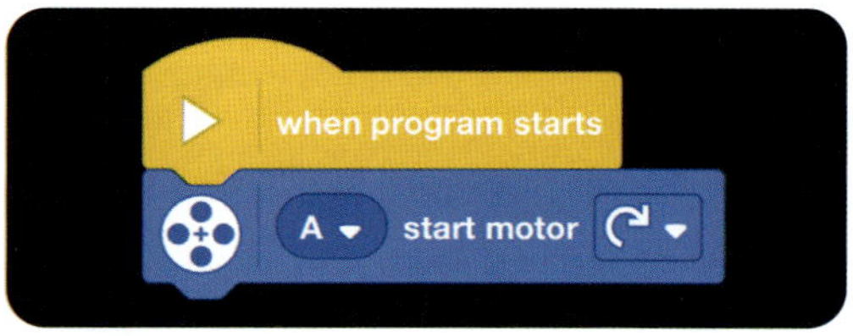

16

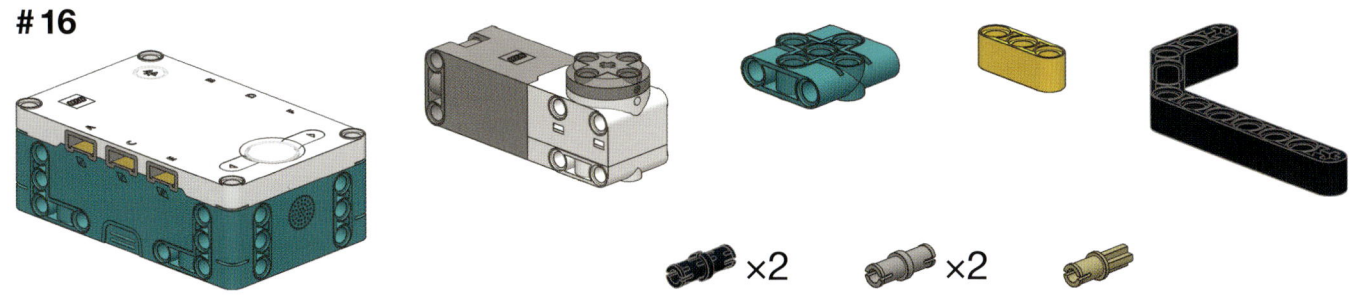

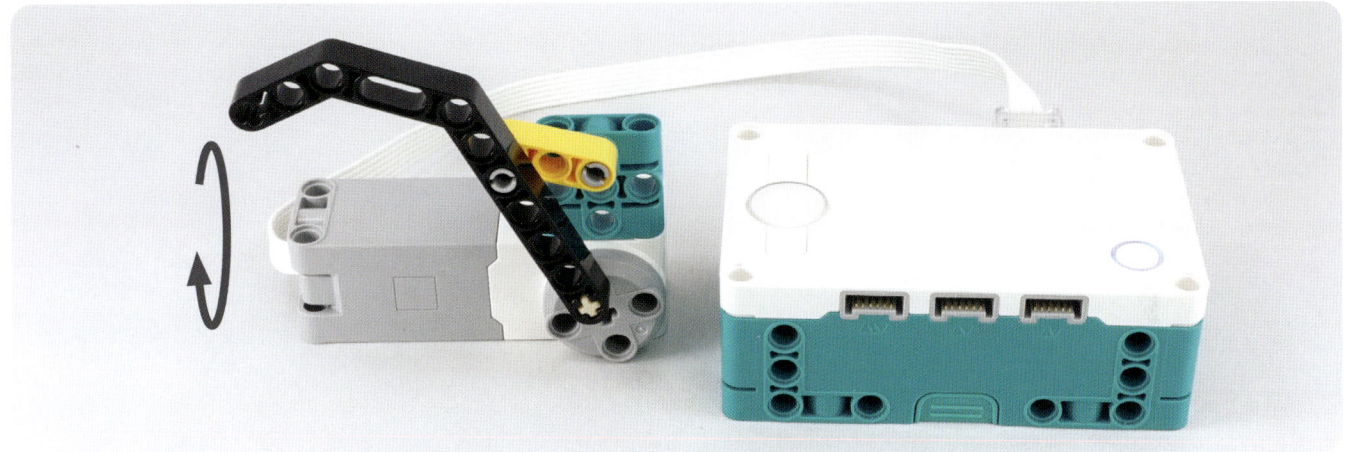

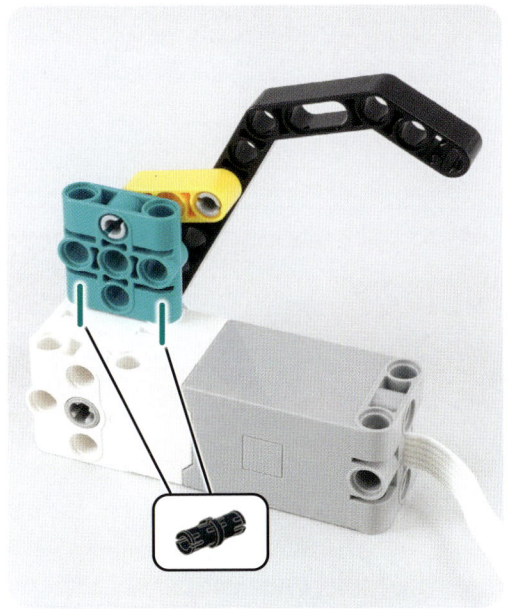

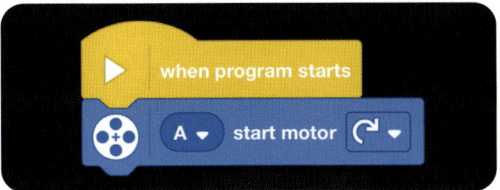

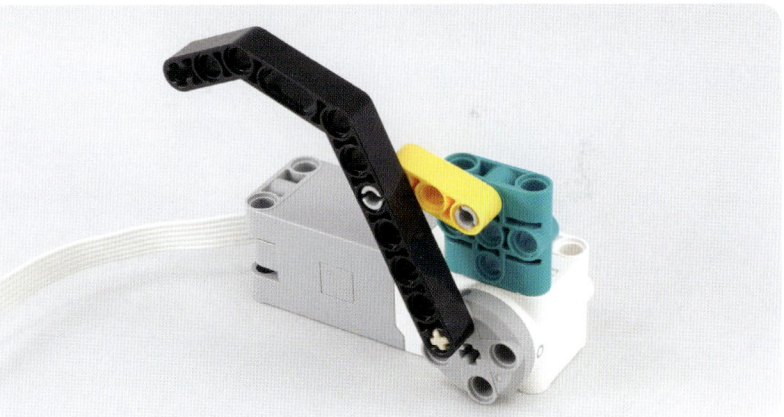

when program starts

A ▾ start motor ↻ ▾

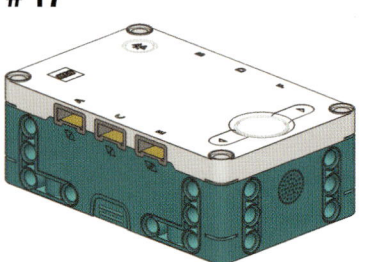
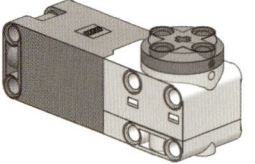
×2

×4 ×2

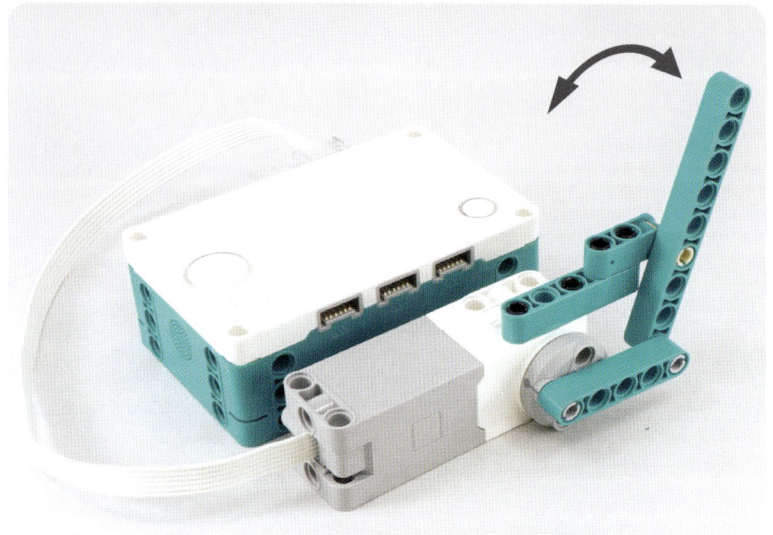

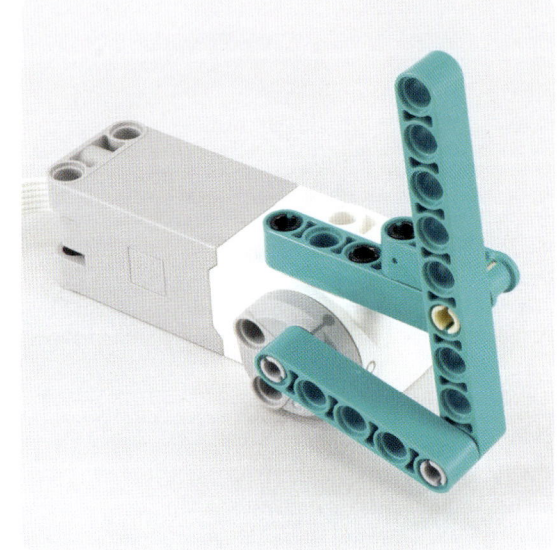

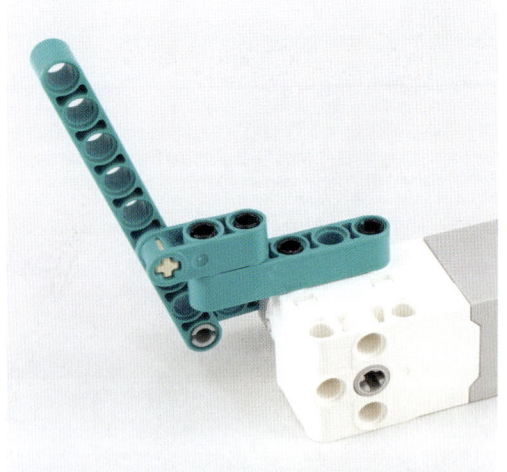

when program starts

A ▼ start motor ↻ ▼

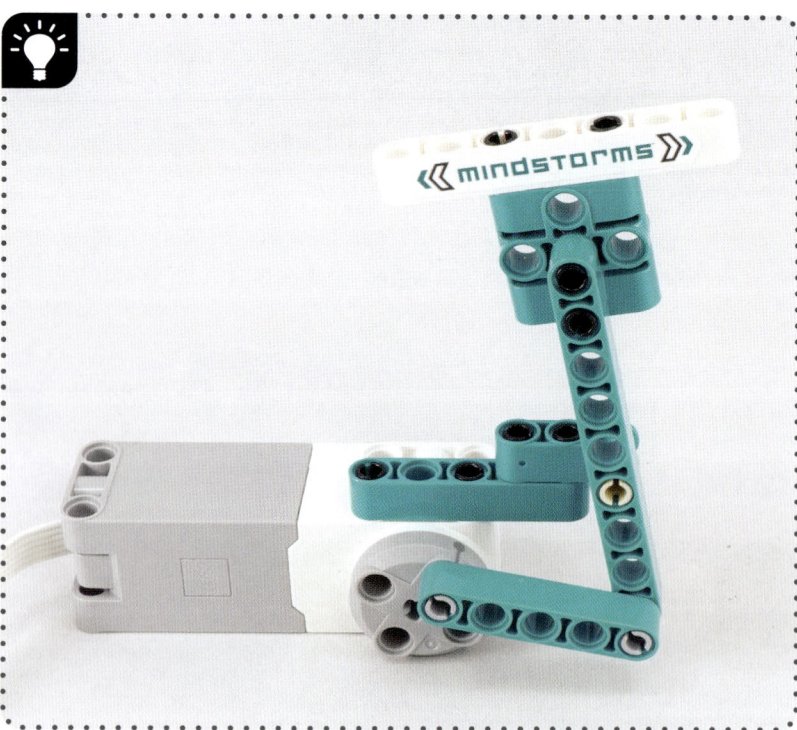

18

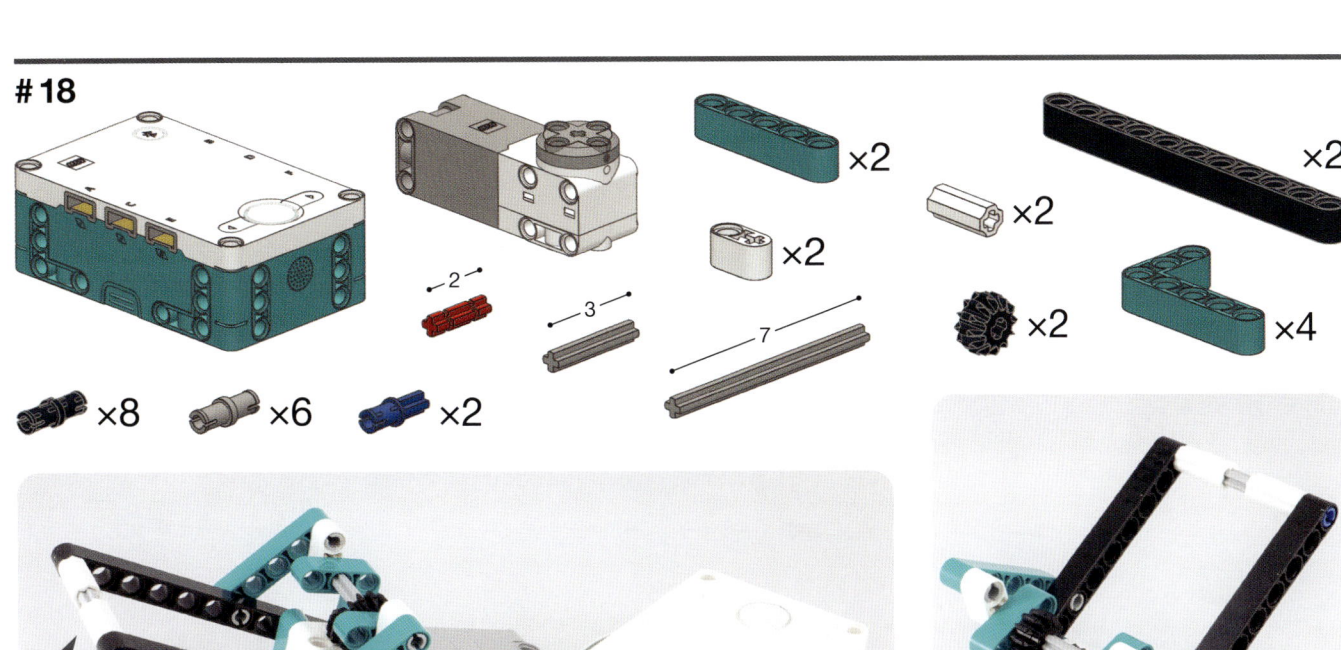

×2 ×2 ×2 ×2 ×2 ×4

×8 ×6 ×2

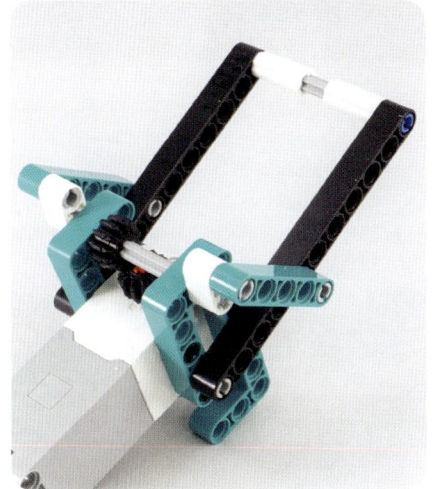

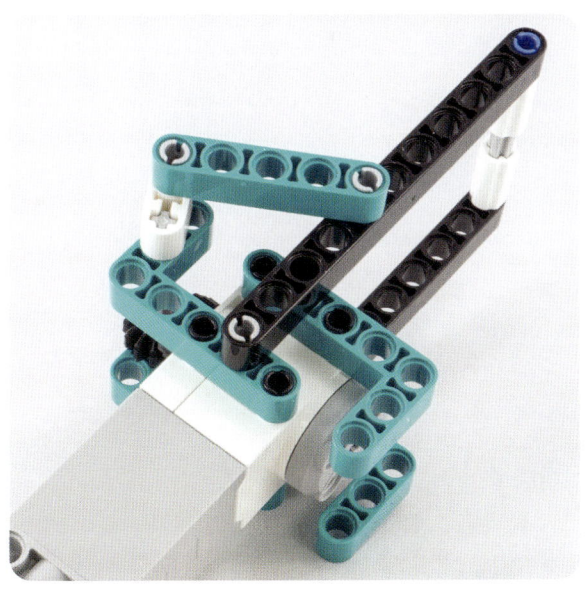

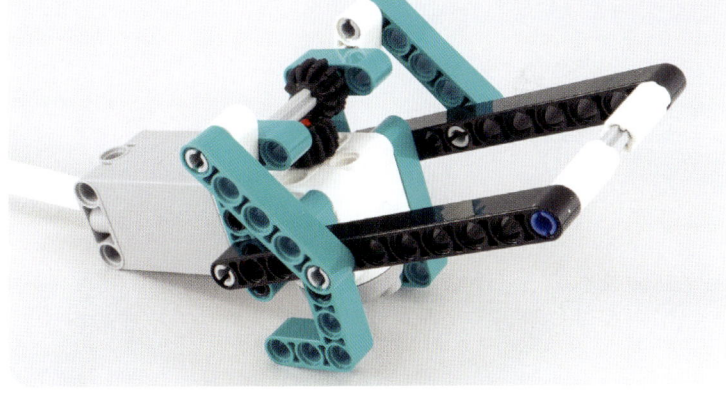

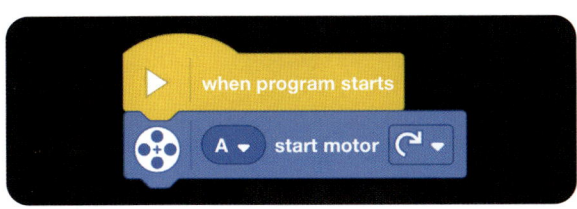

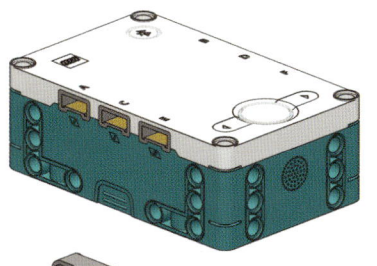

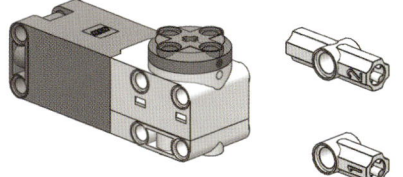

×2

×10

×2

×2

4

5

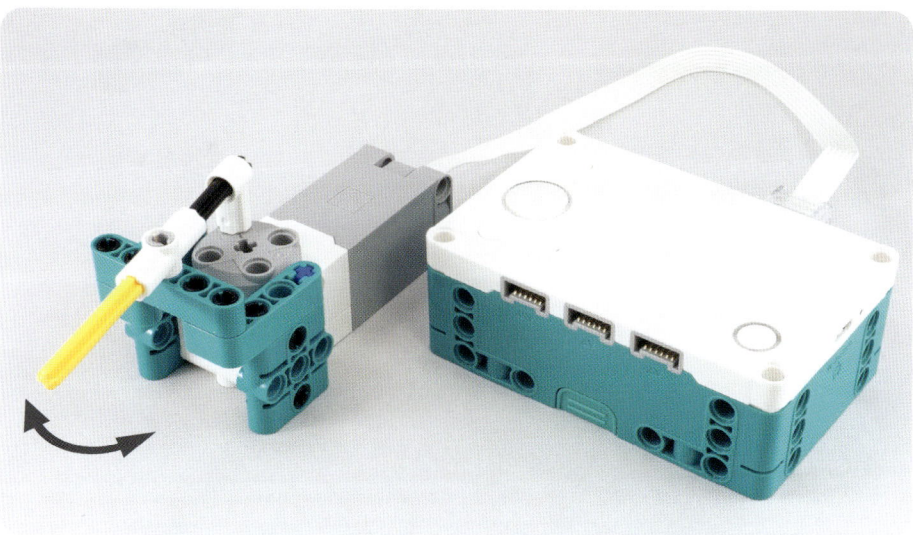

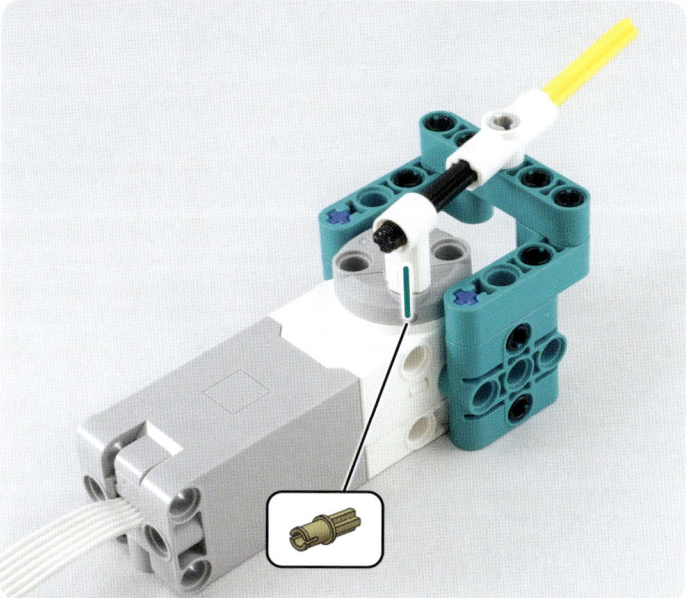

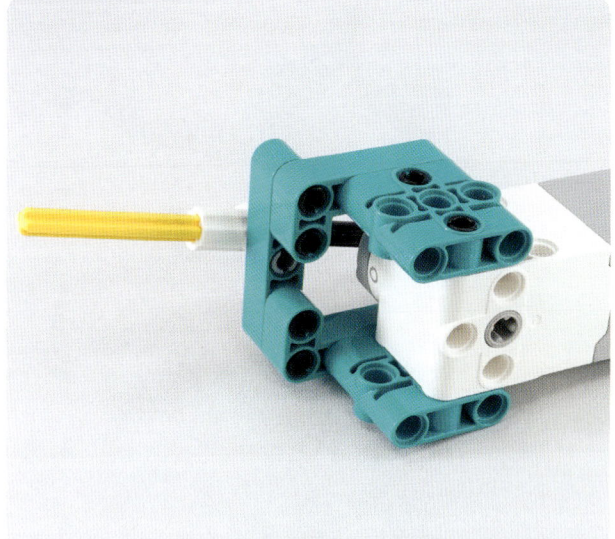

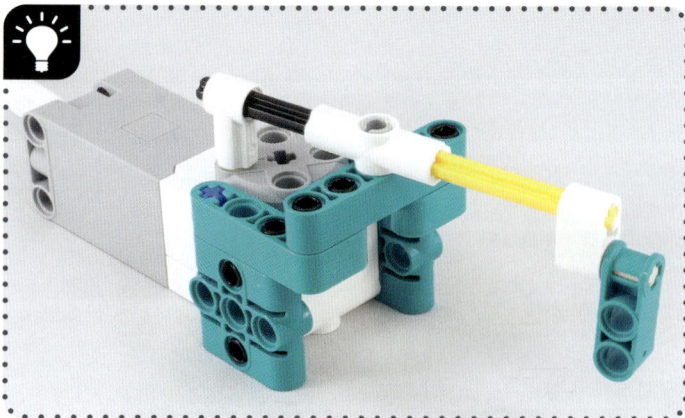

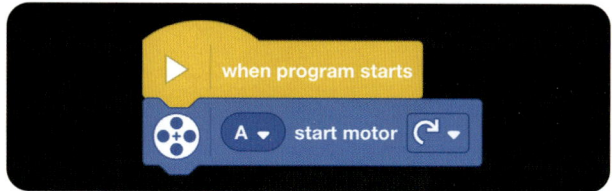

when program starts

A ▾ start motor ↻ ▾

#20

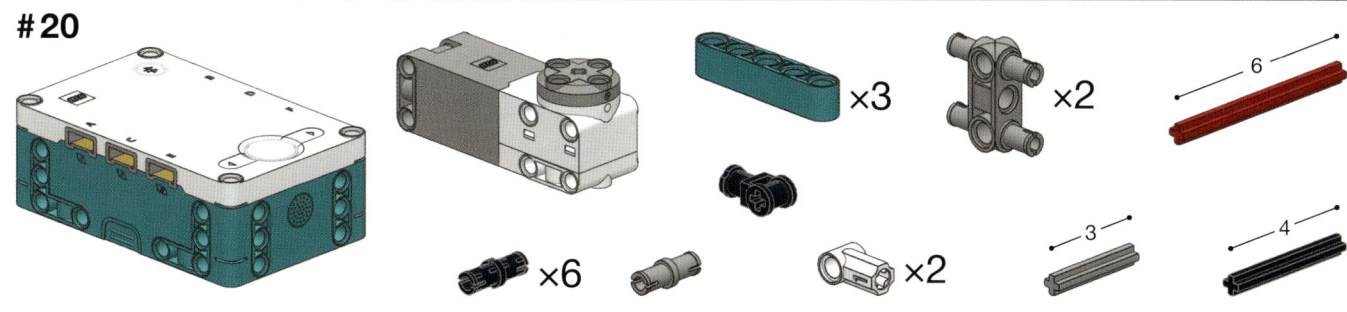

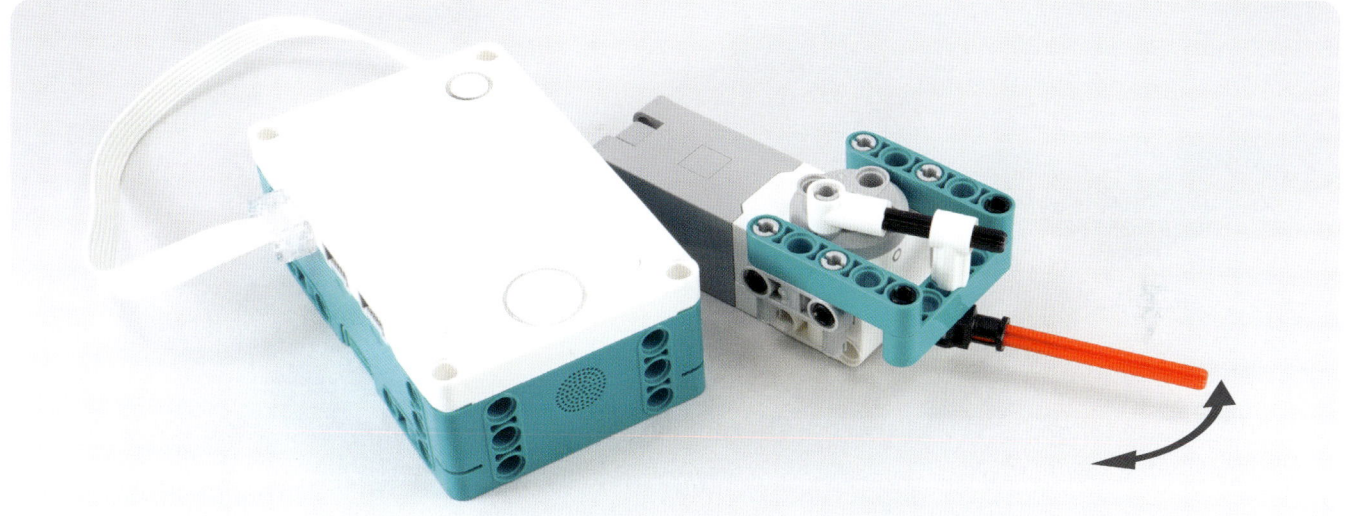

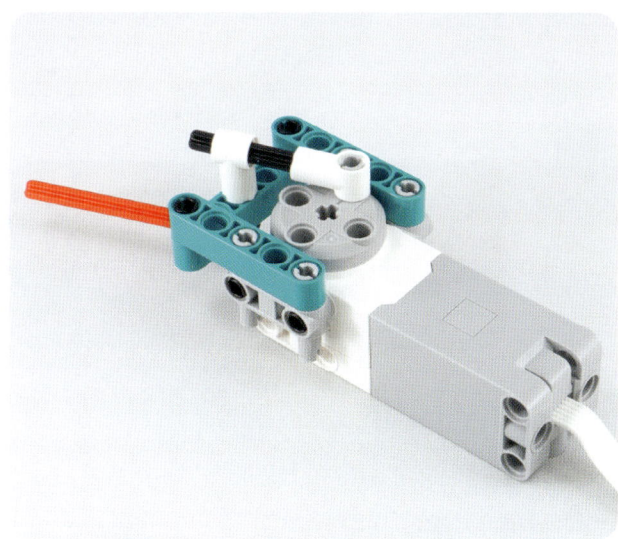

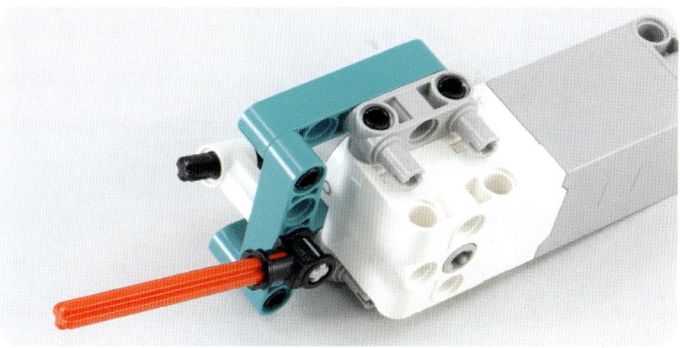

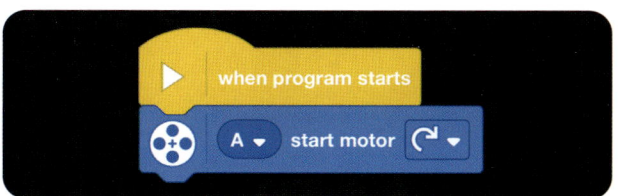

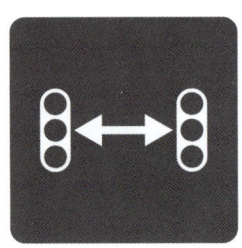

Reciprocating mechanisms

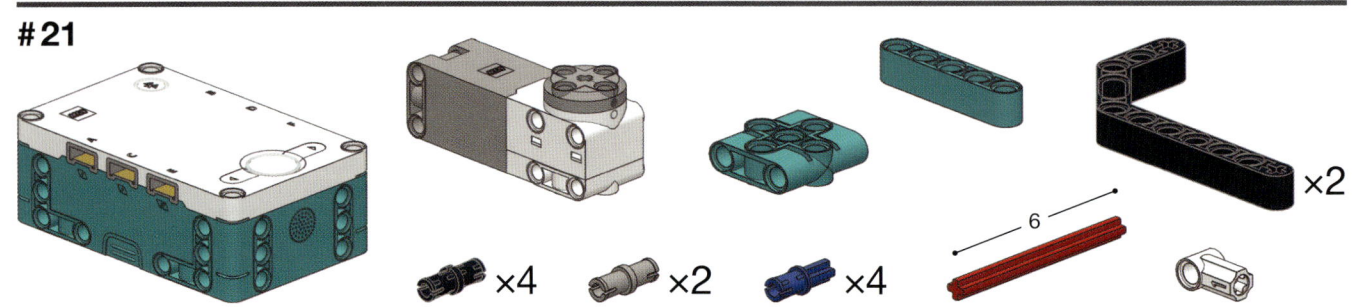

×4   ×2   ×4   6   ×2

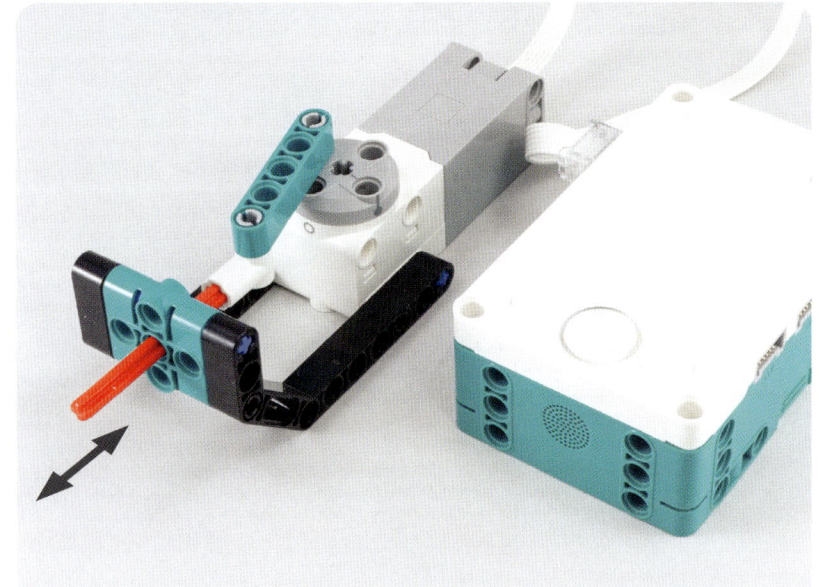

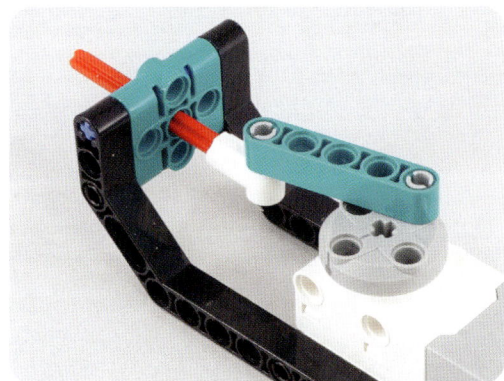

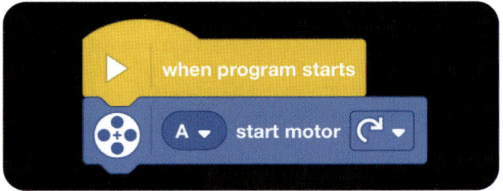

when program starts

A ▾ start motor

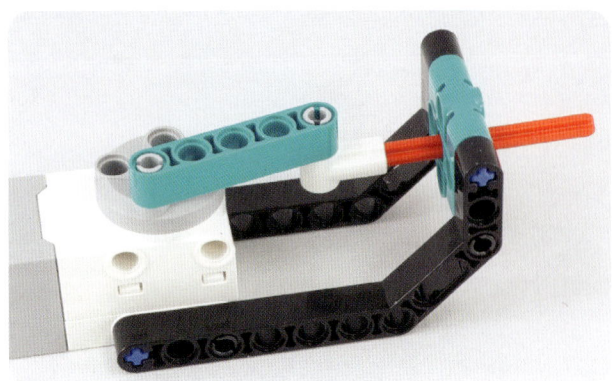

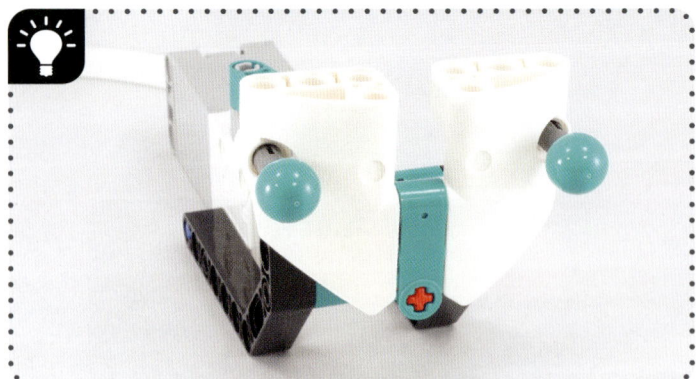

22

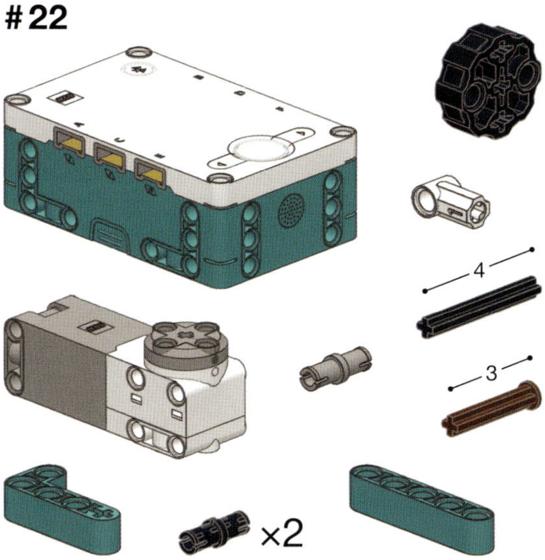

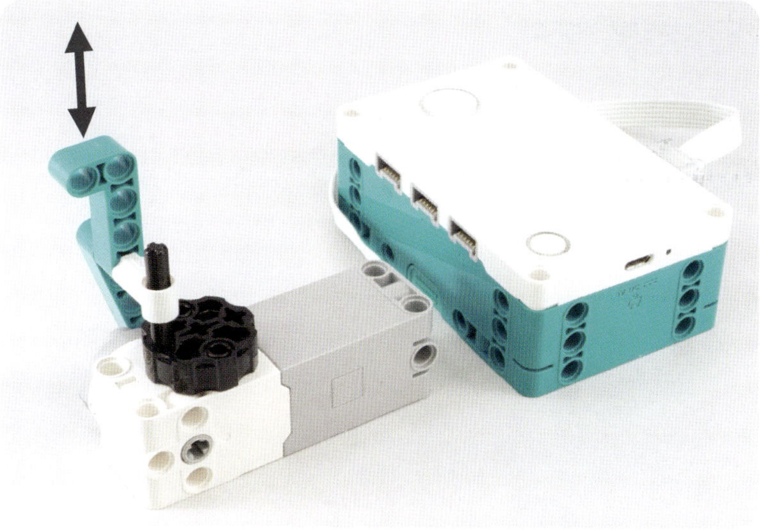

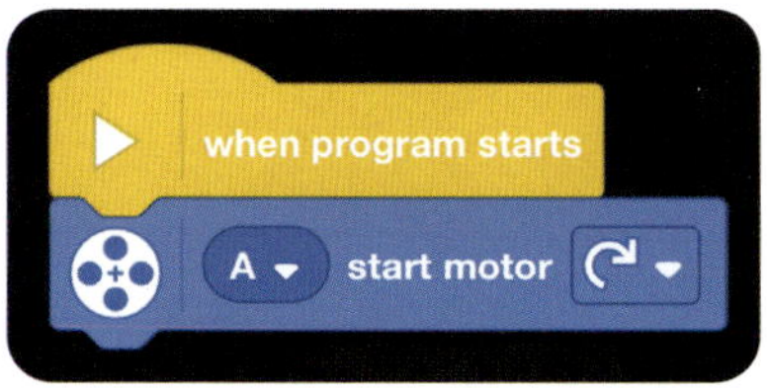

when program starts

A ▼ start motor ↻ ▼

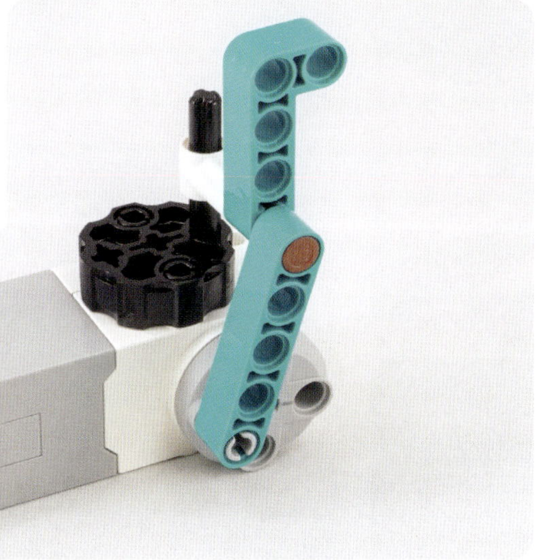

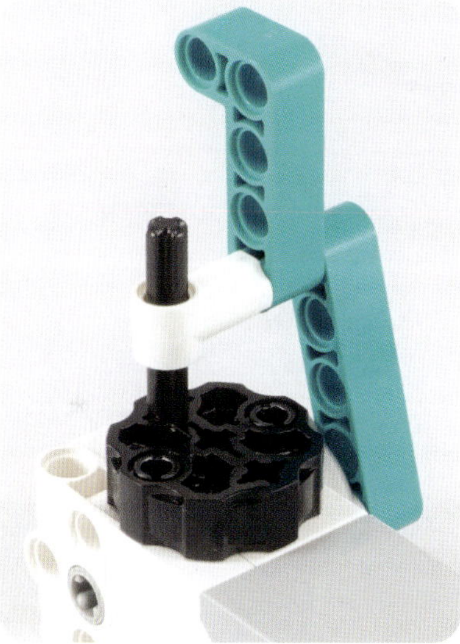

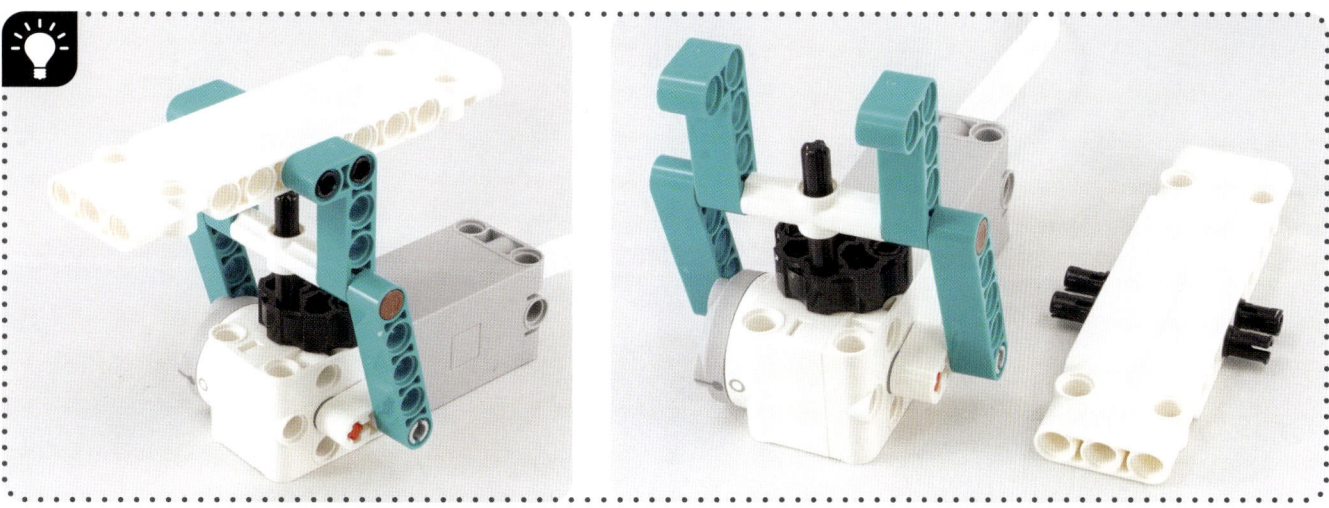

#23

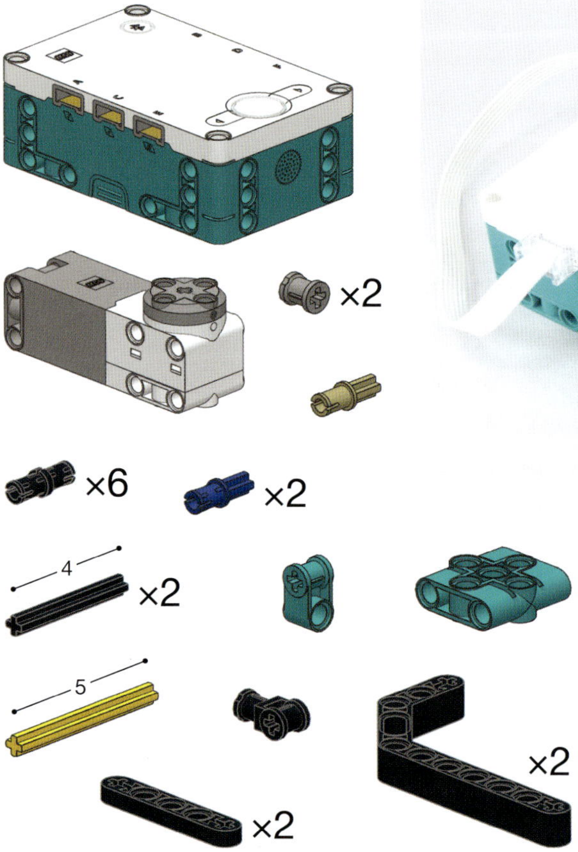

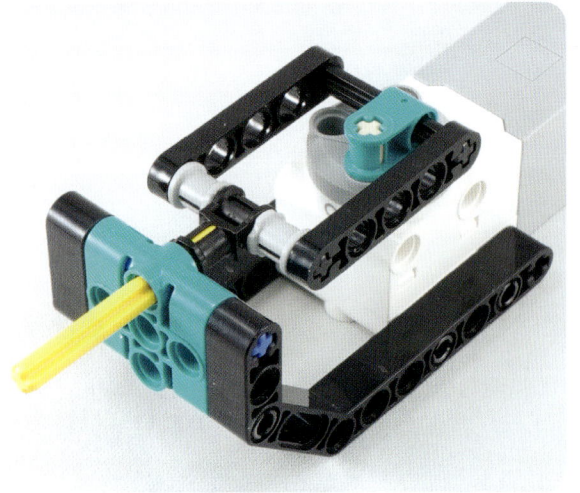

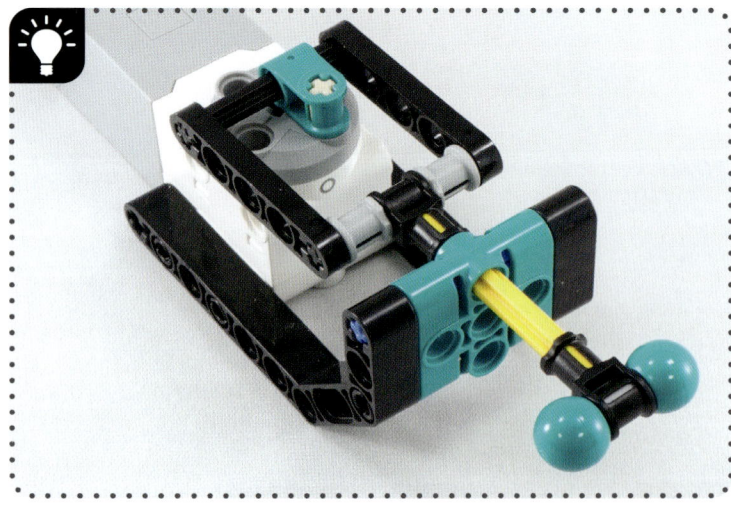

×2

×6 ×2

4 ×2

5 ×2

×2

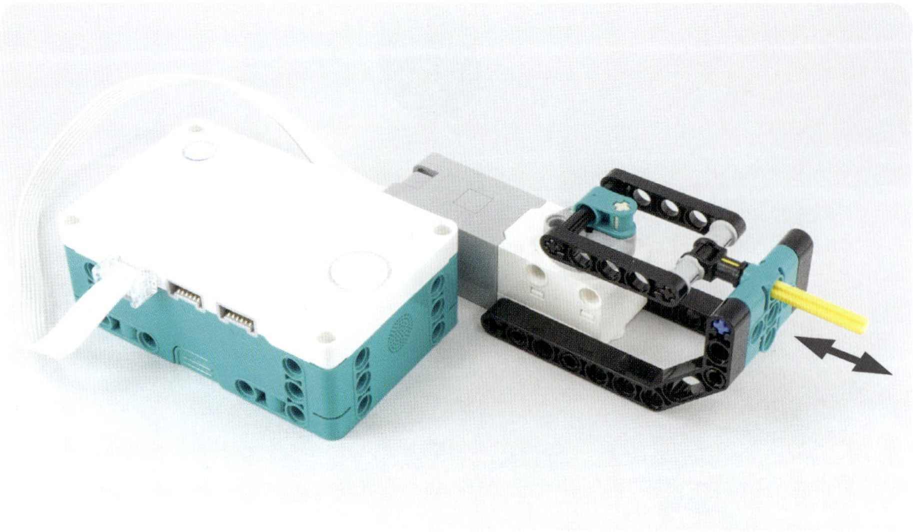

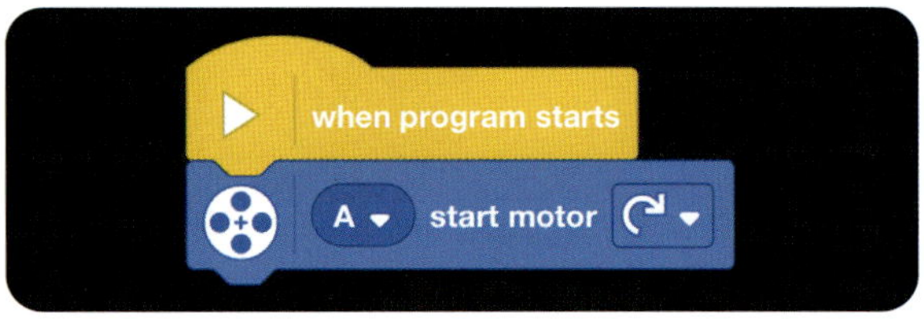

×2

when program starts

A ▾ start motor ↻ ▾

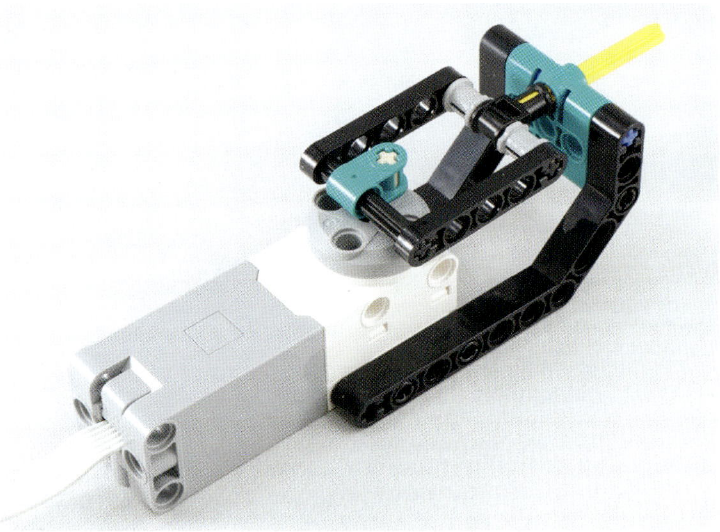

#24

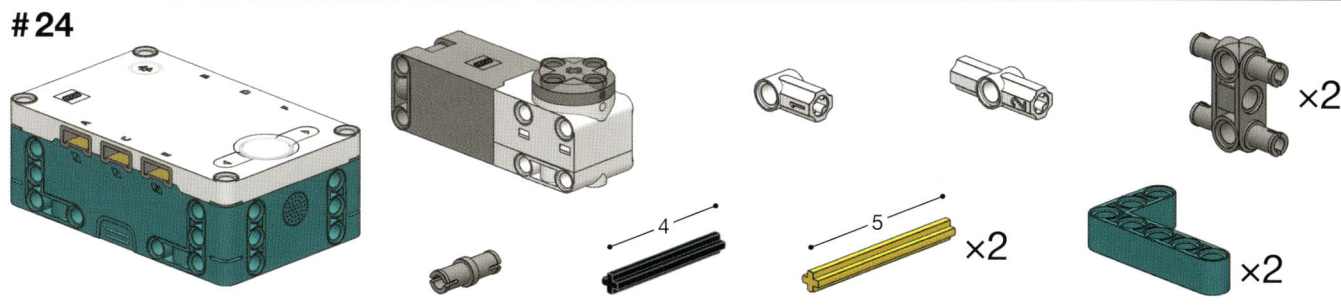

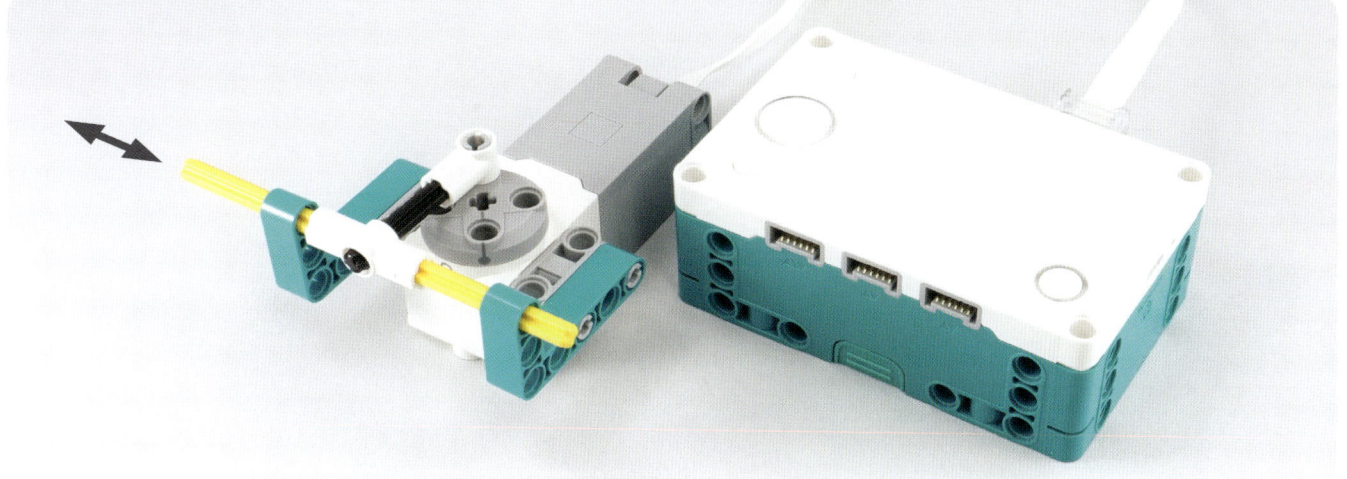

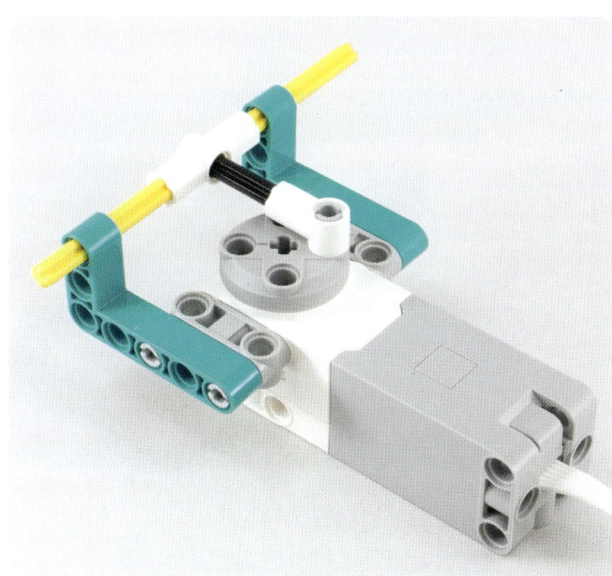

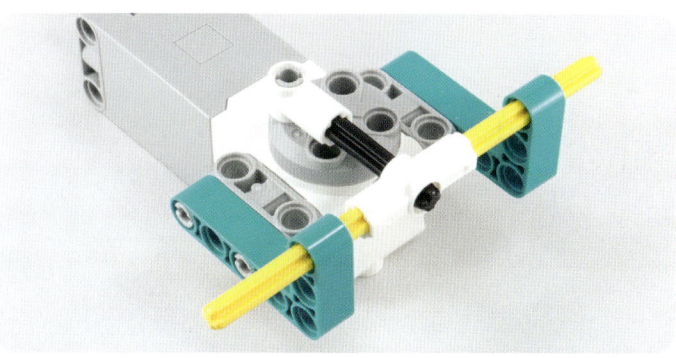

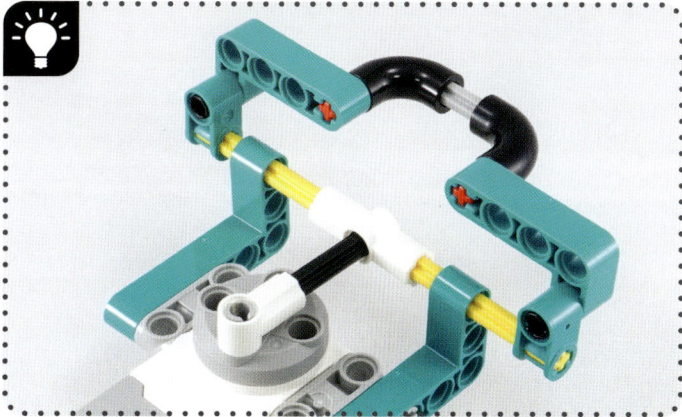

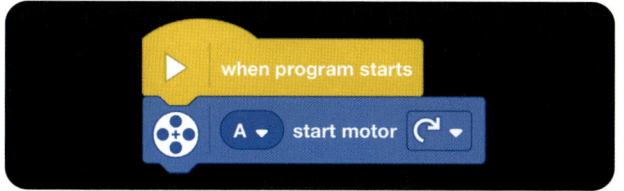

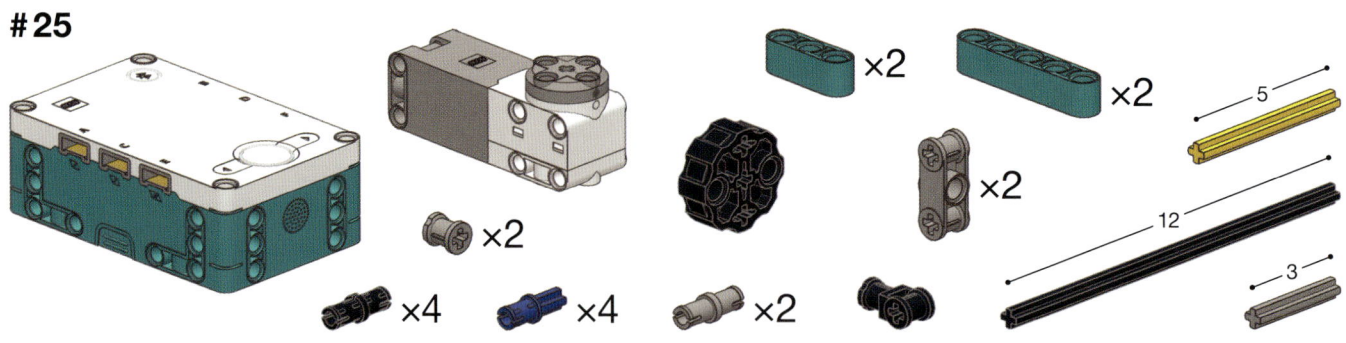

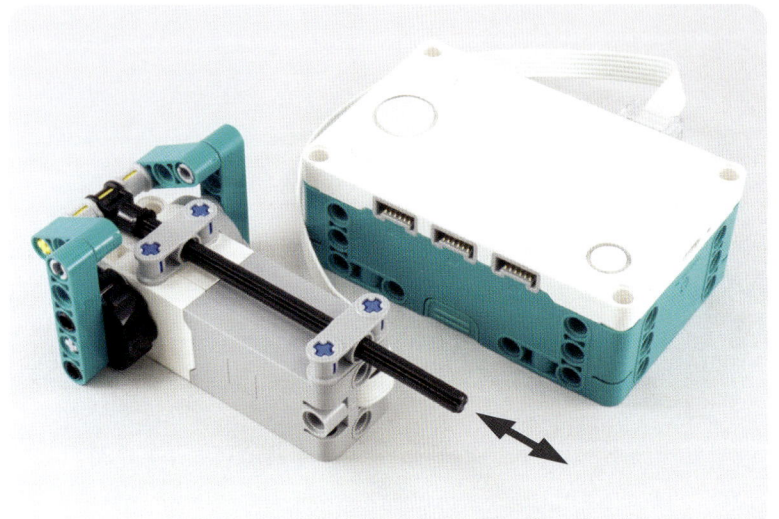

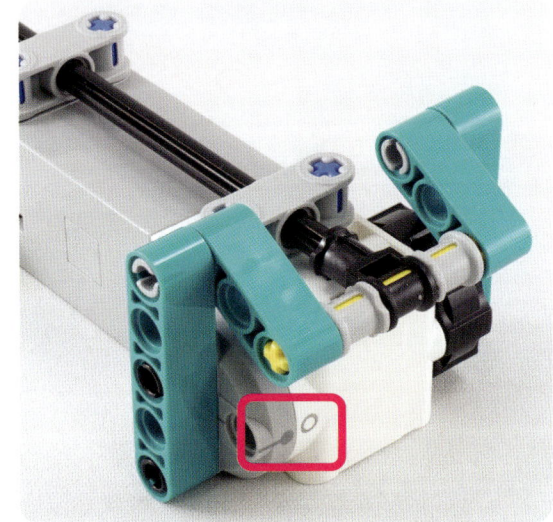

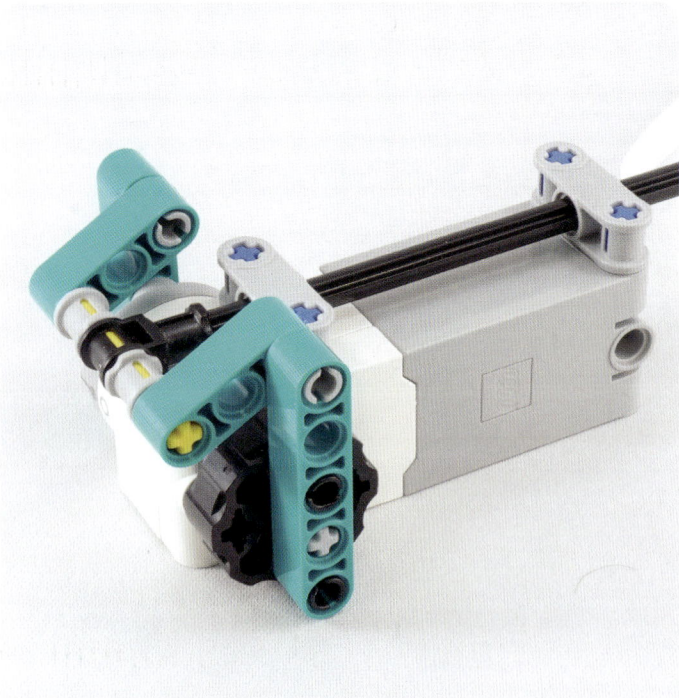

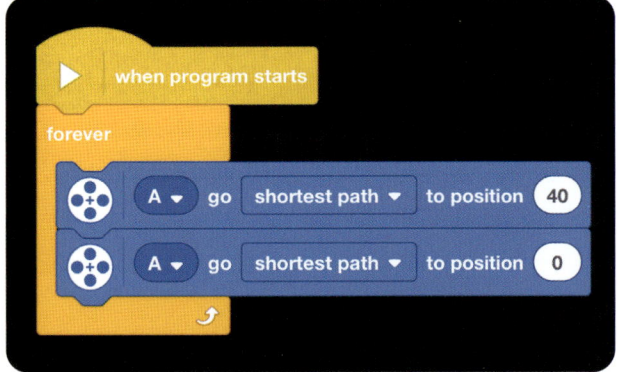

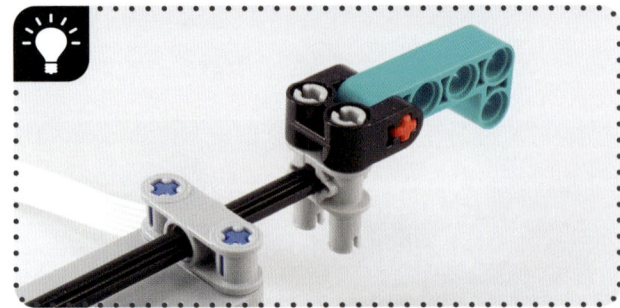

#26

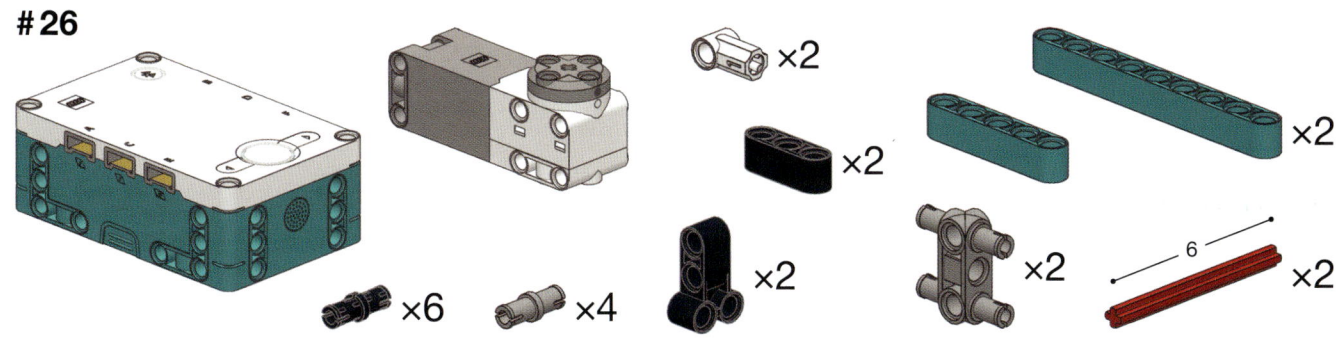

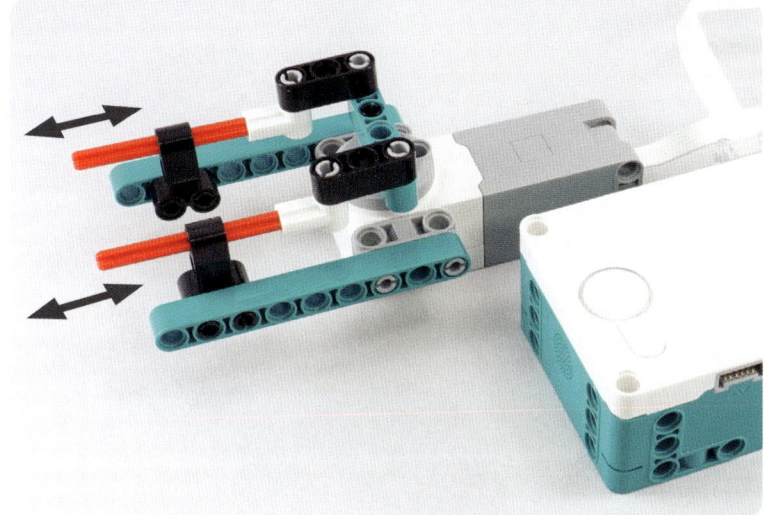

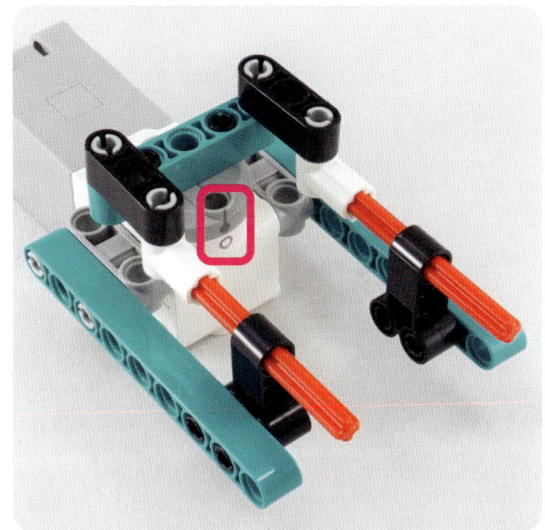

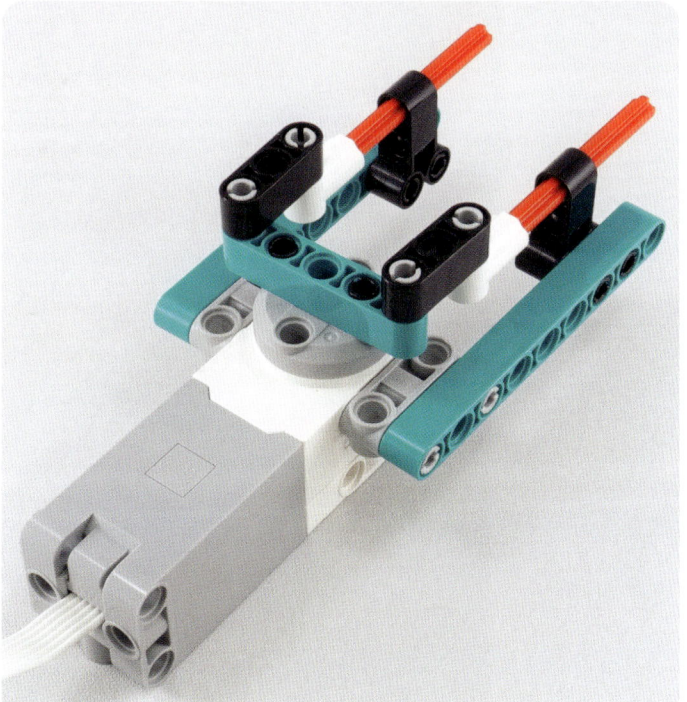

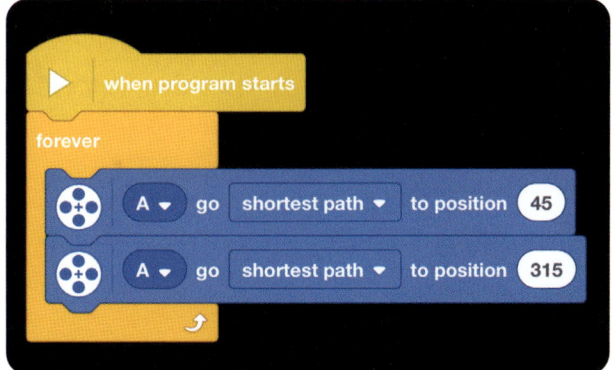

```
when program starts
forever
    A ▼  go  shortest path ▼  to position  45
    A ▼  go  shortest path ▼  to position  315
```

Changing the angle of an axle

 # 27

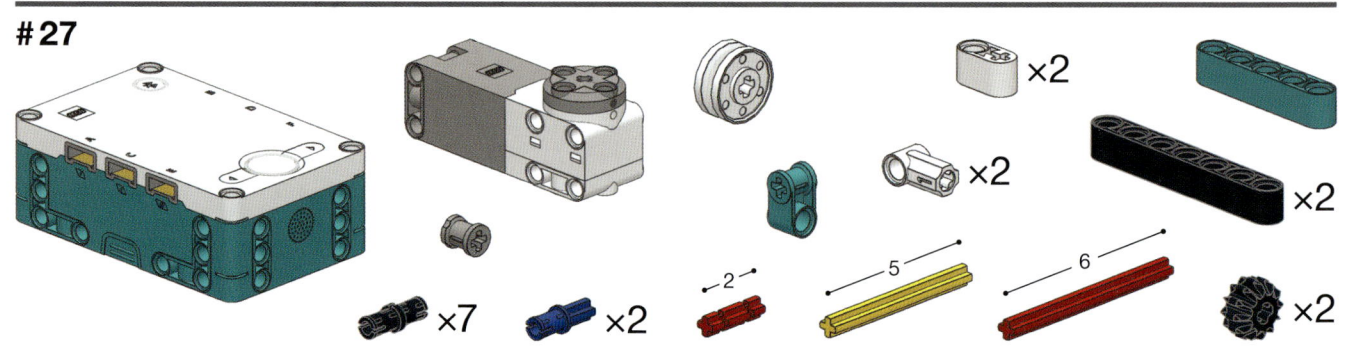

×2 ×2 ×2 ×2

×7 ×2 →2 5 6 ×2

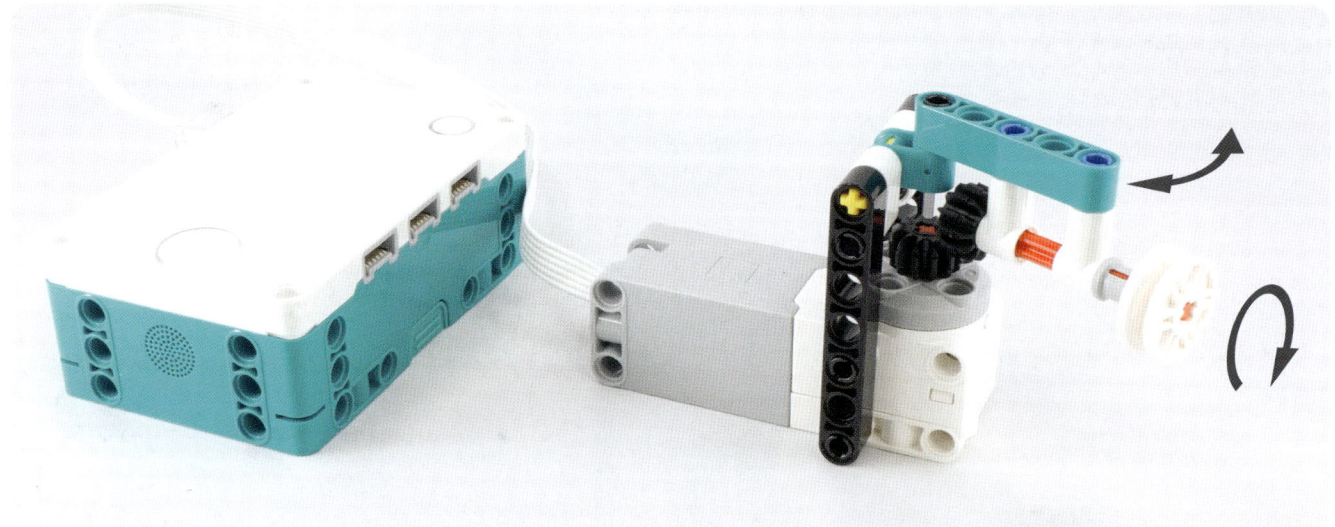

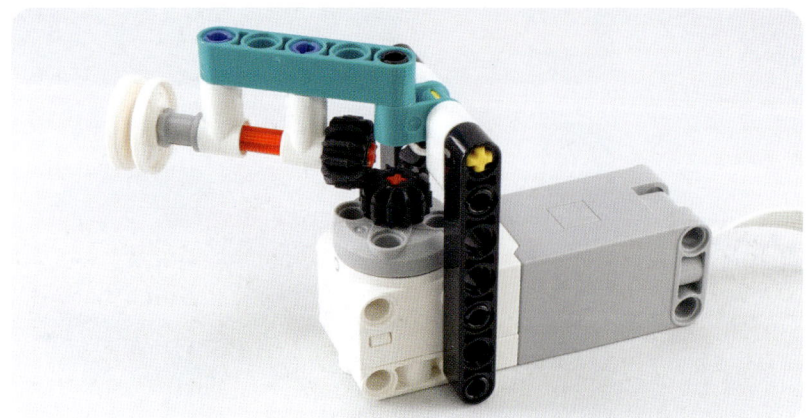

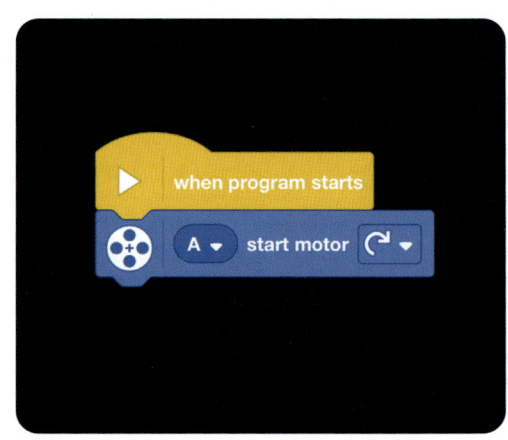

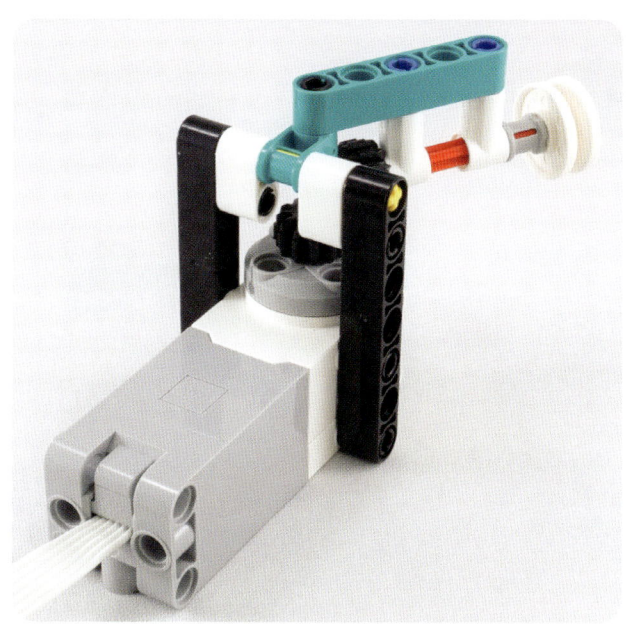

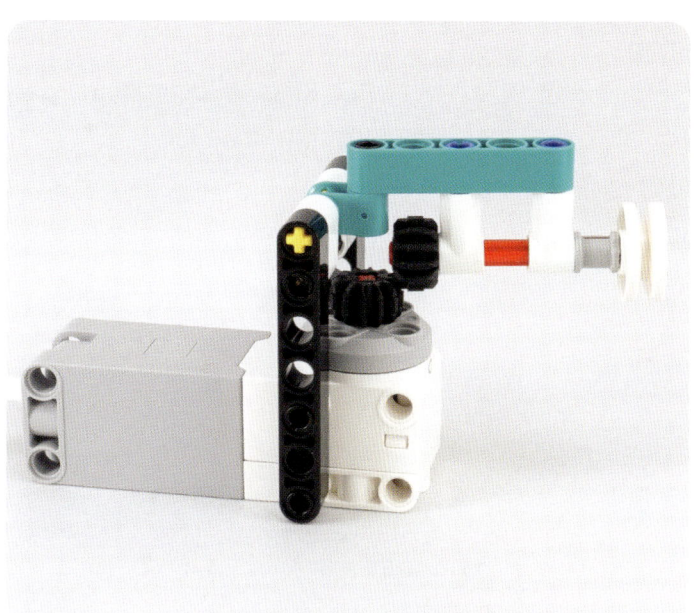

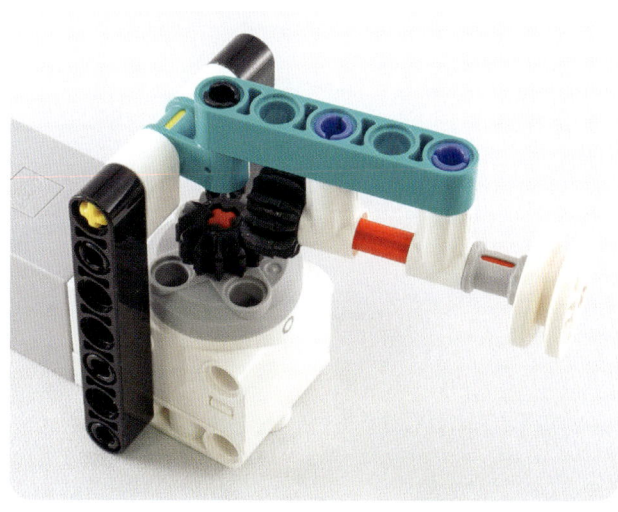

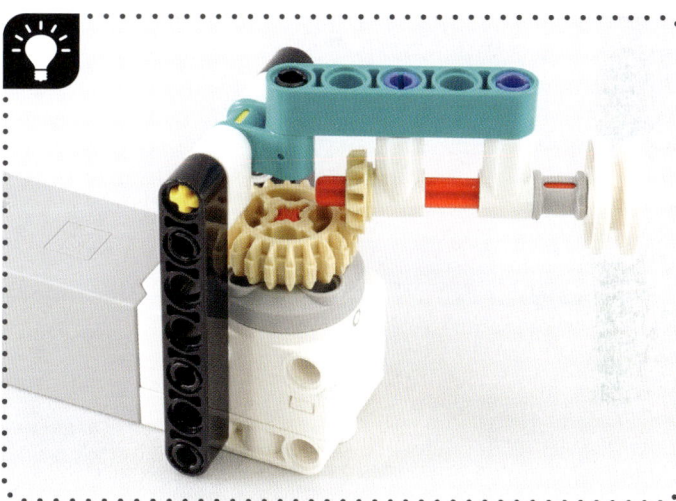

#28

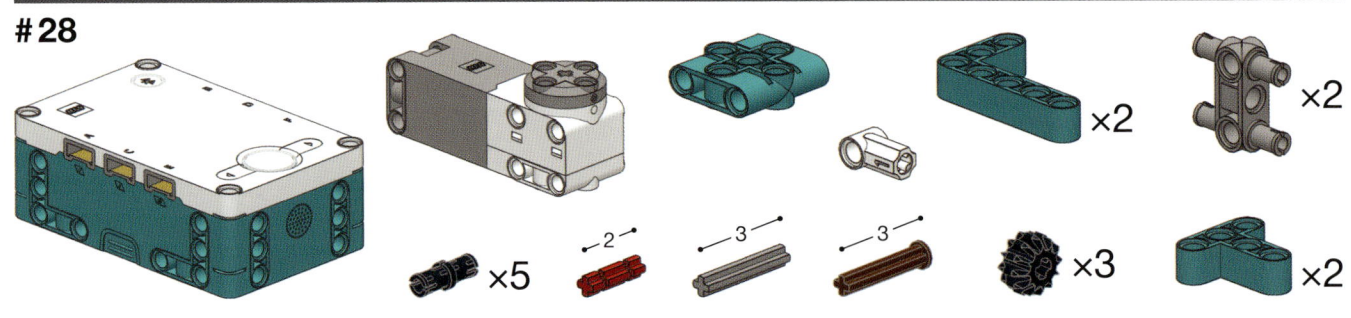

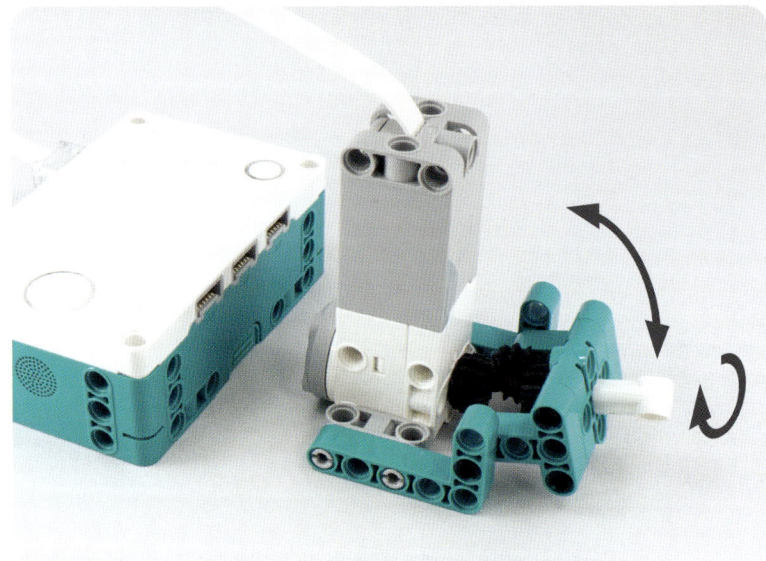

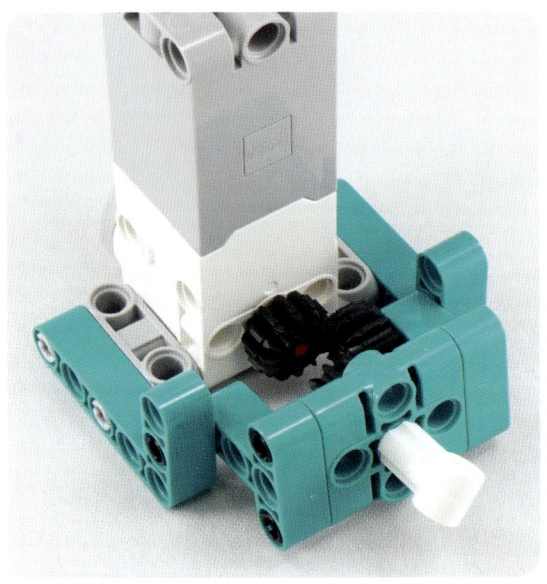

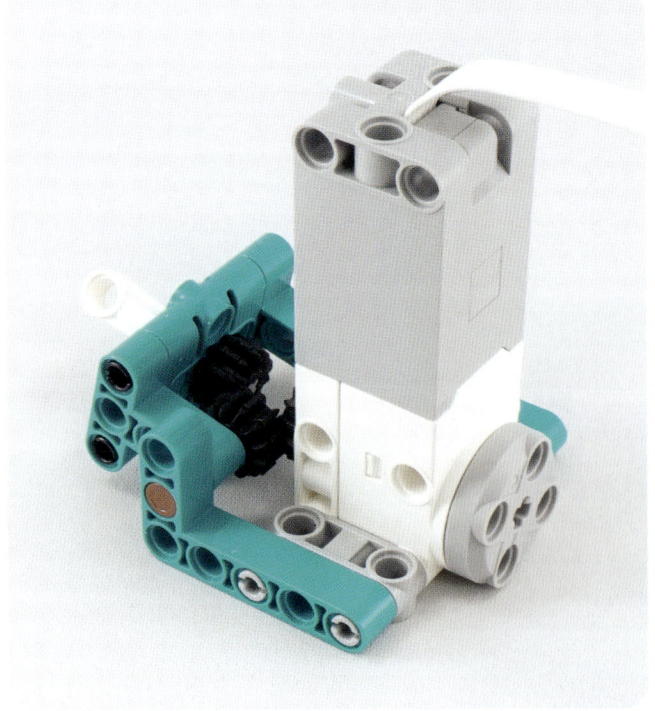

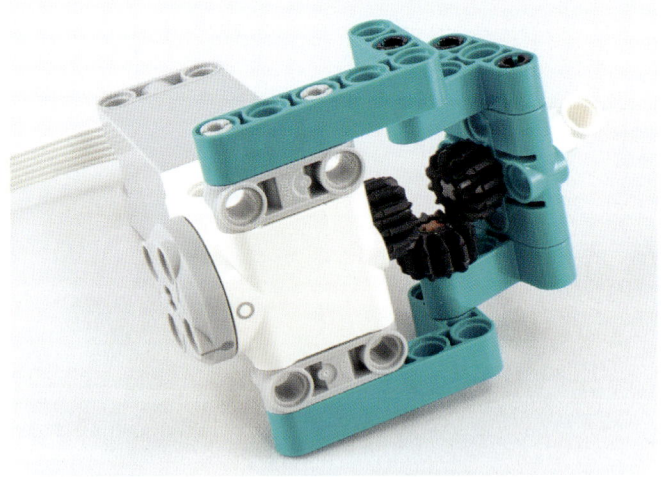

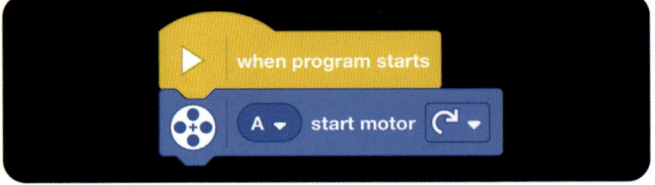

#29

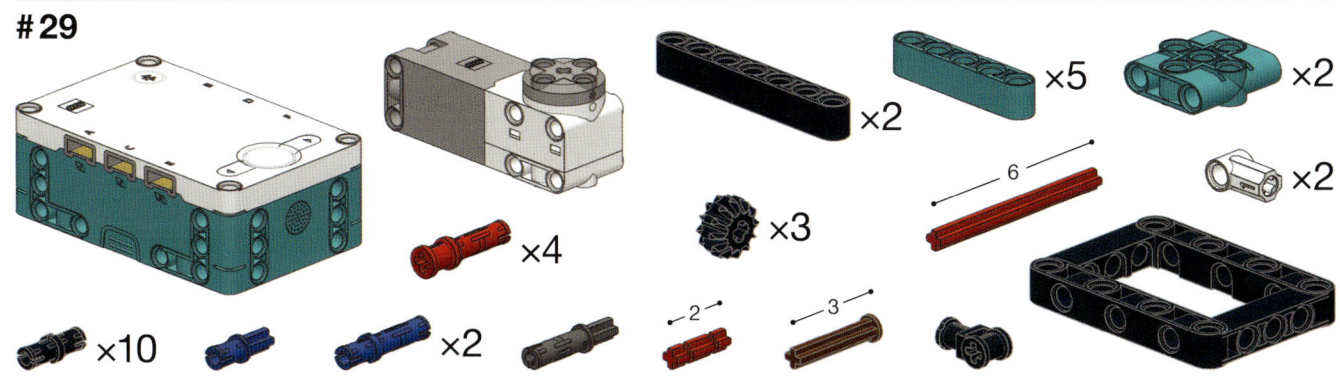

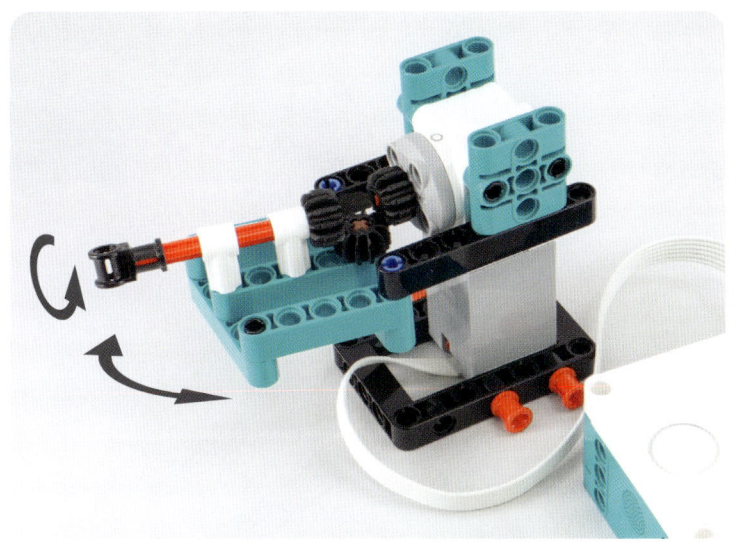

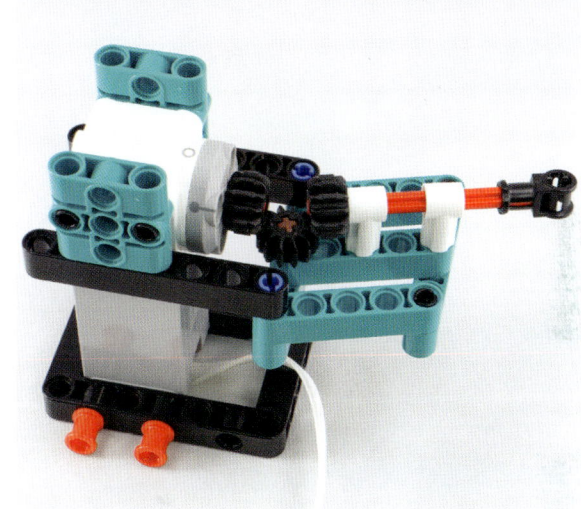

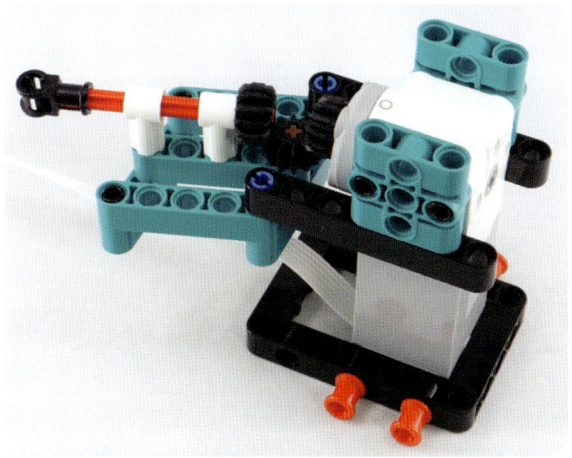

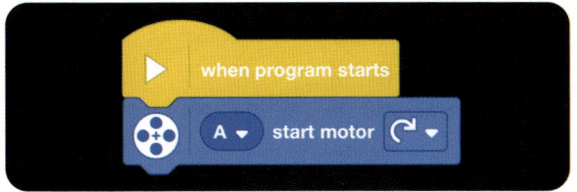

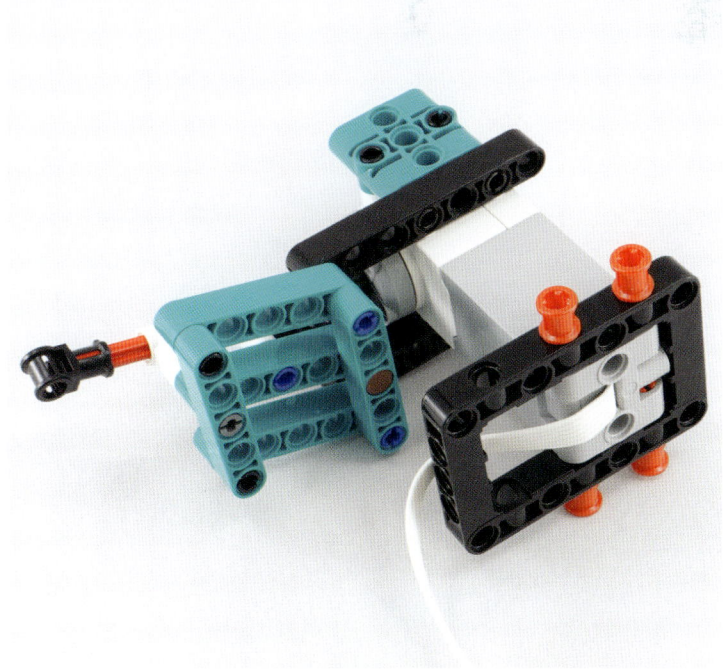

Transmitting rotation with rubber bands

#30

×2

×6   ×2   3   4

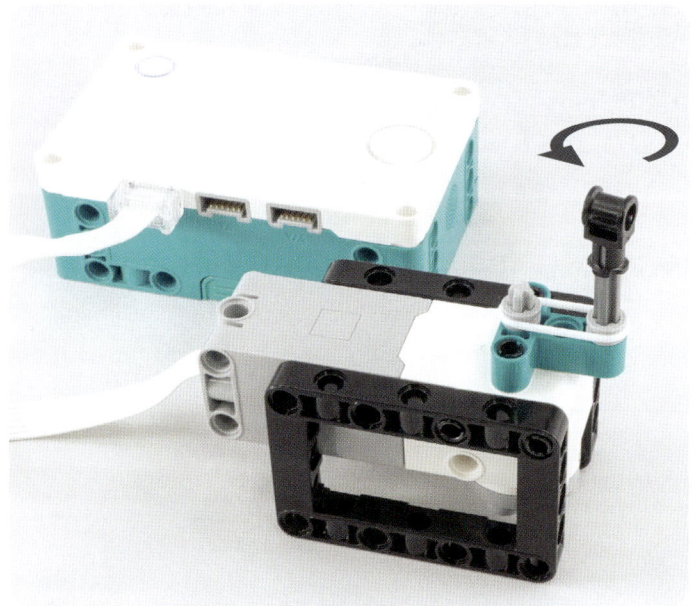

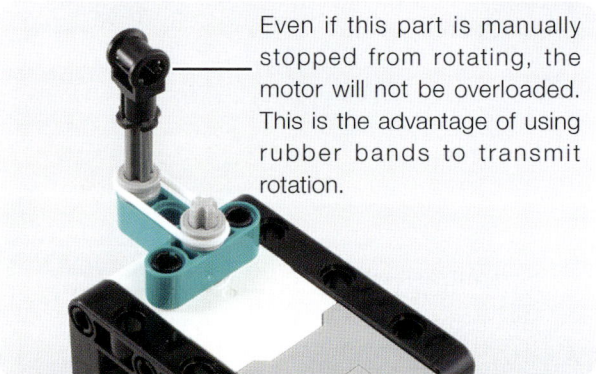

Even if this part is manually stopped from rotating, the motor will not be overloaded. This is the advantage of using rubber bands to transmit rotation.

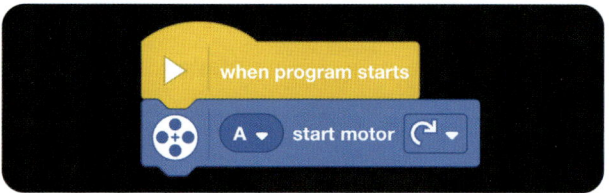

when program starts

A ▾   start motor

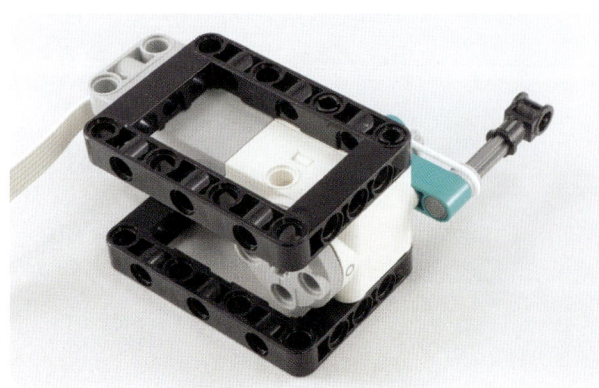

#31

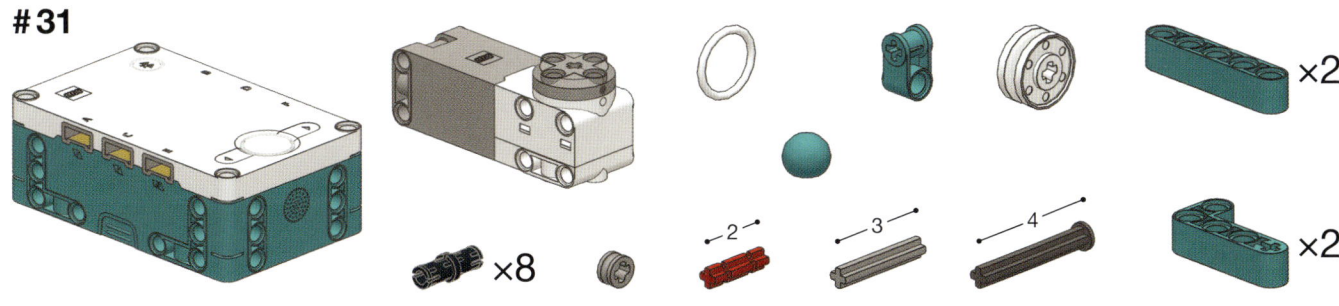

×8   ×2   ×2

- 2
- 3
- 4

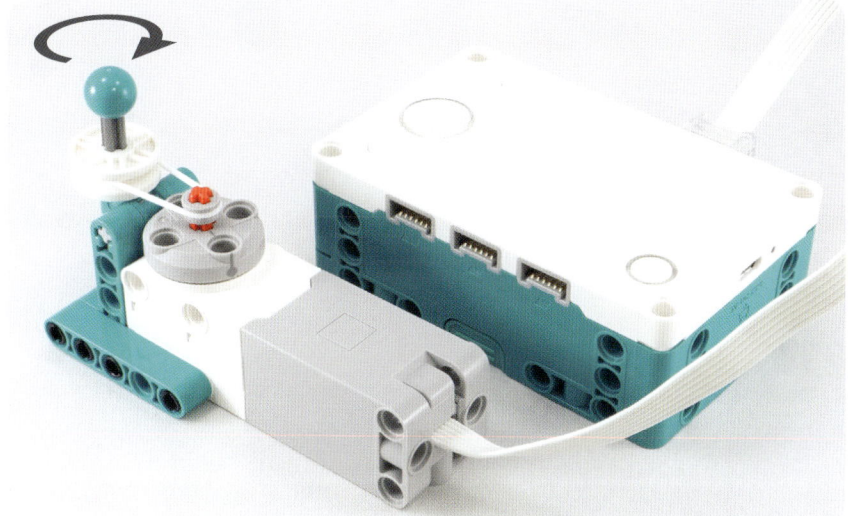

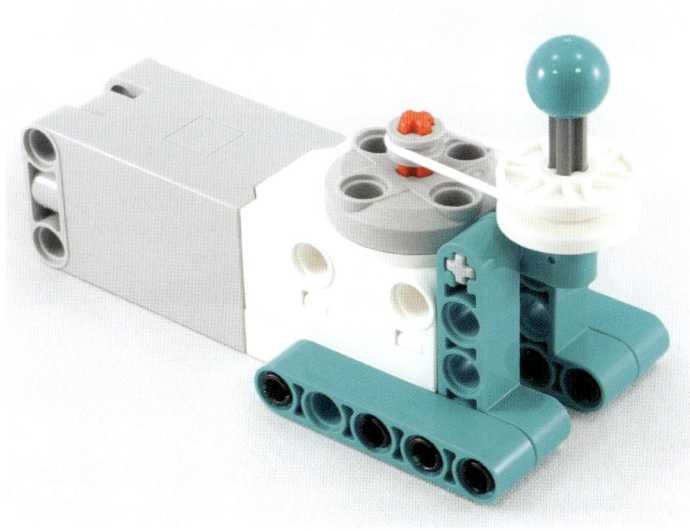

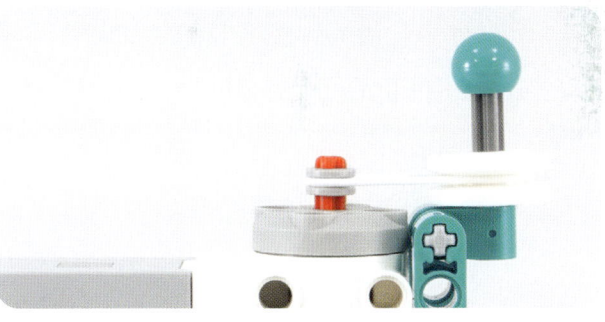

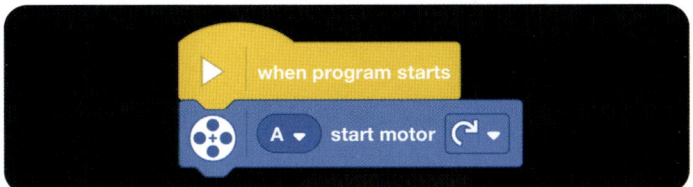

when program starts

A ▼   start motor ↻ ▼

Using cams

#32

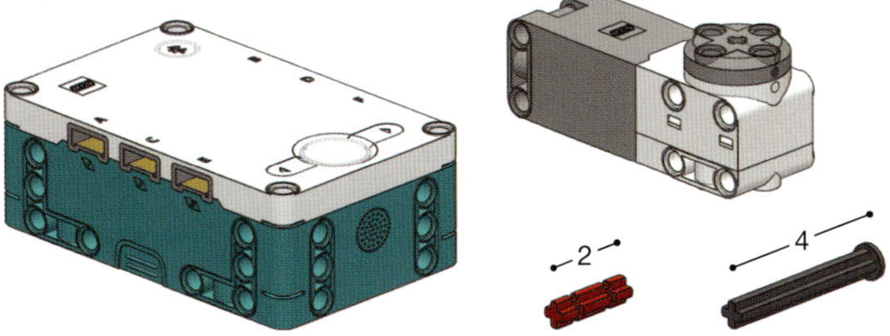

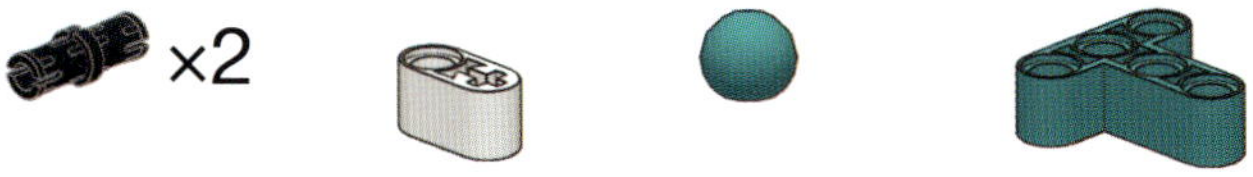

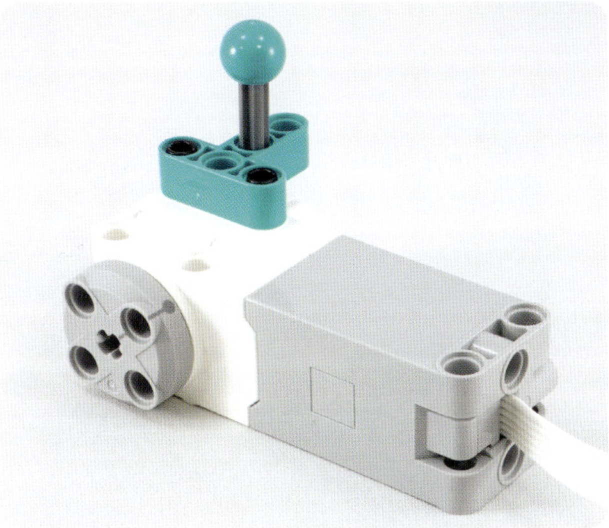

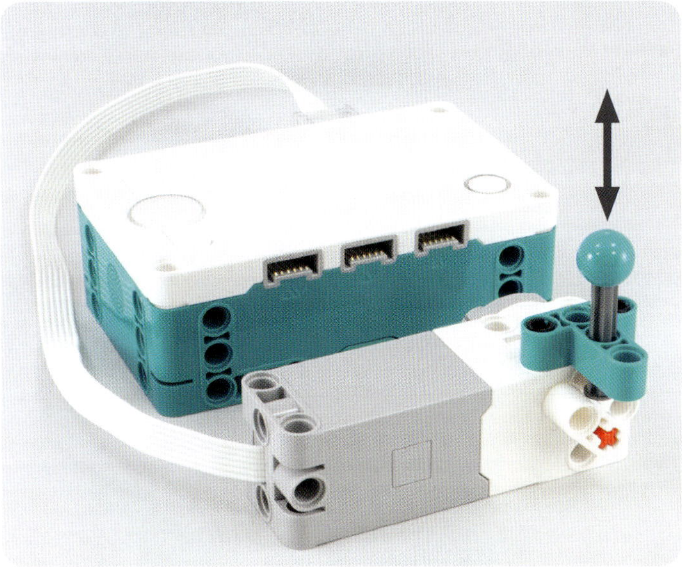

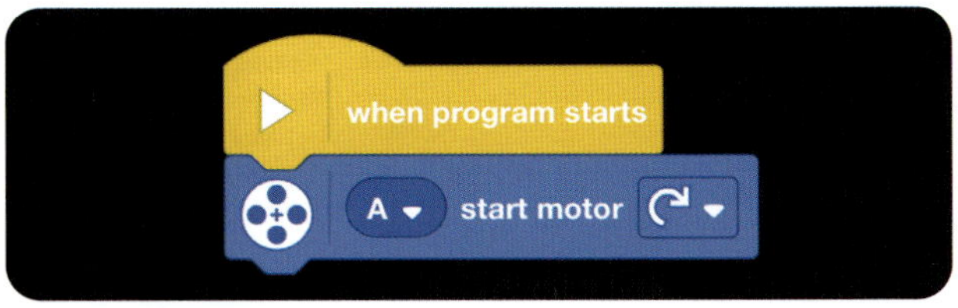

×2

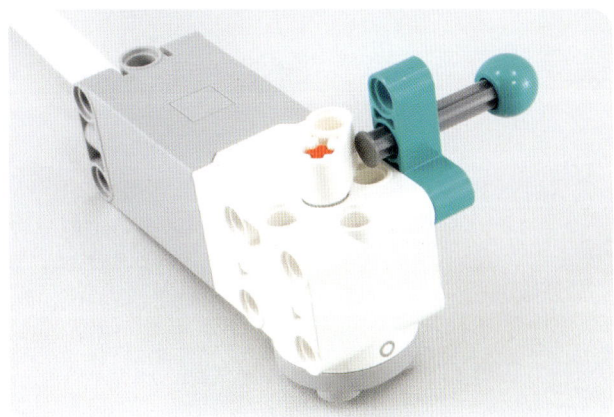

when program starts

A ▾ start motor ↻ ▾

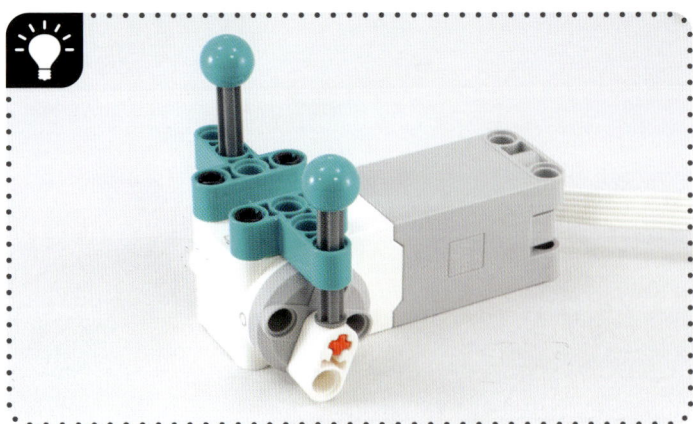

#33

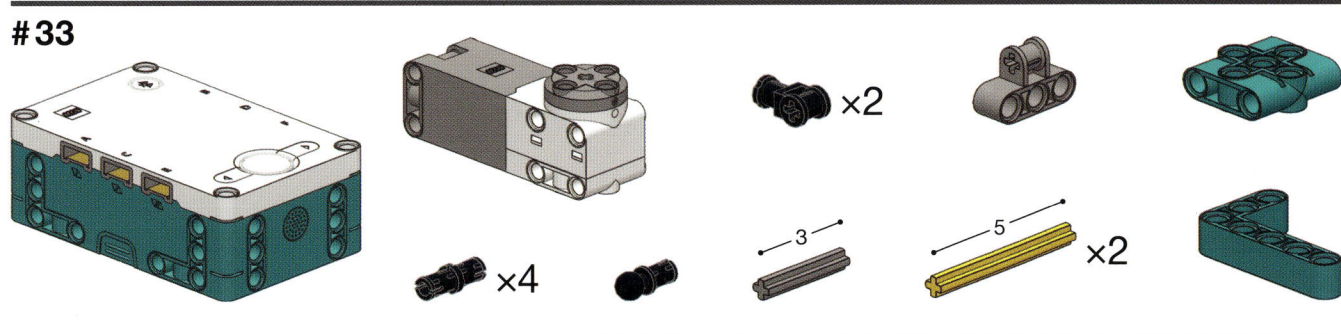

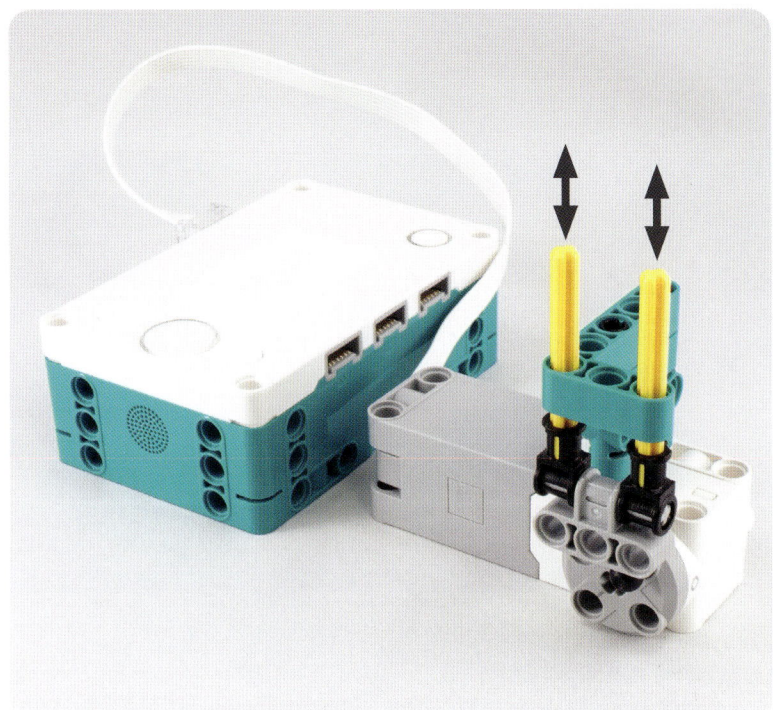

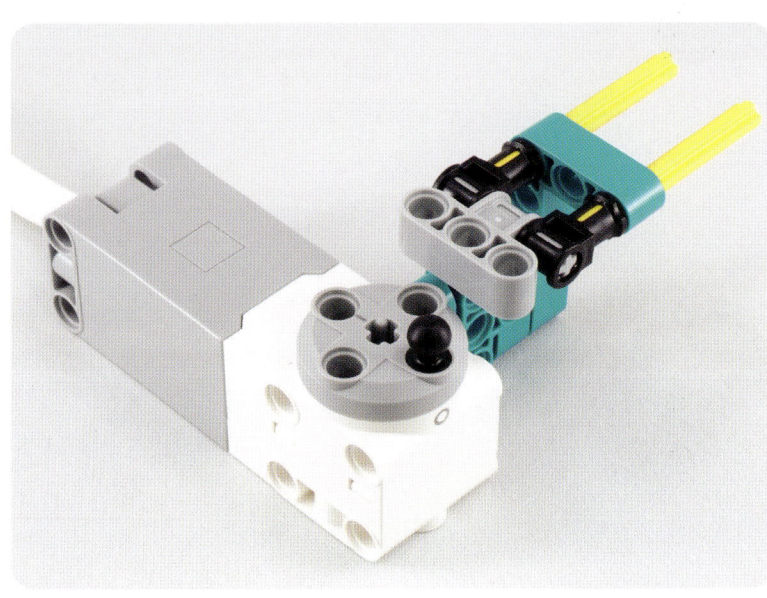

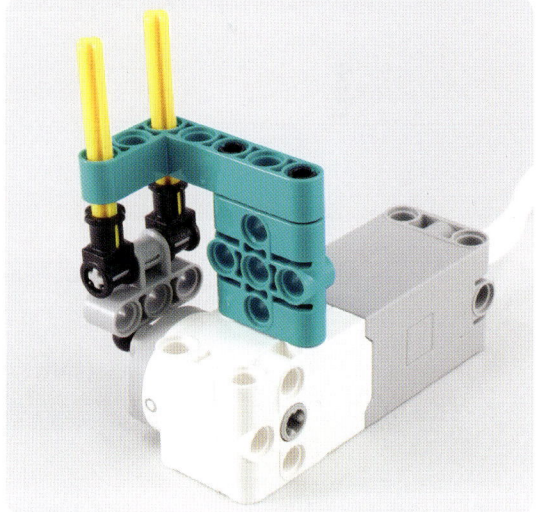

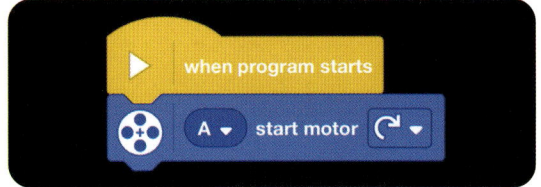

when program starts

A ▾ start motor ↻ ▾

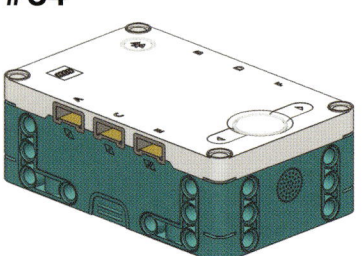
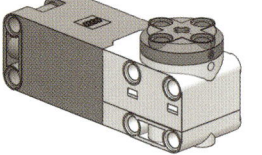

 ×2

×6 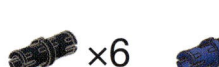 ×2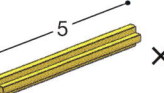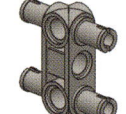

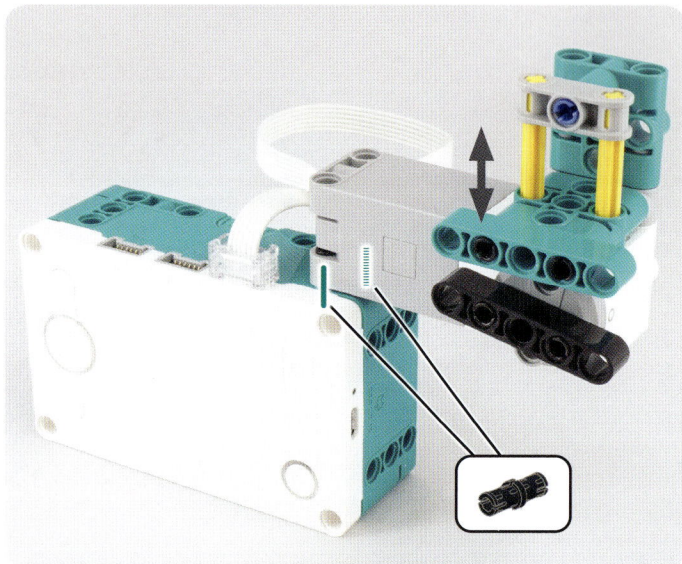

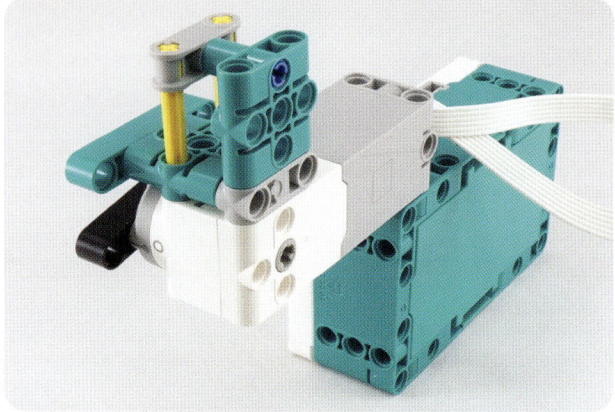

when program starts

A ▾ start motor ↻ ▾

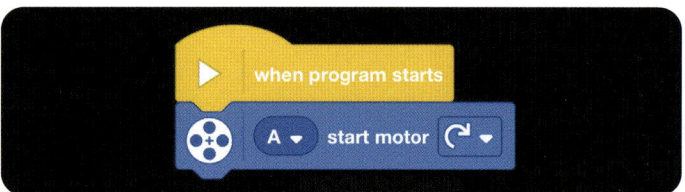

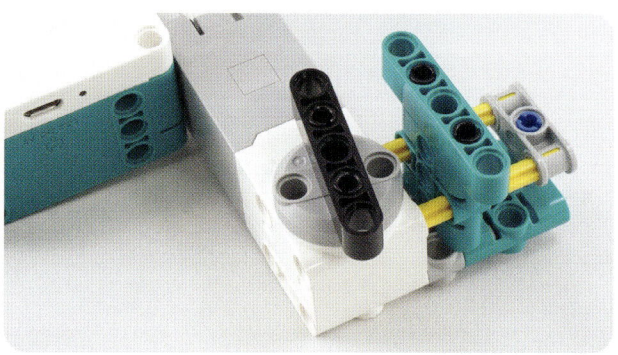

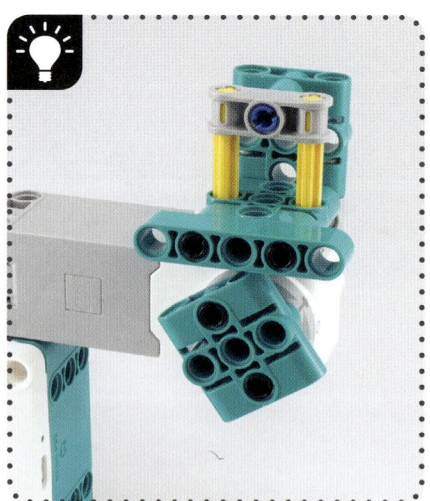

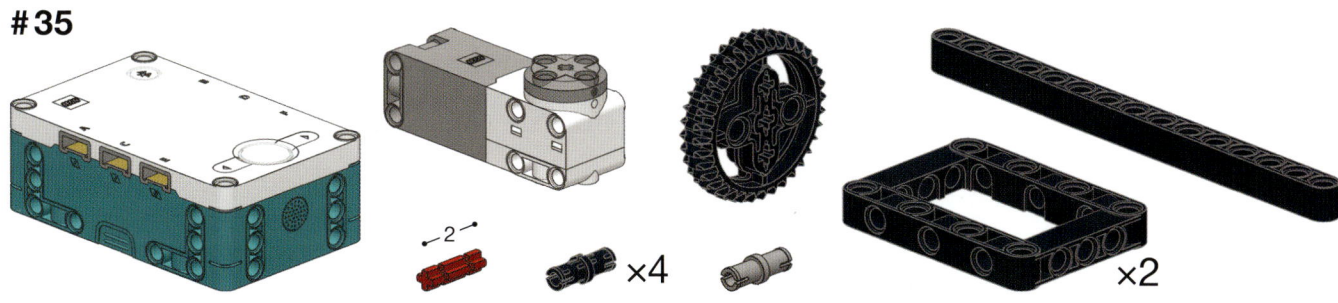

×2   ×4   ×2

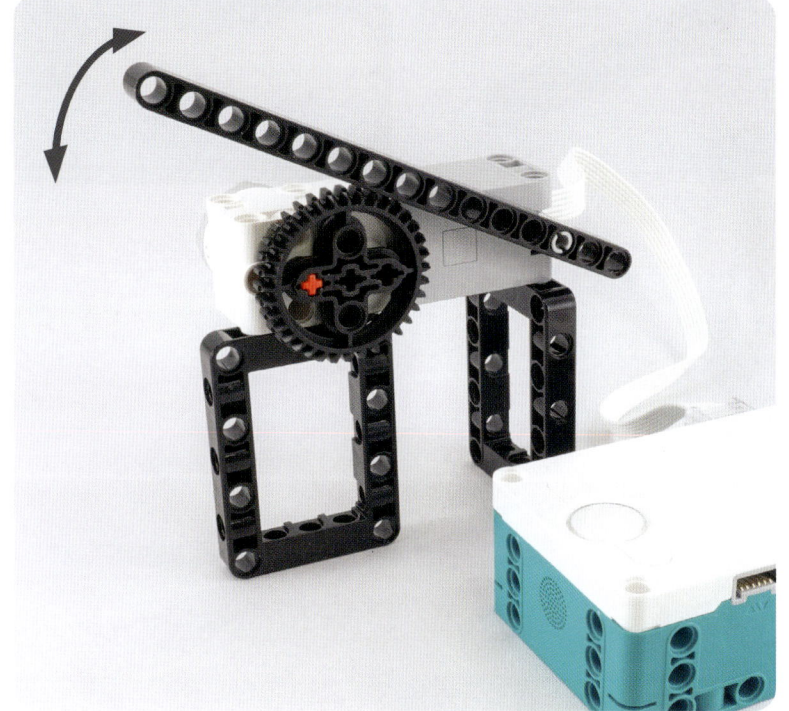

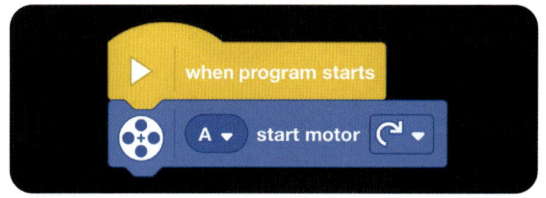

when program starts

A ▾ start motor ↻ ▾

Using turntables

#36

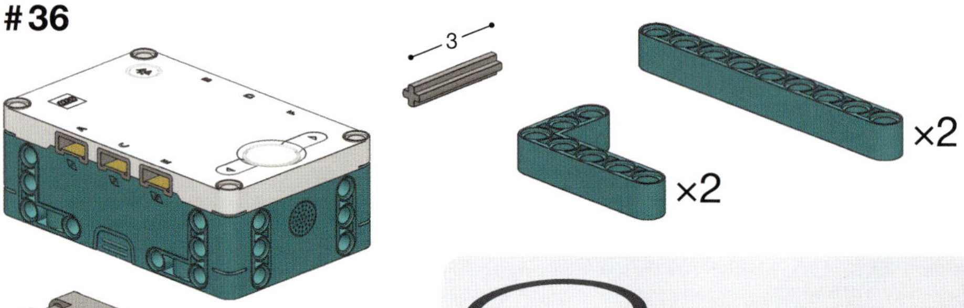

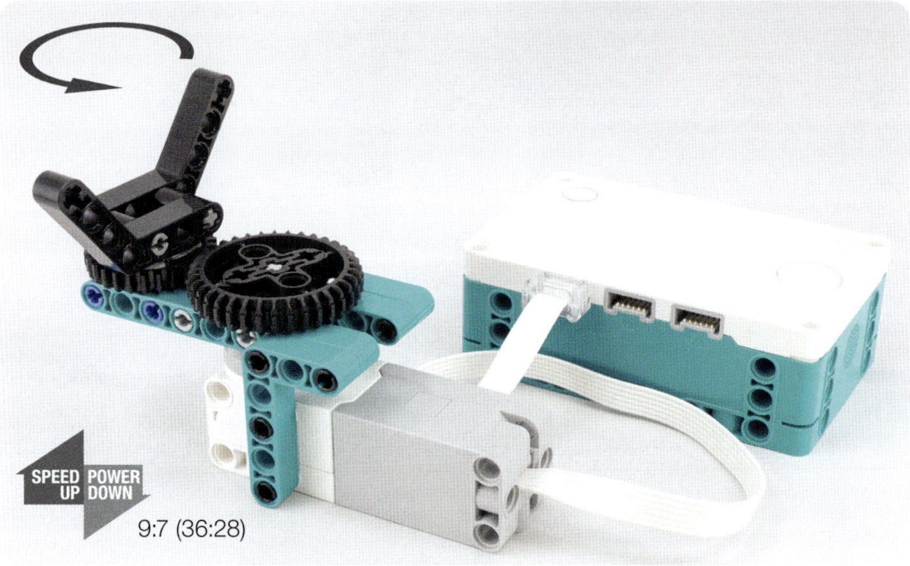

×3

×2

×2

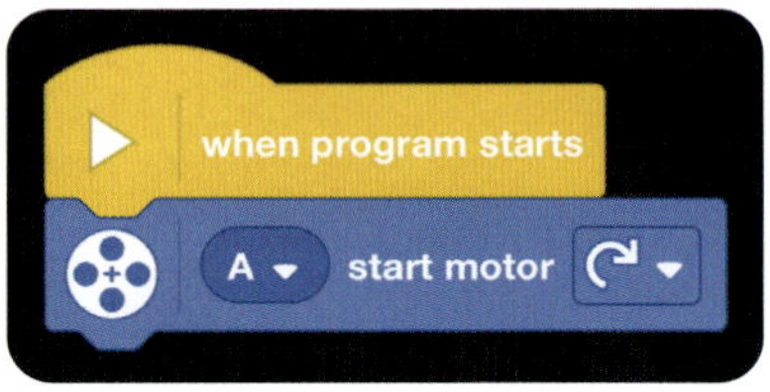

×8

×2

×2

×2

when program starts

A ▾ start motor

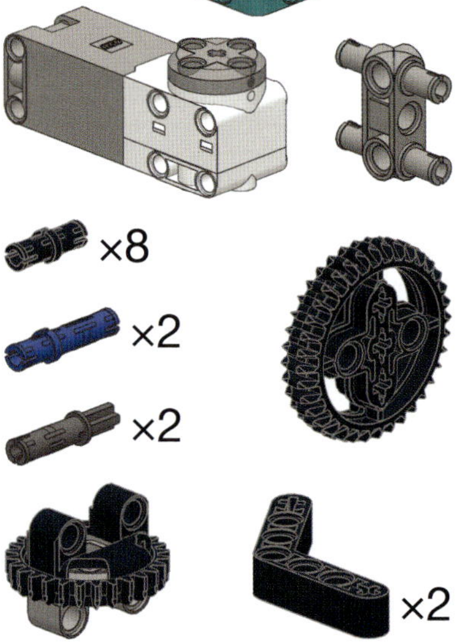

SPEED POWER
UP DOWN

9:7 (36:28)

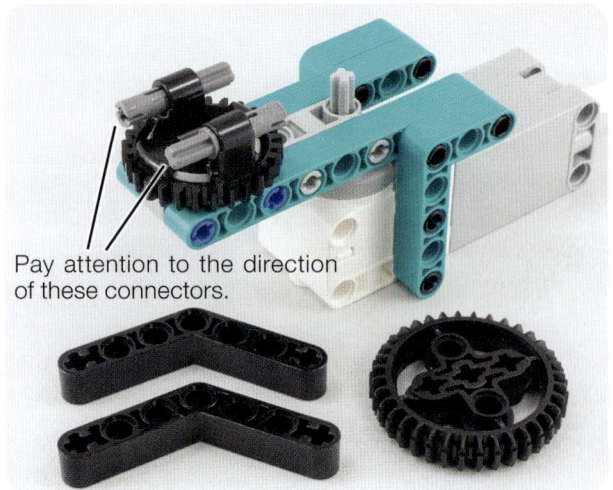

Pay attention to the direction of these connectors.

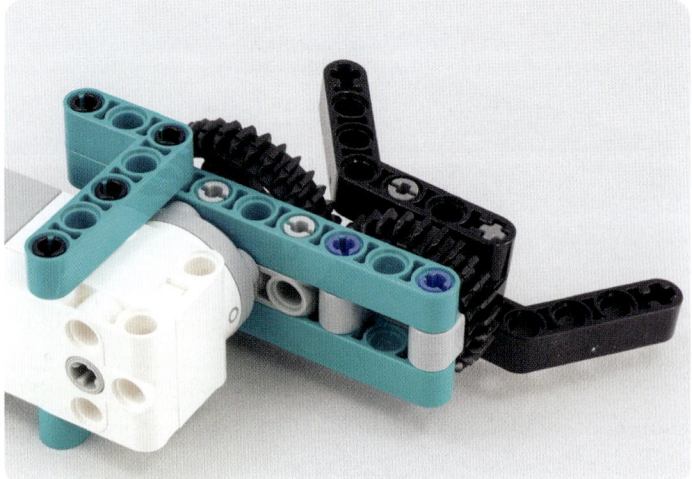

×2

×6

×2

‹—2

5

×2

×2

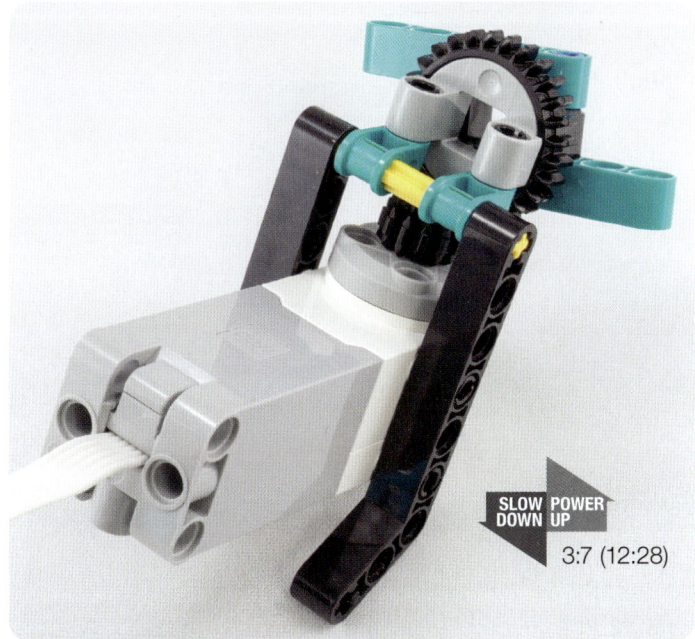

SLOW POWER
DOWN UP

3:7 (12:28)

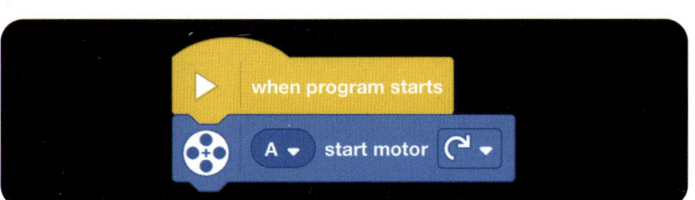

when program starts

A ▾ start motor ↻ ▾

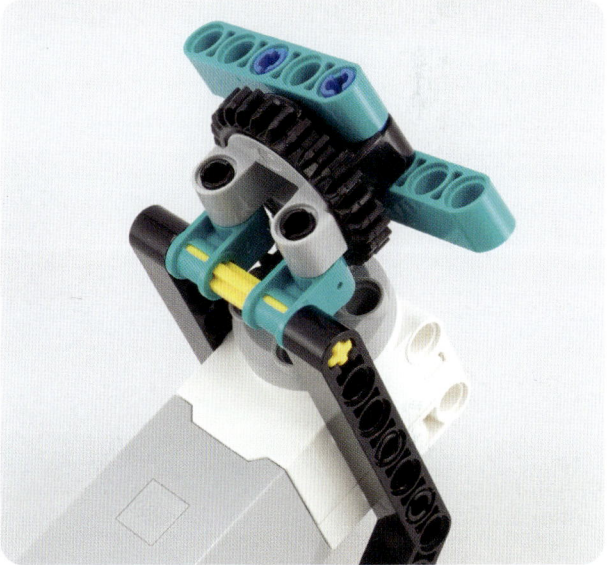

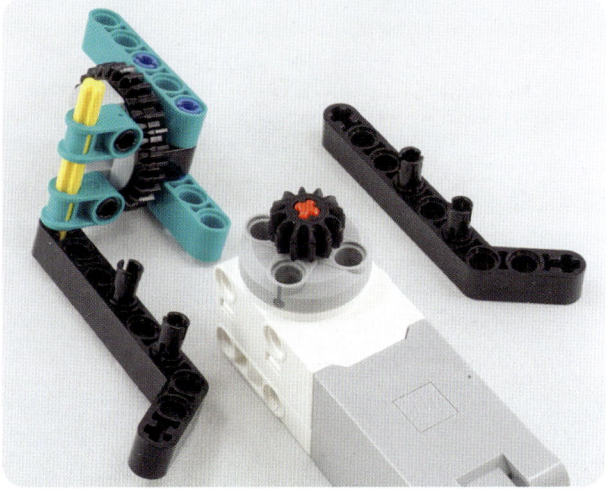

#38

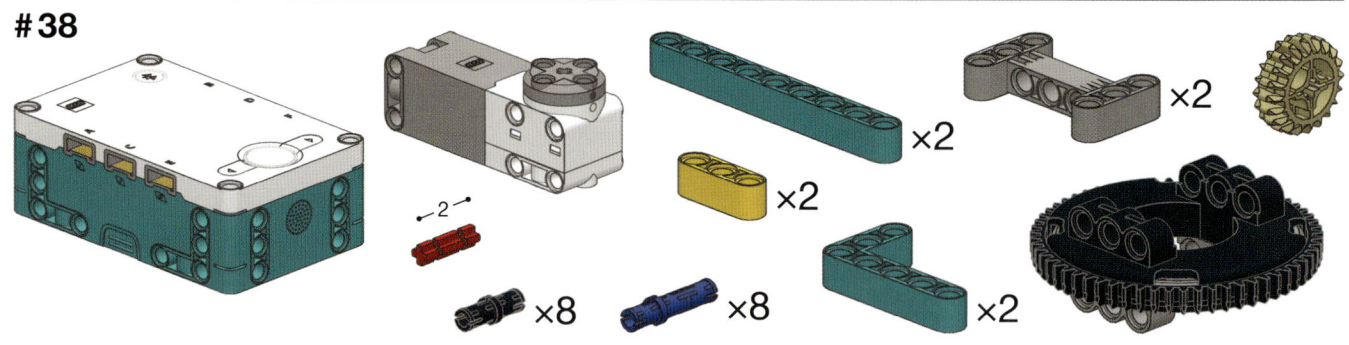

×2 ×2 ×2 ×2 ×8 ×8 ×2

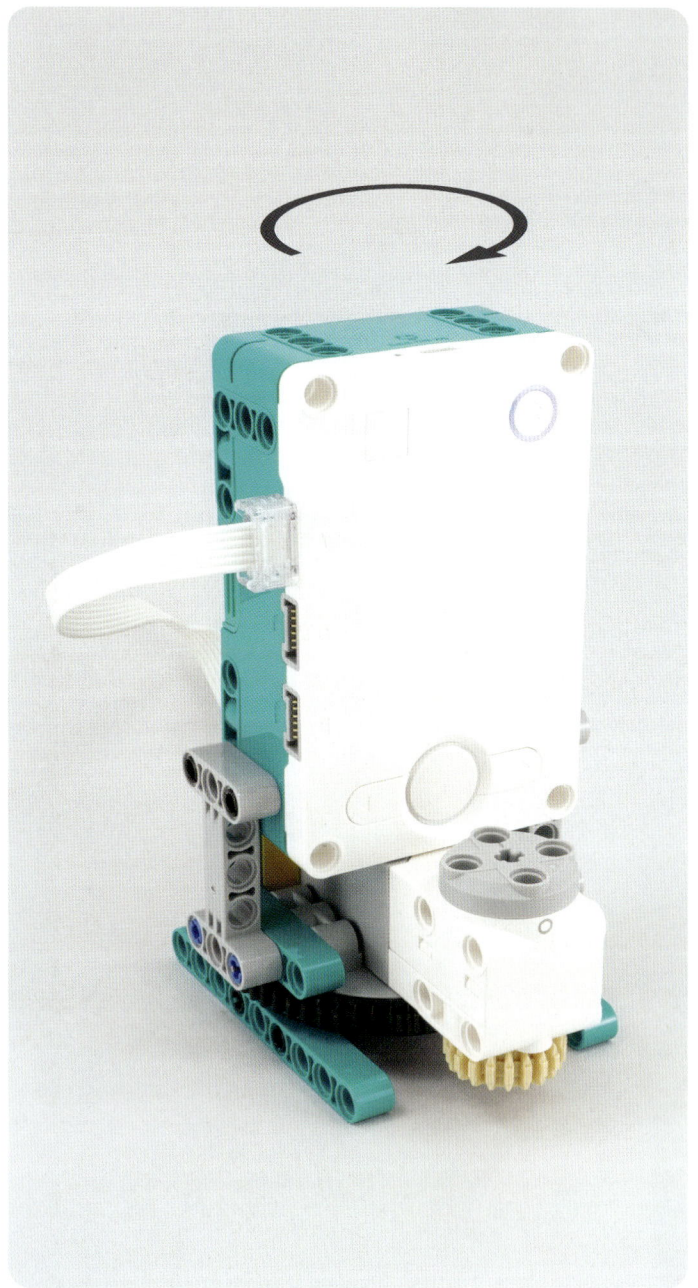

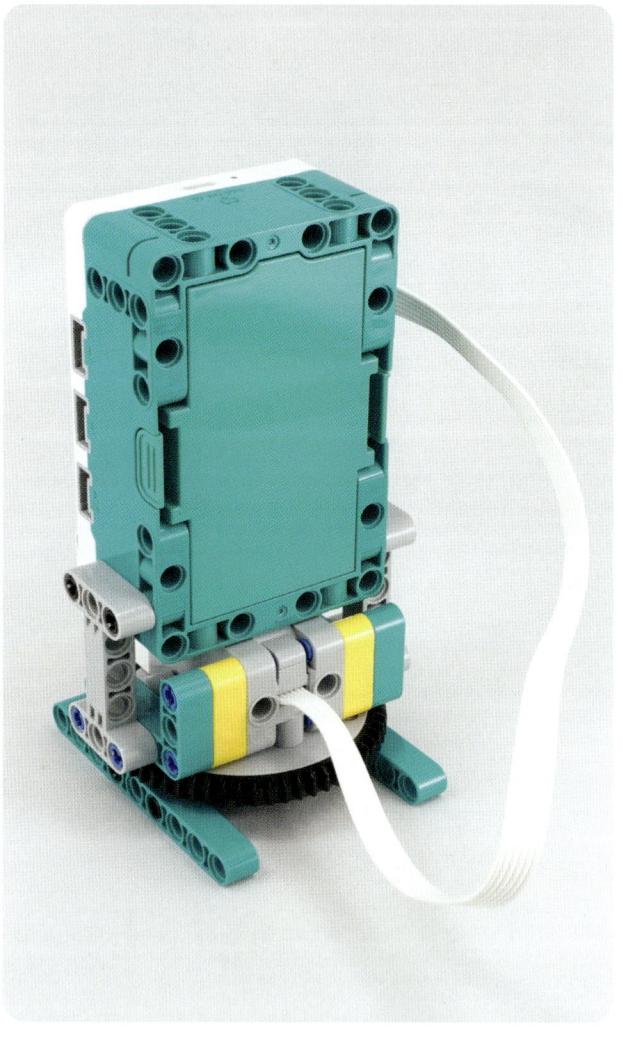

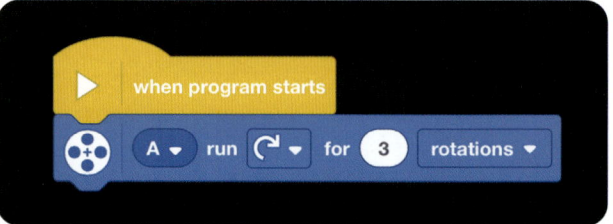

when program starts

A ▾ run ↻ ▾ for 3 rotations ▾

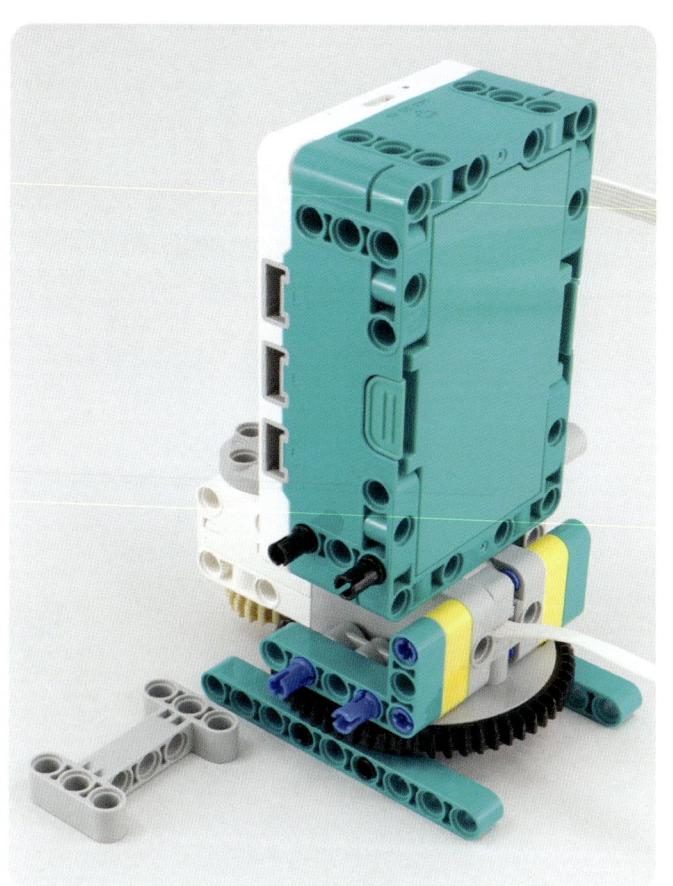

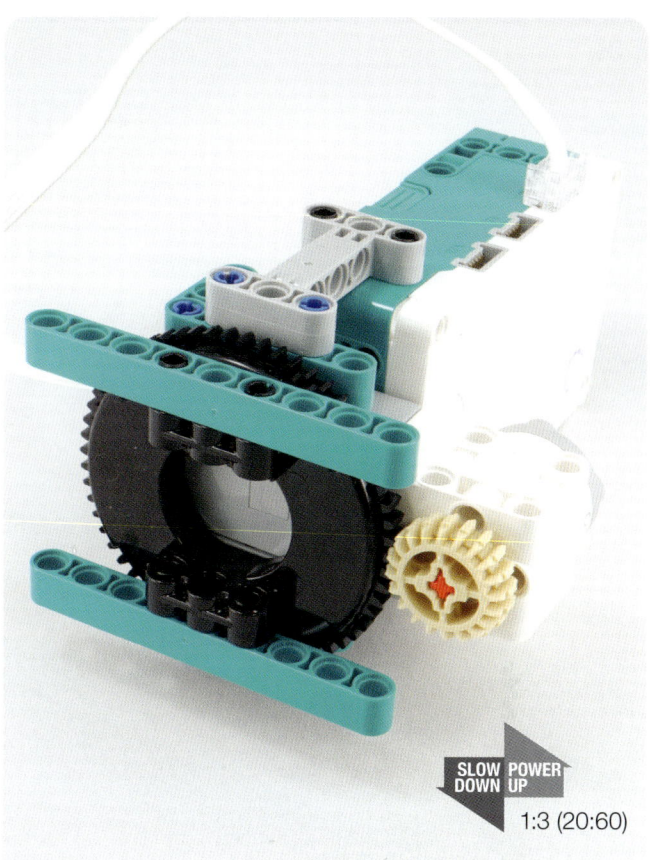

SLOW POWER
DOWN UP
1:3 (20:60)

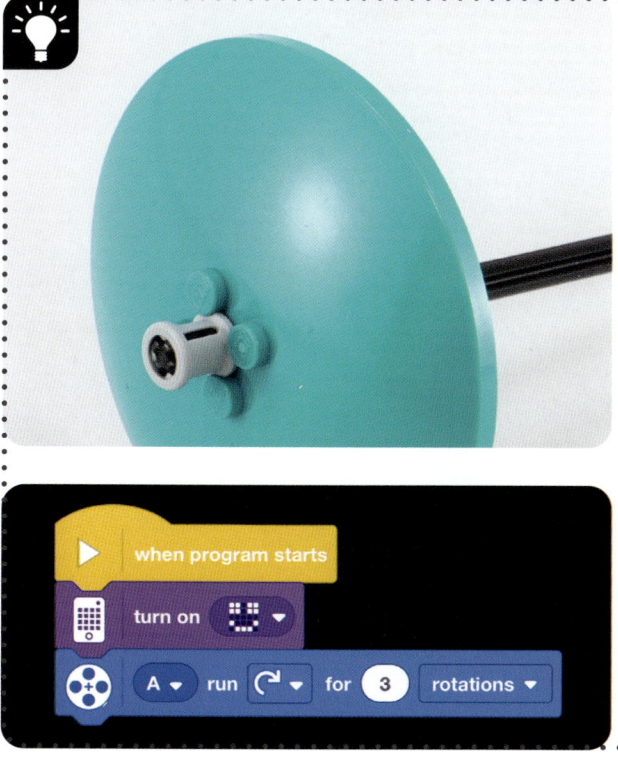

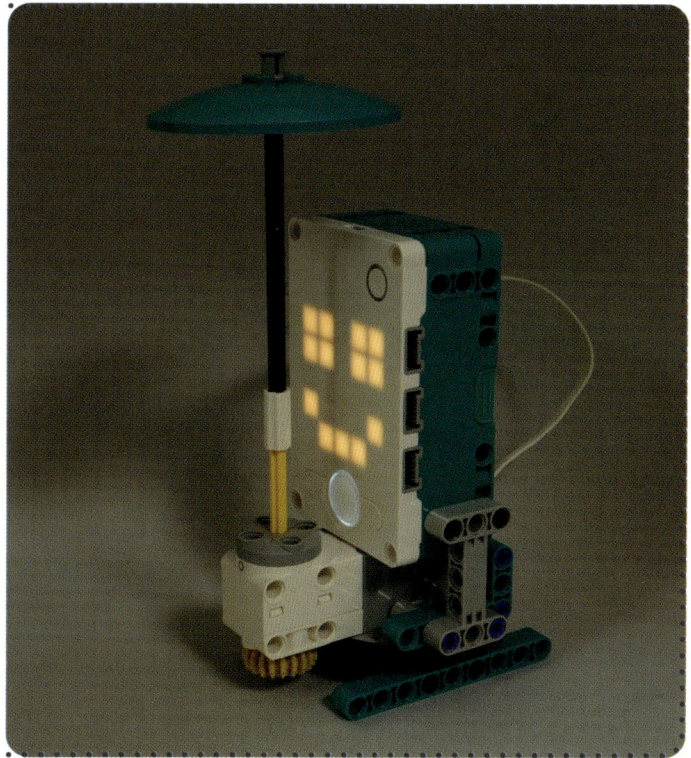

#39

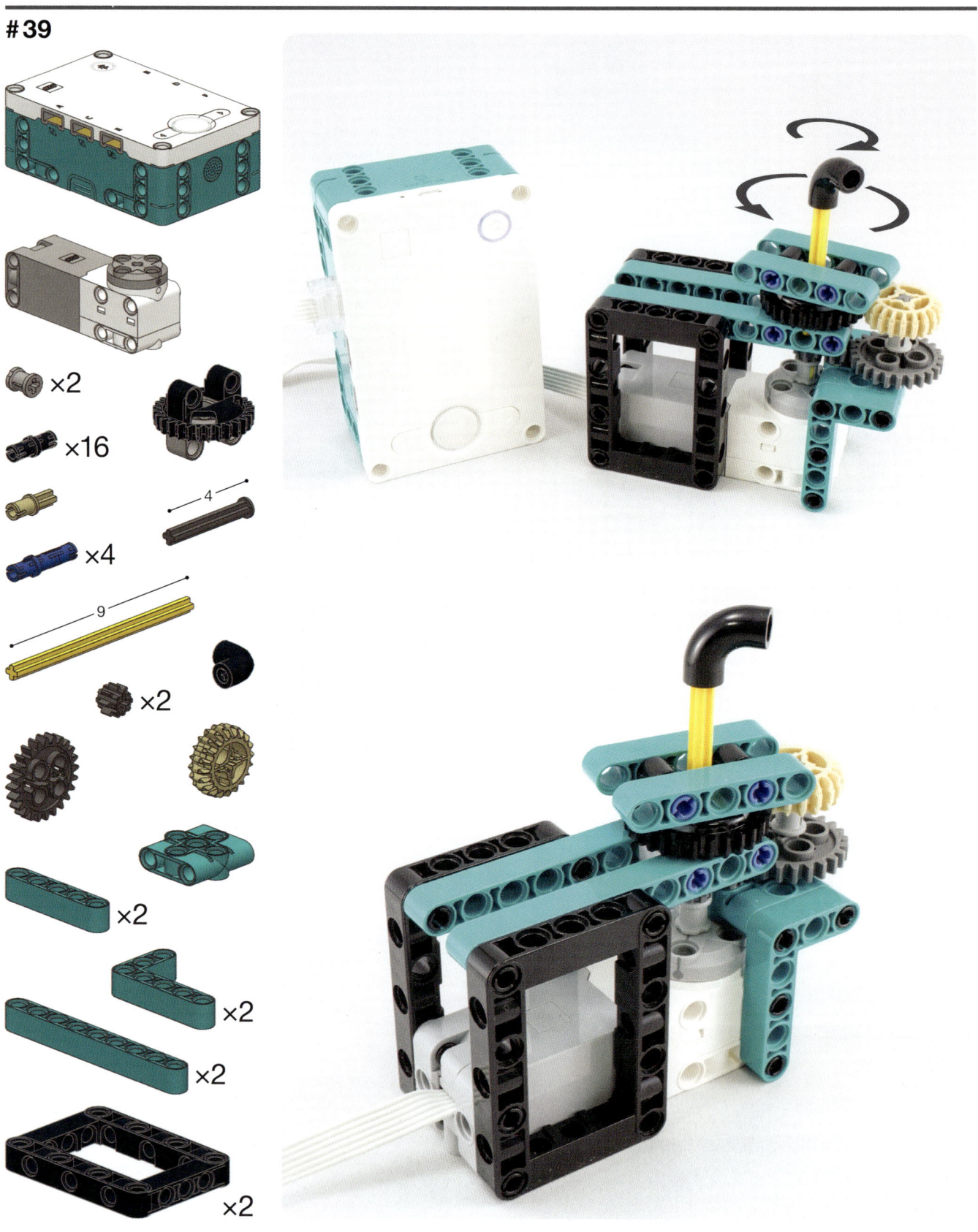

×2

×16

4

×4

9

×2

×2

×2

×2

×2

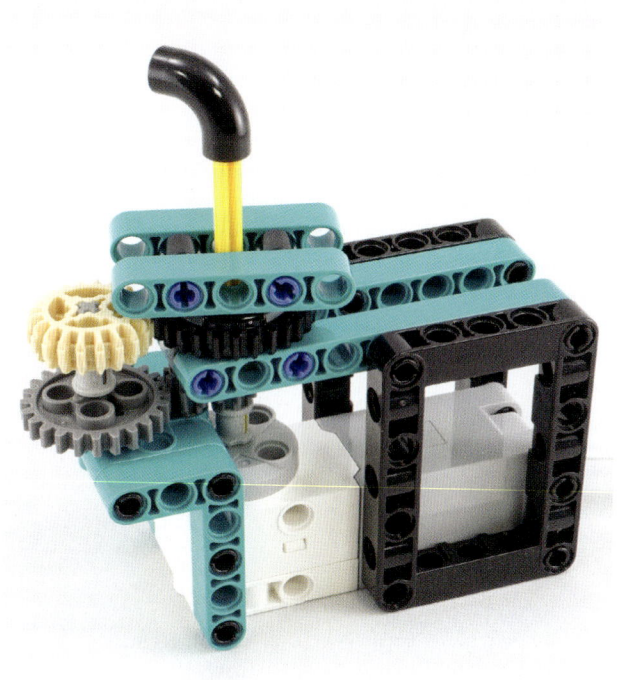

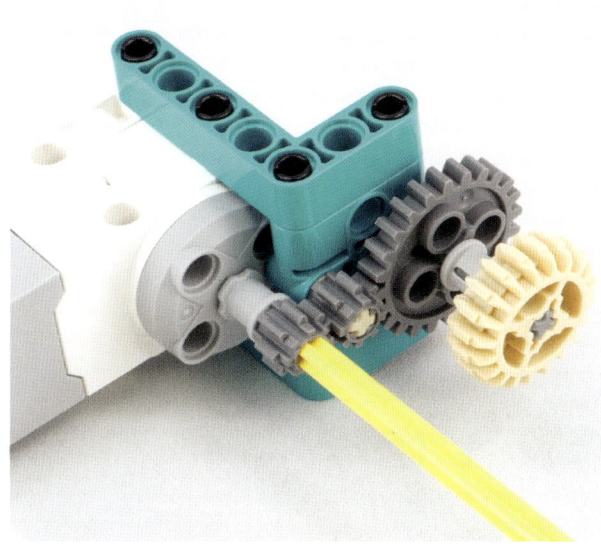

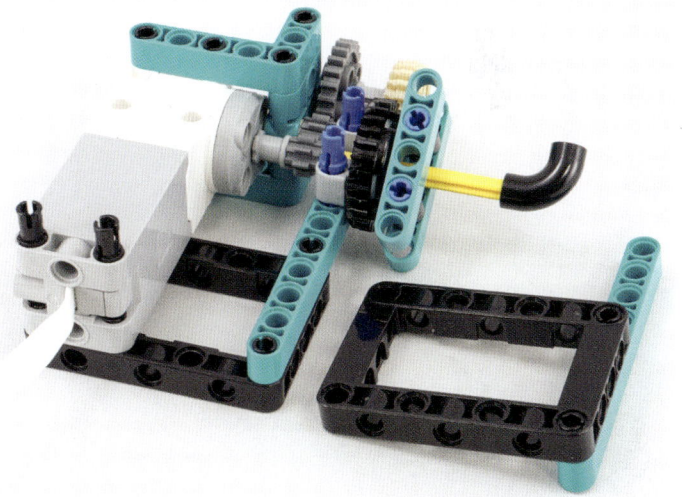

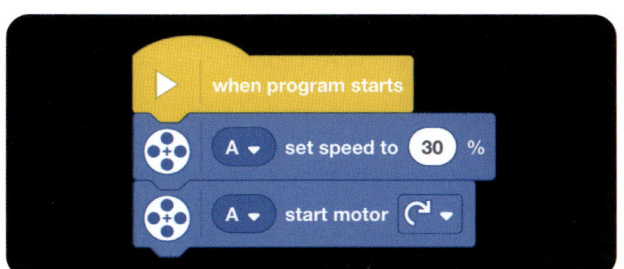

Lighting up the center button

#40

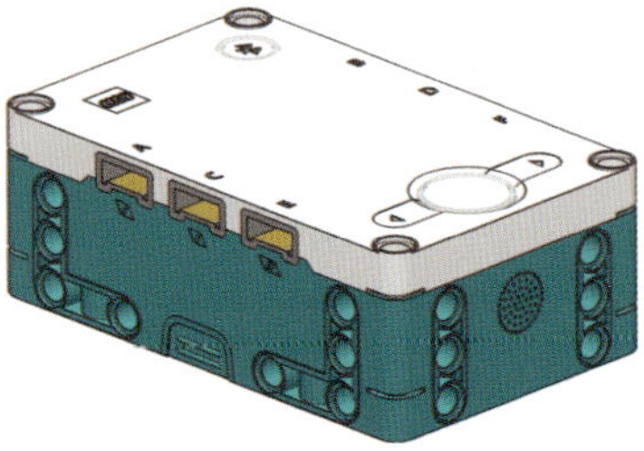

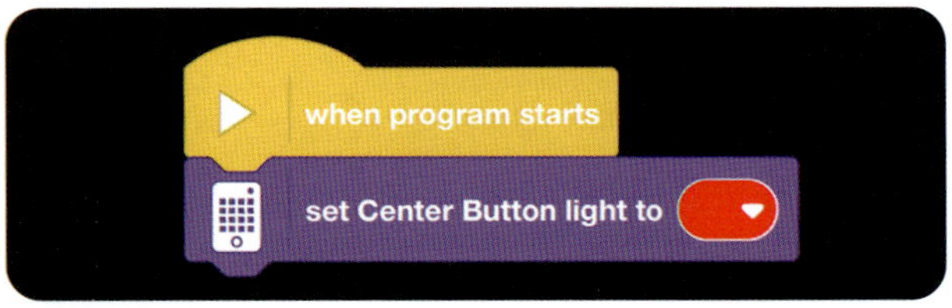

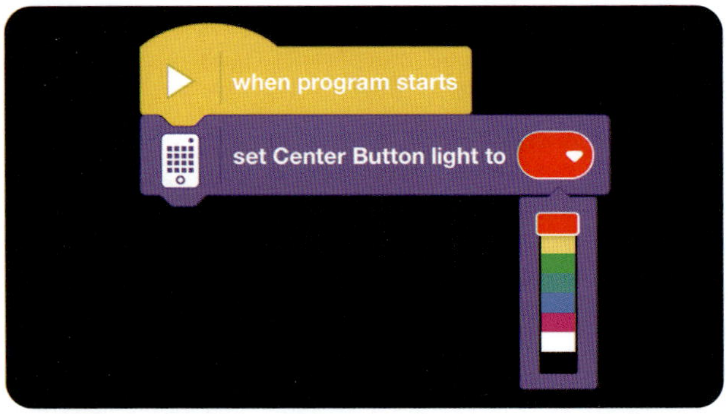

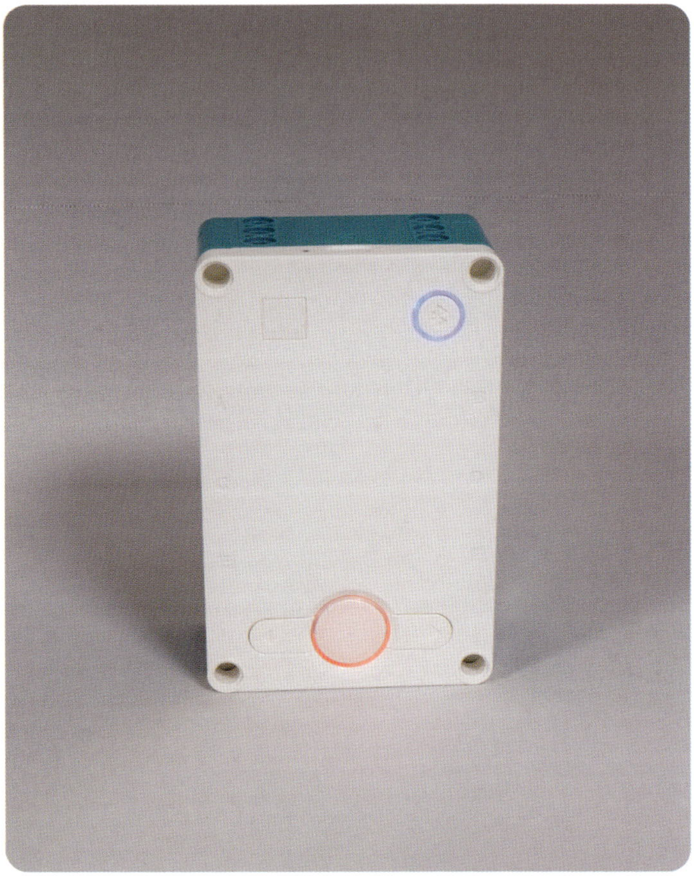

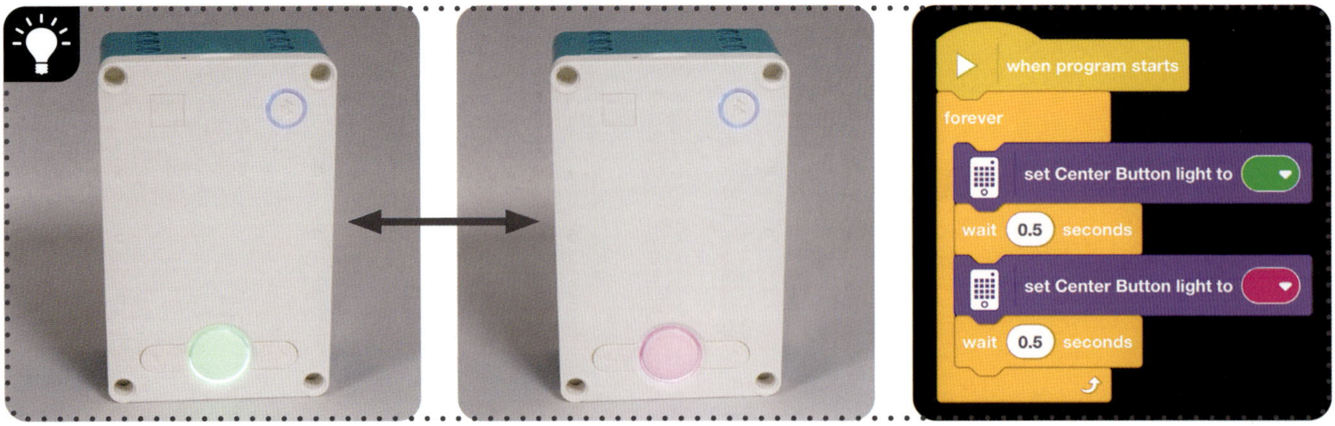

#41

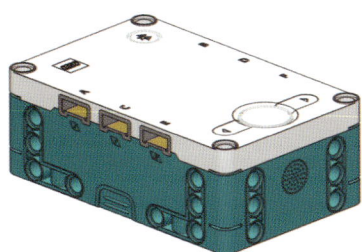

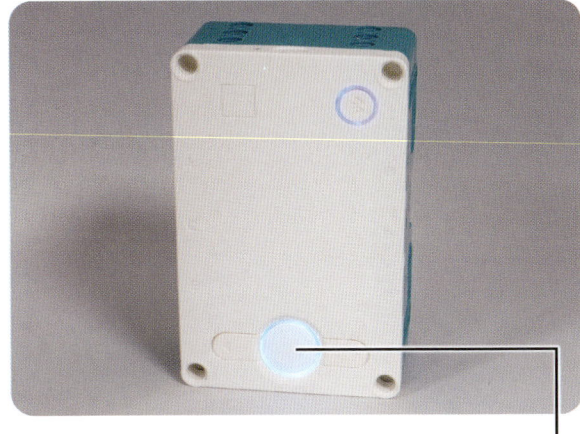

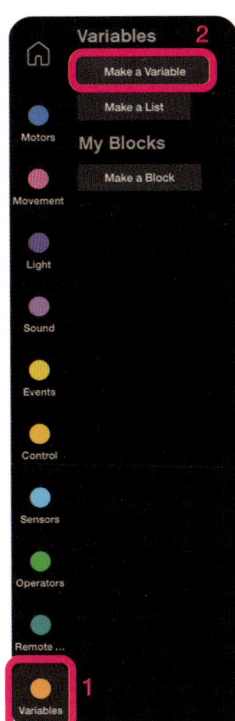

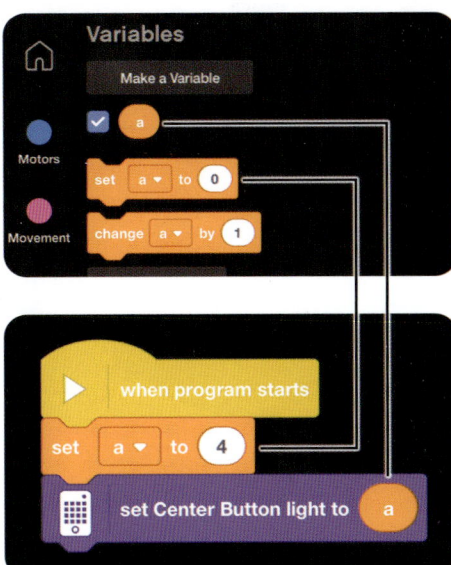

0	1	2	3	4	5	6	7	8	9	10
●	●	●	●	●	●	●	●	●	●	○

This is a program in which the button lights up in different colors other than black in sequence.

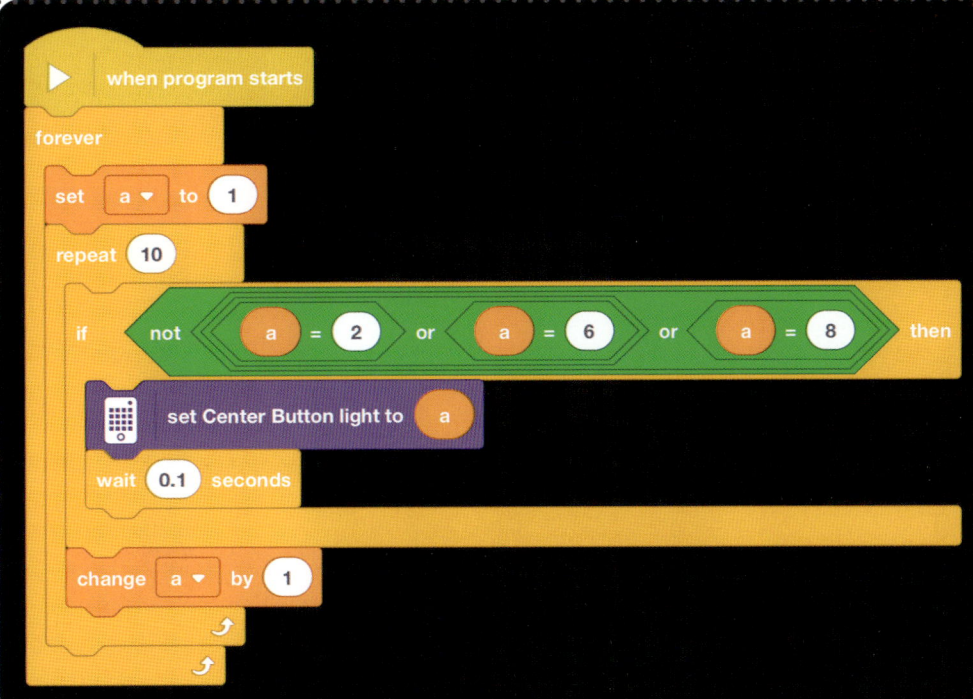

Controlling the LED Matrix

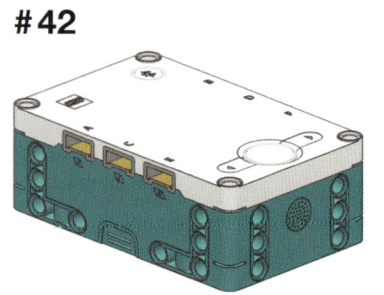

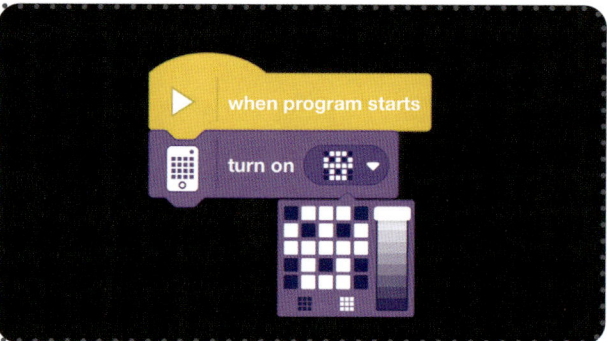

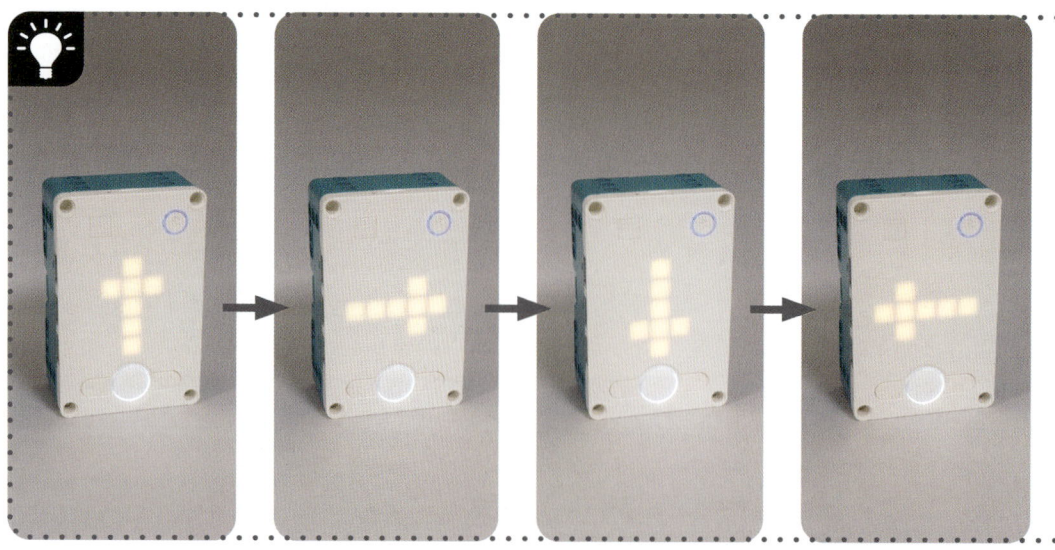

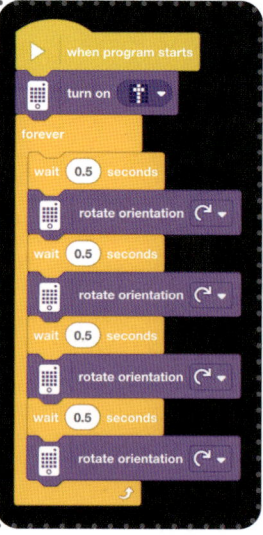

42

#43

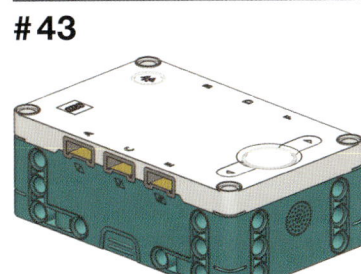

when program starts
write A

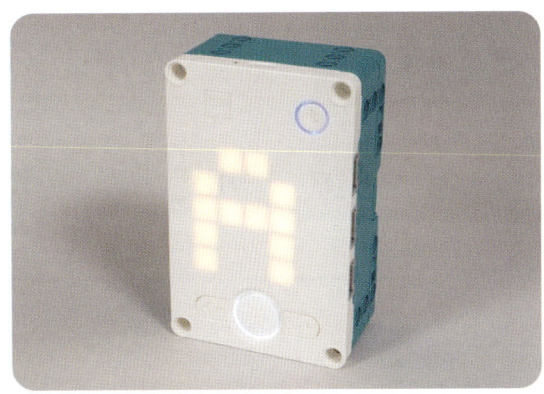

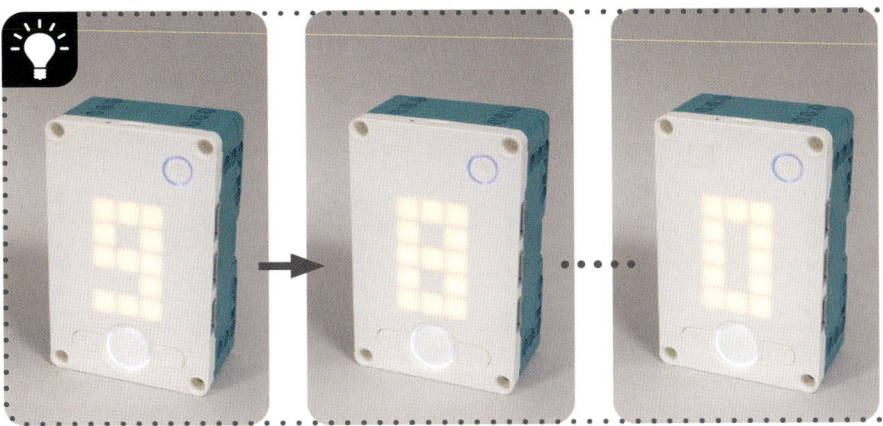

when program starts
set a to 9
repeat 10
  write a
  change a by -1
  wait 1 seconds

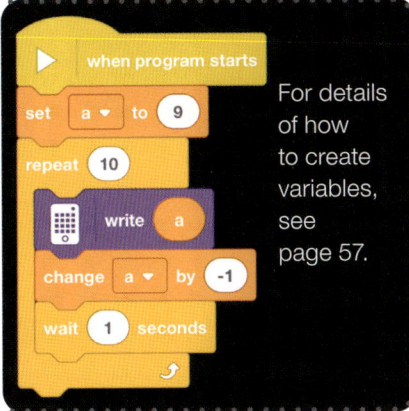

For details of how to create variables, see page 57.

#44

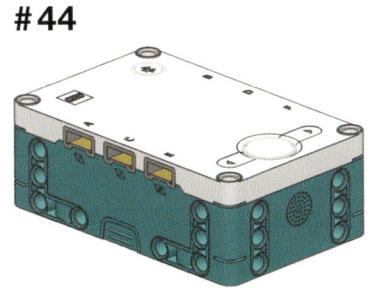

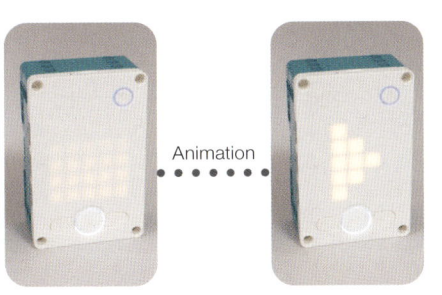

Animation

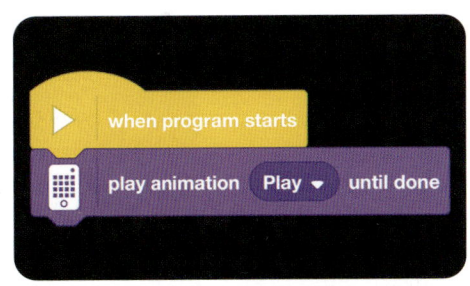

when program starts
play animation Play until done

You can select or edit an animation.

when program starts
play animation Play until done

✓ Play
Animation Editor
Animation Library

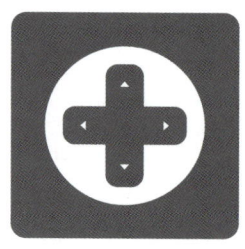

Using the remote control

#45

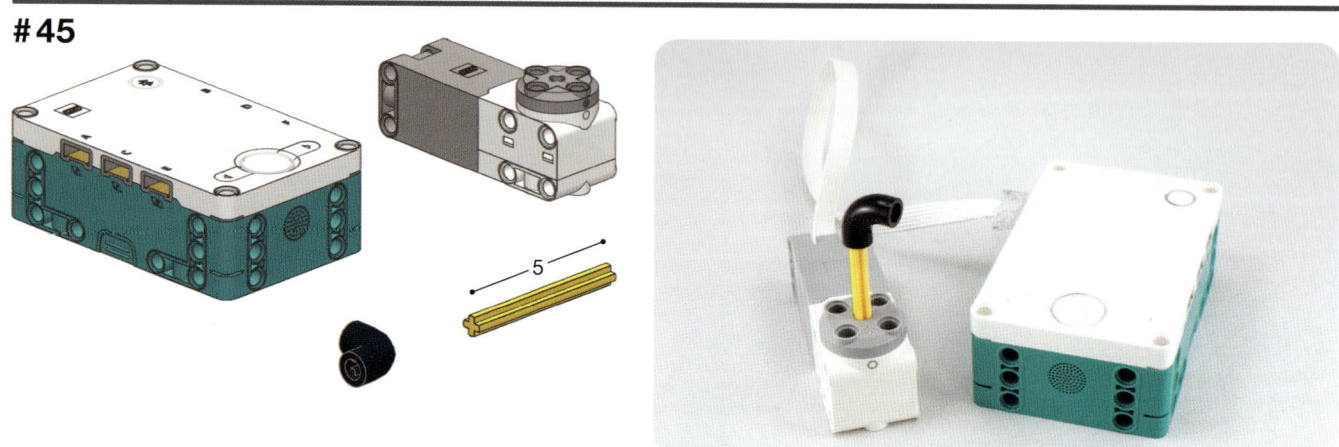

1. Make sure that your tablet or smartphone connects to Bluetooth, and then switch to **Streaming** mode.

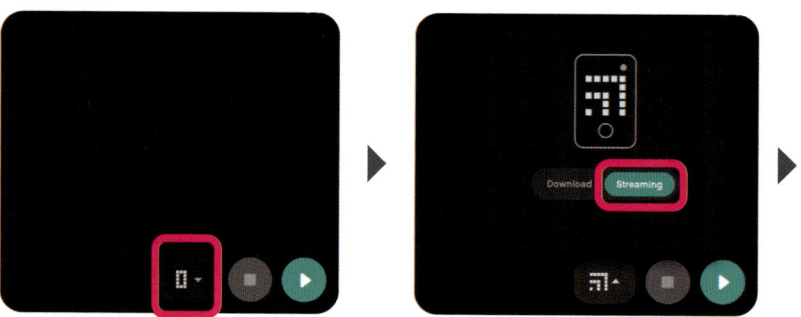

2. Select the controller.

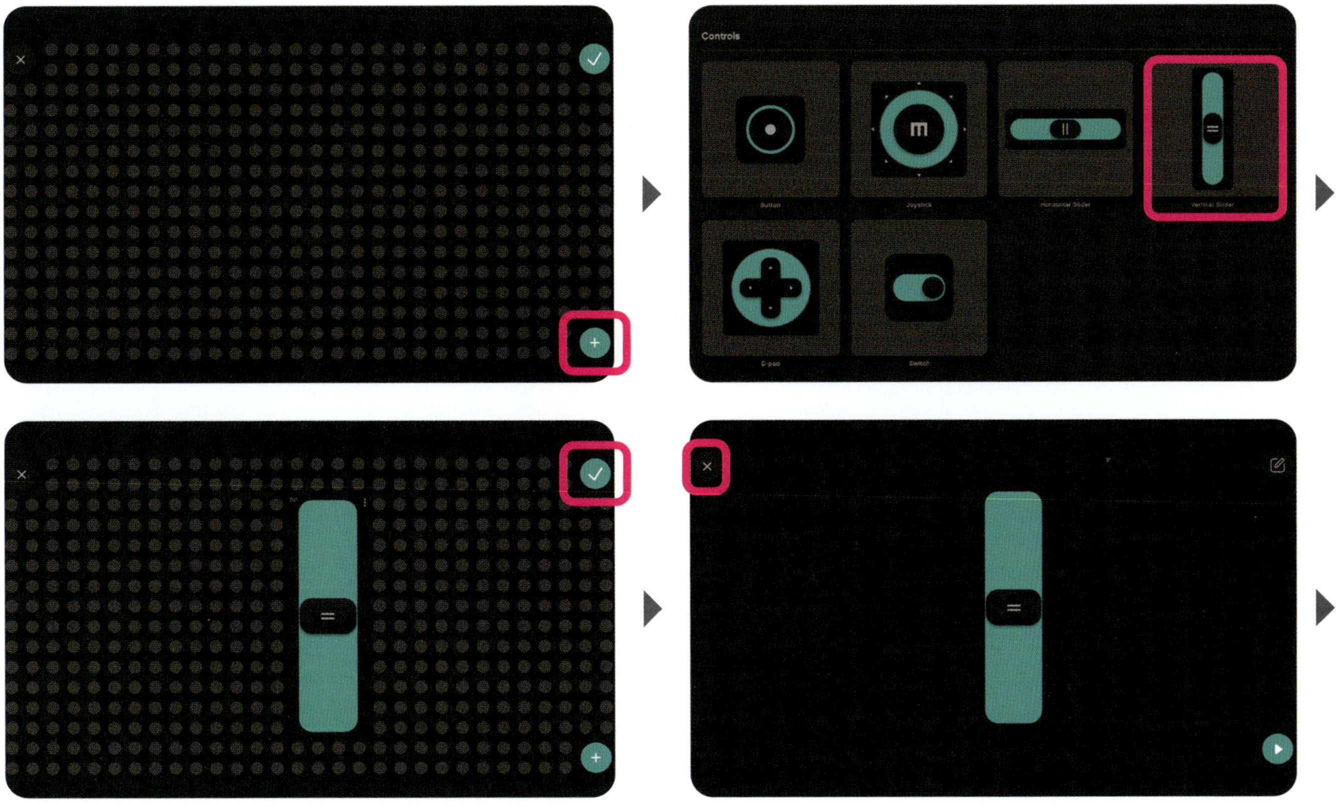

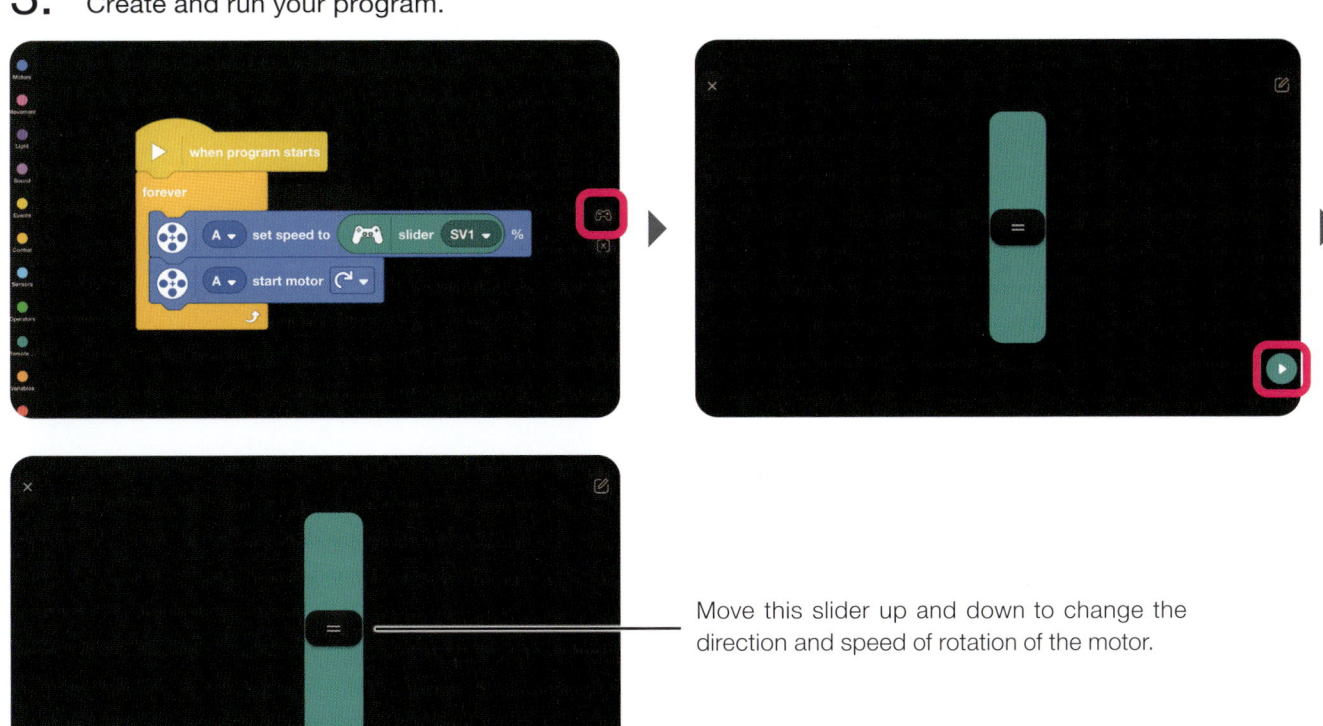

3. Create and run your program.

Move this slider up and down to change the direction and speed of rotation of the motor.

PART 2
Moving
Mechanisms

Vehicles with one motor

#46

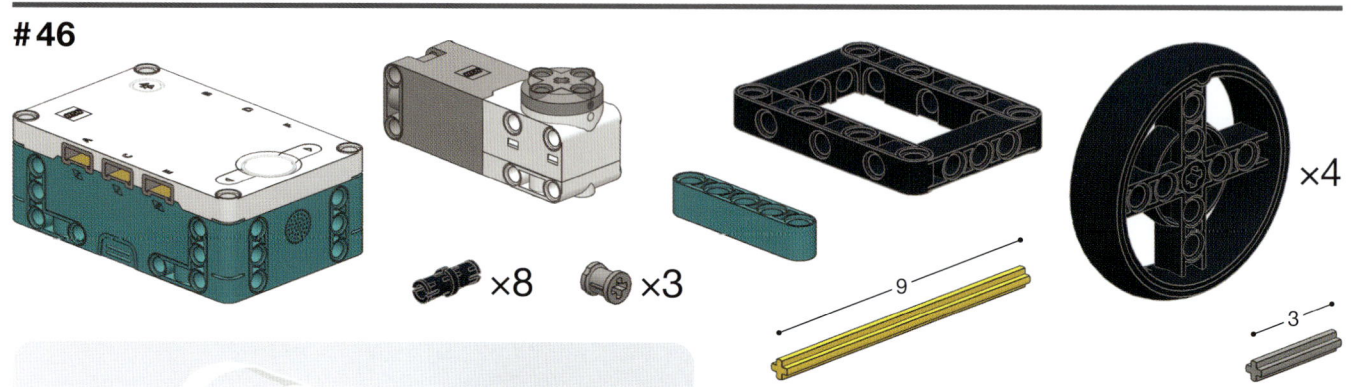

×8 ×3 9 ×4 3

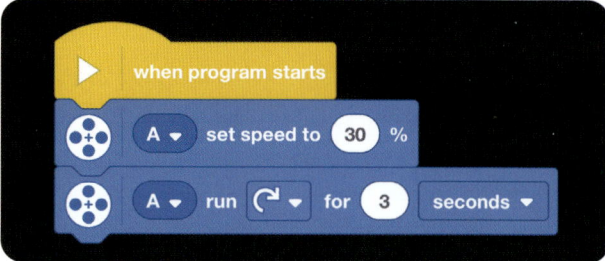

when program starts

A ▾ set speed to 30 %

A ▾ run ↻ ▾ for 3 seconds ▾

#47

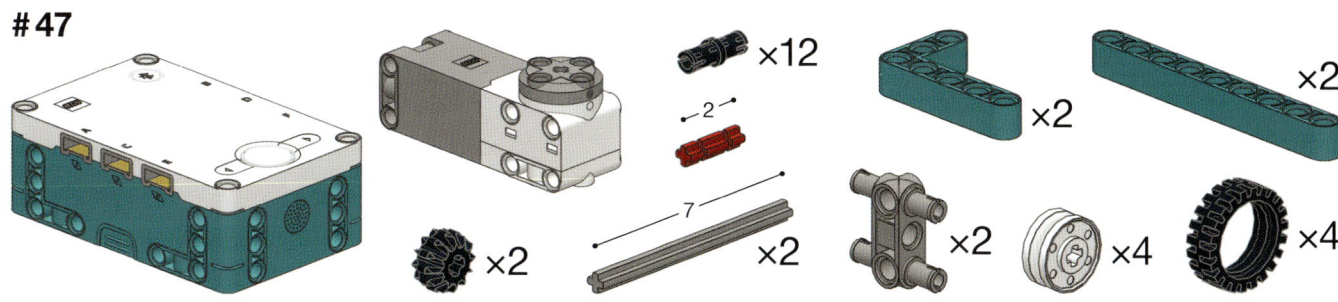

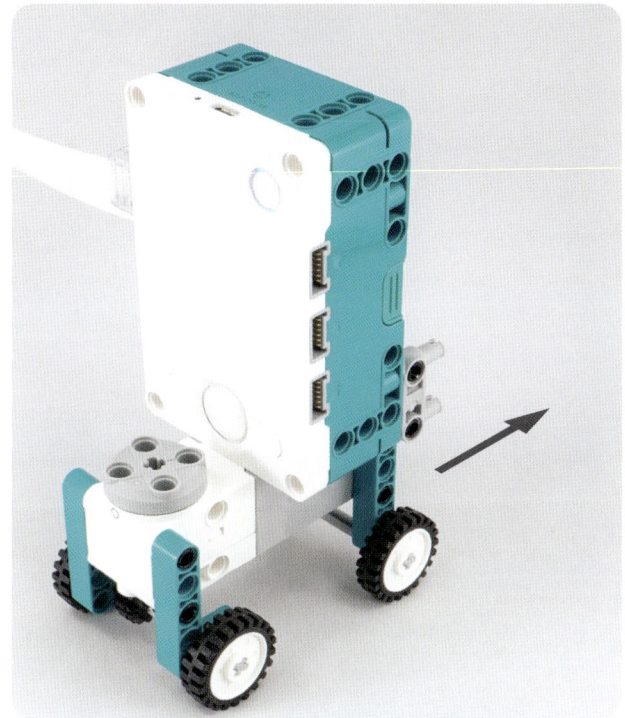

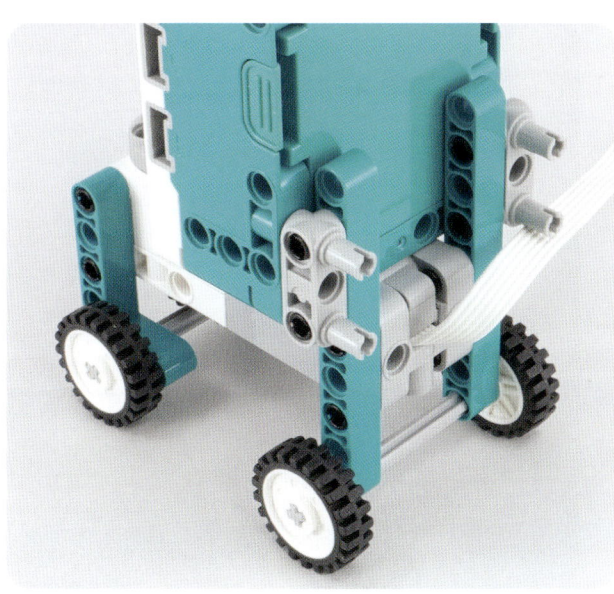

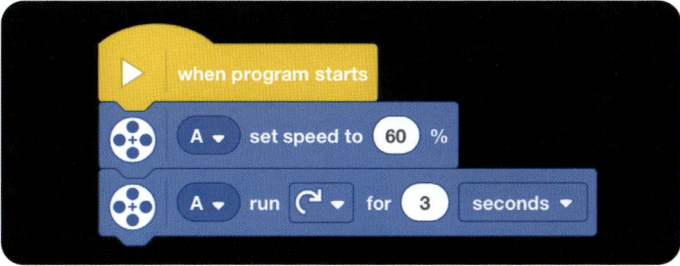

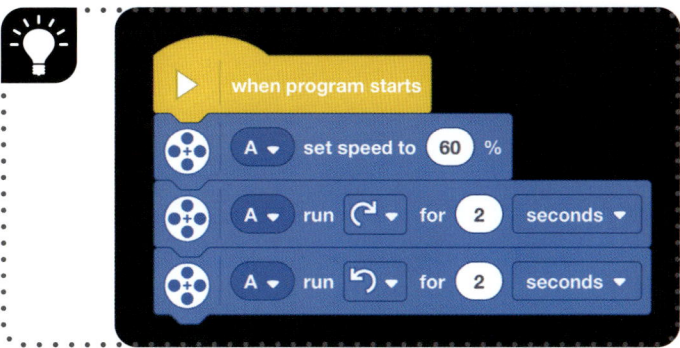

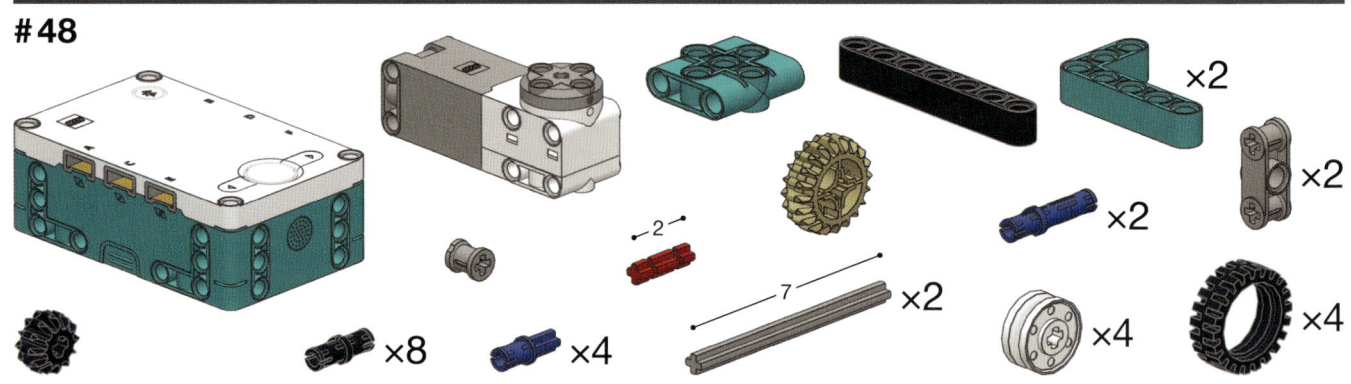

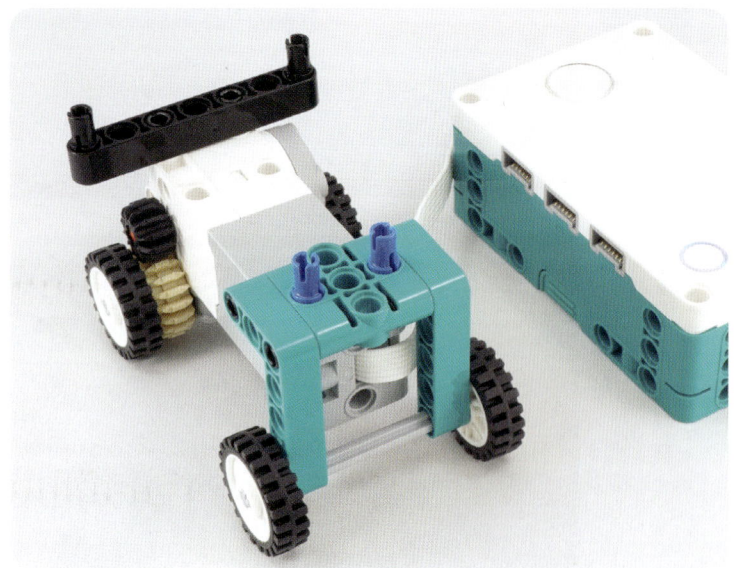

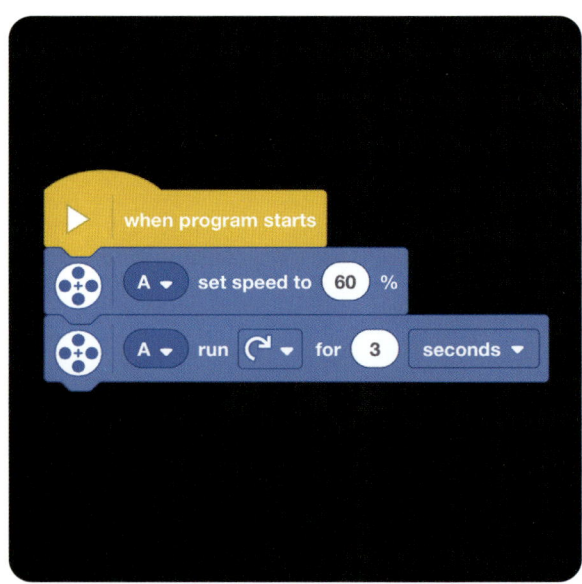

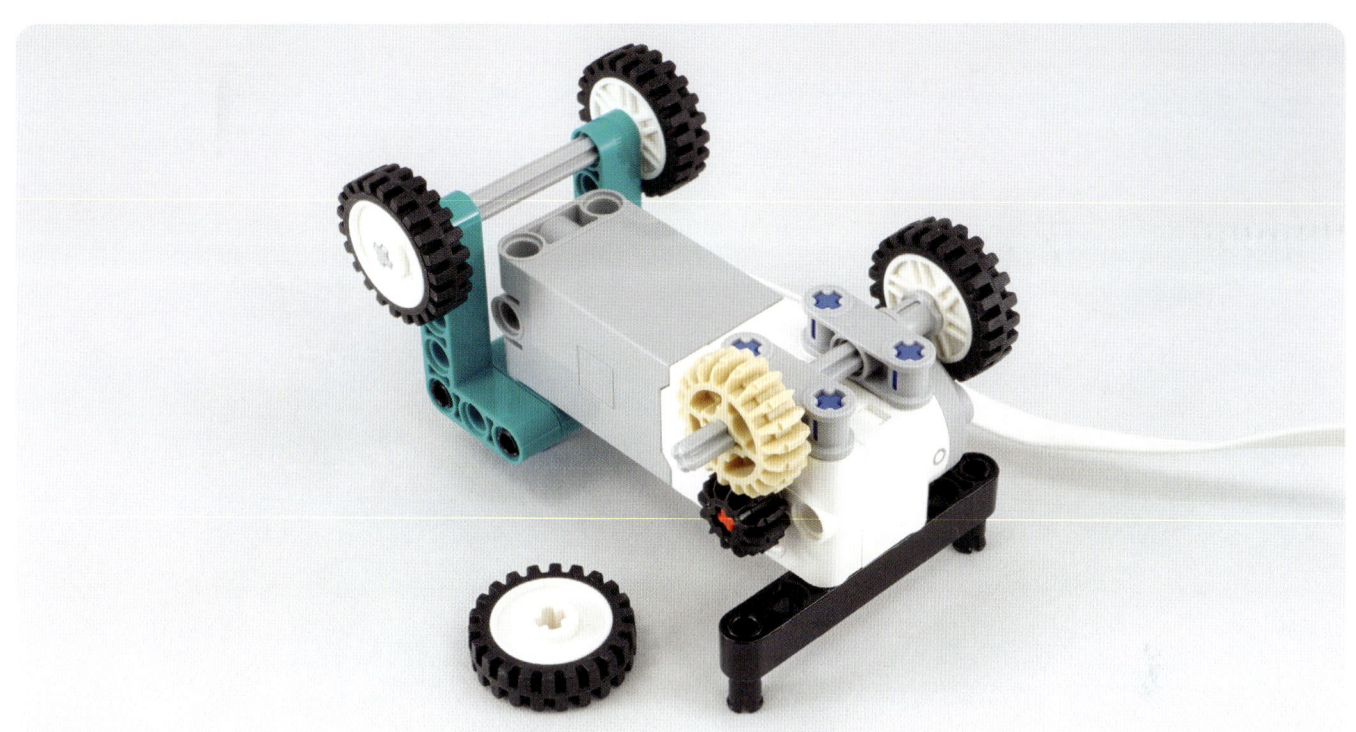

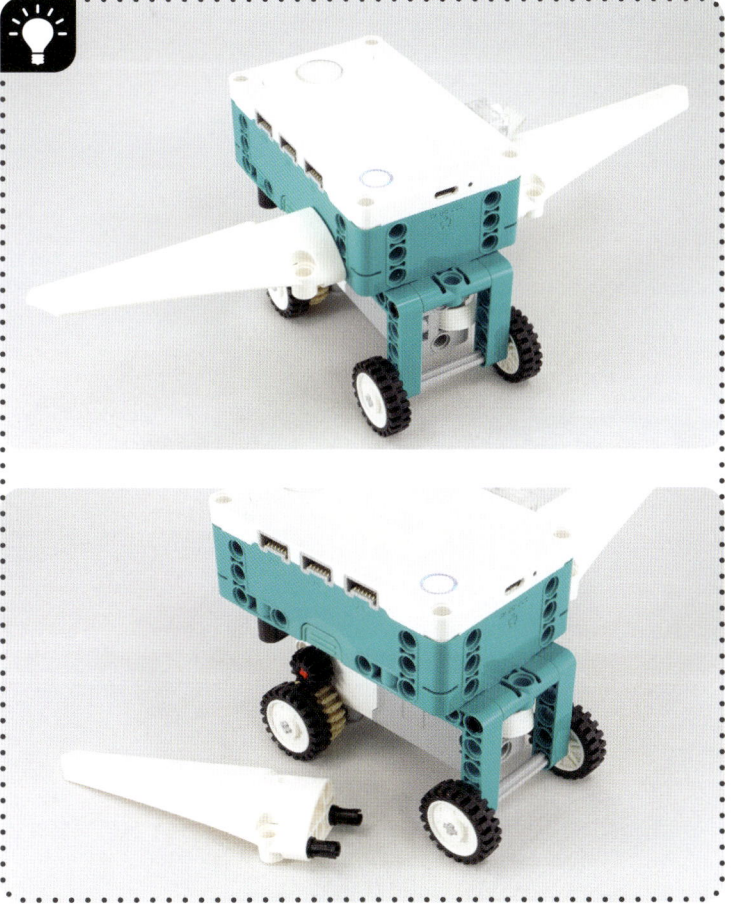

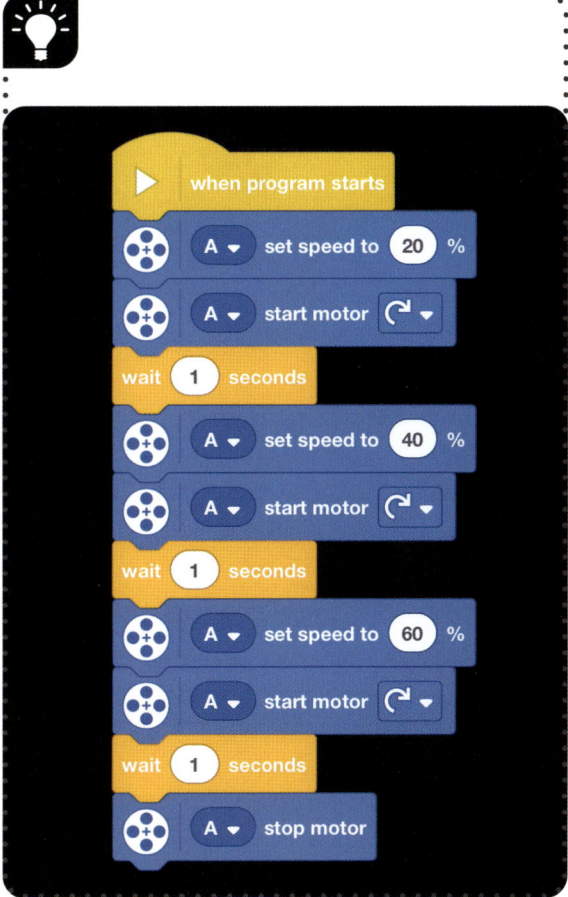

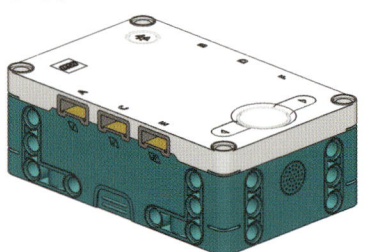

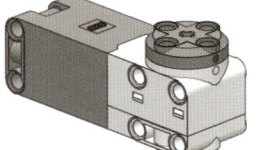 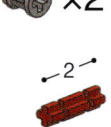 ×2

 2

 ×10 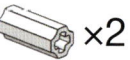 ×2

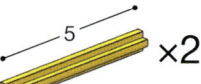 ×2  ×2

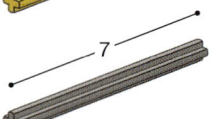 5 ×2 ×2

7

 ×2

×2

 ×2

 ×2

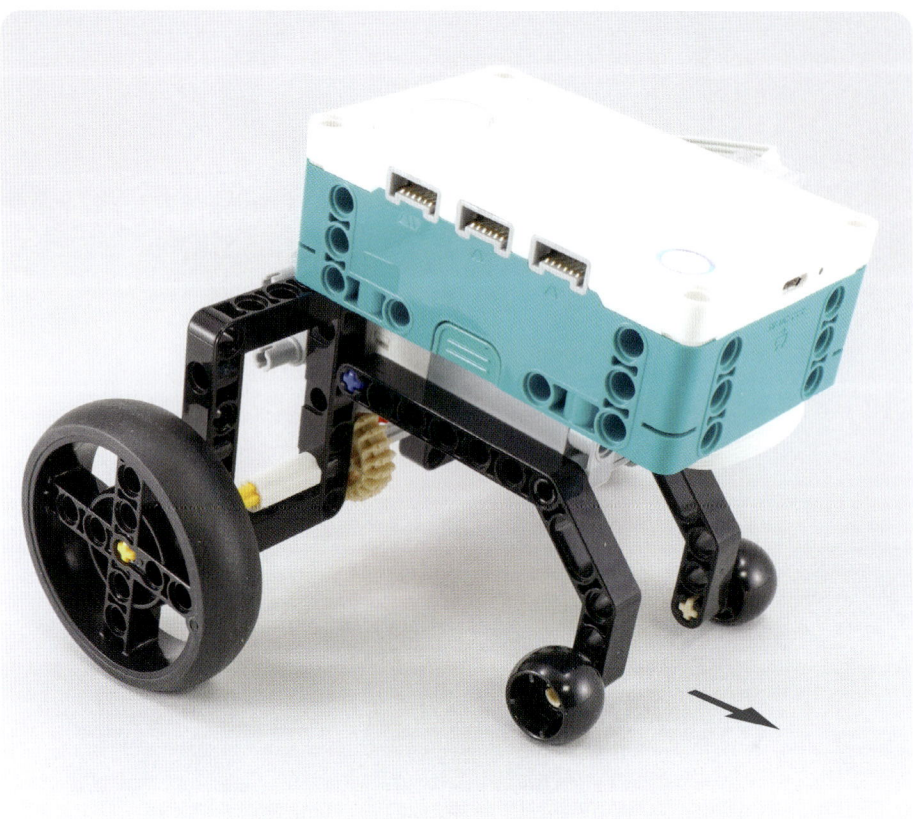

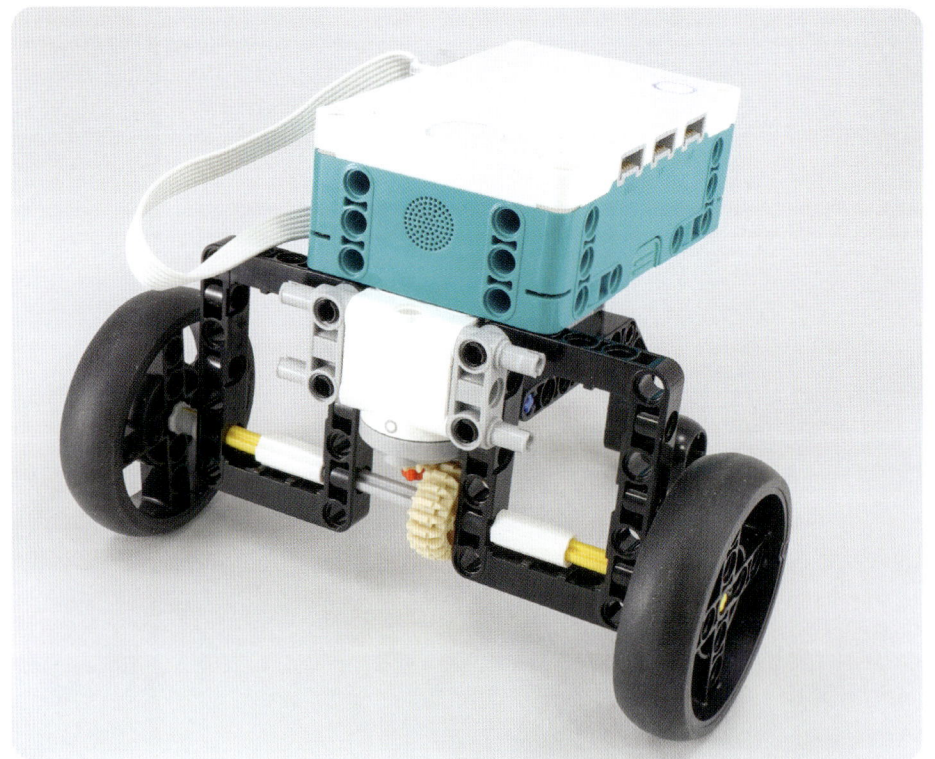

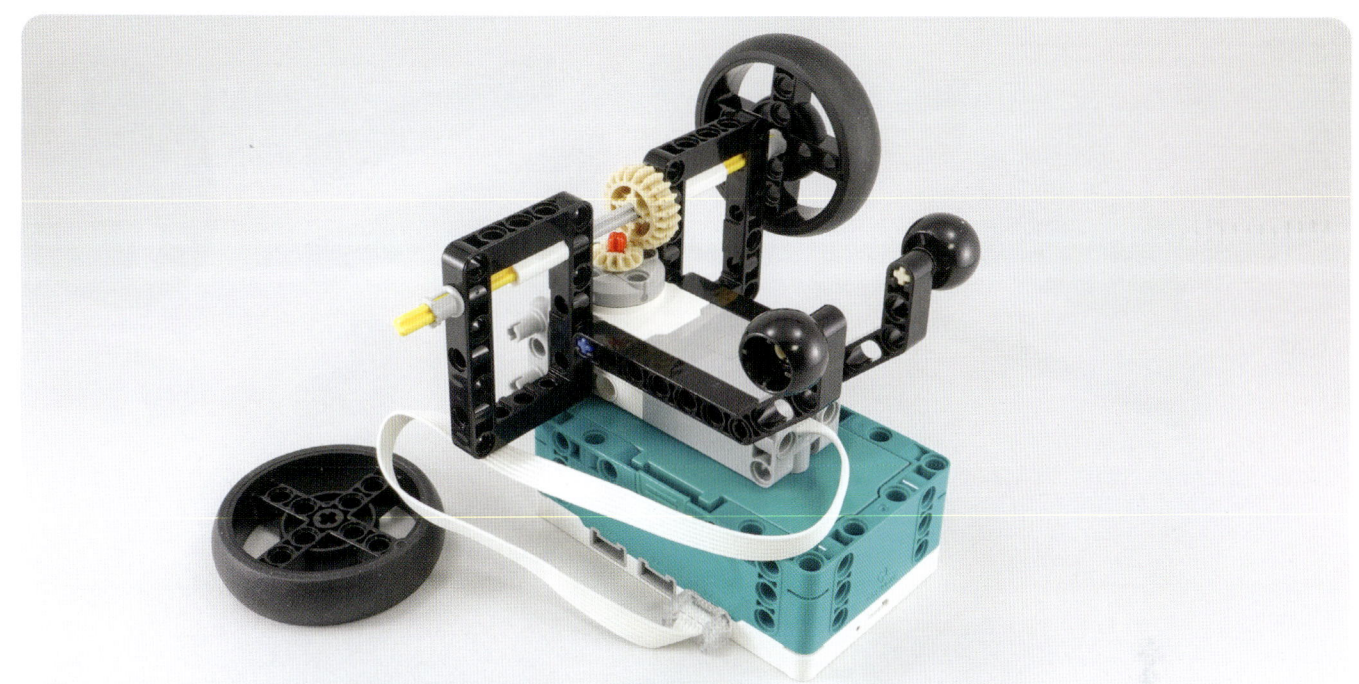

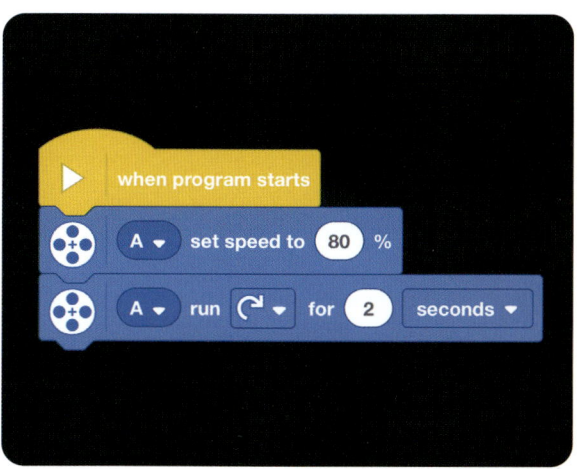

For details about setting up the remote control, see page 60.

Go forward

Stop

Go backward

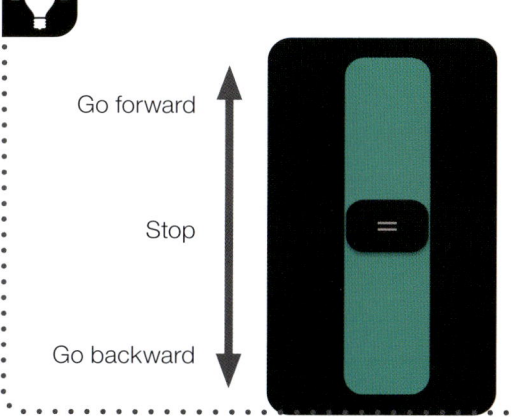

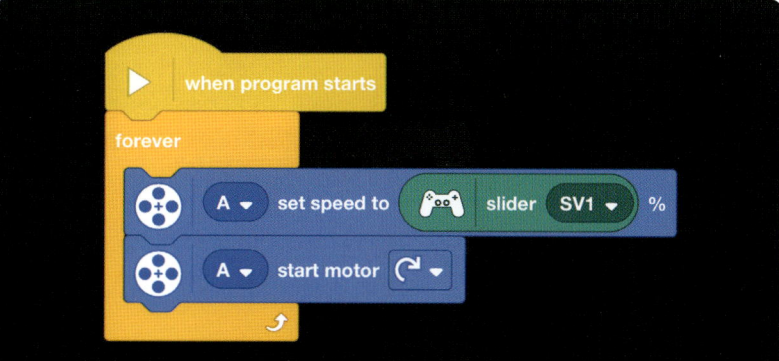

#50

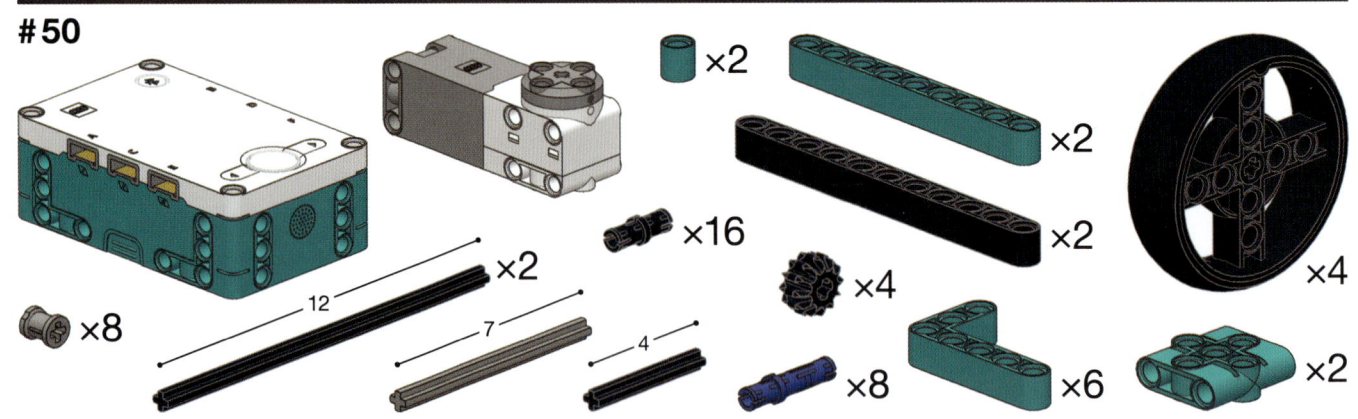

×8   12   ×2   ×16   7   ×4   ×2   ×2   4   ×8   ×6   ×2   ×4

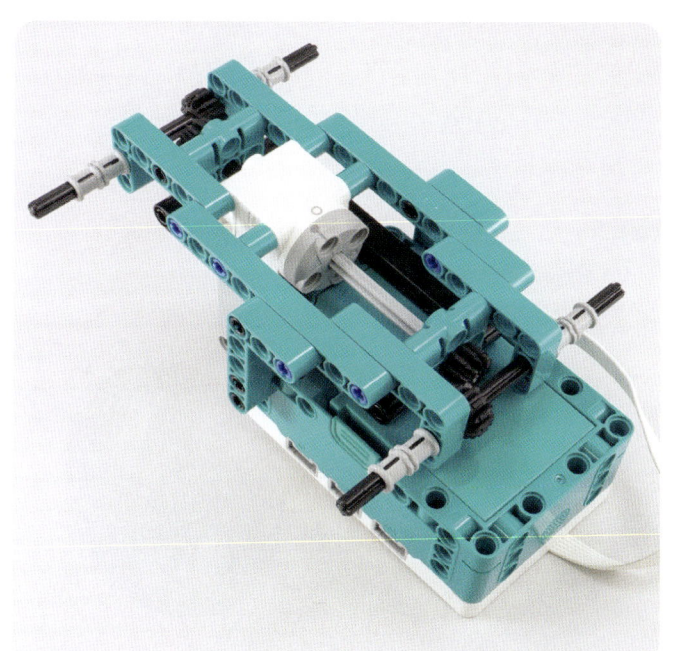

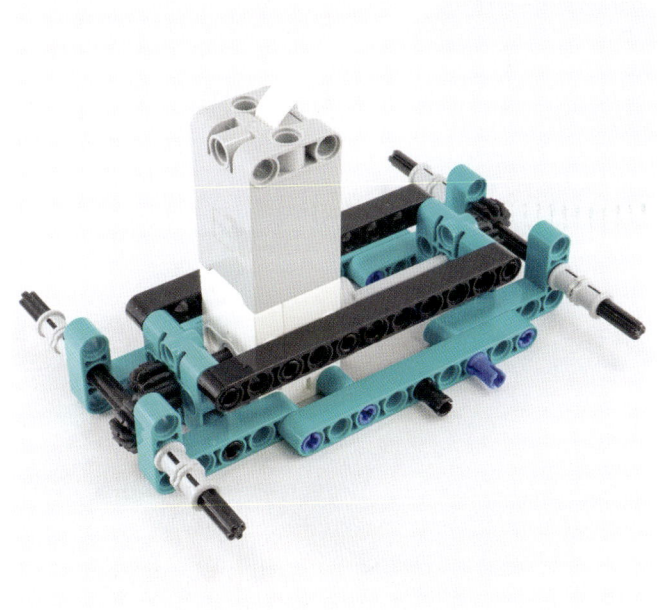

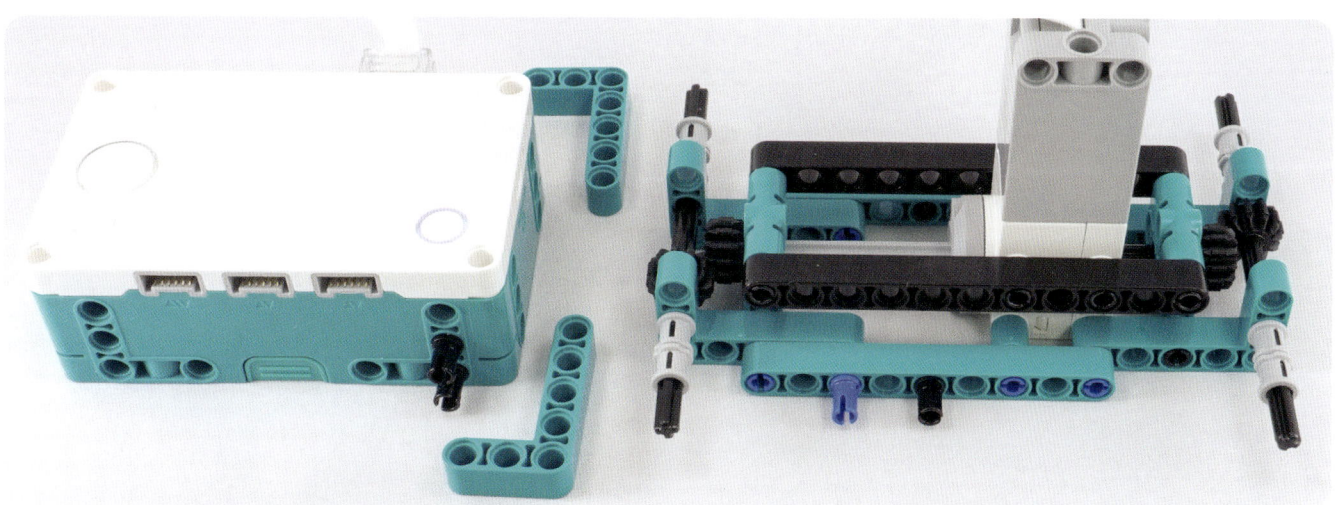

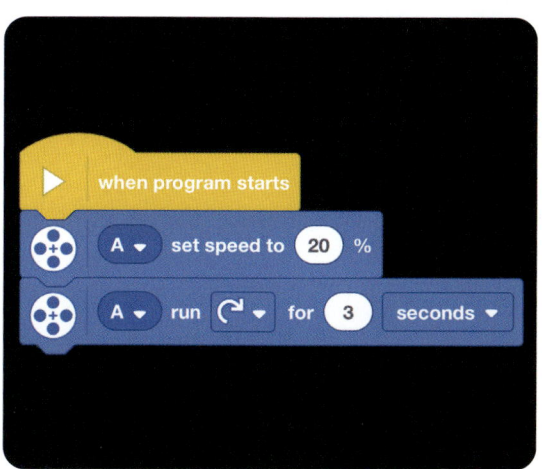

Vehicles with two motors

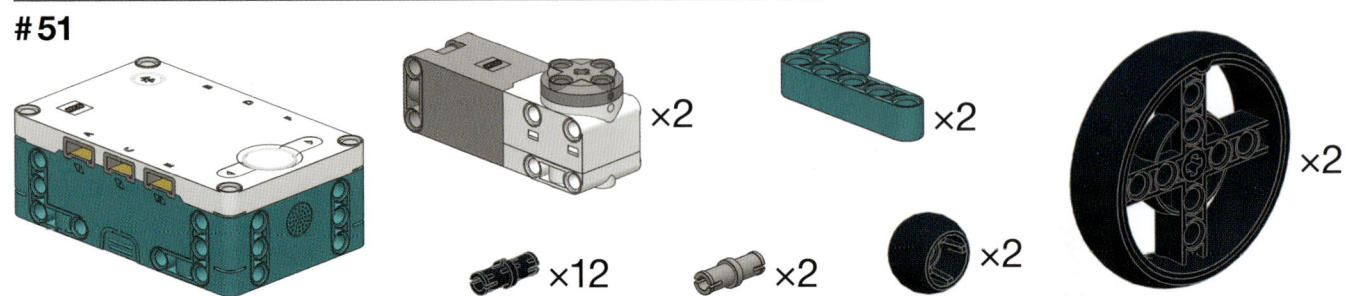

51

×2 ×2 ×2

×12 ×2 ×2

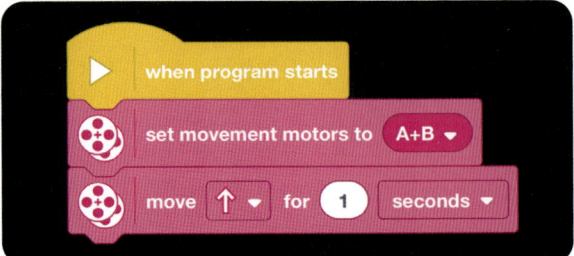

when program starts

set movement motors to A+B

move ↑ ▾ for 1 seconds ▾

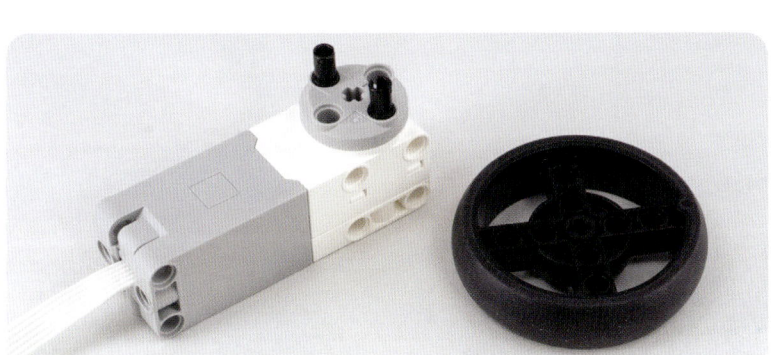

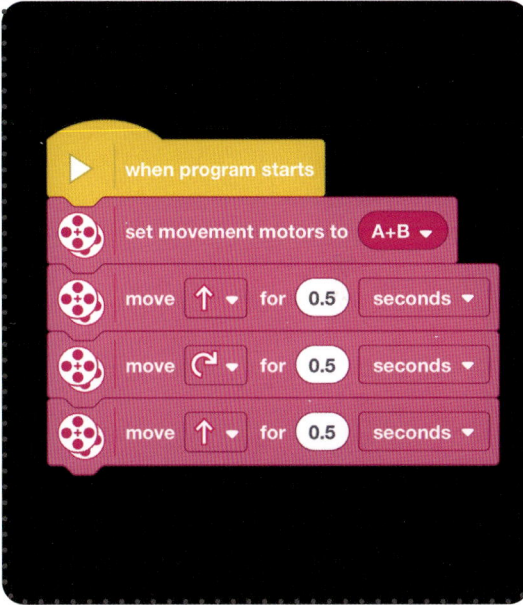

when program starts

set movement motors to A+B ▼

move ↑ ▼ for 0.5 seconds ▼

move ↻ ▼ for 0.5 seconds ▼

move ↑ ▼ for 0.5 seconds ▼

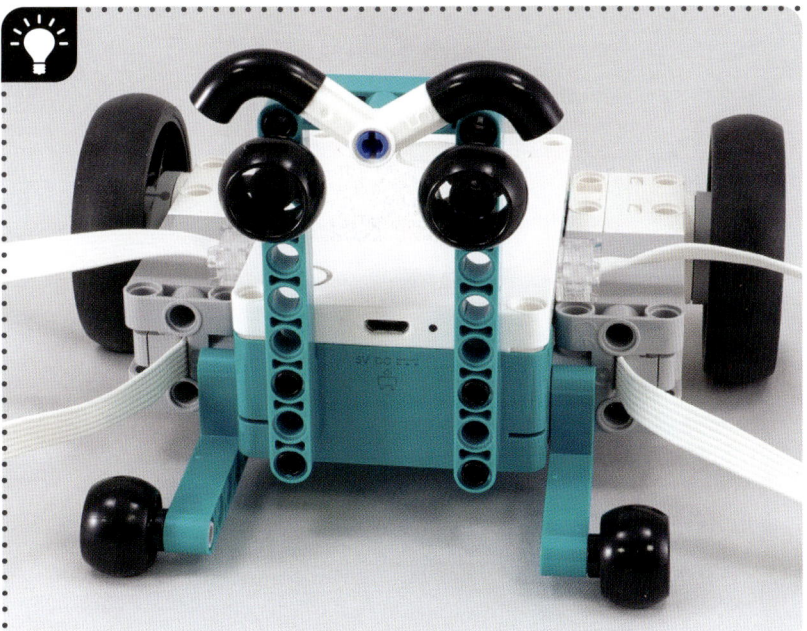

#52

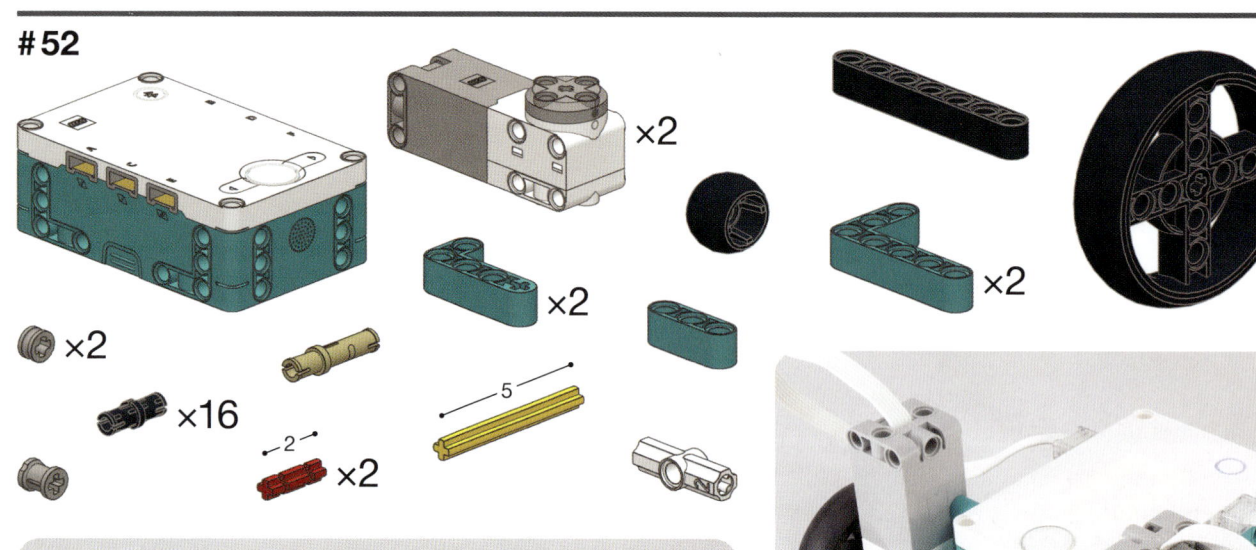

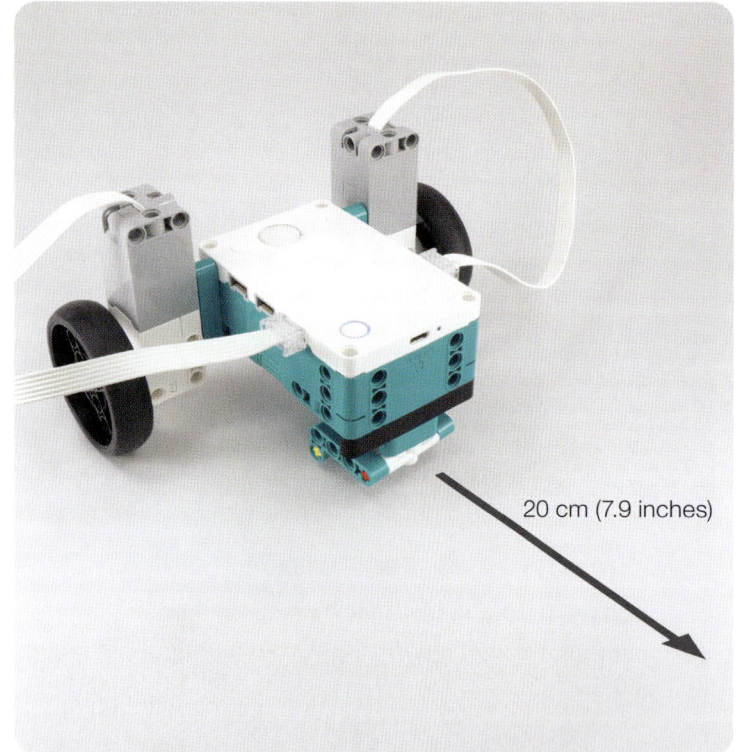

20 cm (7.9 inches)

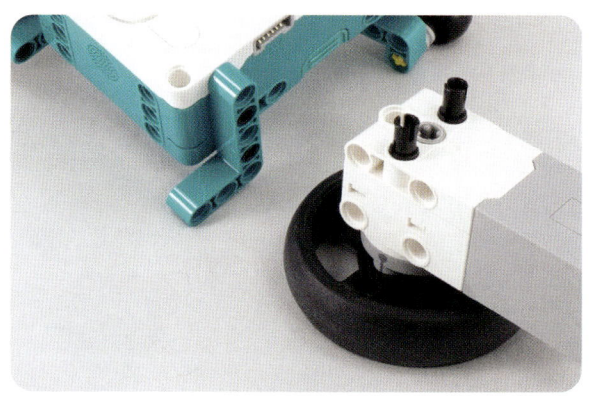

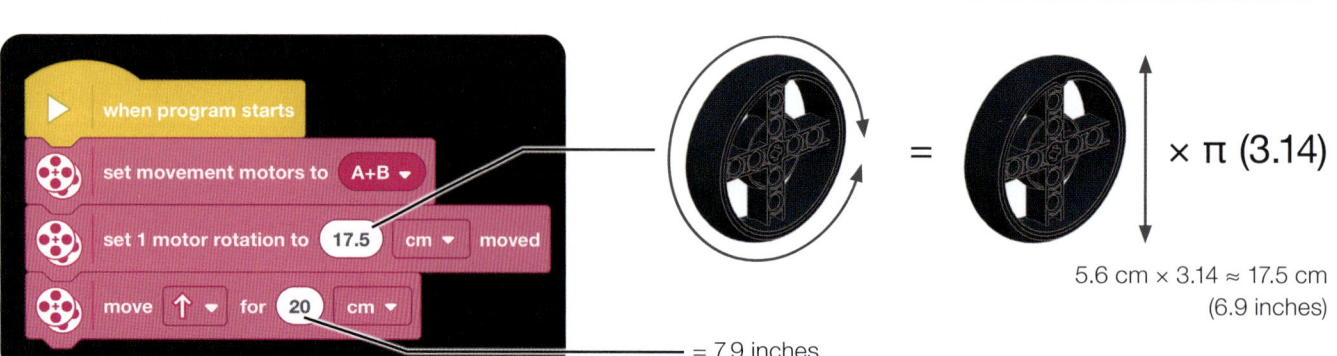

```
when program starts

set movement motors to  A+B ▾

set 1 motor rotation to  17.5  cm ▾  moved

move  ↑ ▾  for  20  cm ▾
```

= 7.9 inches

= × π (3.14)

5.6 cm × 3.14 ≈ 17.5 cm
(6.9 inches)

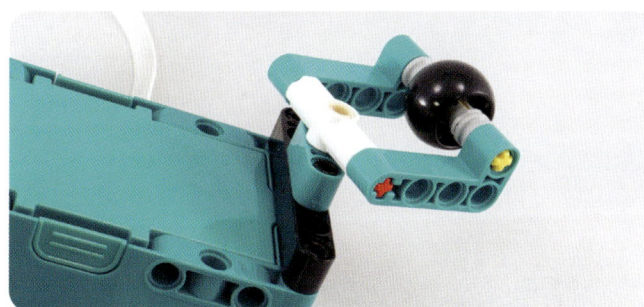

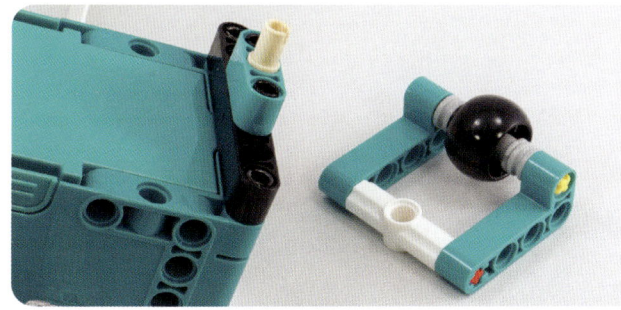

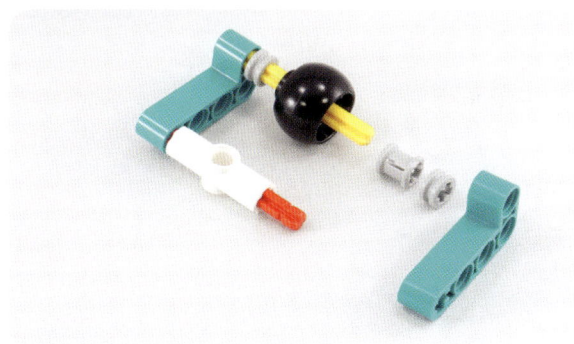

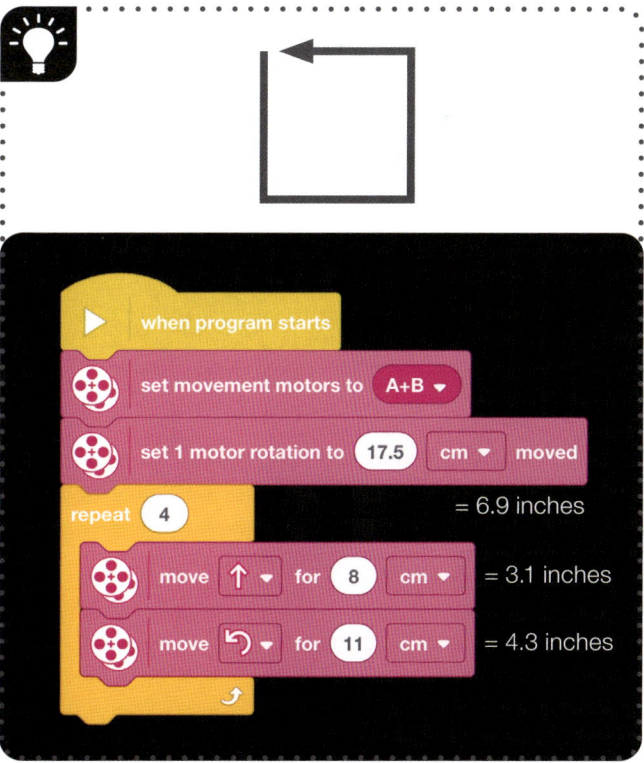

when program starts

set movement motors to A+B

set 1 motor rotation to 17.5 cm moved   = 6.9 inches

repeat 4

move ↑ for 8 cm   = 3.1 inches

move ↺ for 11 cm   = 4.3 inches

#53

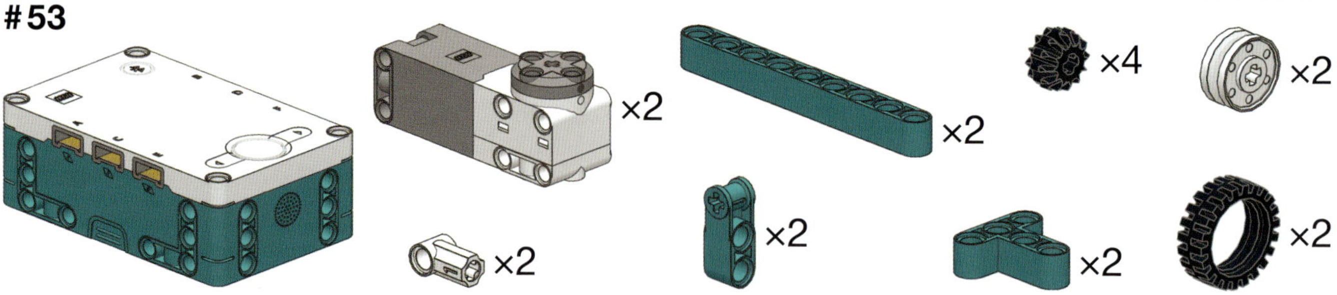

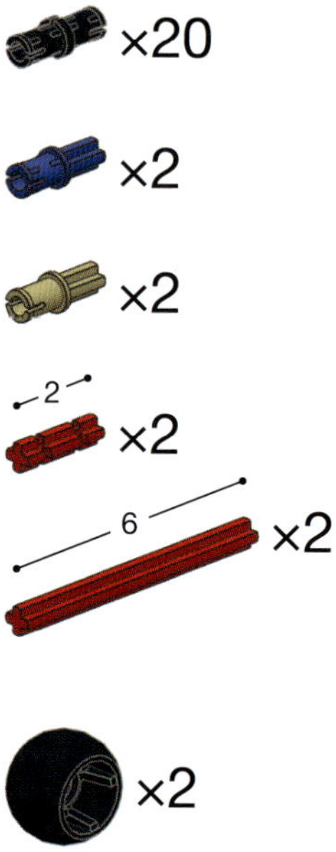

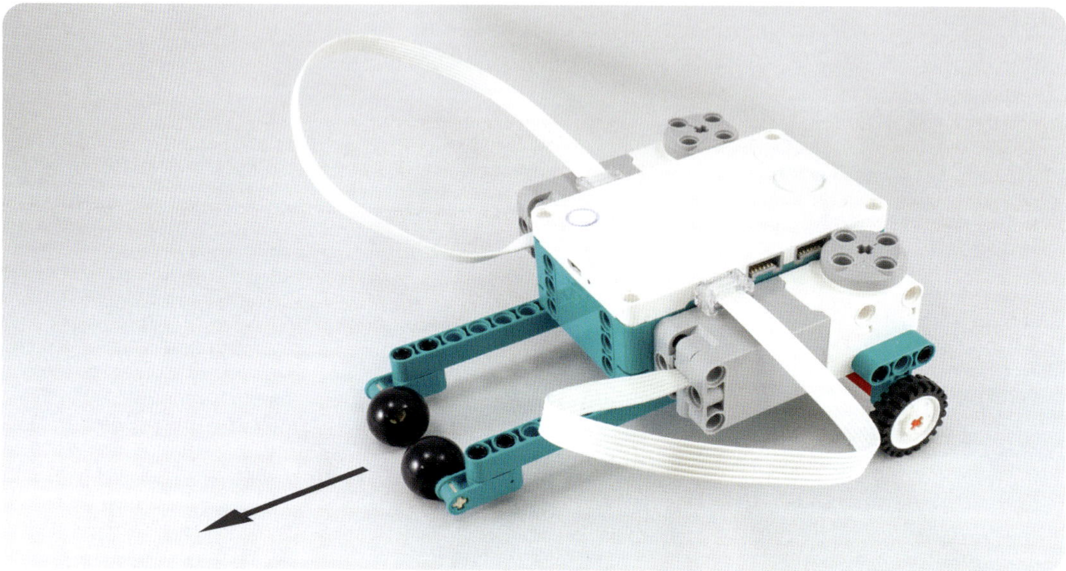

×2

×2

×4

×2

×2

×2

×2

×20

×2

×2

2
×2

6
×2

×2

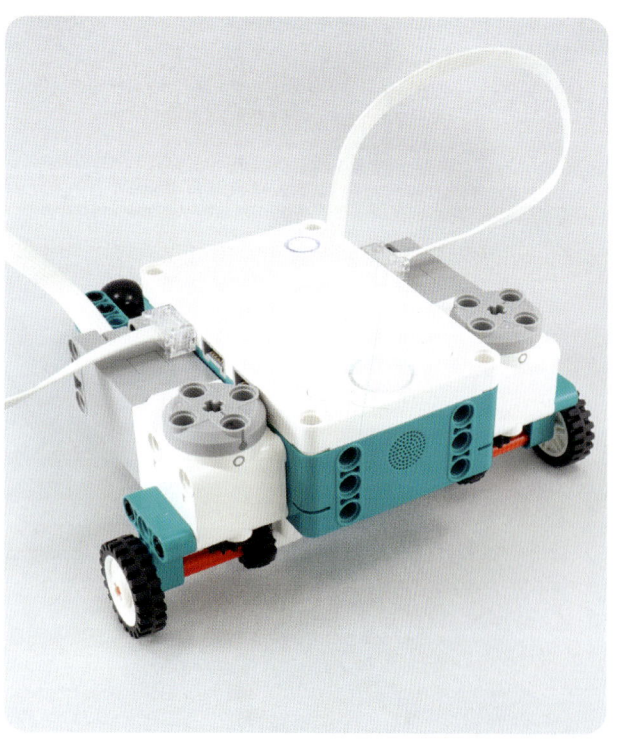

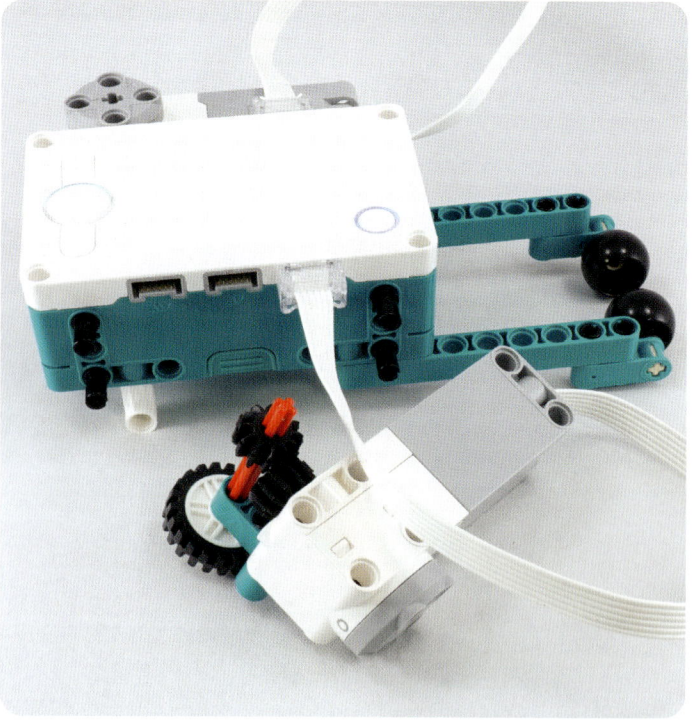

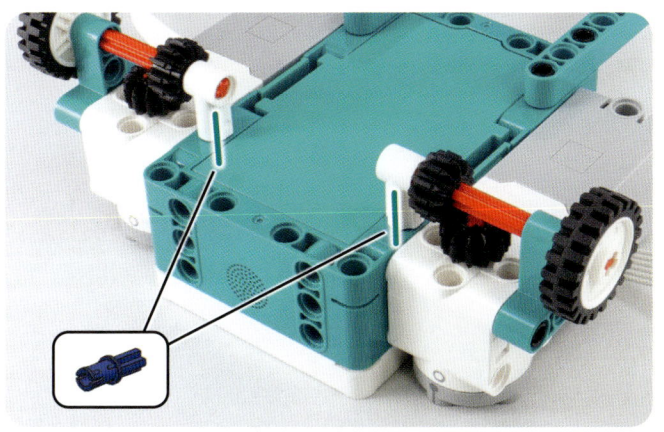

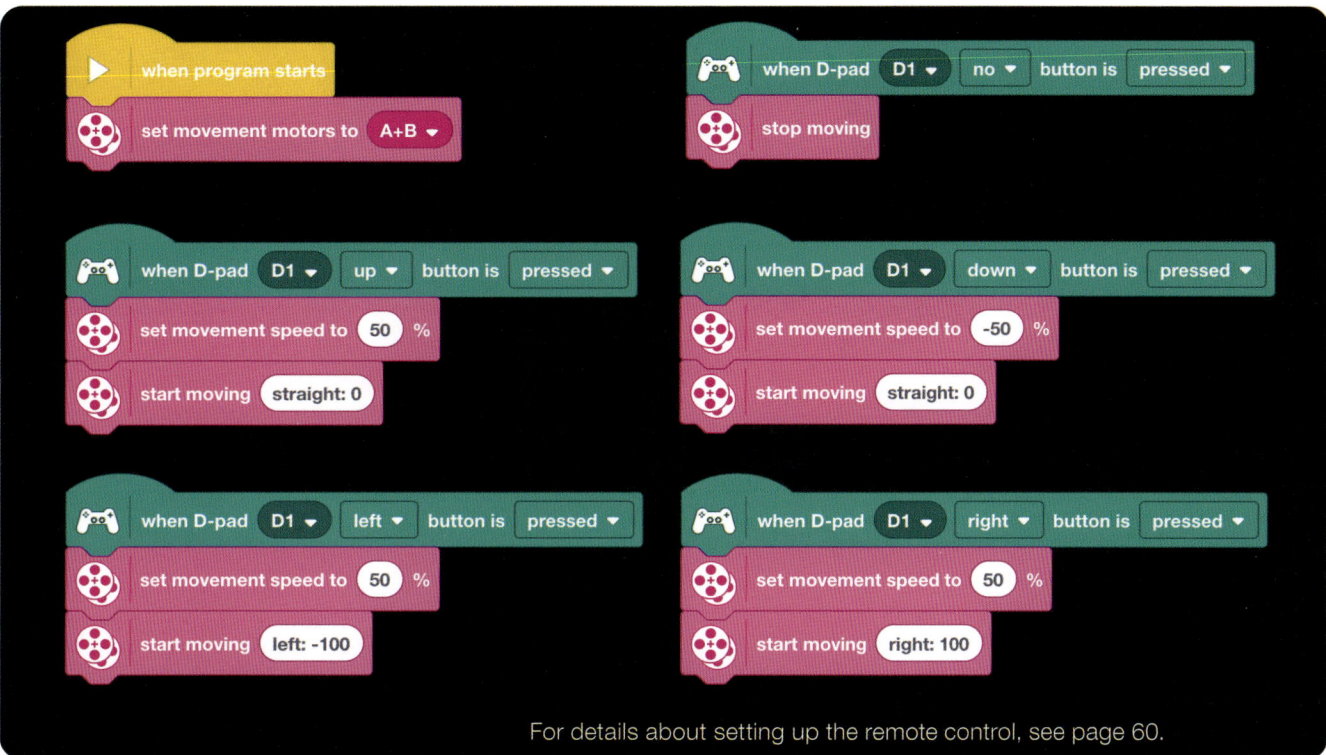

when program starts

set movement motors to A+B

when D-pad D1 ▾ no ▾ button is pressed ▾
stop moving

when D-pad D1 ▾ up ▾ button is pressed ▾
set movement speed to 50 %
start moving straight: 0

when D-pad D1 ▾ down ▾ button is pressed ▾
set movement speed to -50 %
start moving straight: 0

when D-pad D1 ▾ left ▾ button is pressed ▾
set movement speed to 50 %
start moving left: -100

when D-pad D1 ▾ right ▾ button is pressed ▾
set movement speed to 50 %
start moving right: 100

For details about setting up the remote control, see page 60.

In the above program, the D-pad is used as a controller.

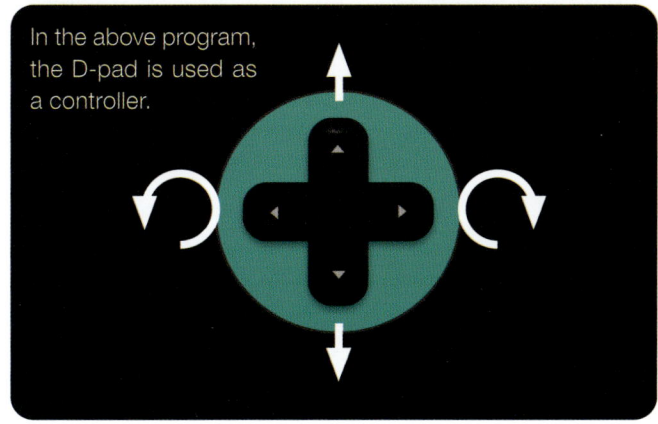

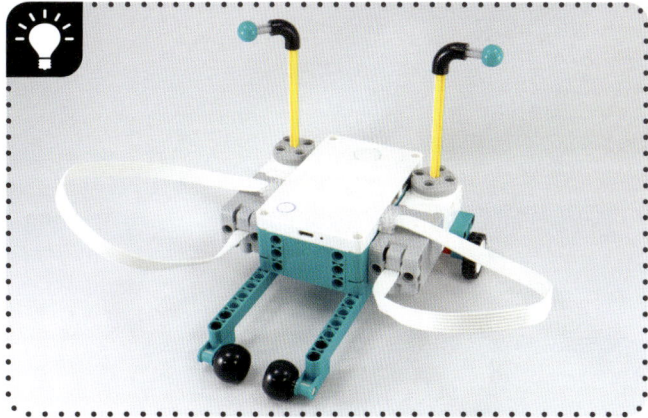

54

×2
×10
×2
×4

×2
×2
×2
4
4
×2

×2

×4

×2

×2
×2

×2
×2

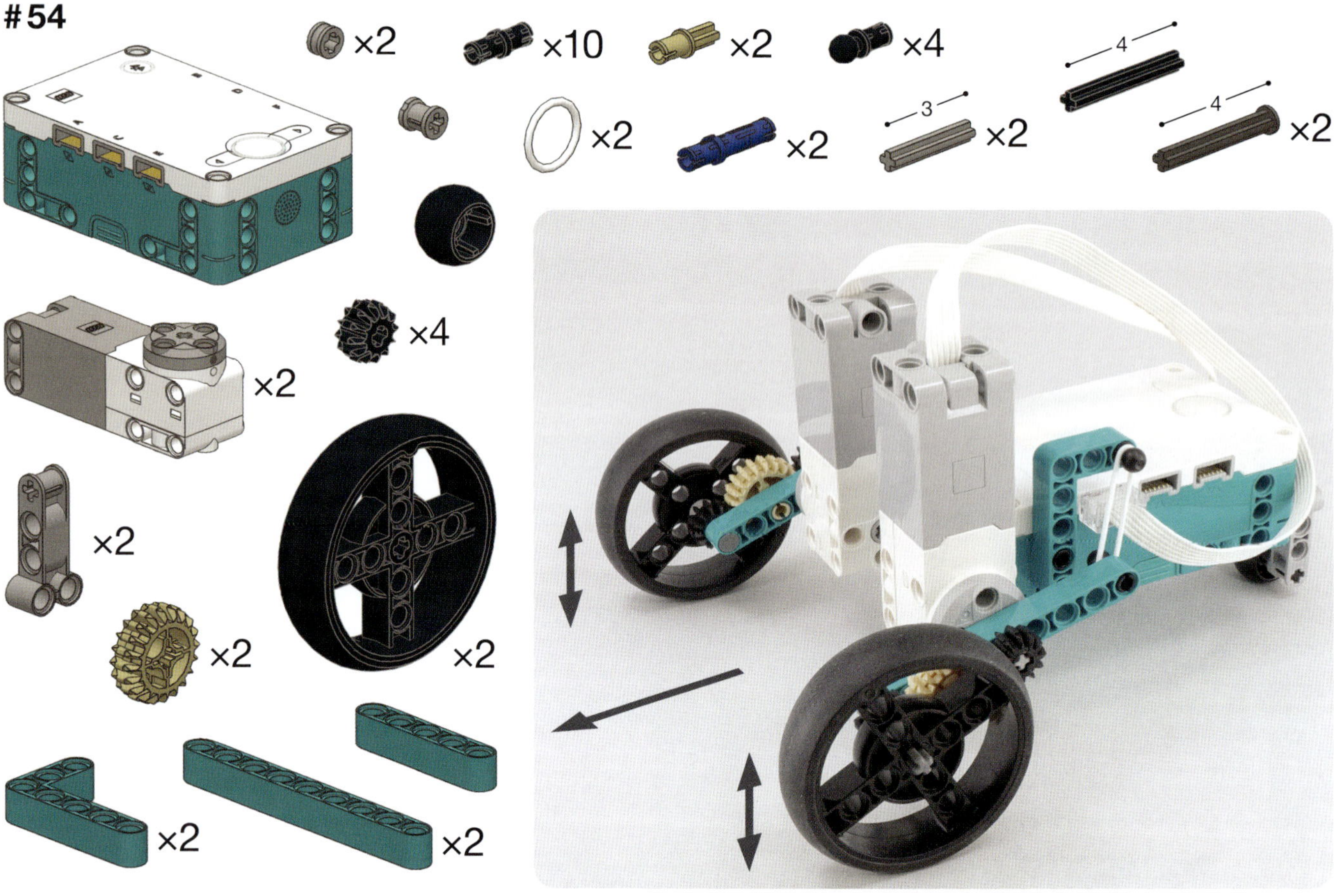

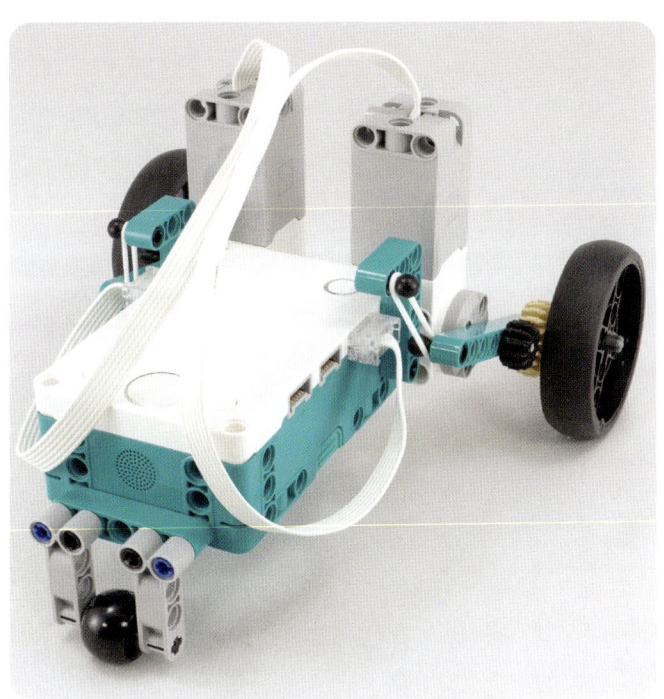

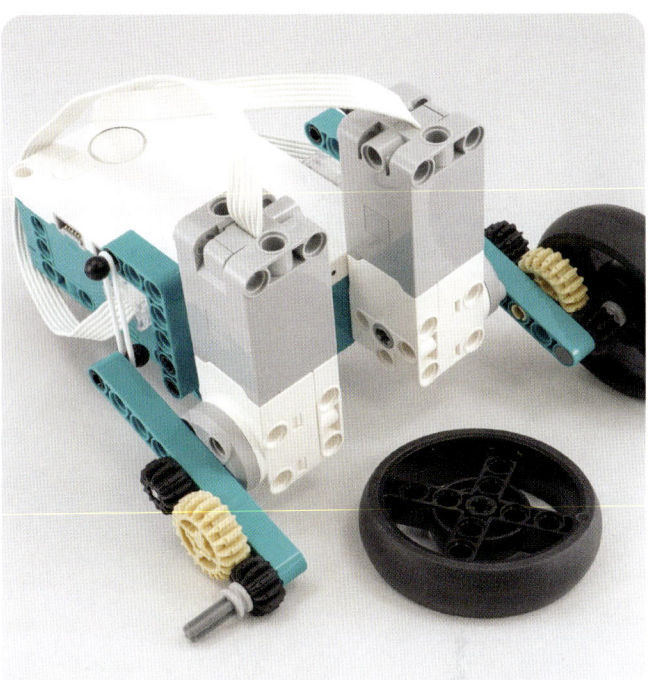

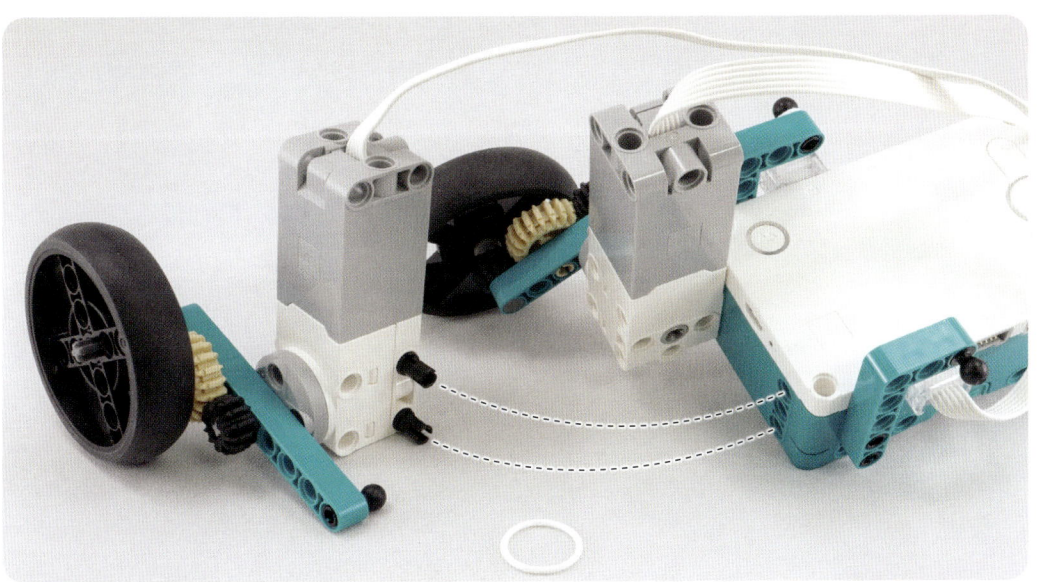

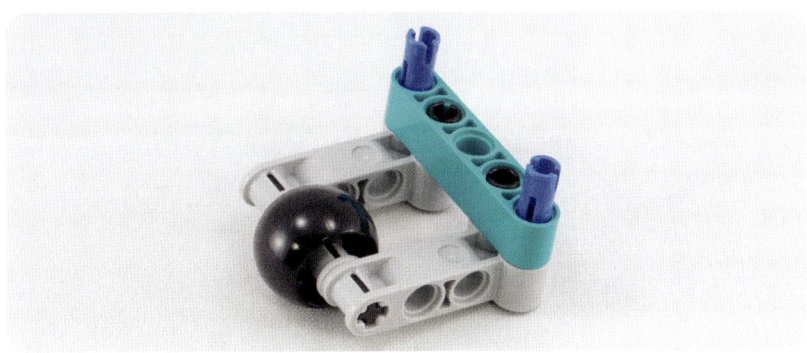

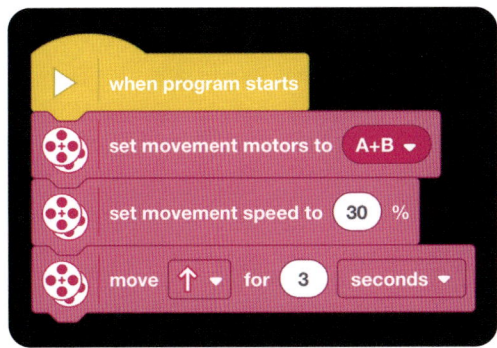

Turning with a steering wheel

#55

×2 ×5 ×2

×22 ×3

×2

~2

~3 ×2

×2

~4

×2

×4

×2

×3

×2 ×3

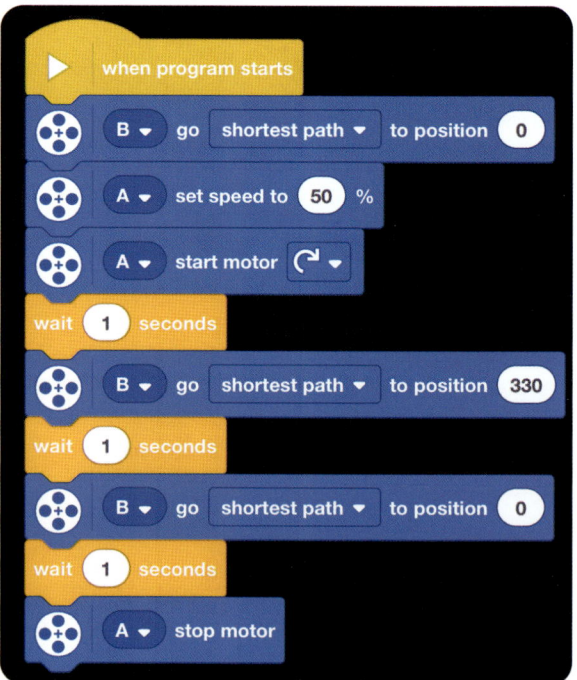

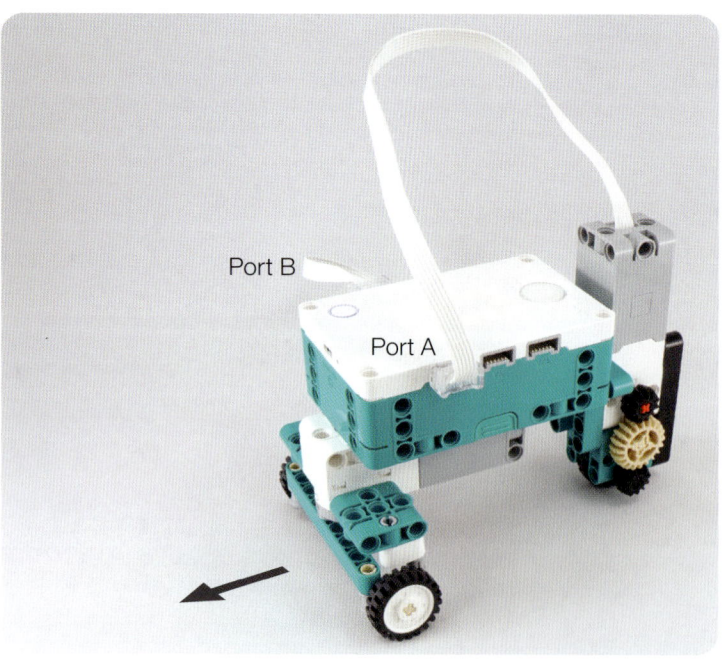

Port B

Port A

when program starts

B ▾ go shortest path ▾ to position 0

A ▾ set speed to 50 %

A ▾ start motor ↻ ▾

wait 1 seconds

B ▾ go shortest path ▾ to position 330

wait 1 seconds

B ▾ go shortest path ▾ to position 0

wait 1 seconds

A ▾ stop motor

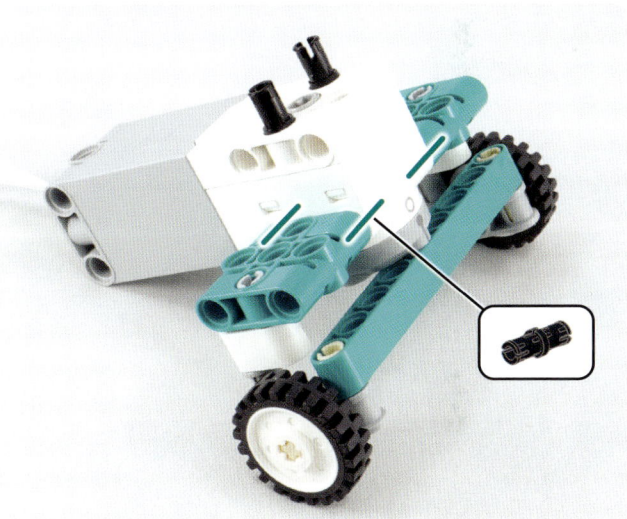

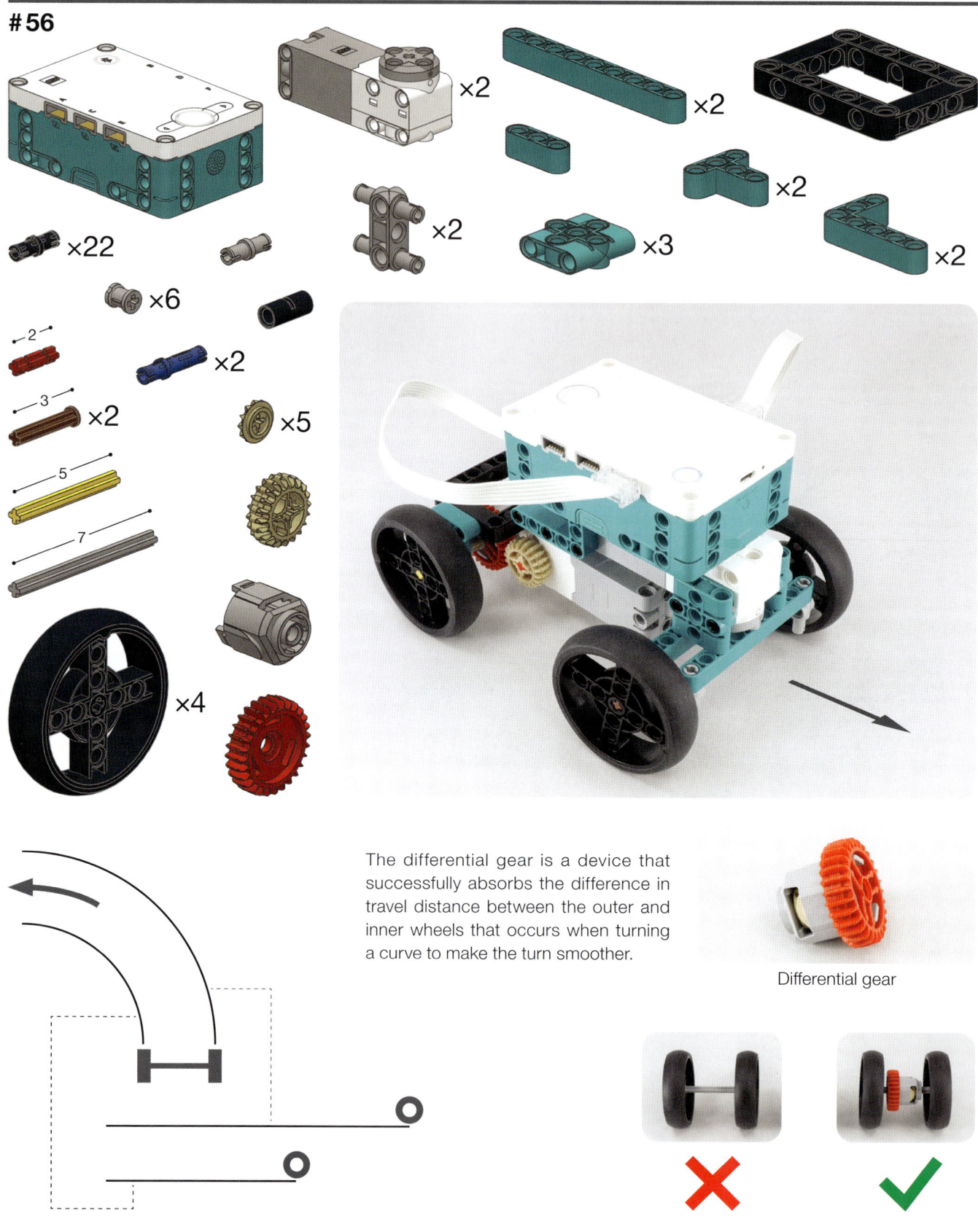

56

×22
×6
2
×2
3
×2
5
7
×4

×2
×2
×5

×2
×3
×2
×2
×2

The differential gear is a device that successfully absorbs the difference in travel distance between the outer and inner wheels that occurs when turning a curve to make the turn smoother.

Differential gear

❌ ✅

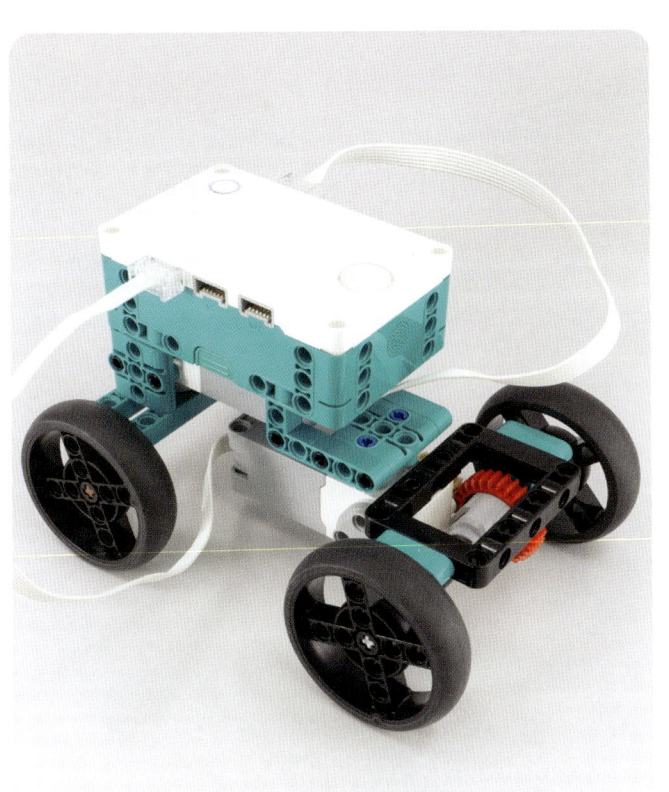

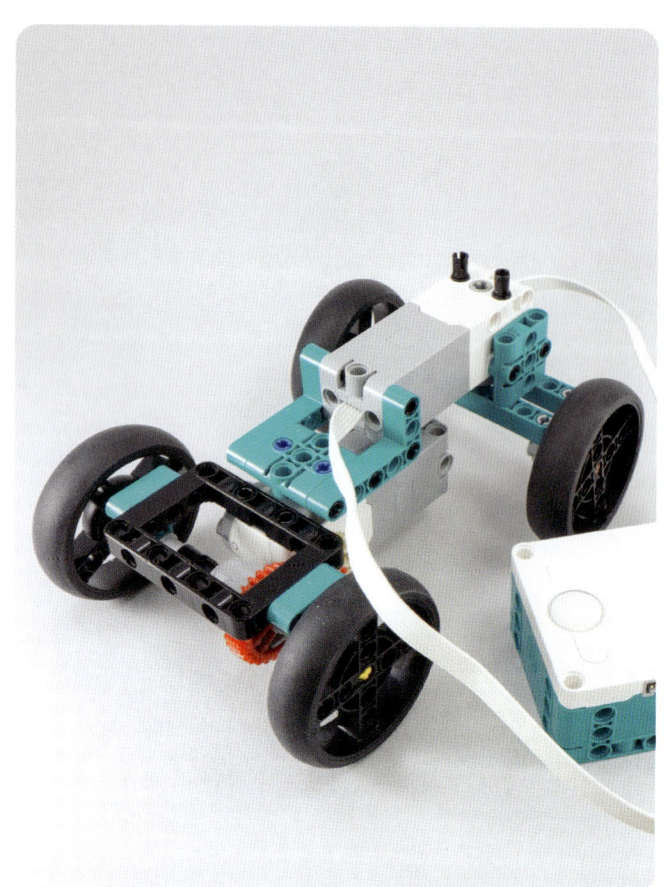

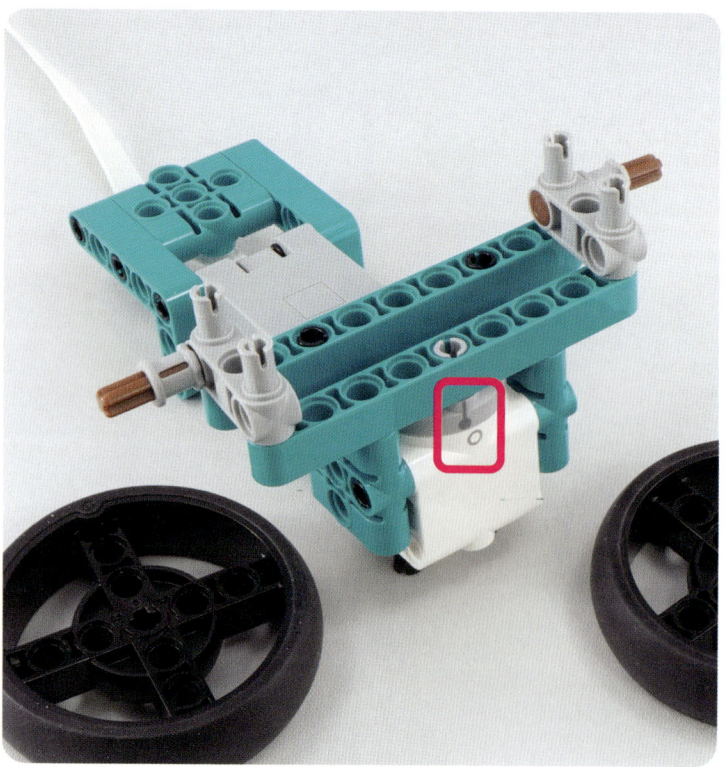

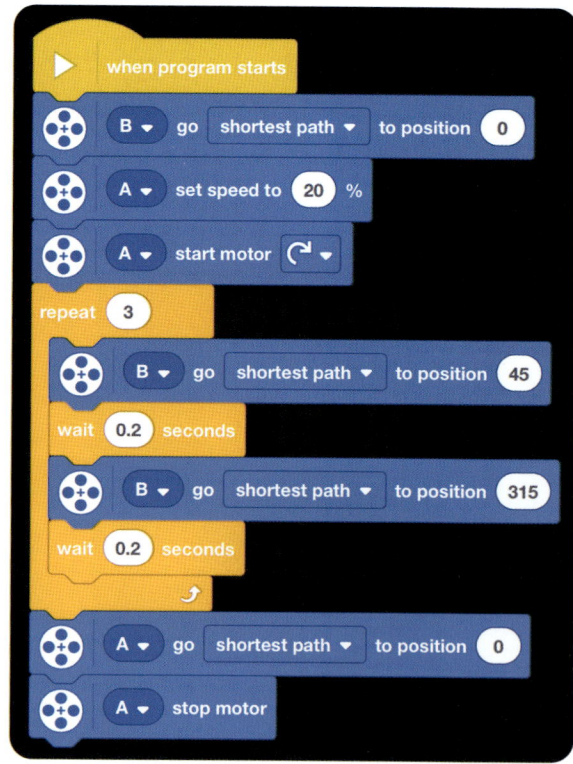

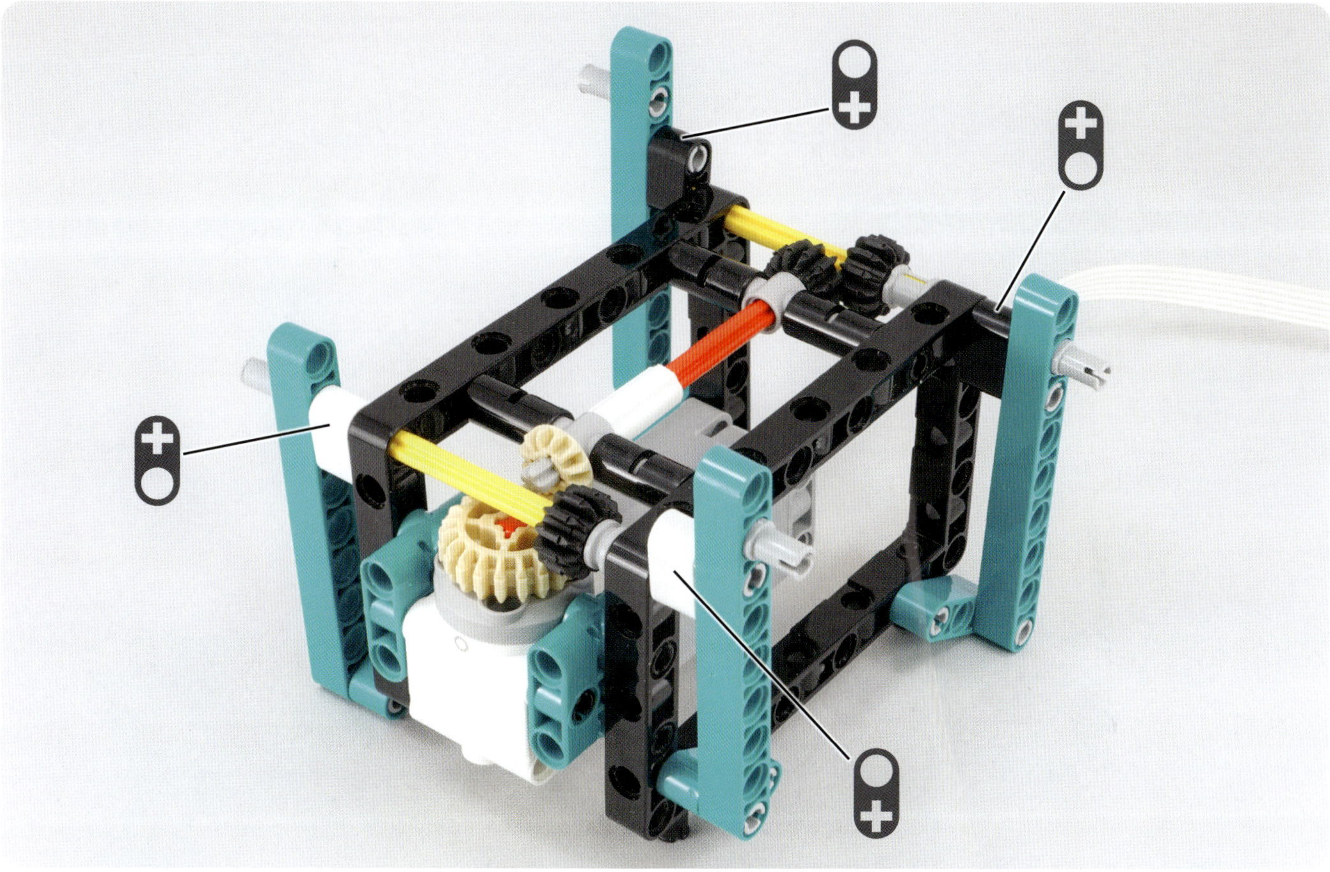

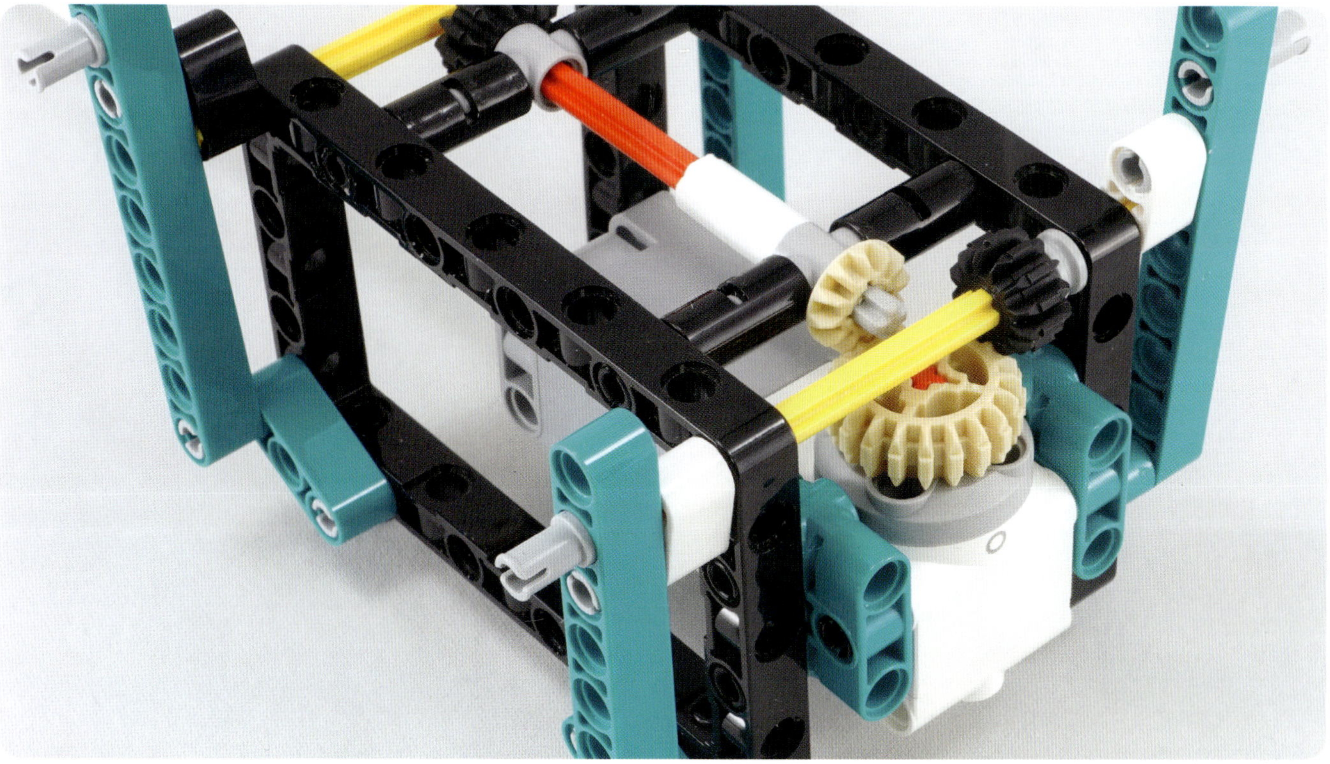

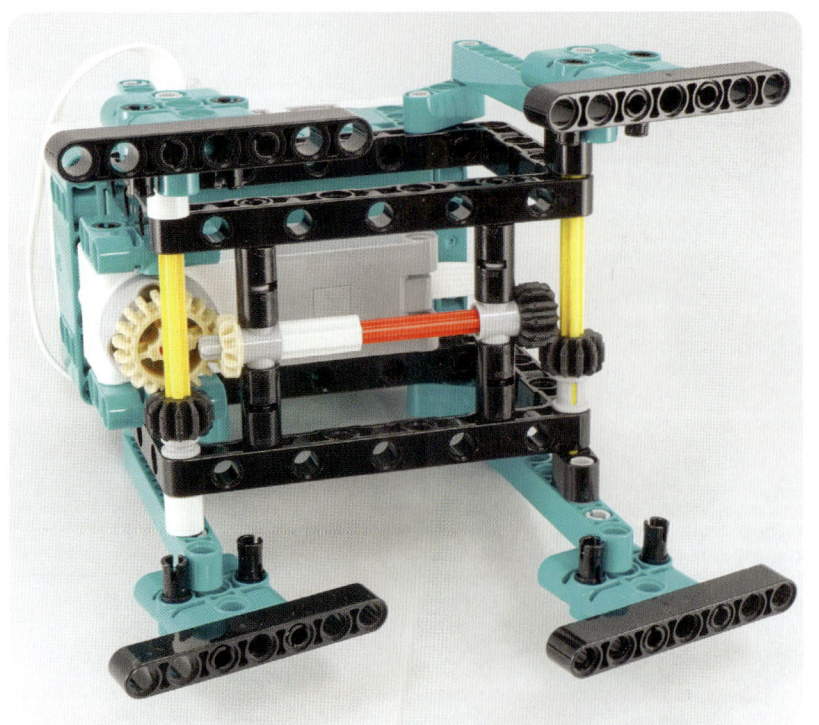

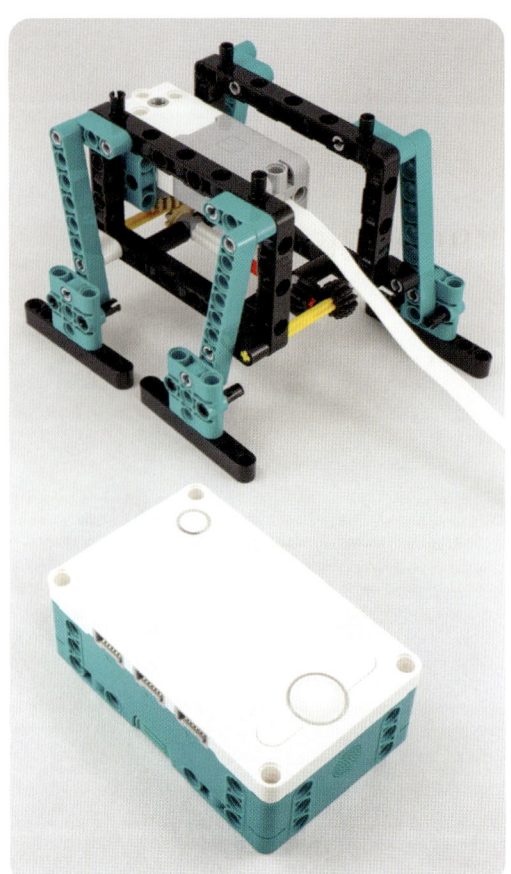

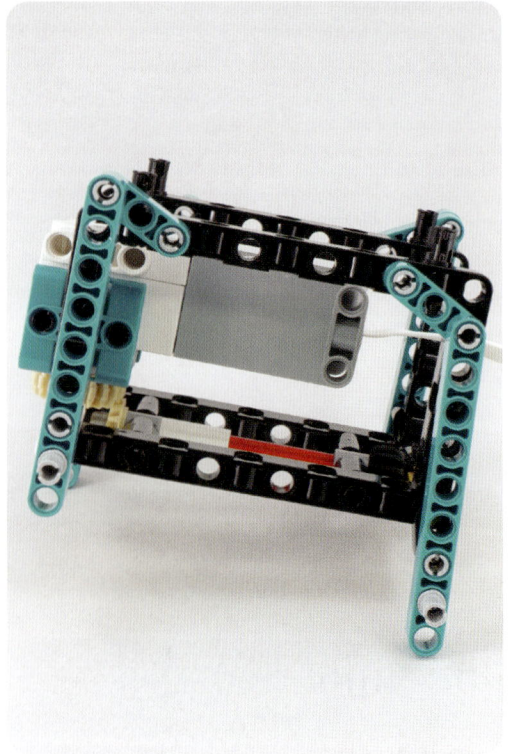

#59

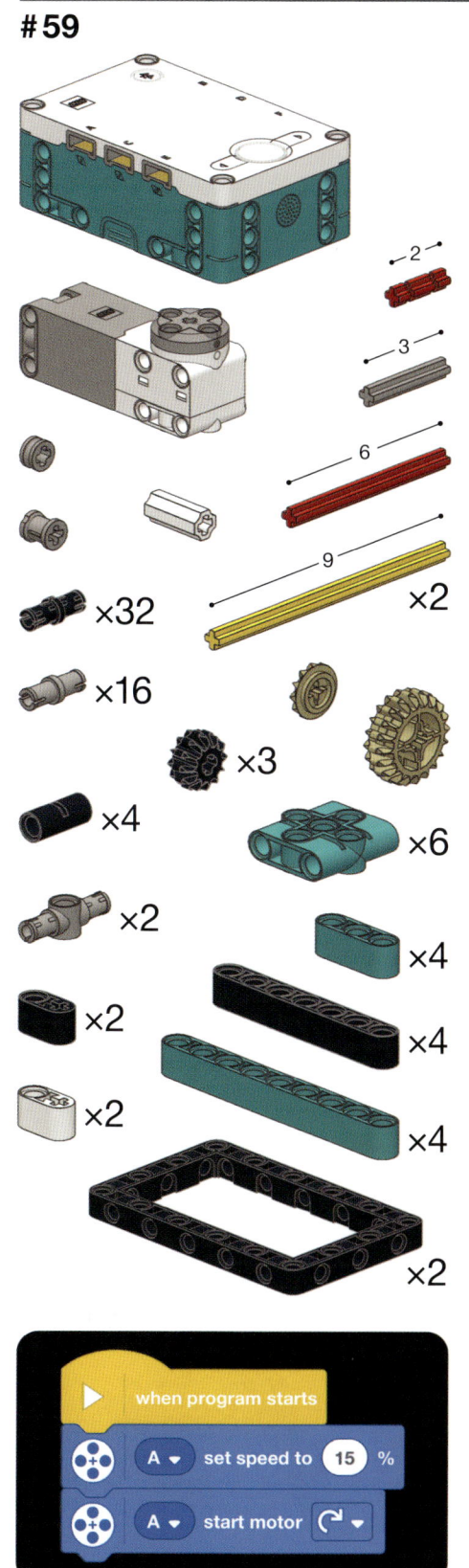

×2

×3

×6

×2

×9

×2

×32

×16

×3

×4

×6

×2

×4

×2

×4

×2

×4

×2

```
when program starts
A ▾  set speed to  15 %
A ▾  start motor ↻ ▾
```

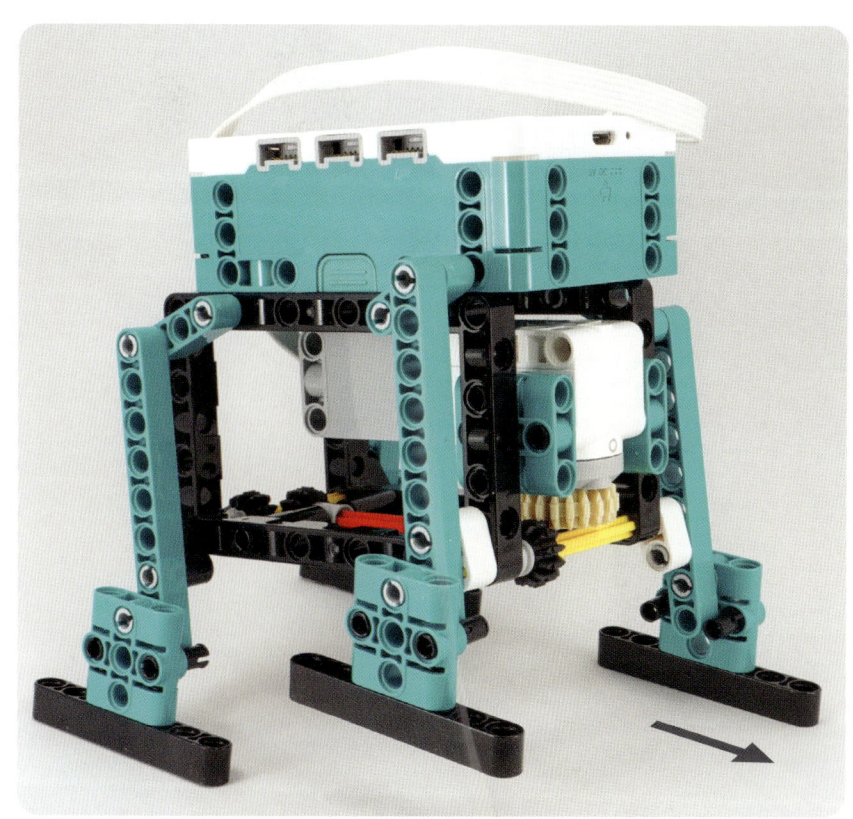

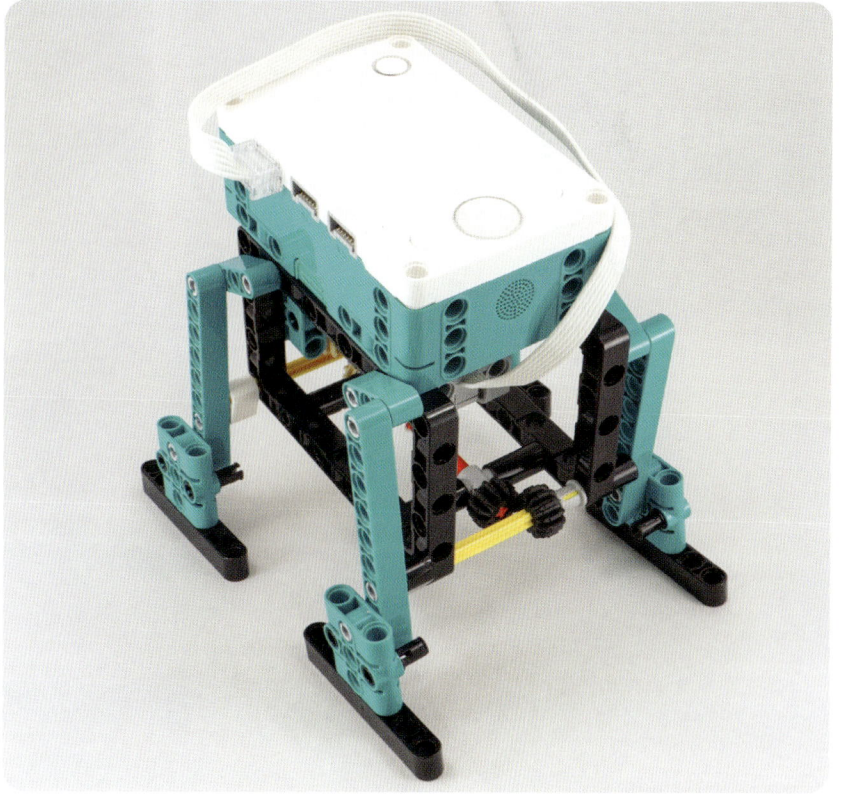

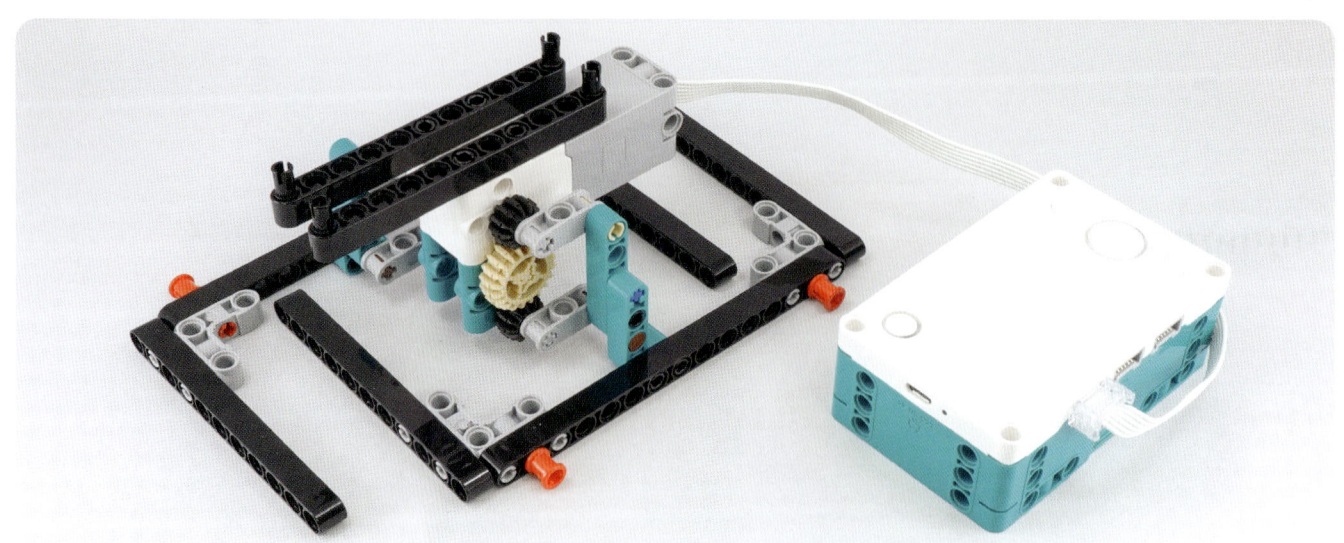

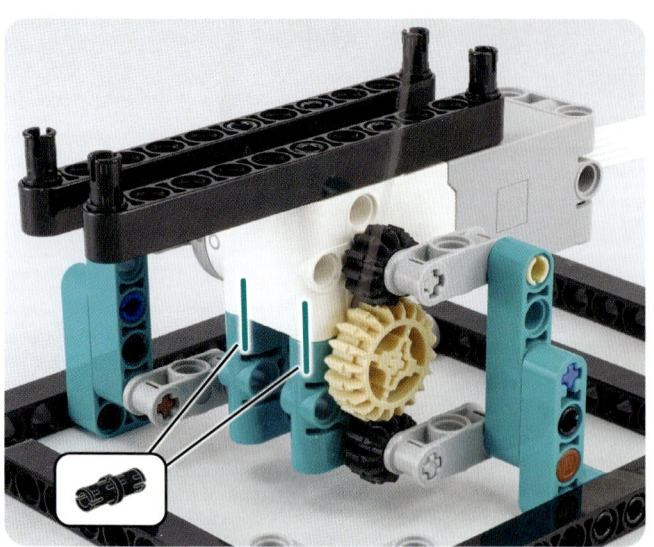

These two parts should be assembled so that they face opposite directions on the left and right.

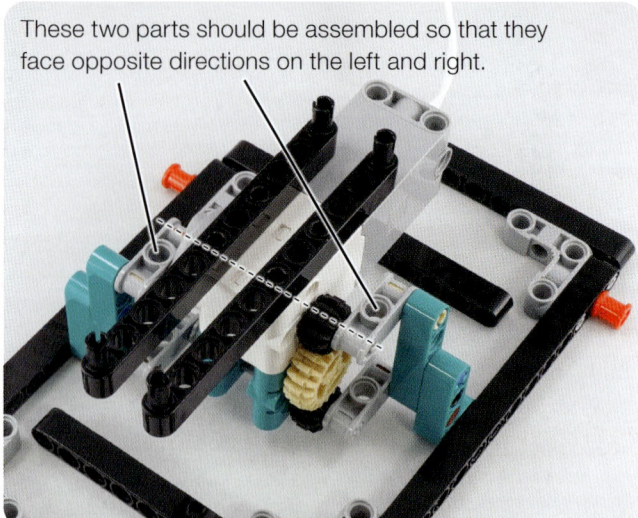

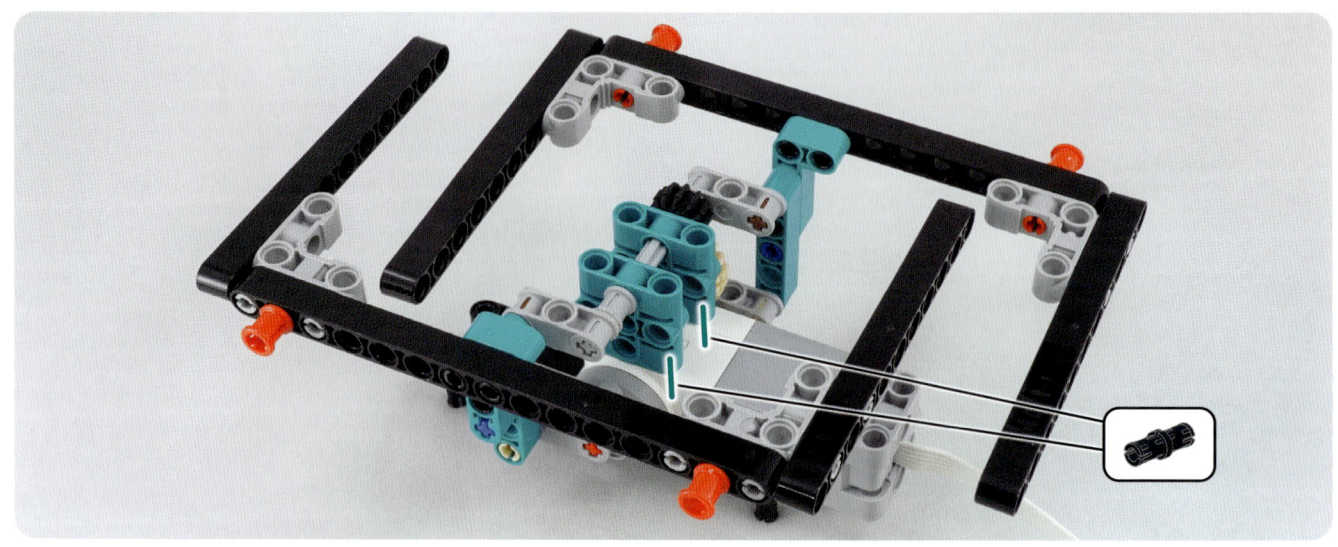

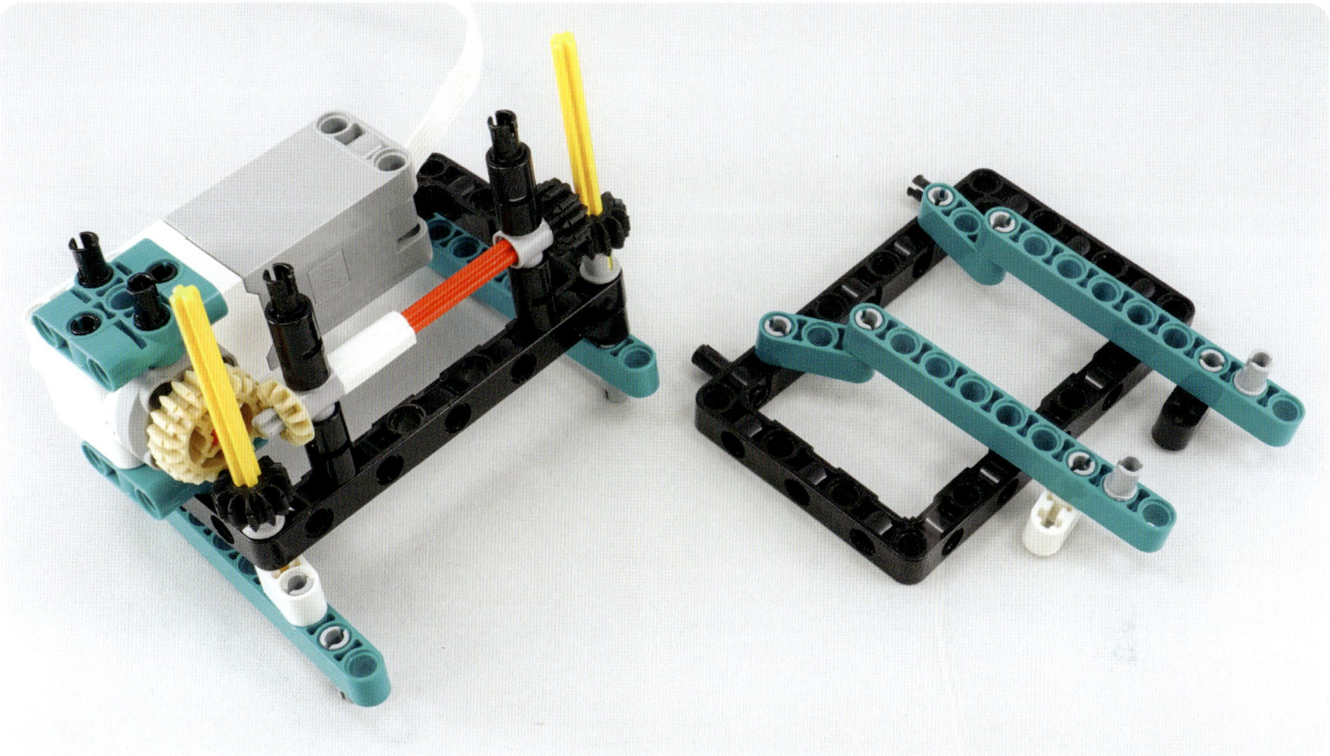

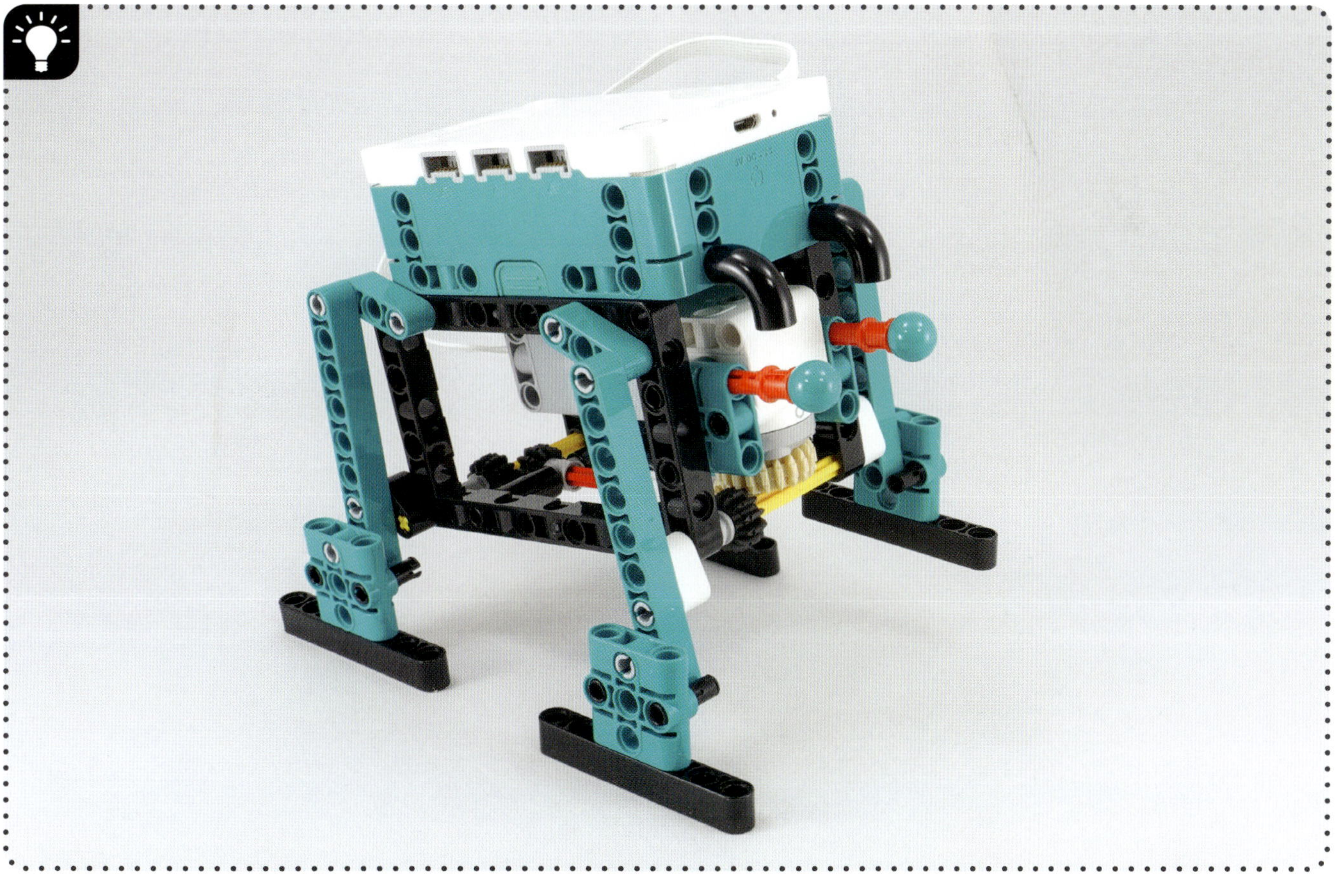

PART 3
Practical
Mechanisms

Gripping tools

#60

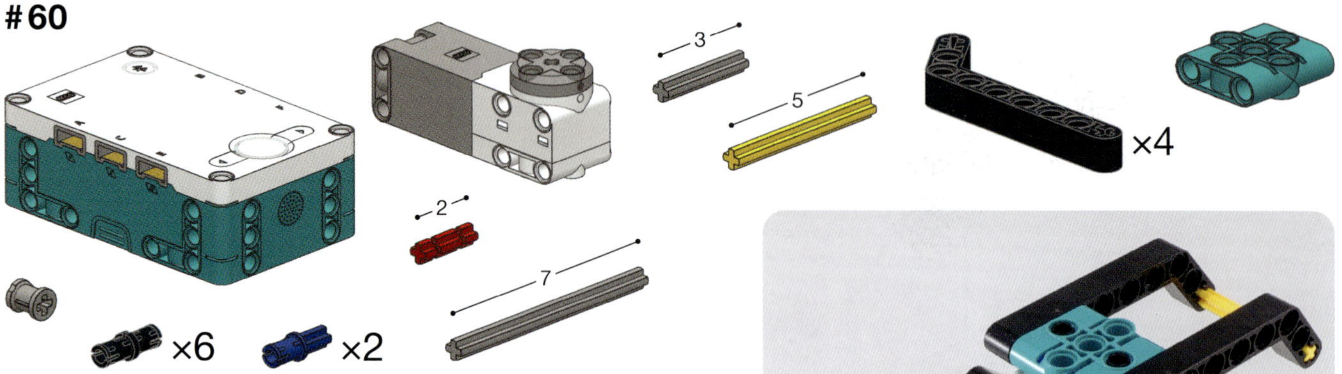

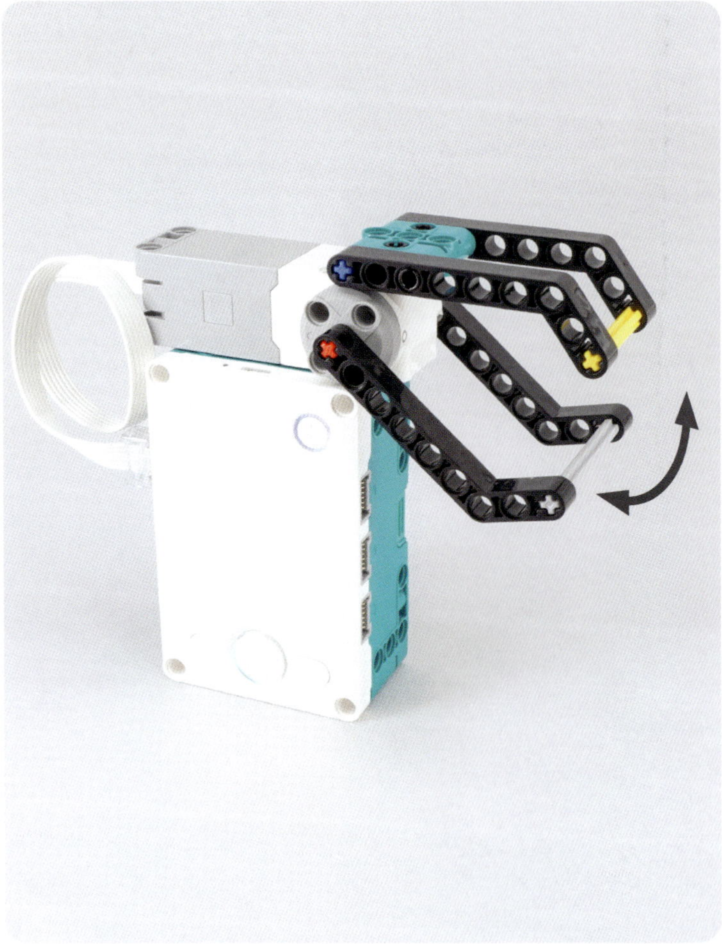

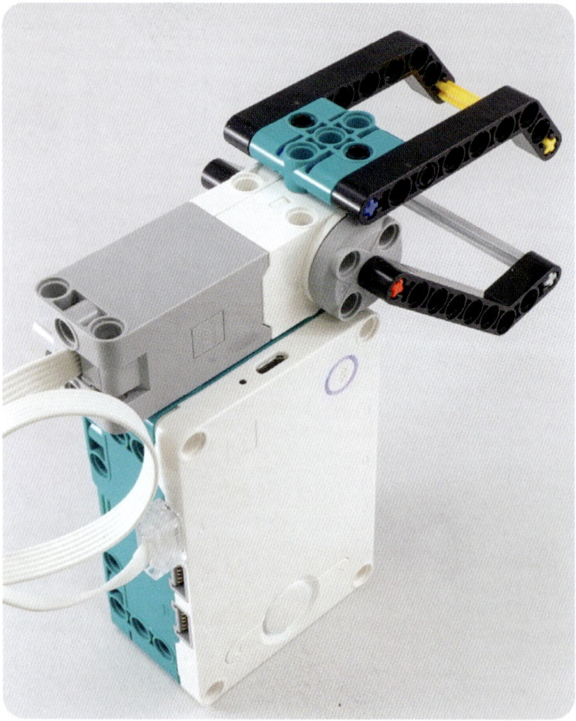

×6   ×2

3

5

7

2

×4

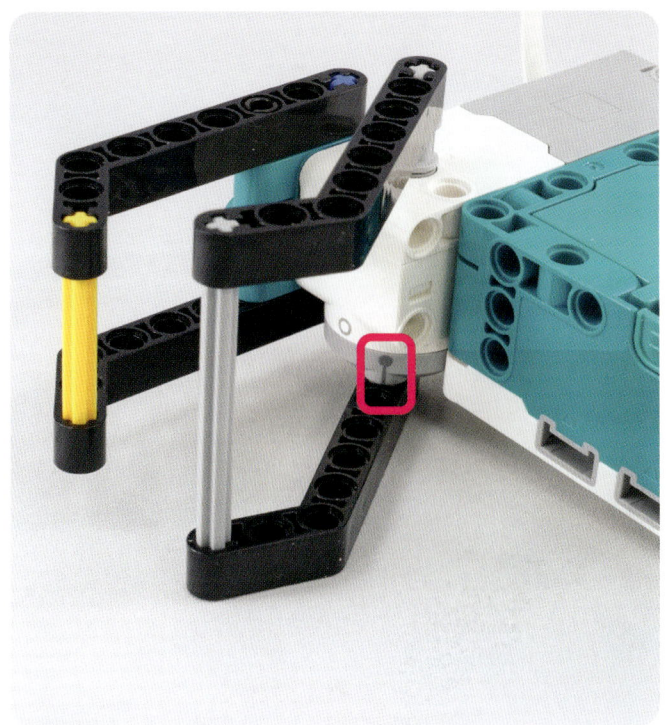

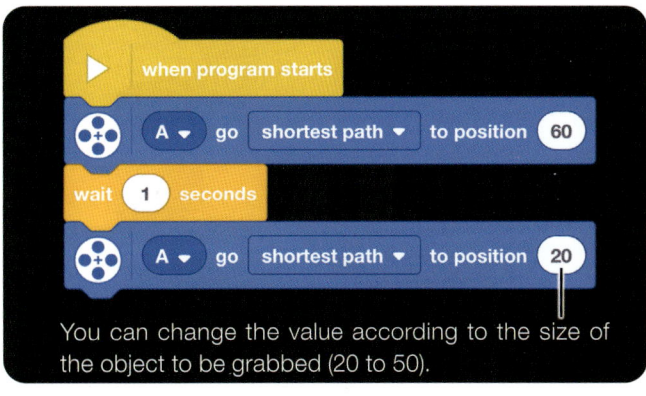

You can change the value according to the size of the object to be grabbed (20 to 50).

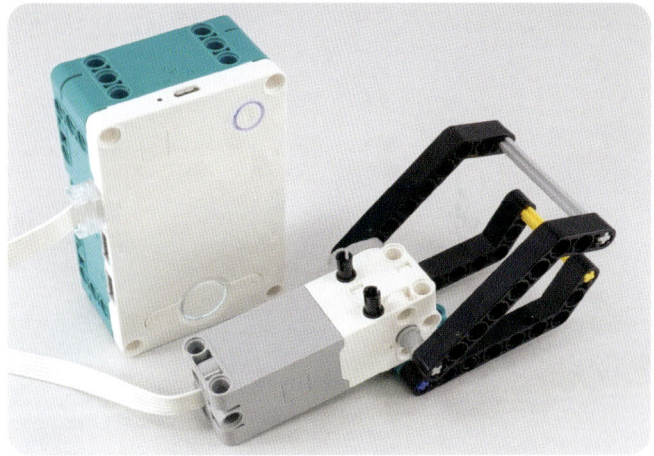

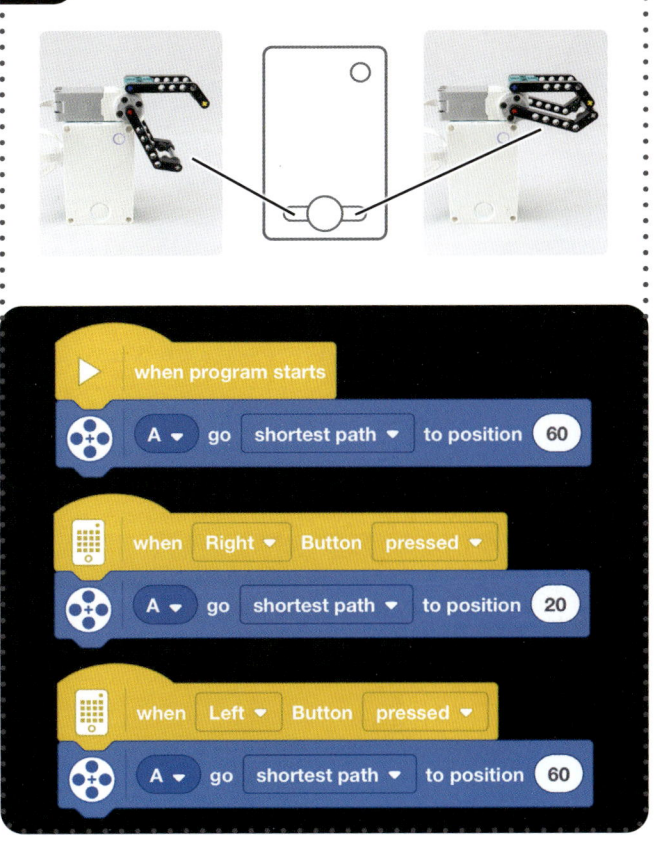

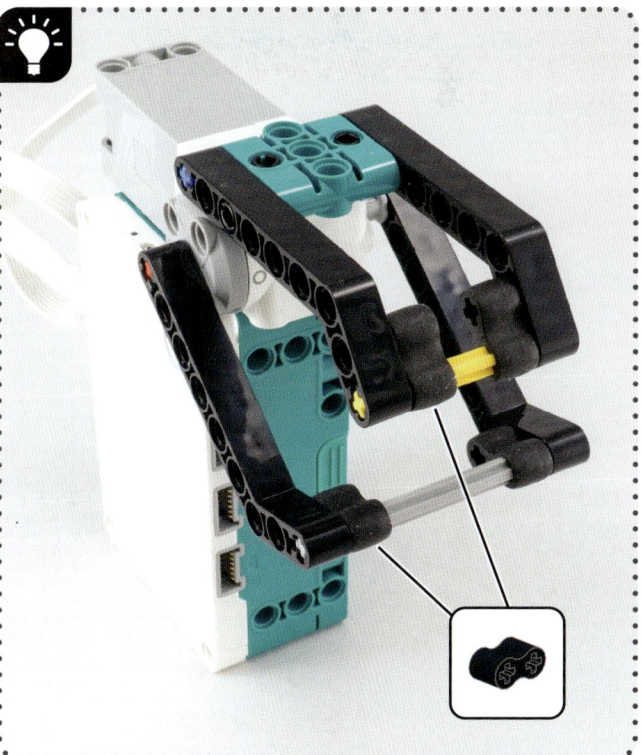

#61

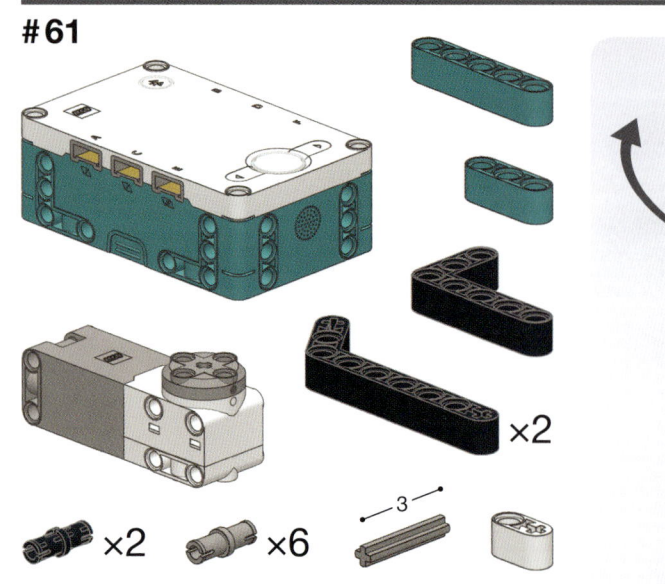

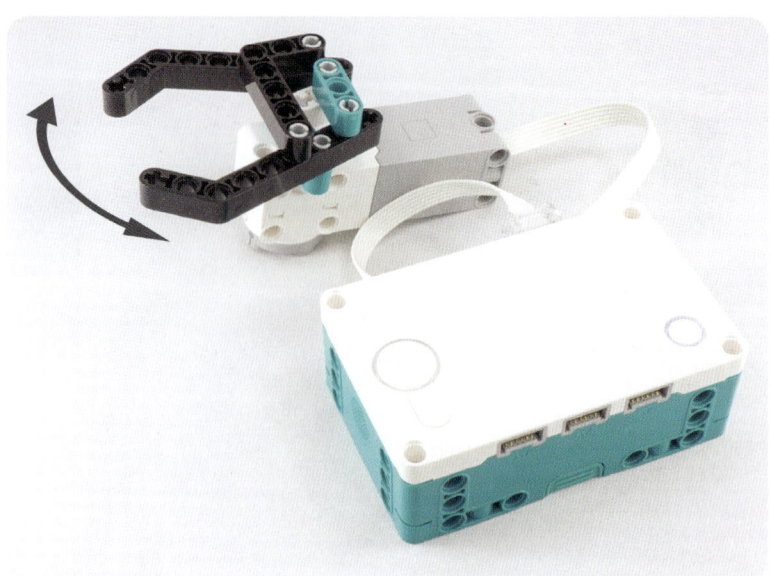

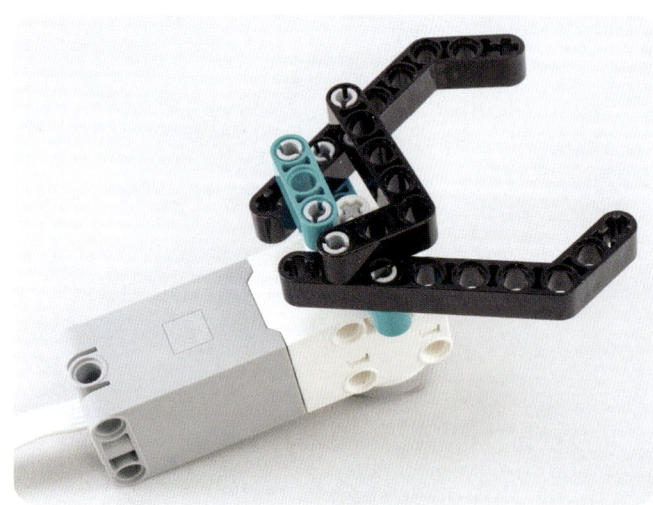

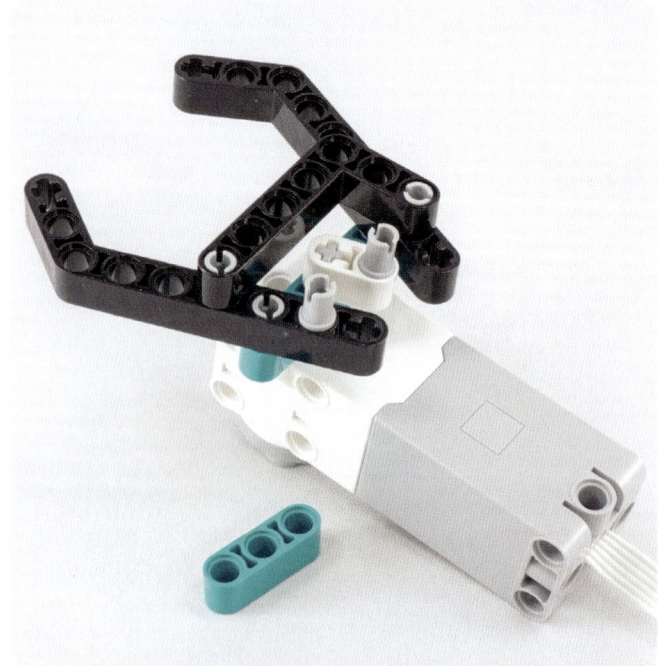

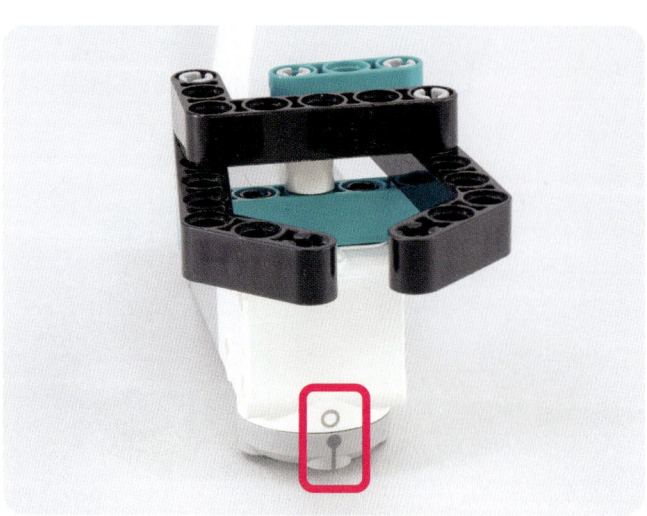

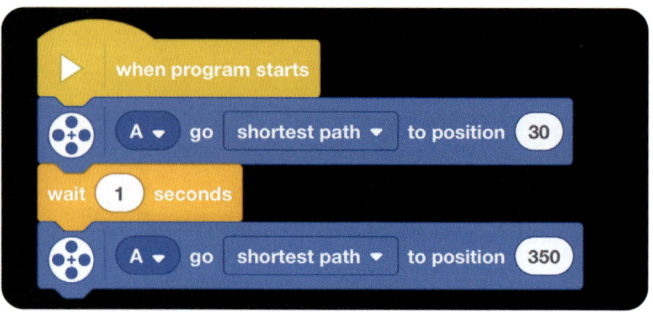

#62

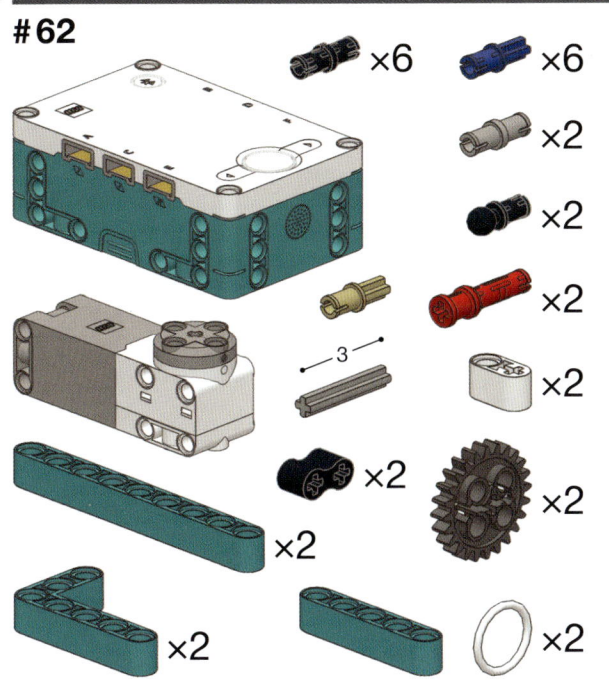

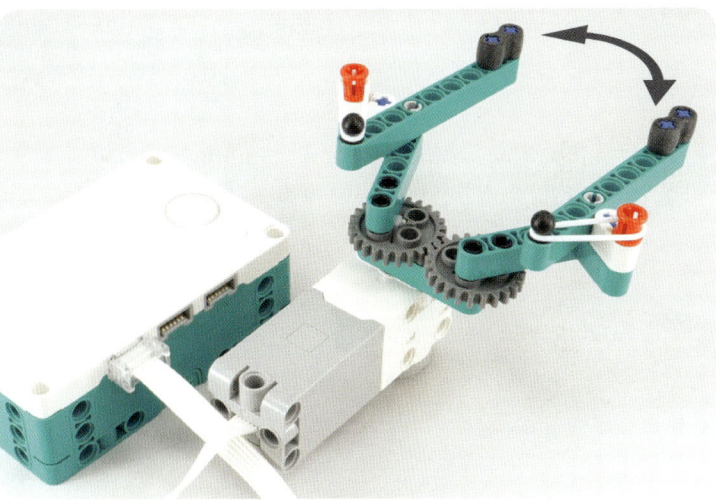

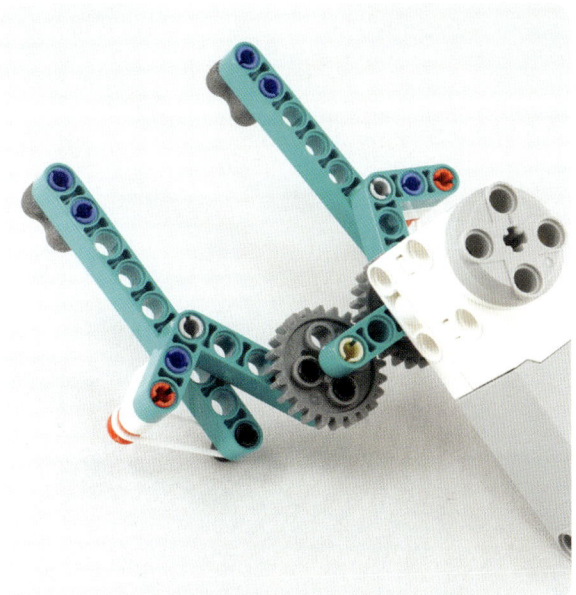

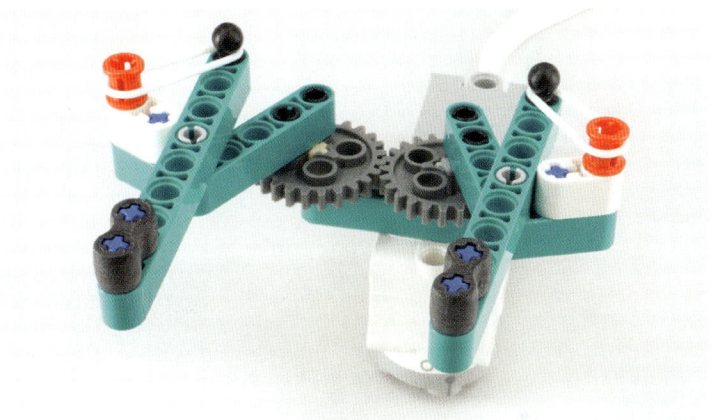

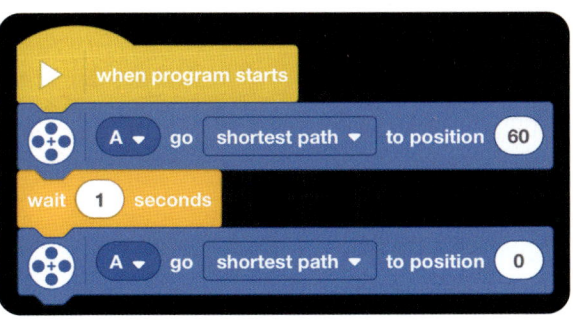

Match the positions of these two marks.

Then adjust the arms so that their tips are touching.

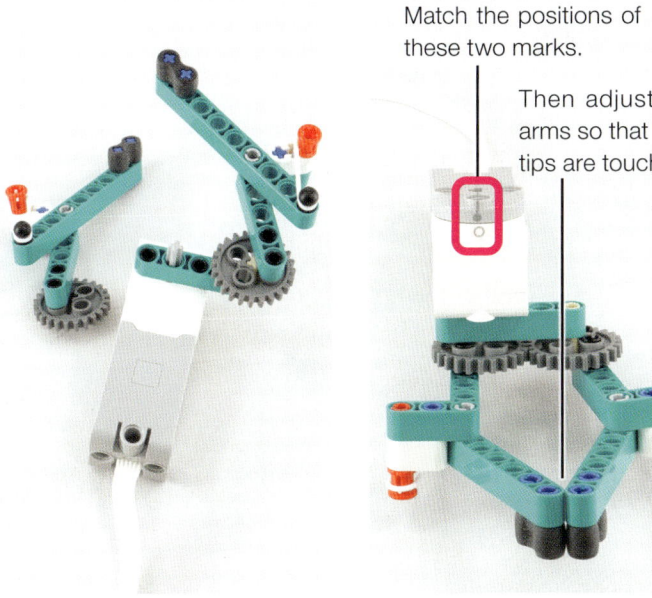

```
▶ when program starts

⊕  A ▾  go  shortest path ▾  to position  60

wait  1  seconds

⊕  A ▾  go  shortest path ▾  to position  0
```

#63

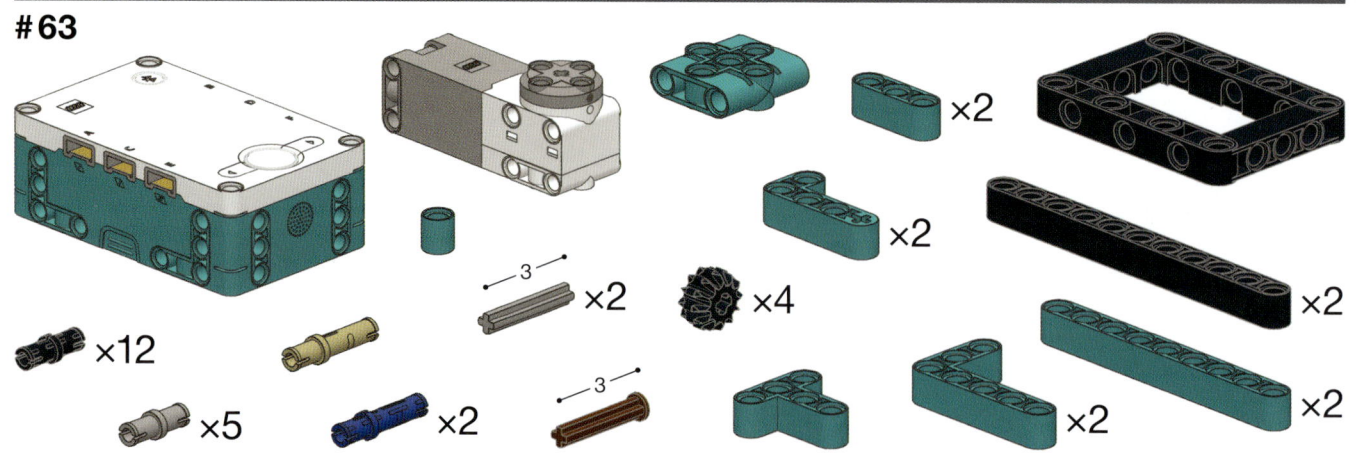

×12   ×5   ×2   3 ×2   ×4   3 ×2   ×2   ×2   ×2   ×2

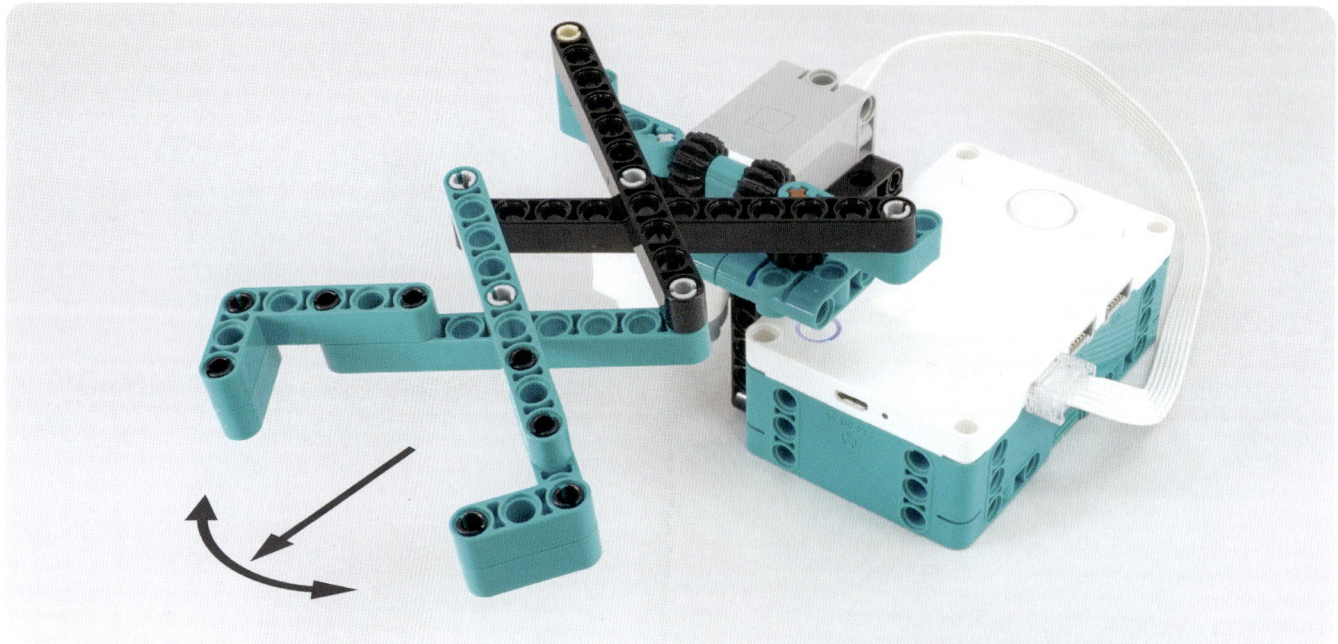

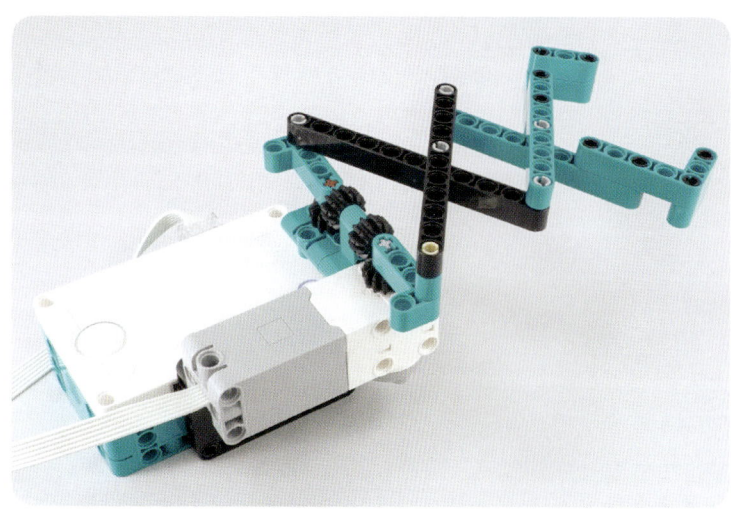

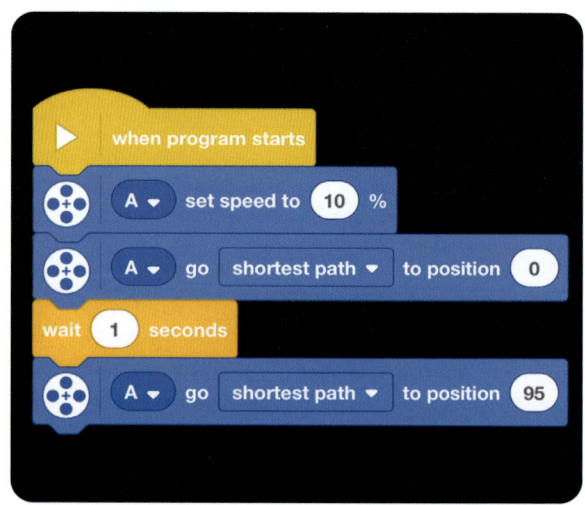

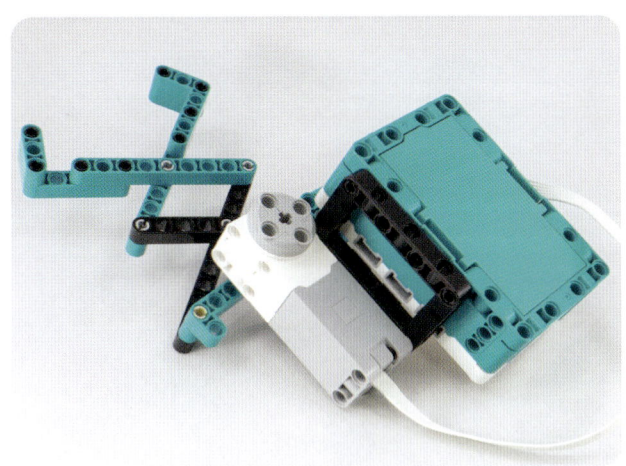

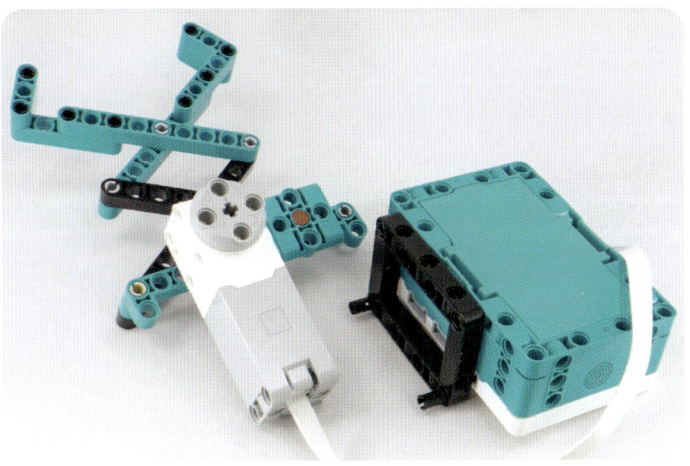

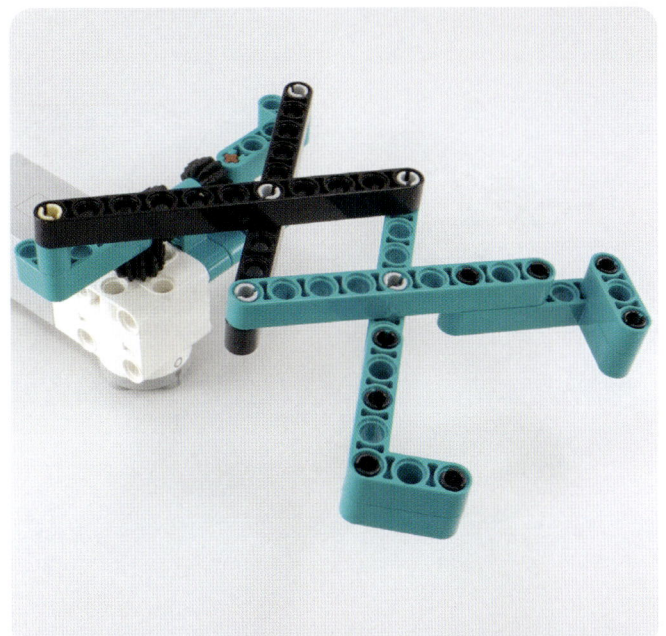

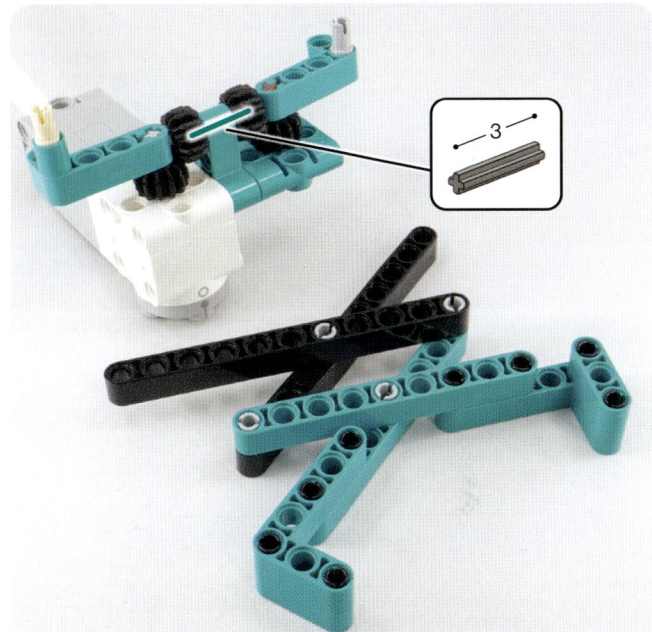

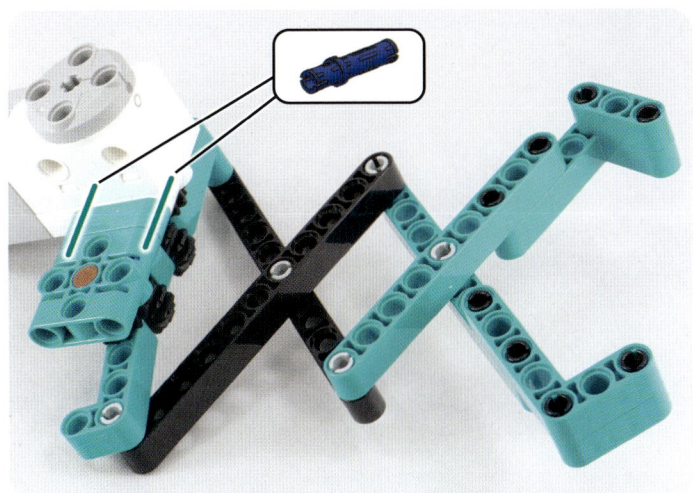

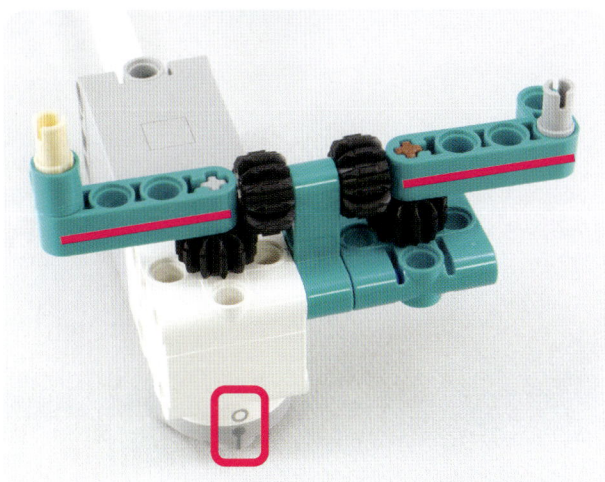

#64

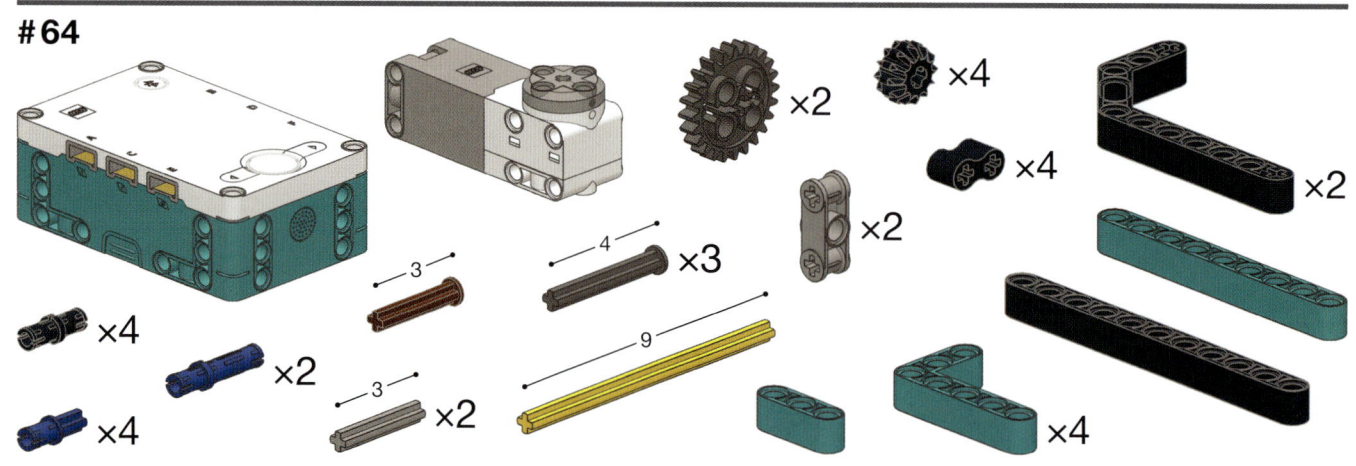

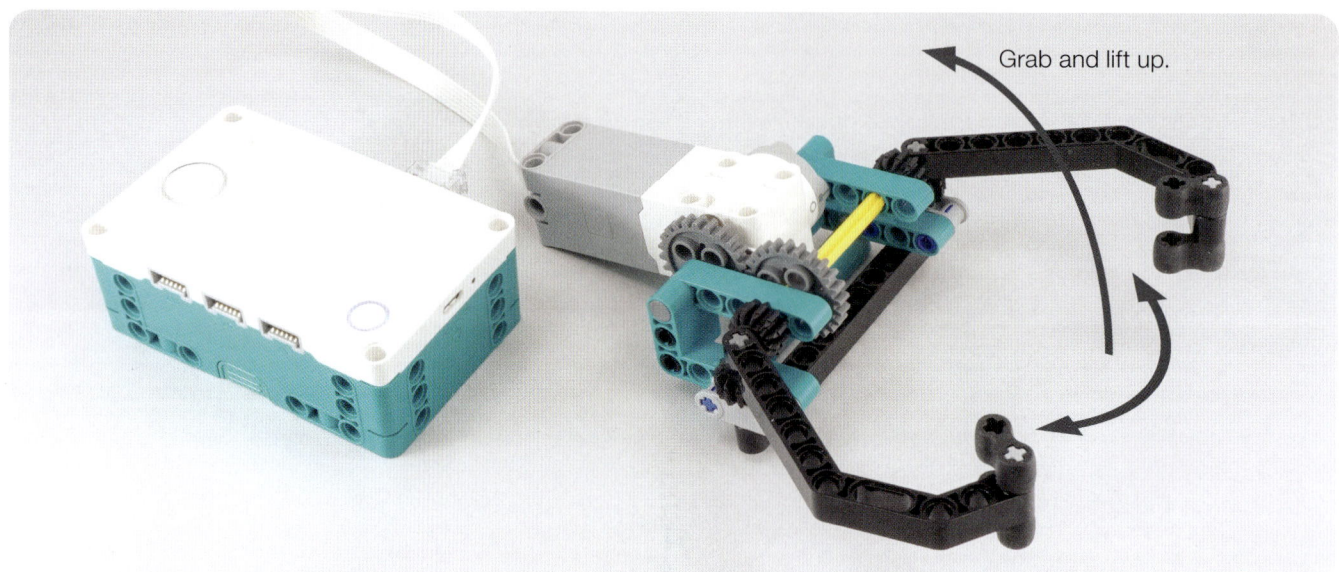

Grab and lift up.

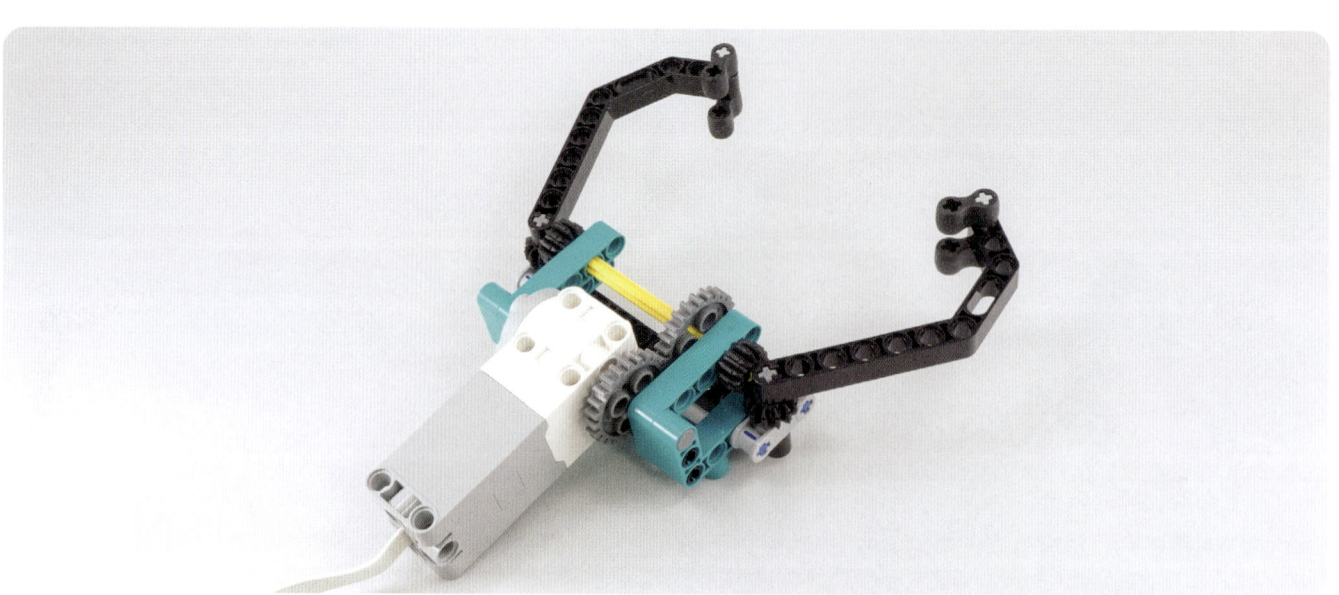

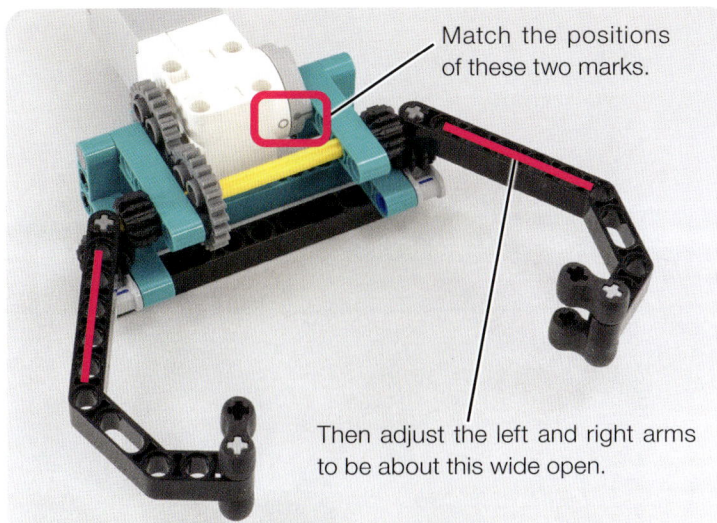

Match the positions of these two marks.

Then adjust the left and right arms to be about this wide open.

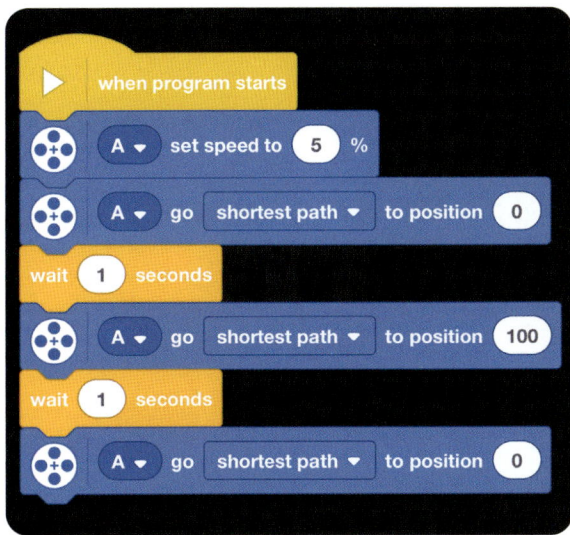

```
when program starts
A ▾ set speed to 5 %
A ▾ go shortest path ▾ to position 0
wait 1 seconds
A ▾ go shortest path ▾ to position 100
wait 1 seconds
A ▾ go shortest path ▾ to position 0
```

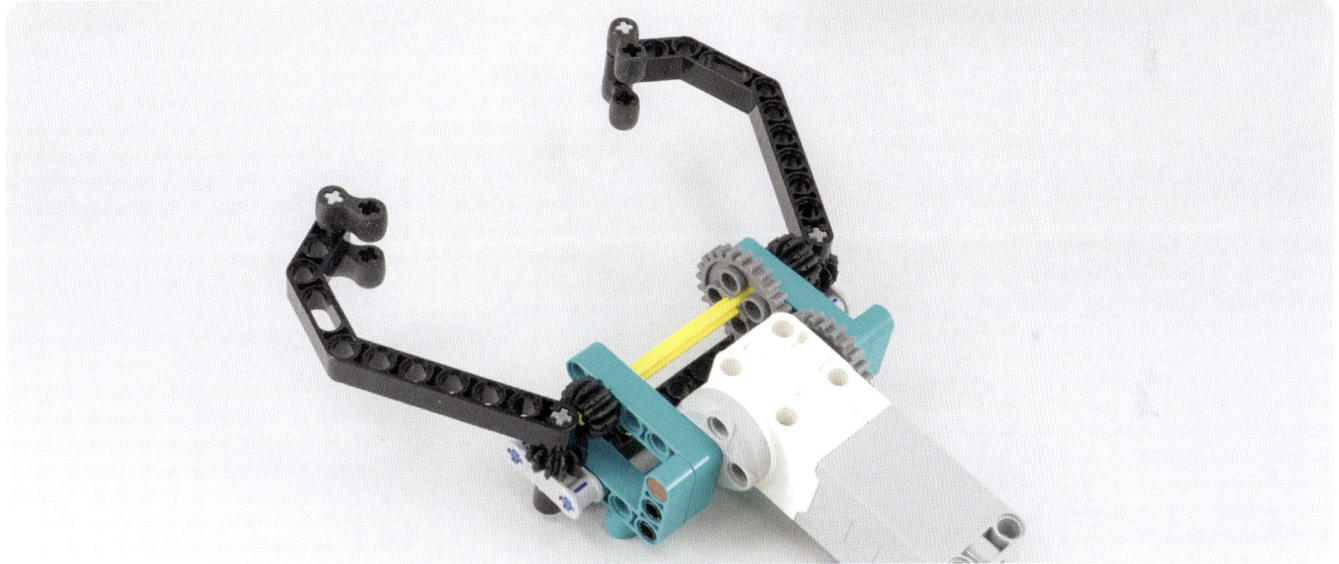

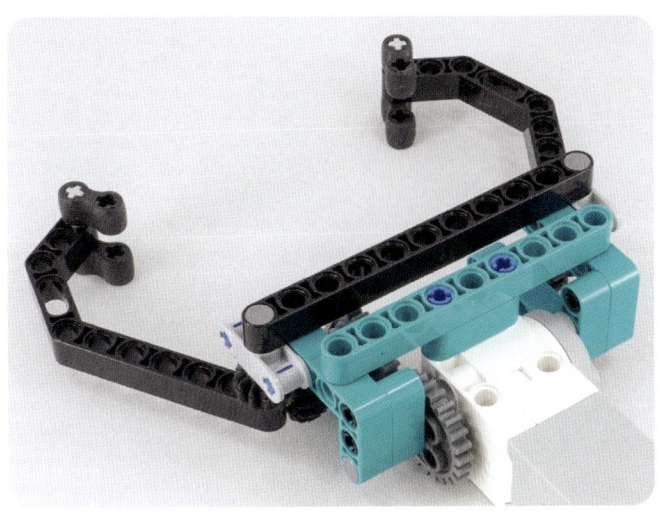

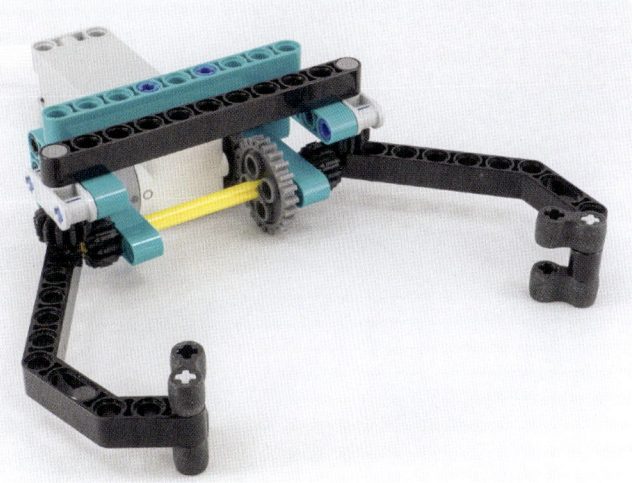

Lifting devices

#65

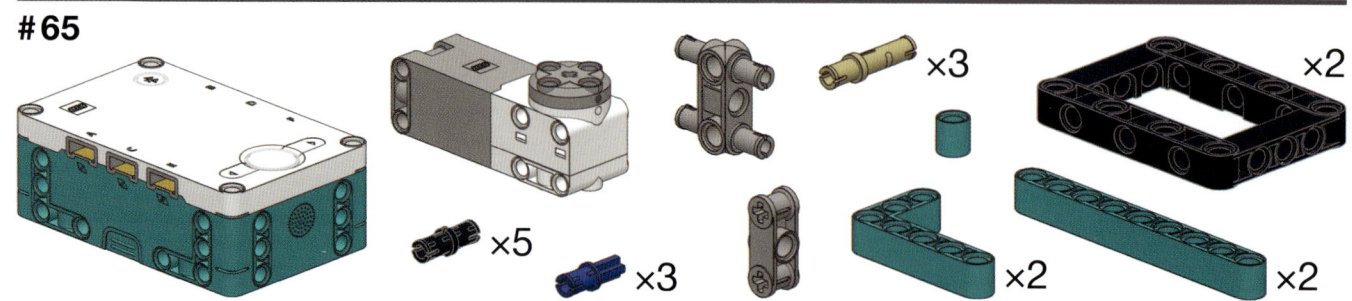

×5 ×3 ×3 ×2 ×2 ×2

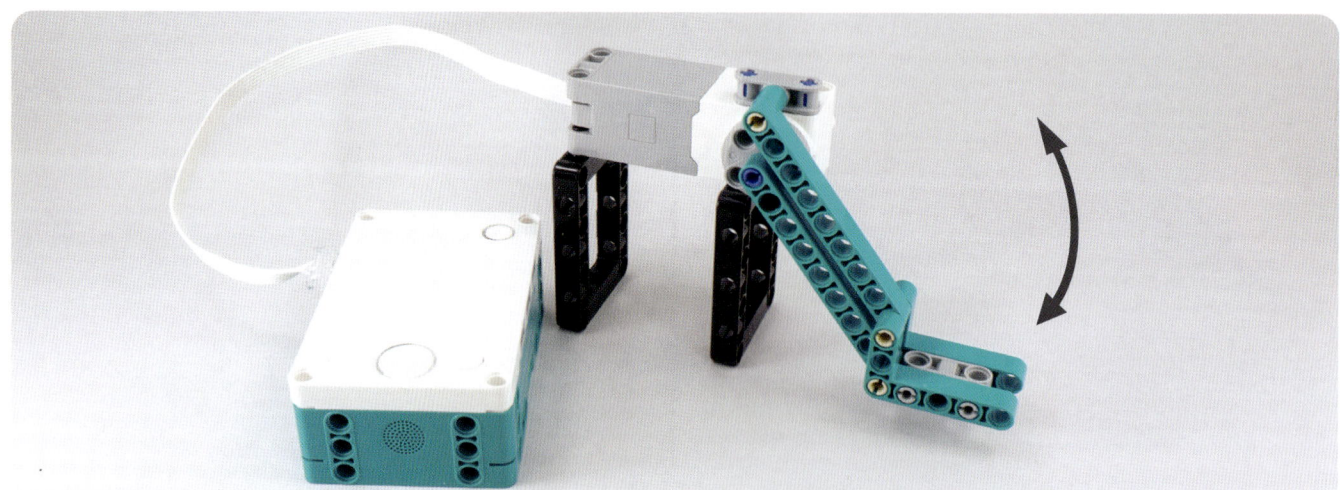

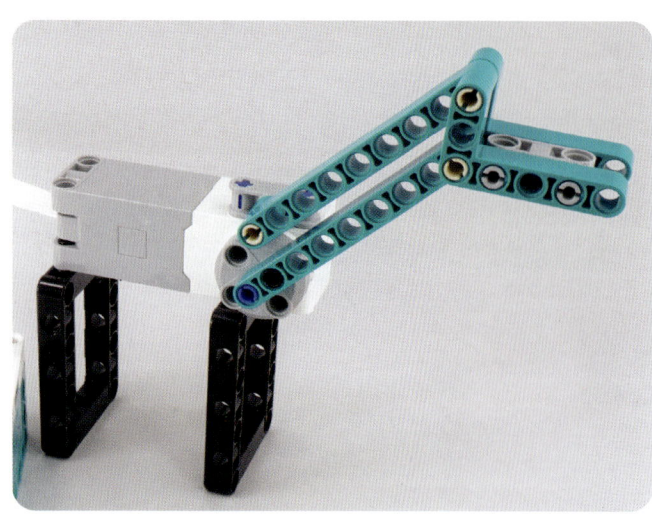

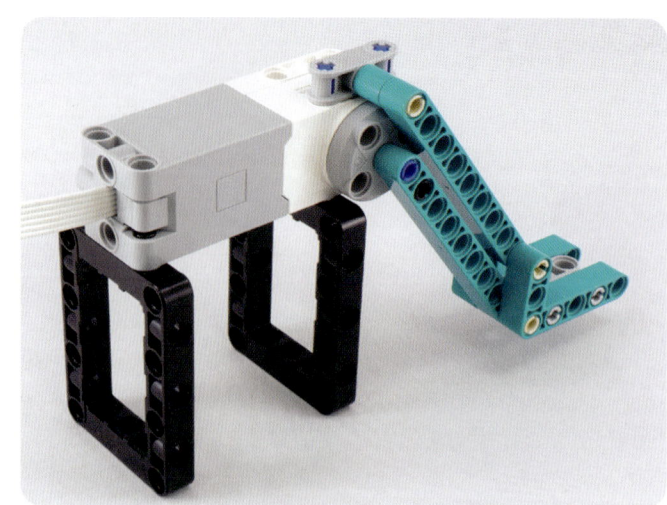

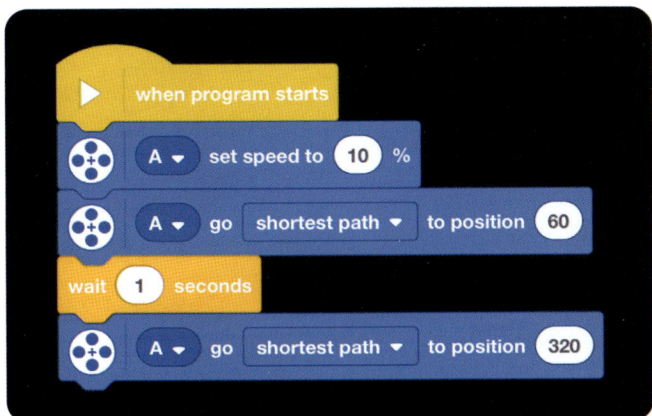

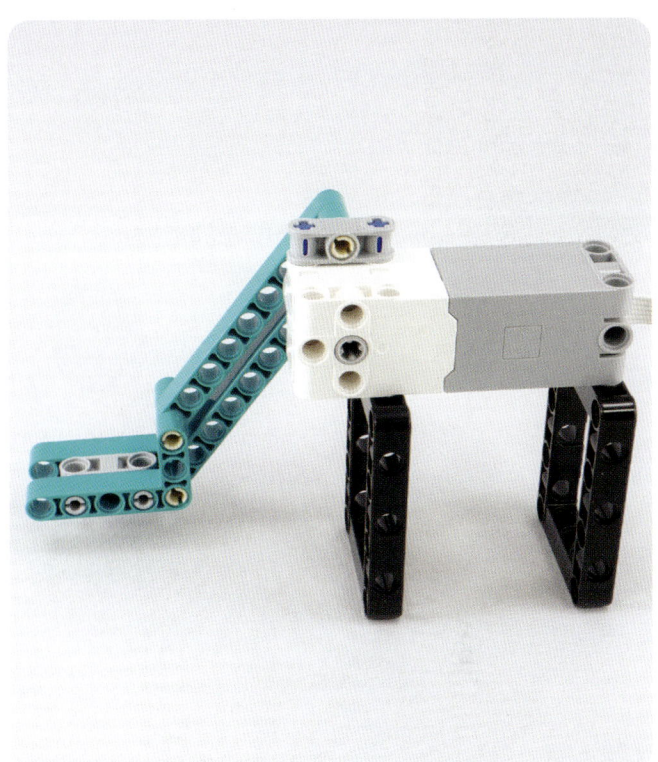

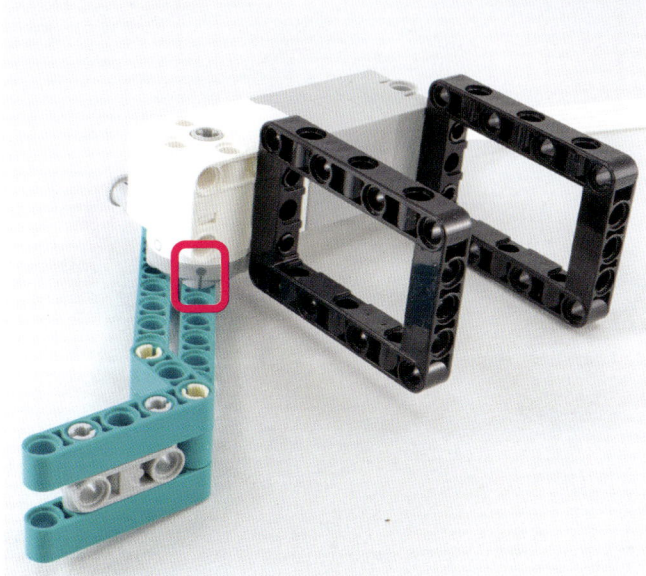

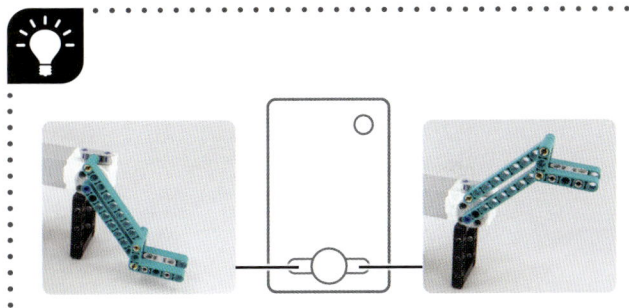

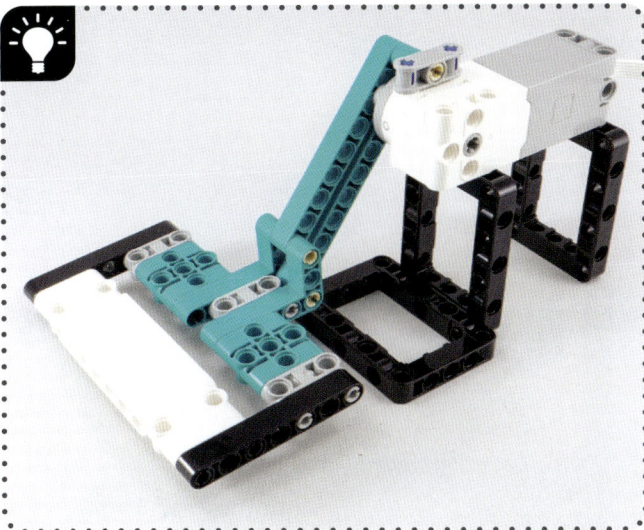

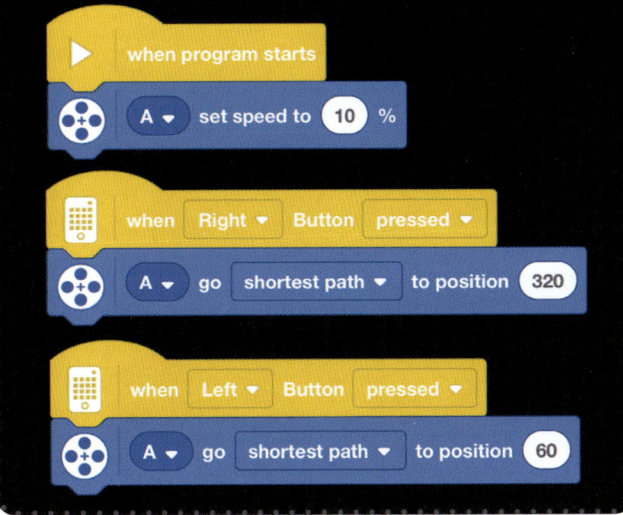

#66

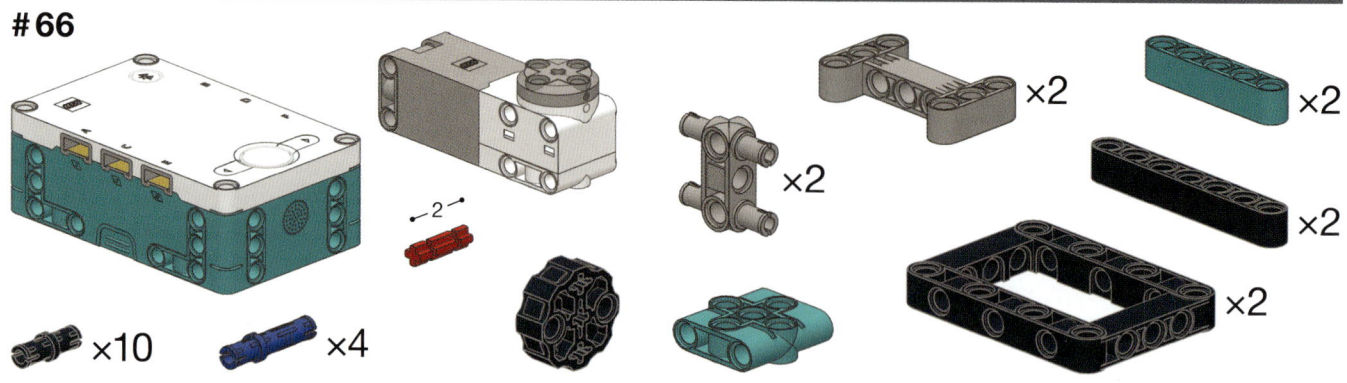

×10   ×4

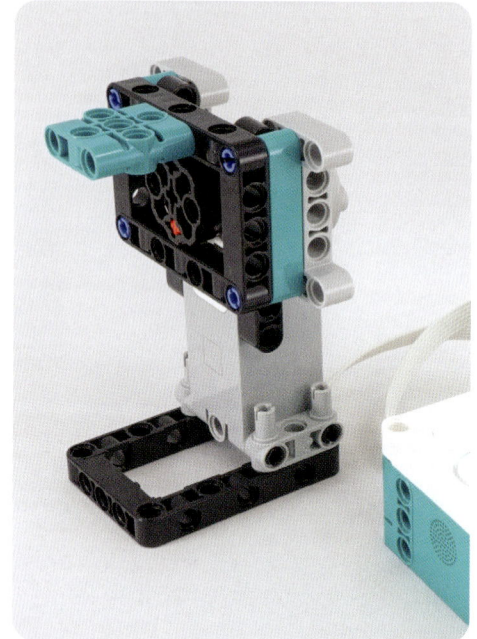

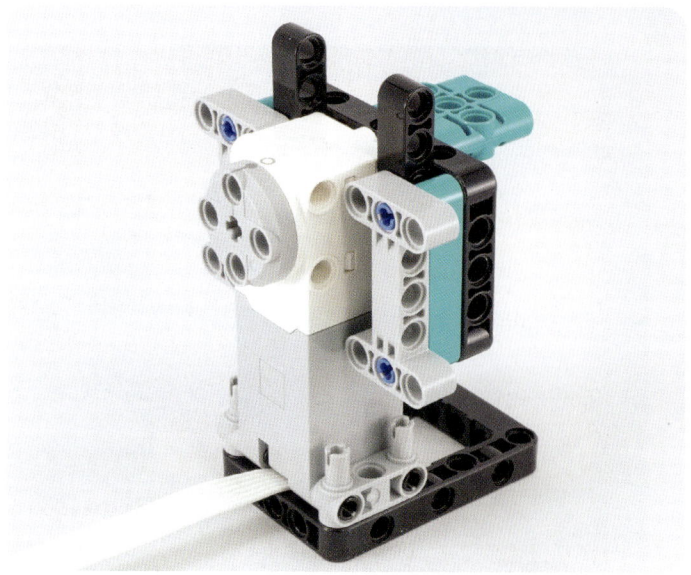

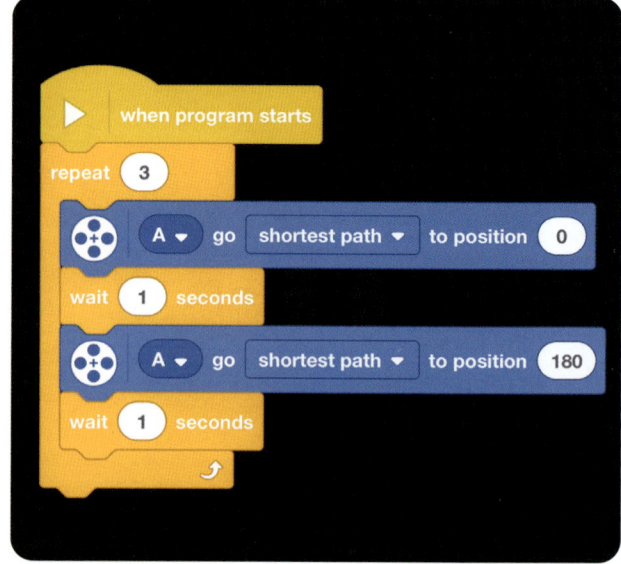

when program starts

repeat 3
  A ▼ go shortest path ▼ to position 0
  wait 1 seconds
  A ▼ go shortest path ▼ to position 180
  wait 1 seconds

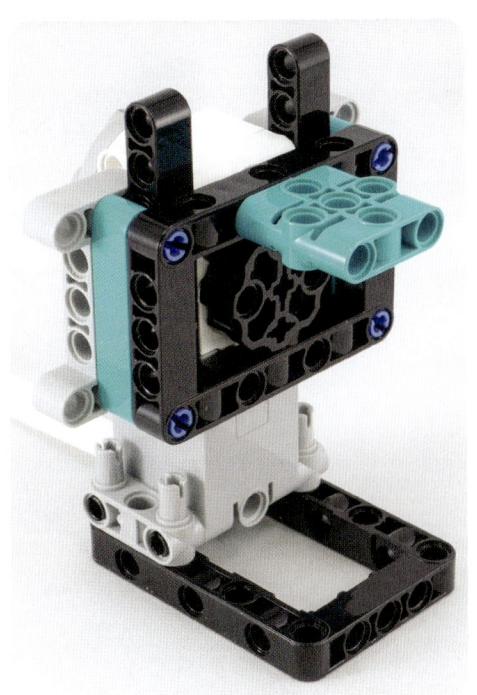

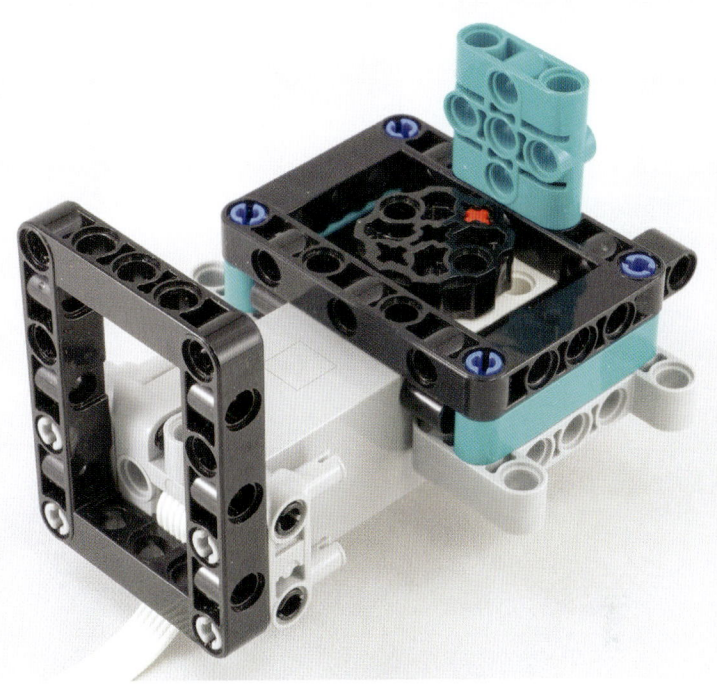

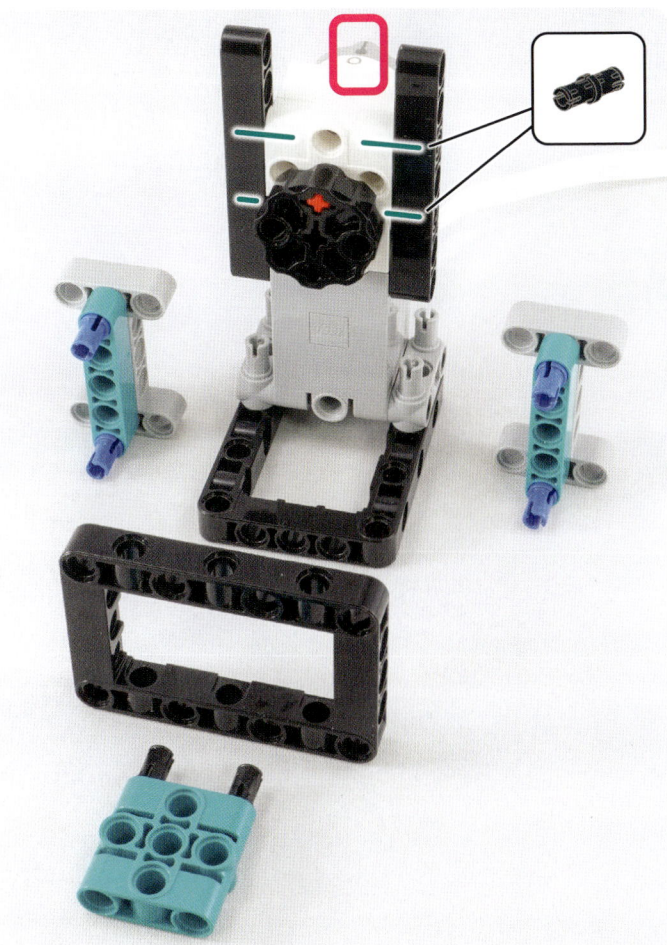

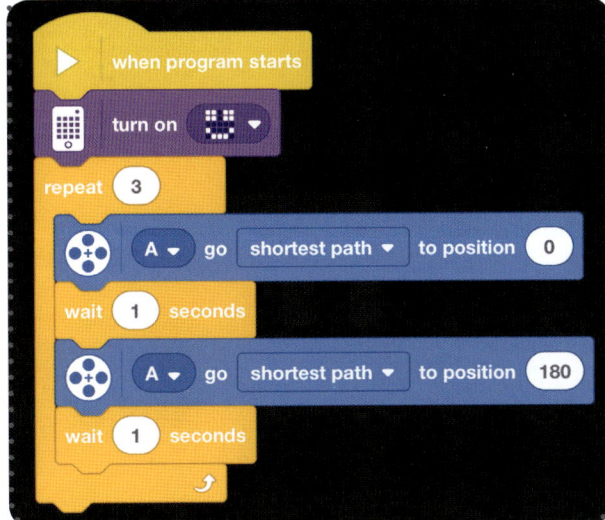

#67

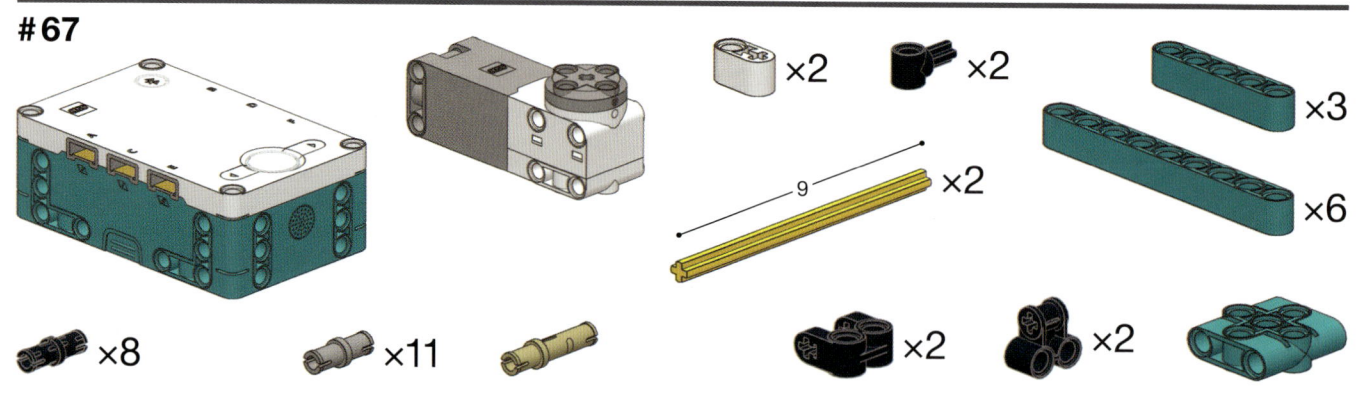

×2 ×2 ×3 ×6

9 ×2

×8 ×11 ×2 ×2

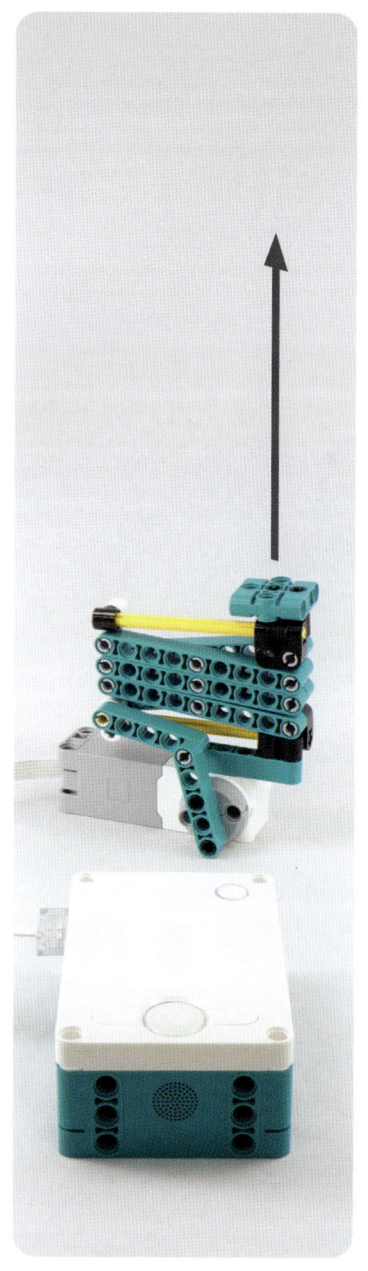

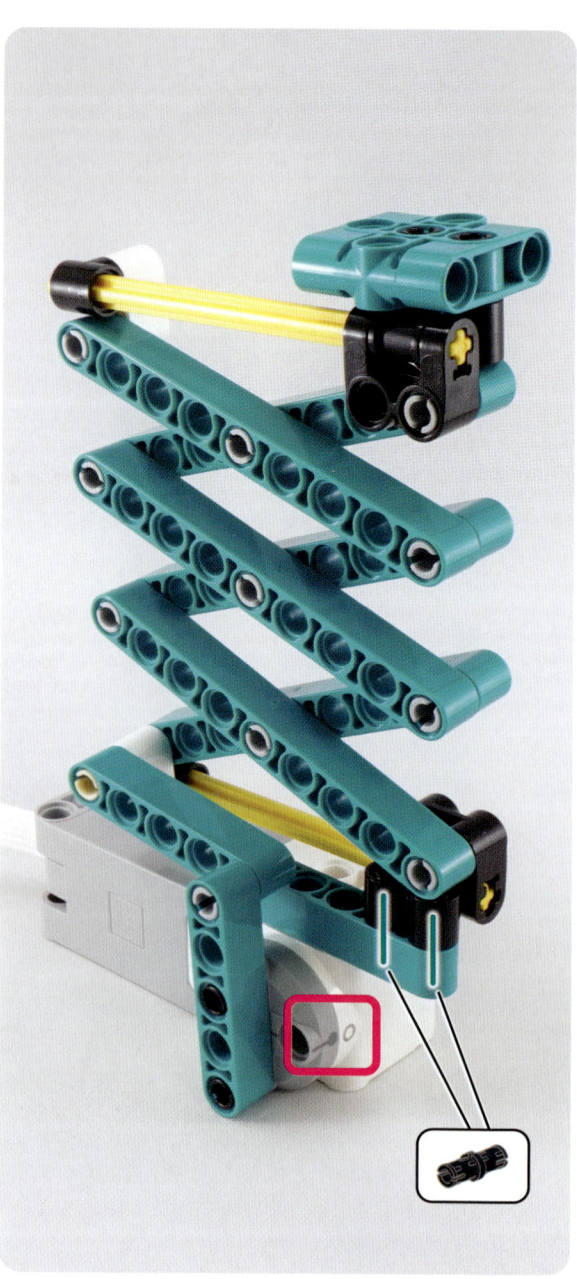

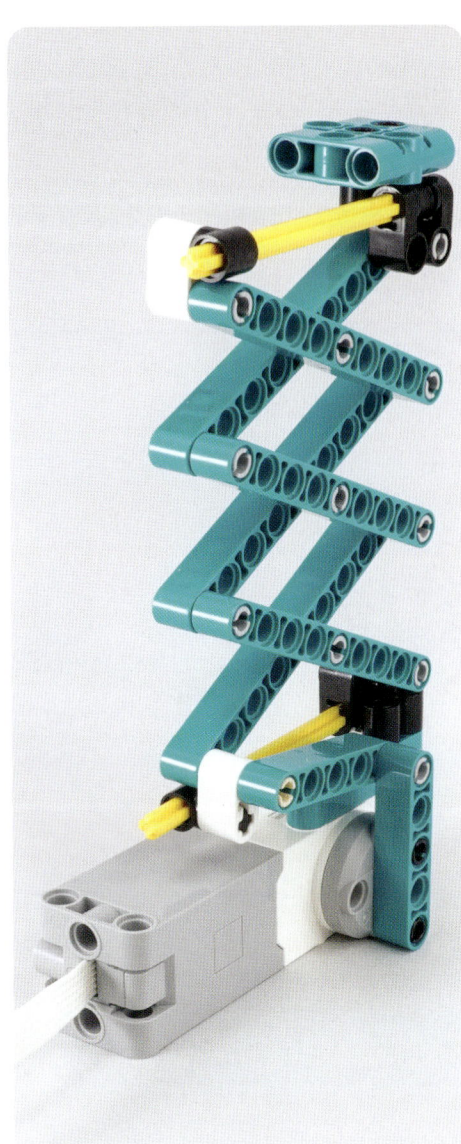

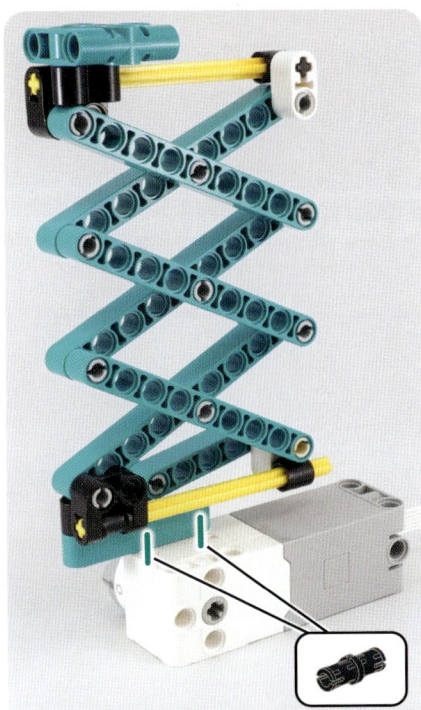

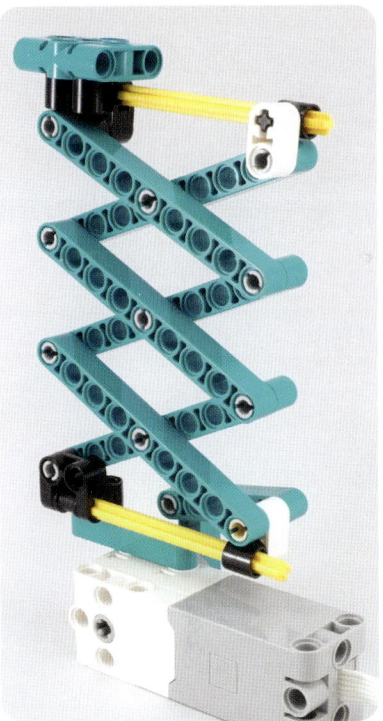

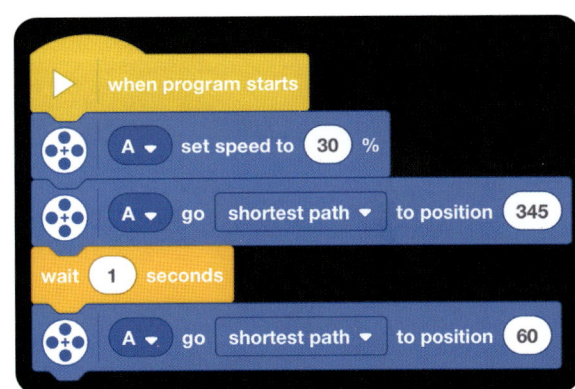

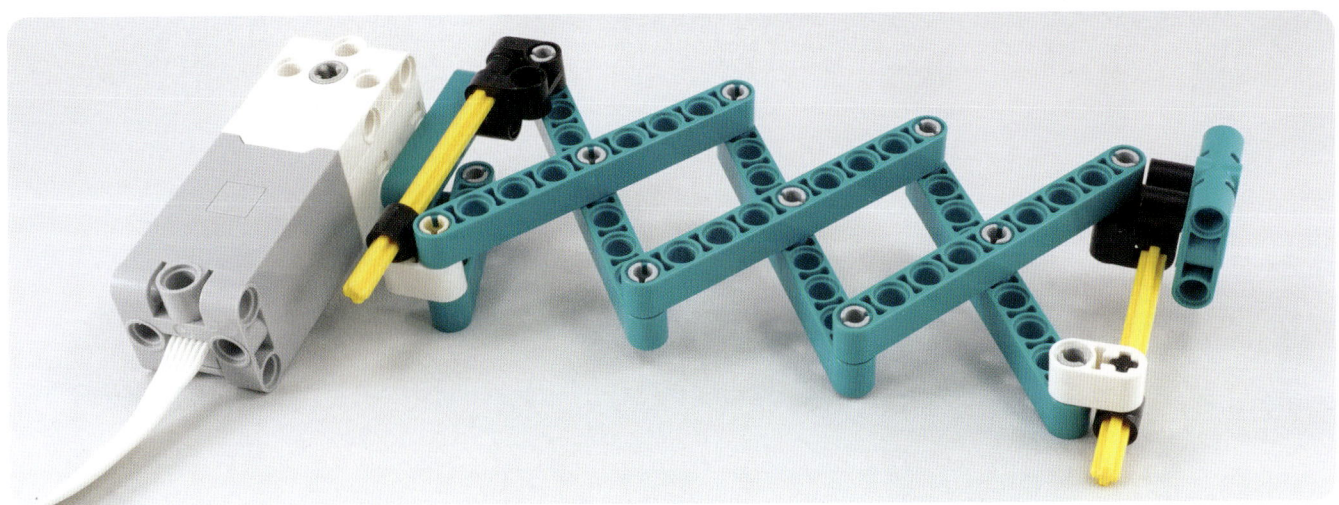

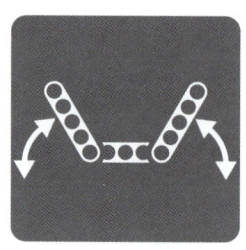

Flapping wings

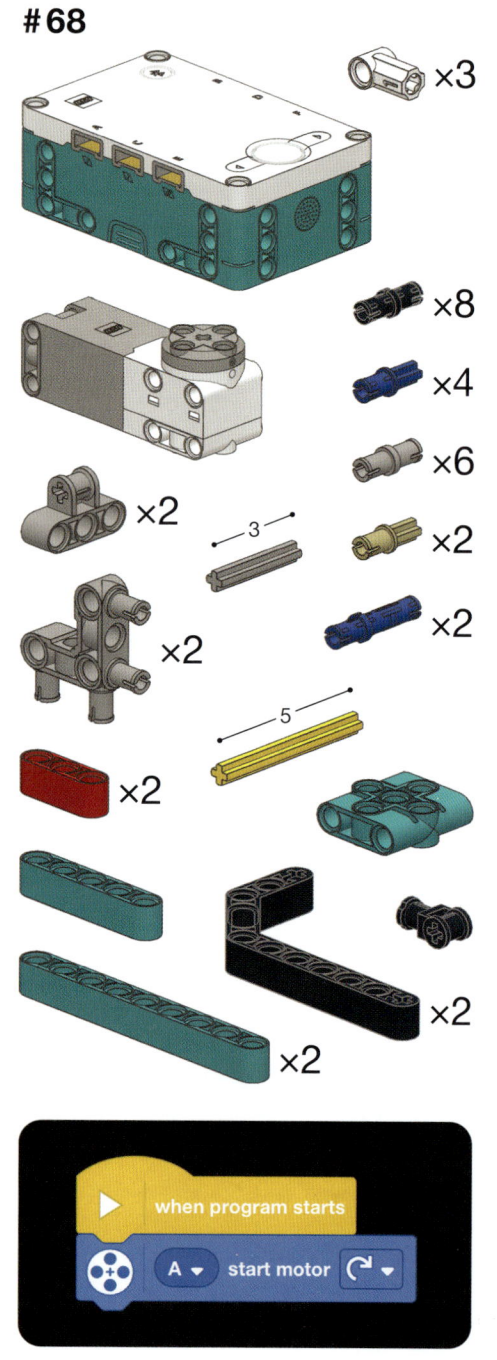

#68

×3
×8
×4
×6
×2
3
×2
×2
×2
5
×2
×2
×2

when program starts
A
start motor

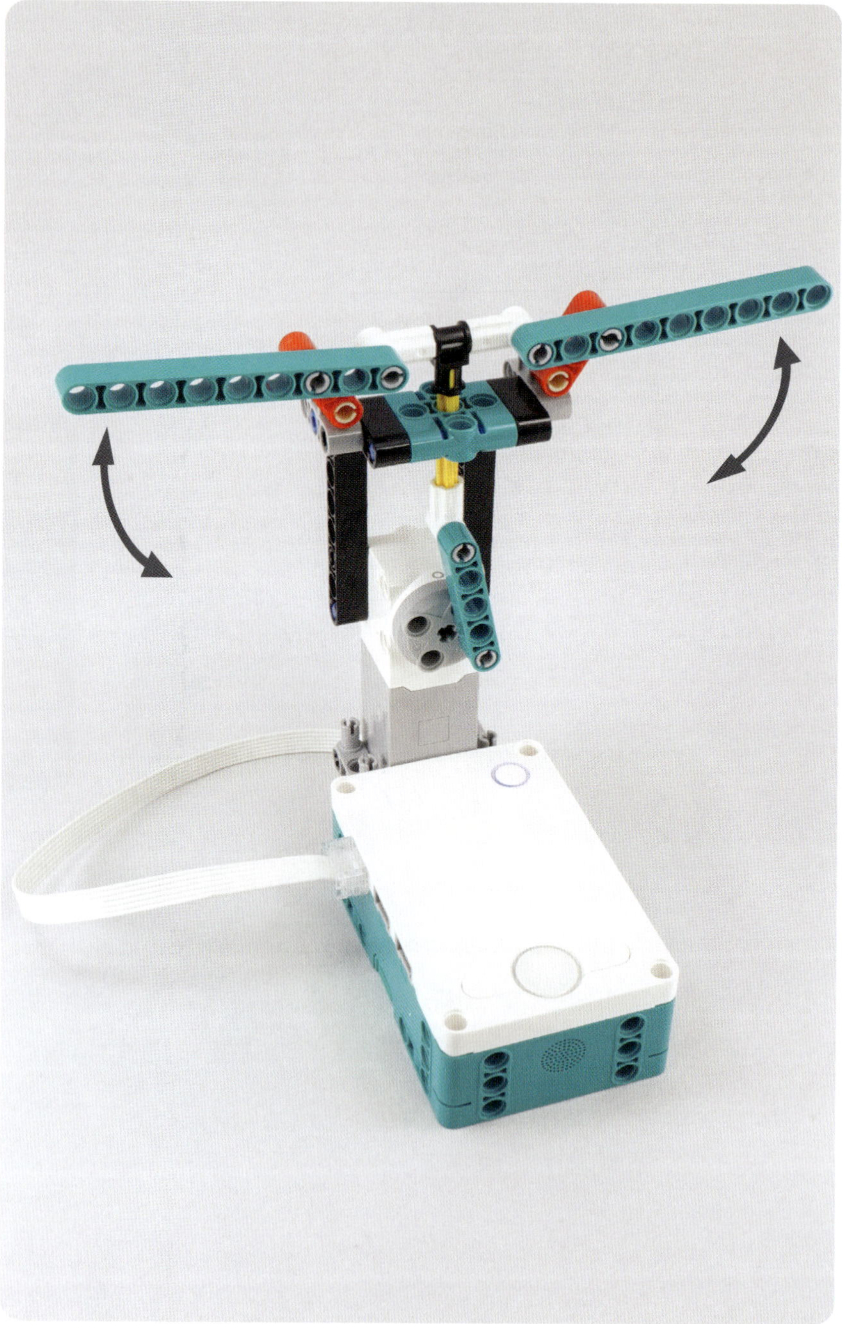

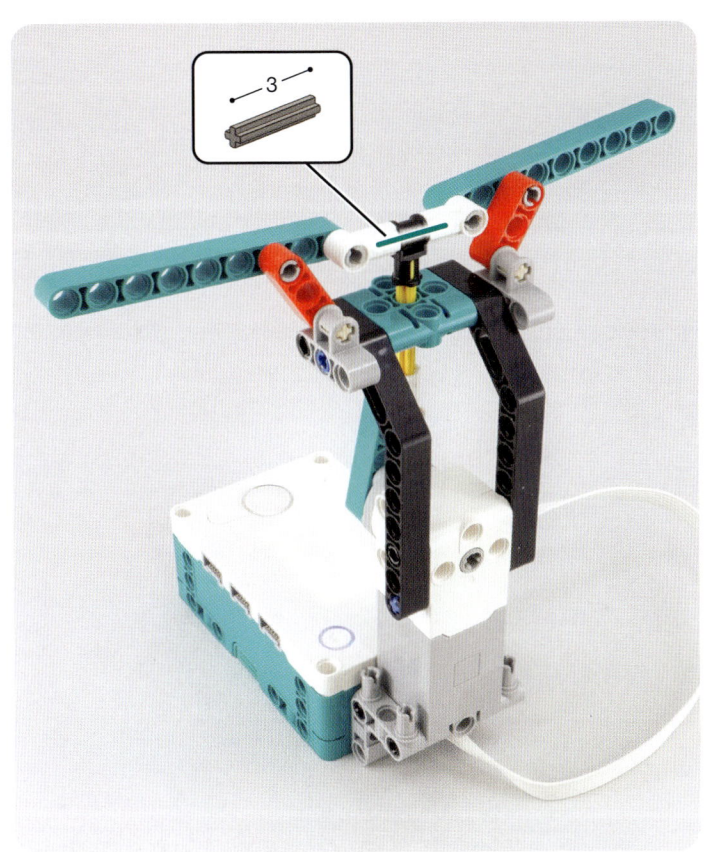

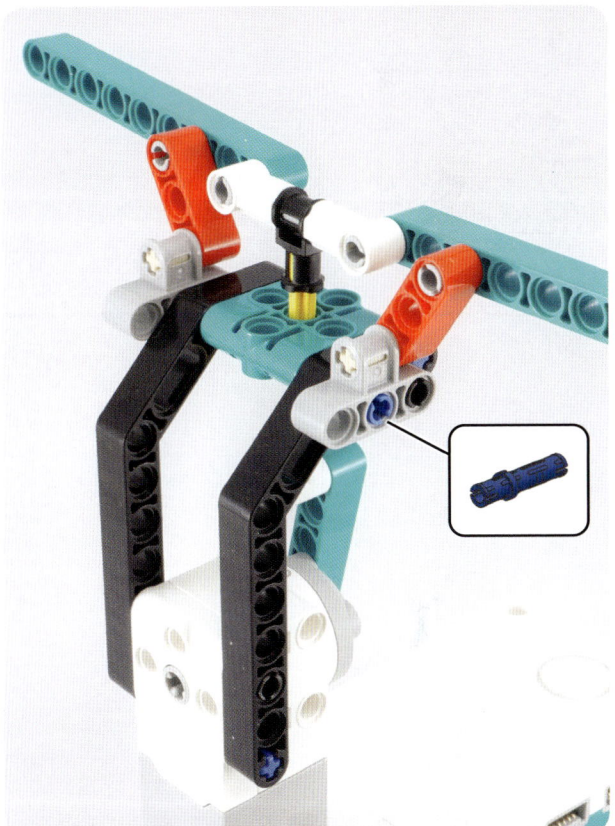

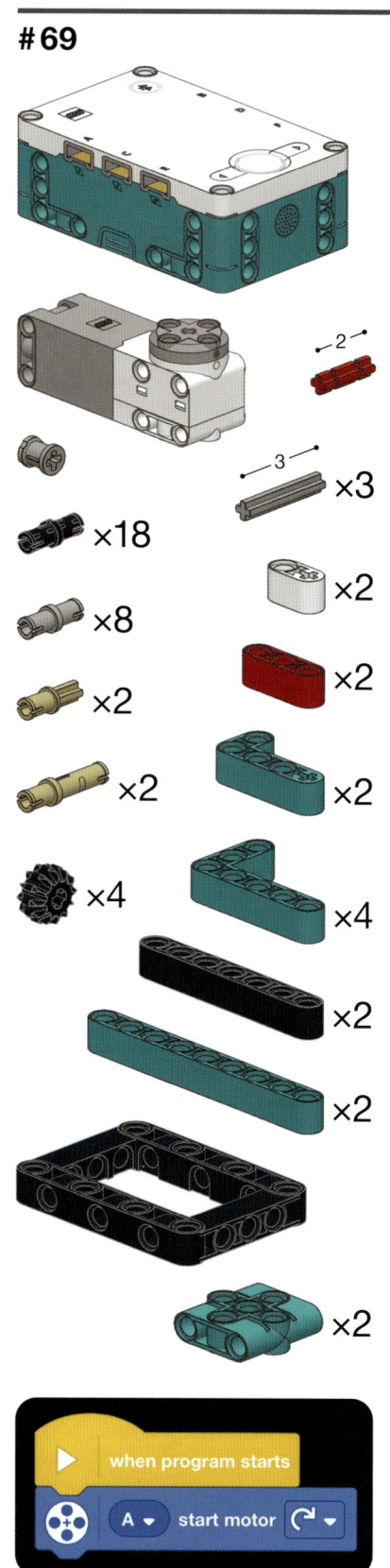

69

×2

×3

×18

×2

×8

×2

×2

×2

×2

×4

×4

×2

×2

×2

when program starts

A ▾ start motor ↻ ▾

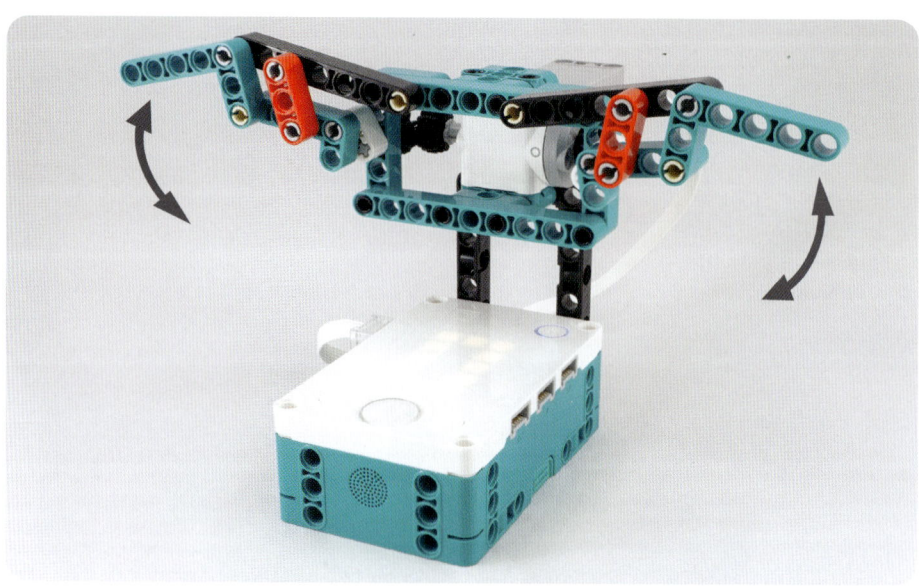

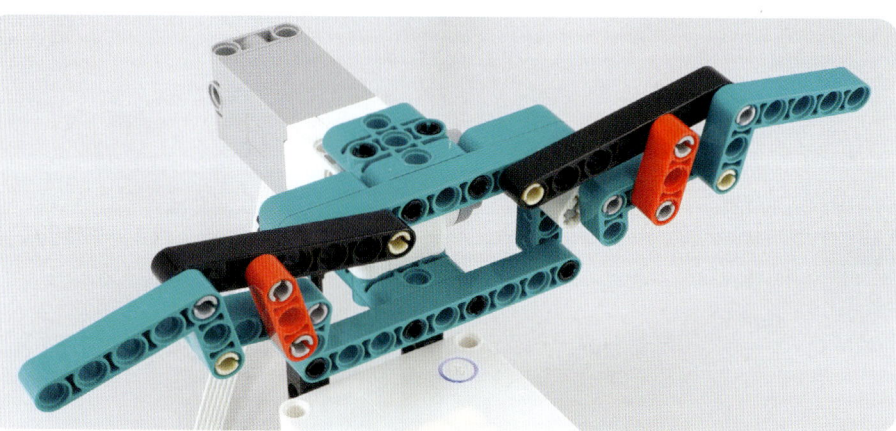

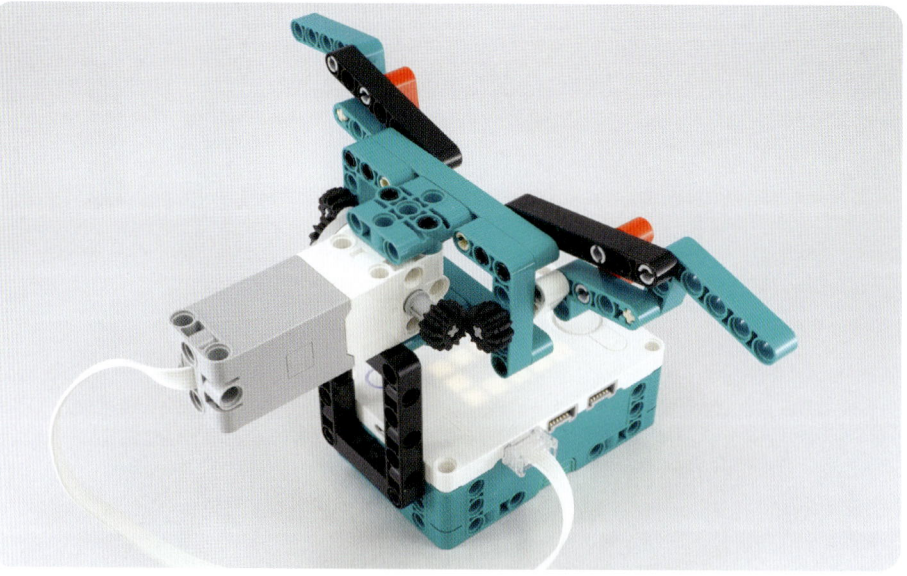

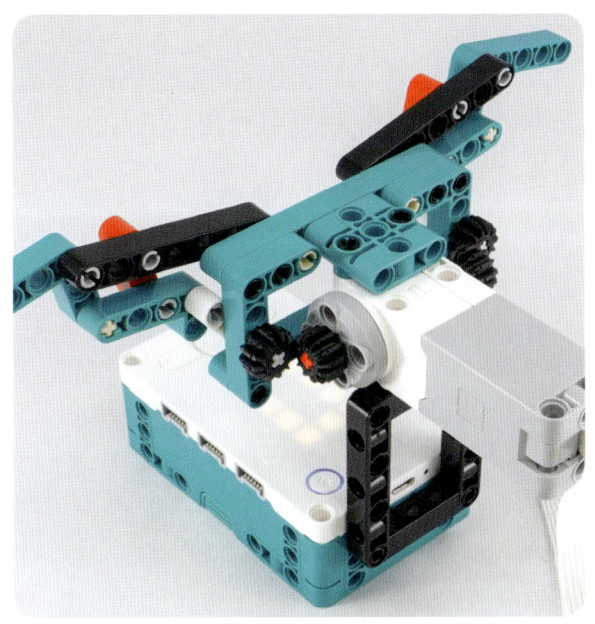

Make sure that these parts are symmetrical.

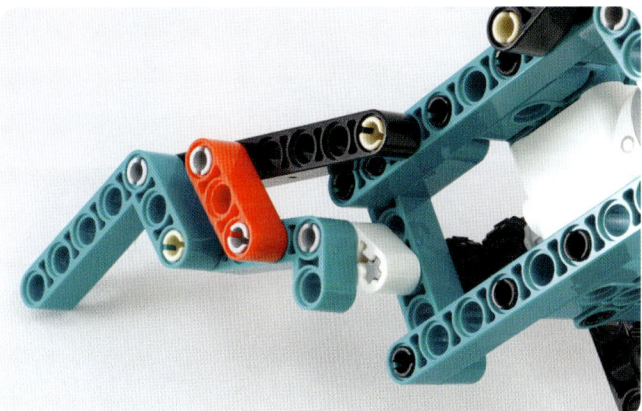

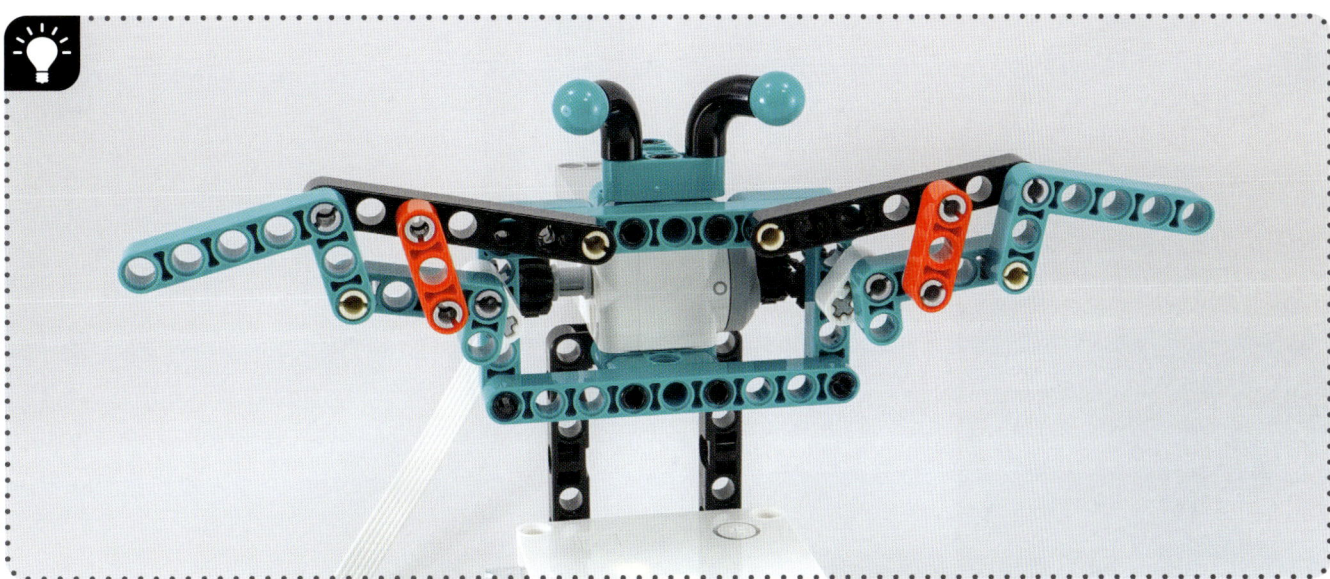

#70

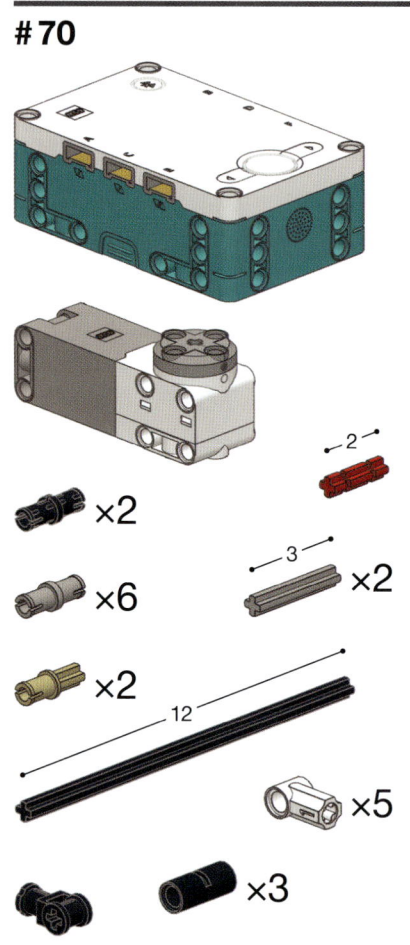

×2

×2

×6

×3

12

×5

×3

×2

×3

when program starts

A ▾ set speed to 50 %

A ▾ start motor ↻ ▾

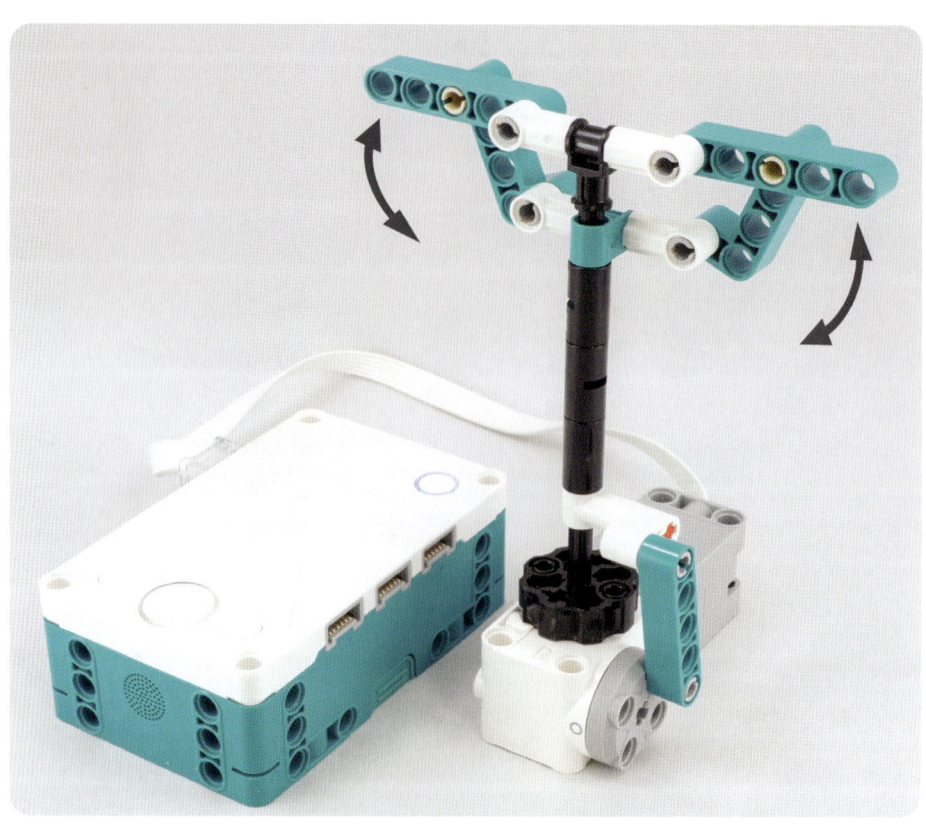

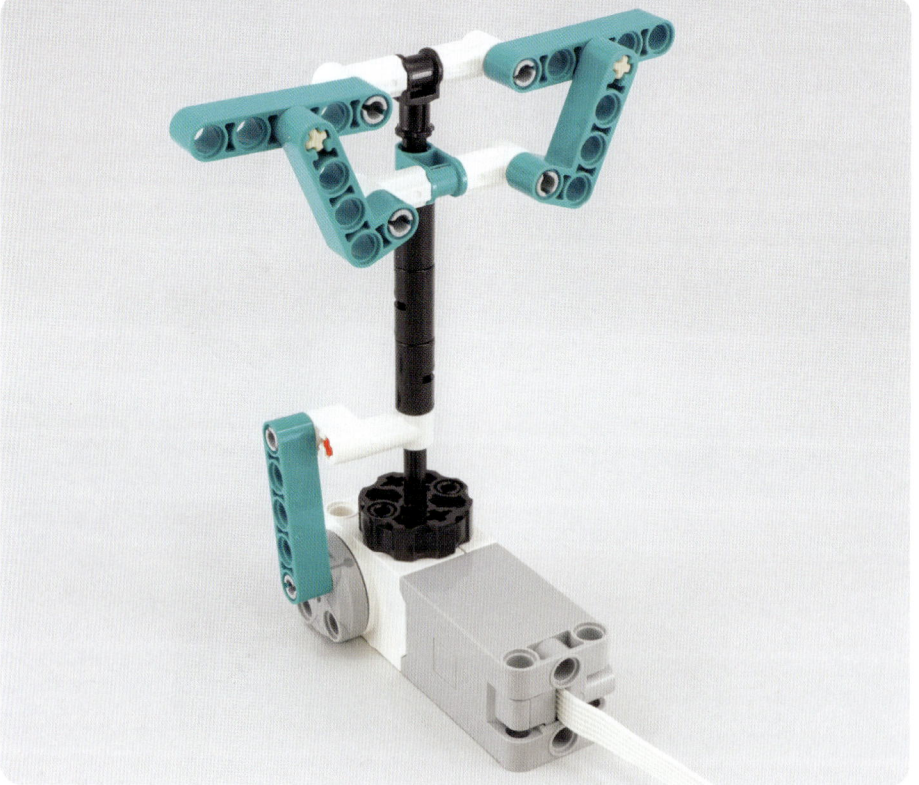

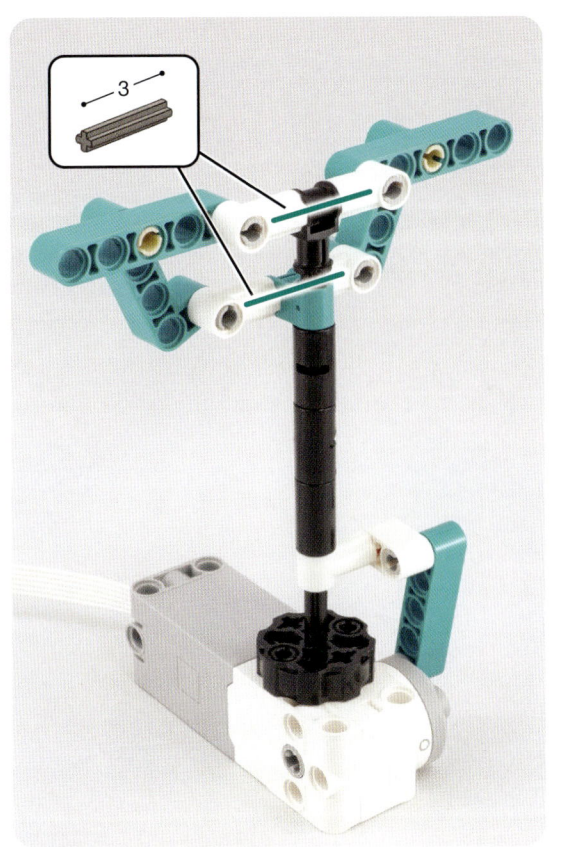

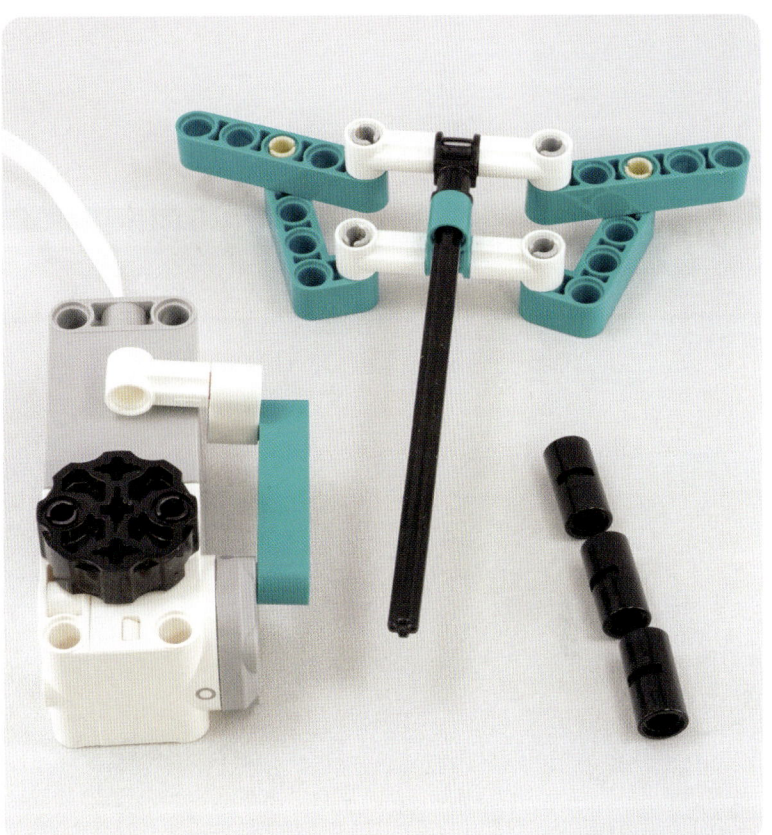

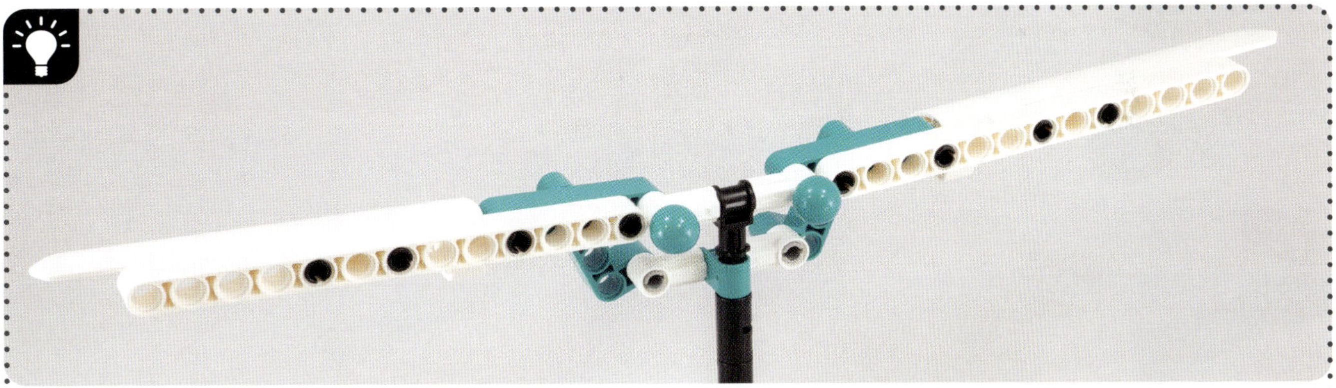

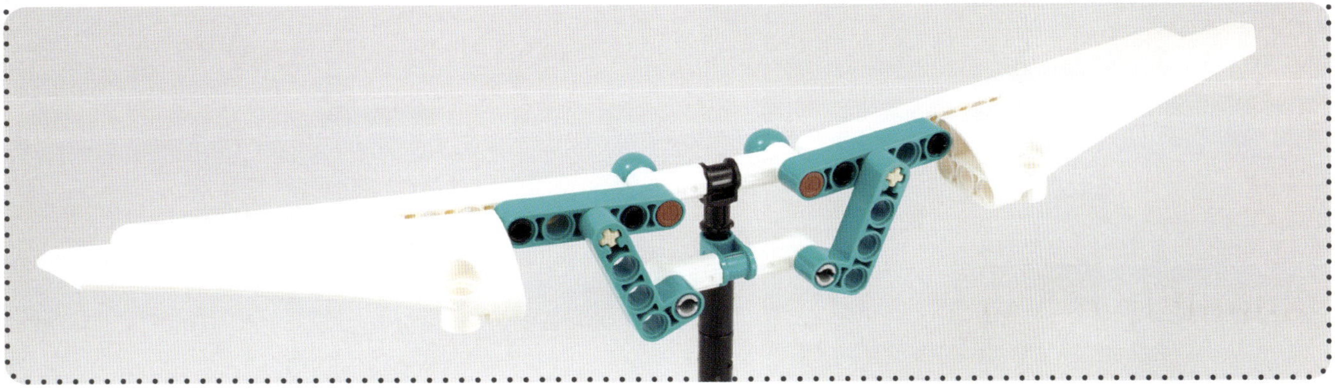

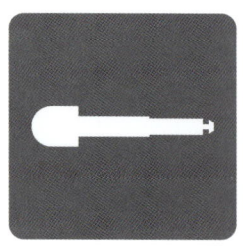

Using the projectile launcher

#71

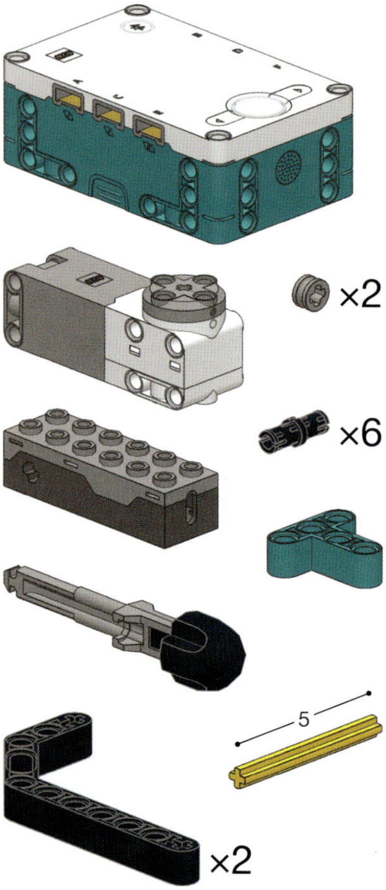

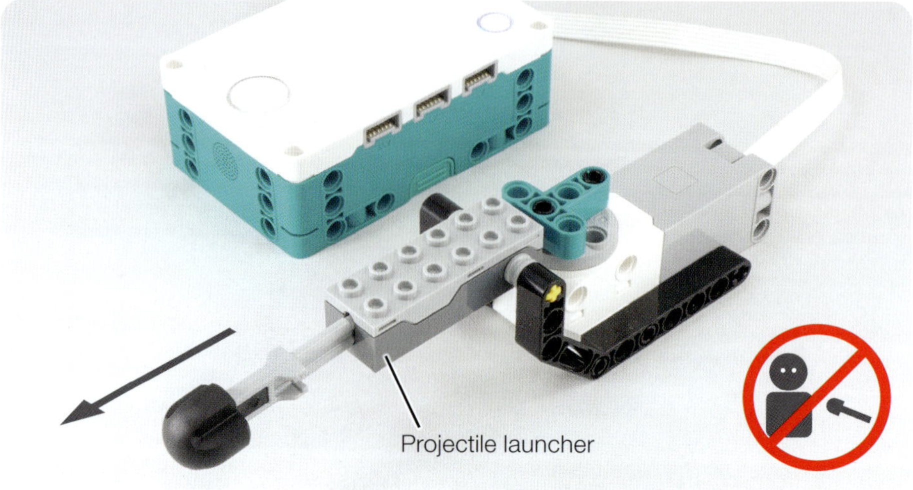

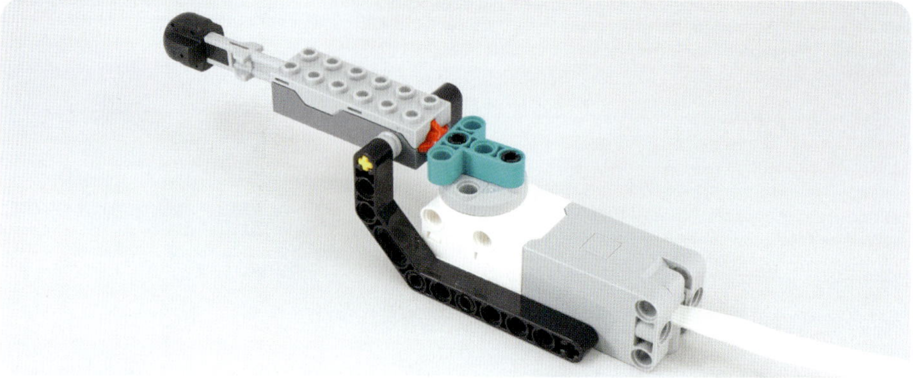

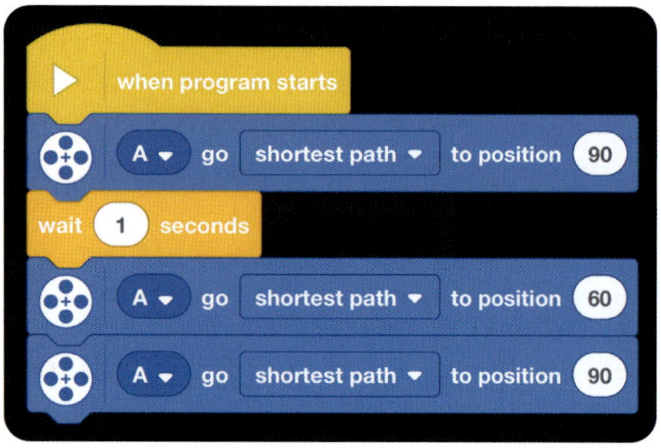

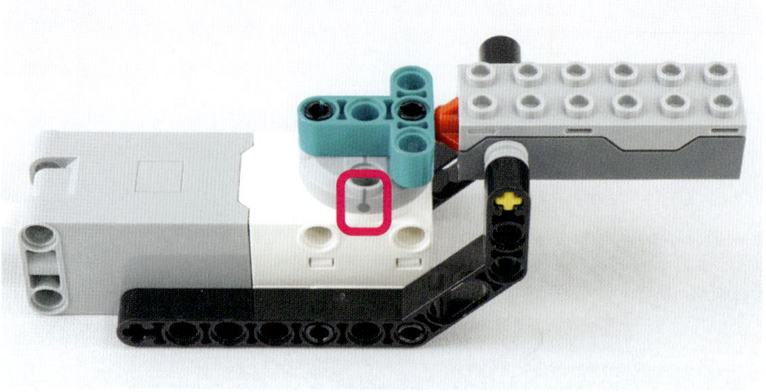

×2

×6

×2

5

Projectile launcher

when program starts

A ▾ go shortest path ▾ to position 90

wait 1 seconds

A ▾ go shortest path ▾ to position 60

A ▾ go shortest path ▾ to position 90

#72

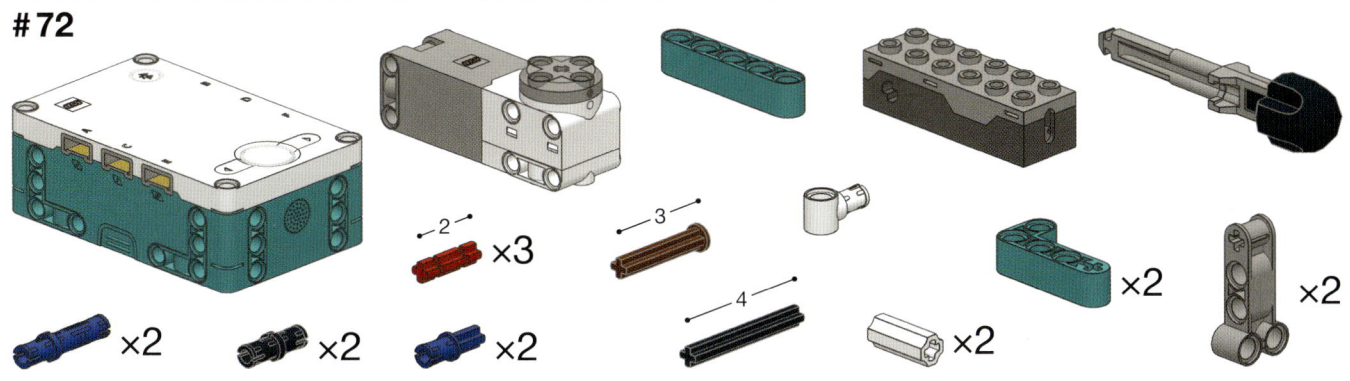

×3 ×2
×2 ×3 ×4 ×2 ×2

×2 ×2 ×2

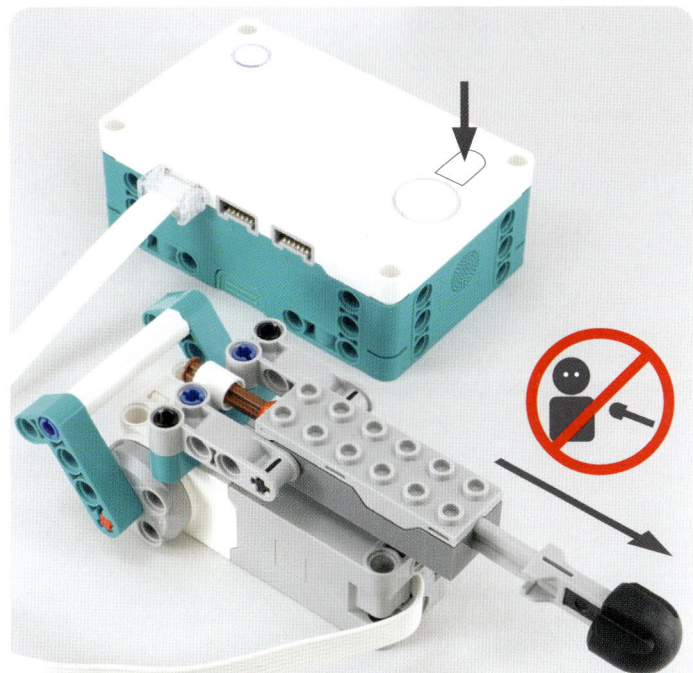

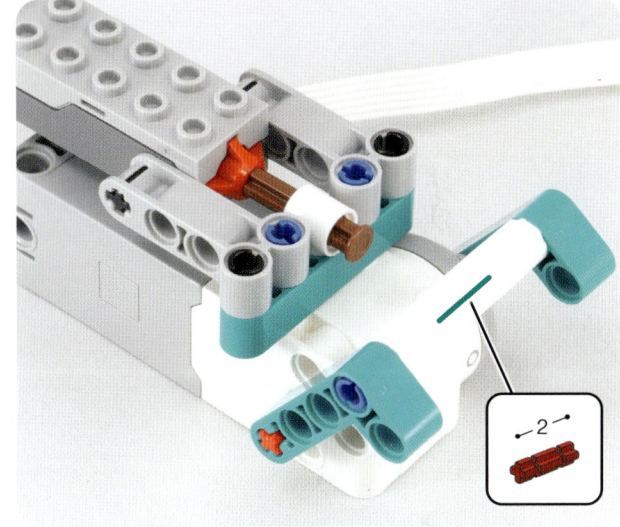

×2

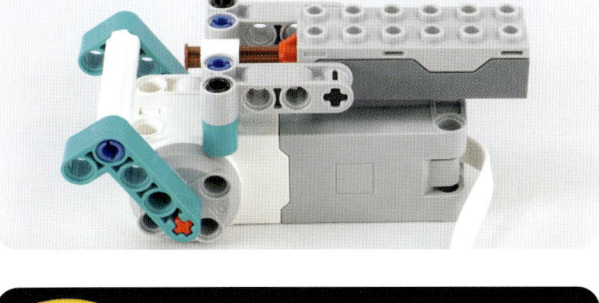

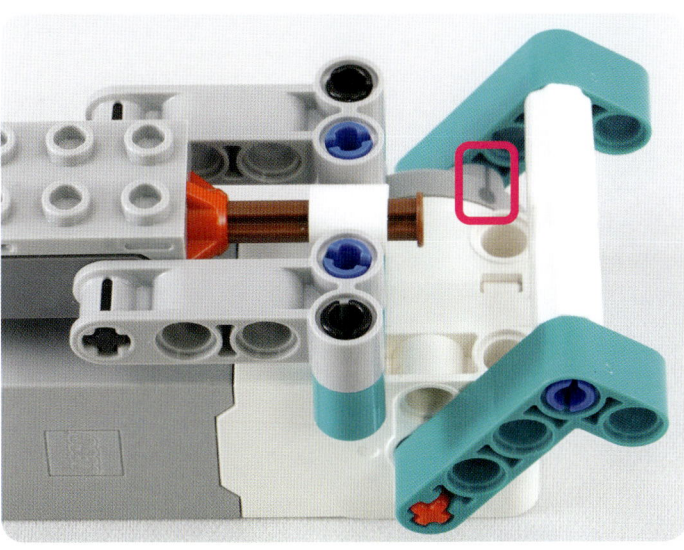

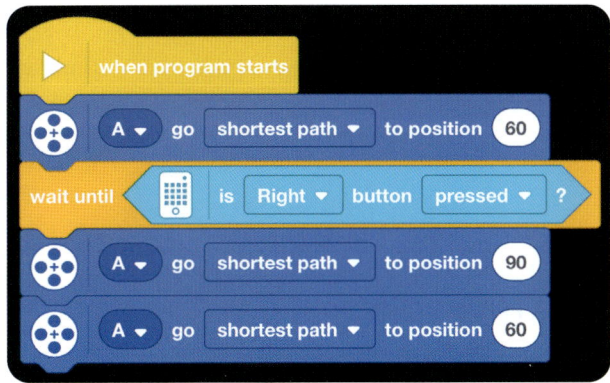

when program starts

A ▾ go shortest path ▾ to position 60

wait until 🔢 is Right ▾ button pressed ▾ ?

A ▾ go shortest path ▾ to position 90

A ▾ go shortest path ▾ to position 60

#73

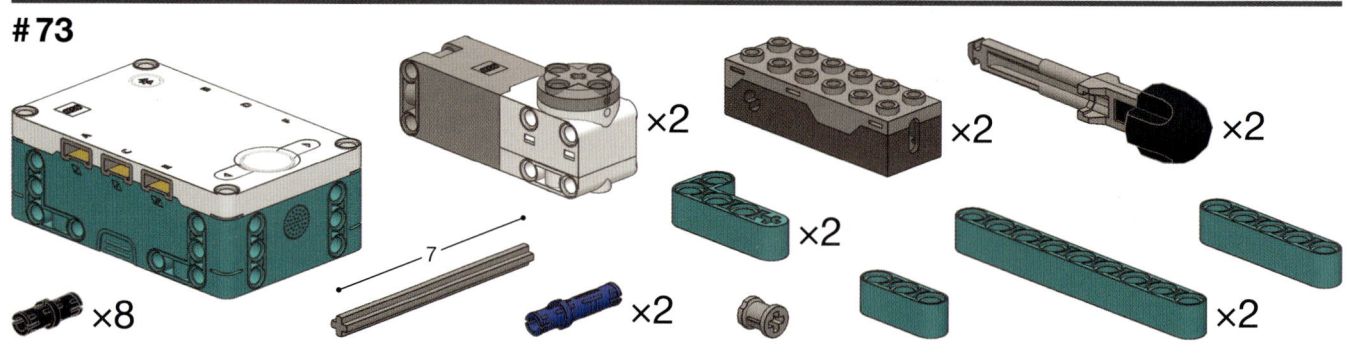

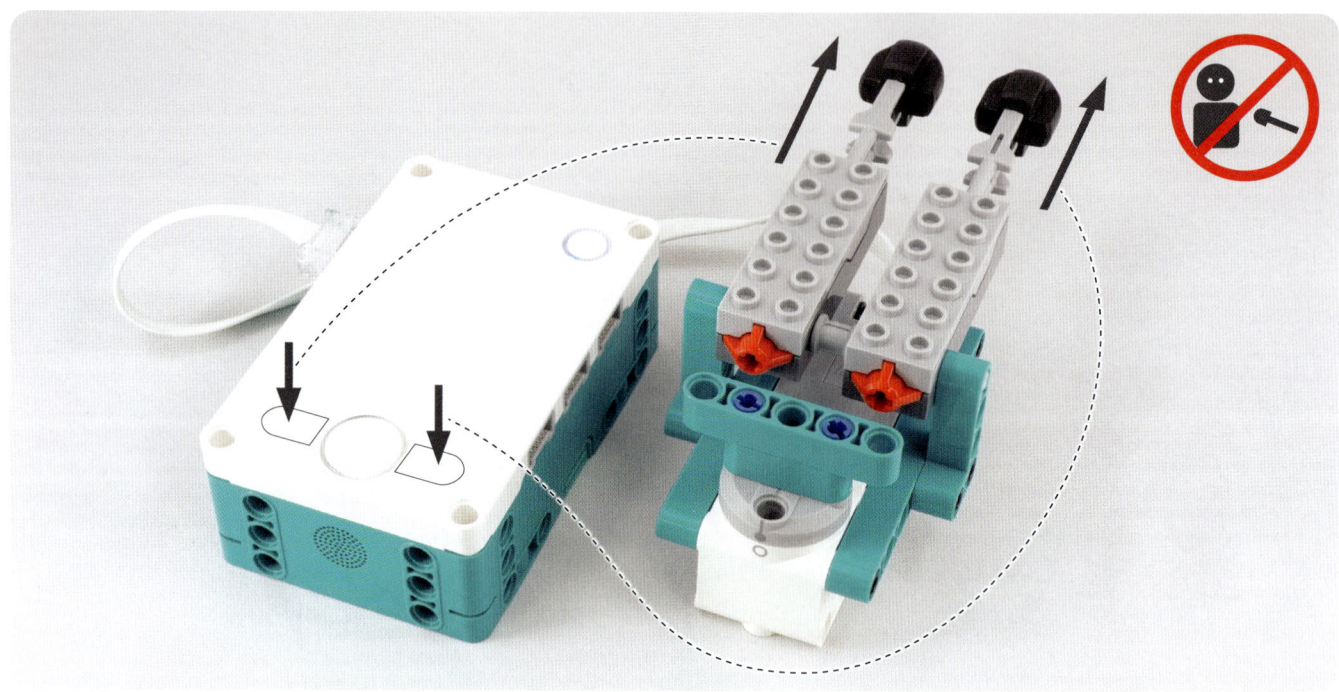

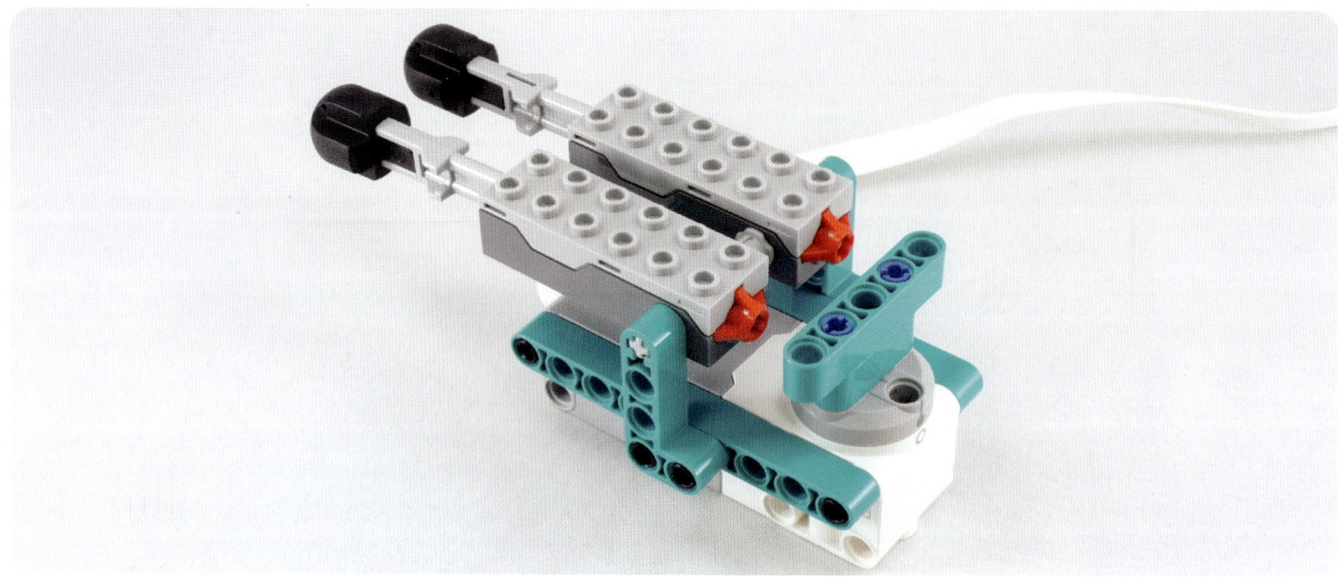

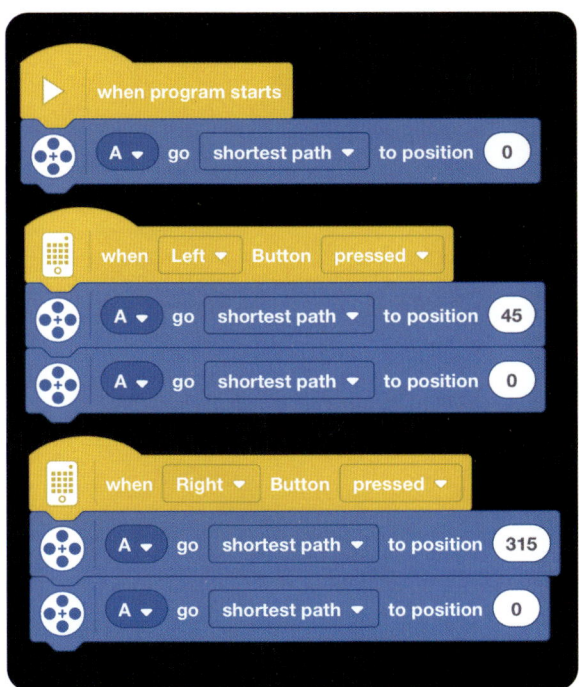

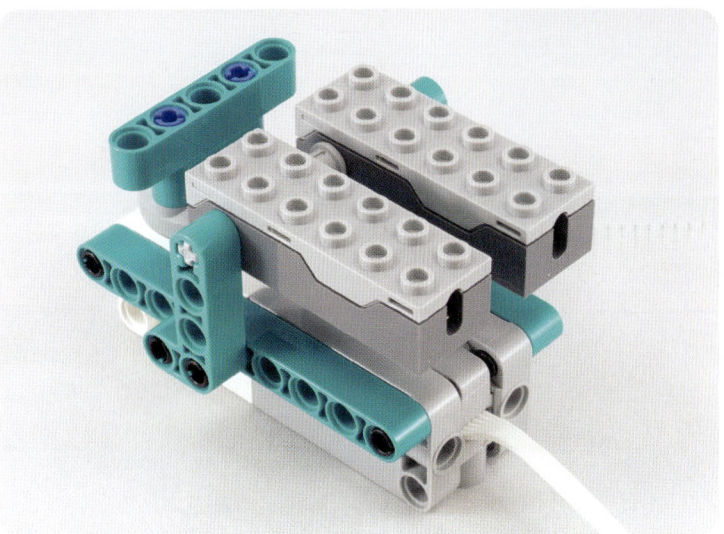

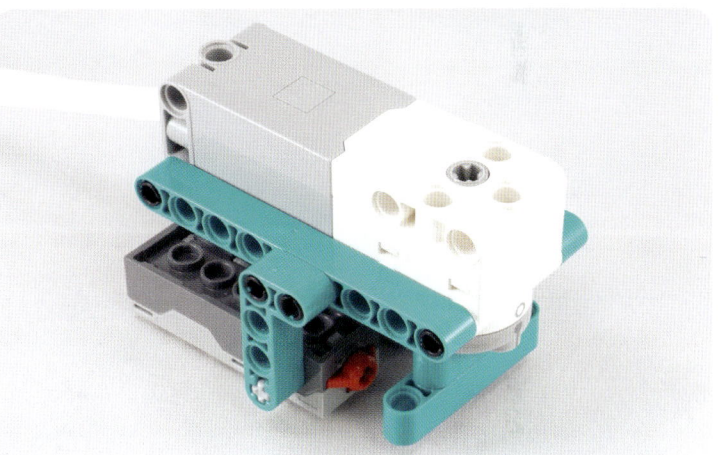

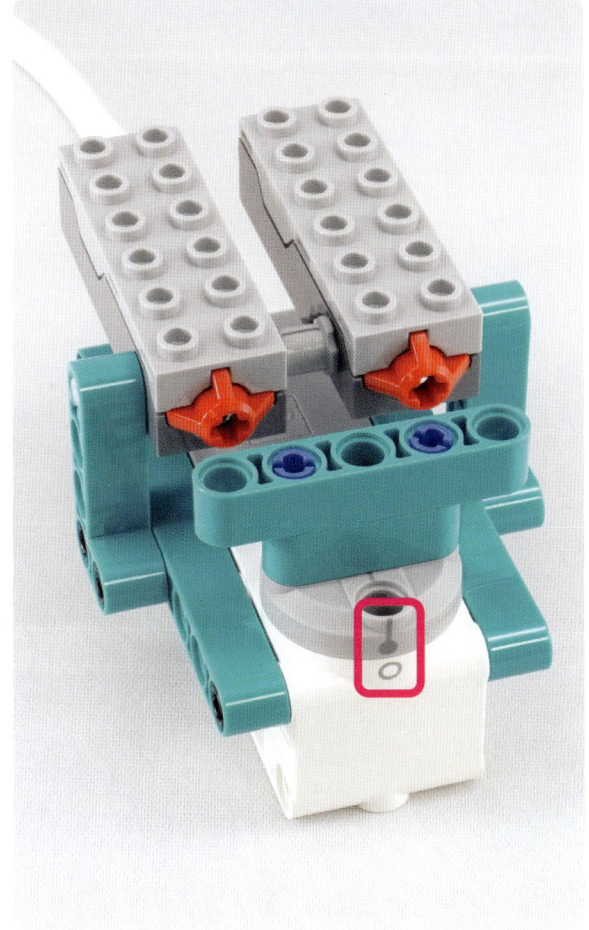

Shooting devices

#74

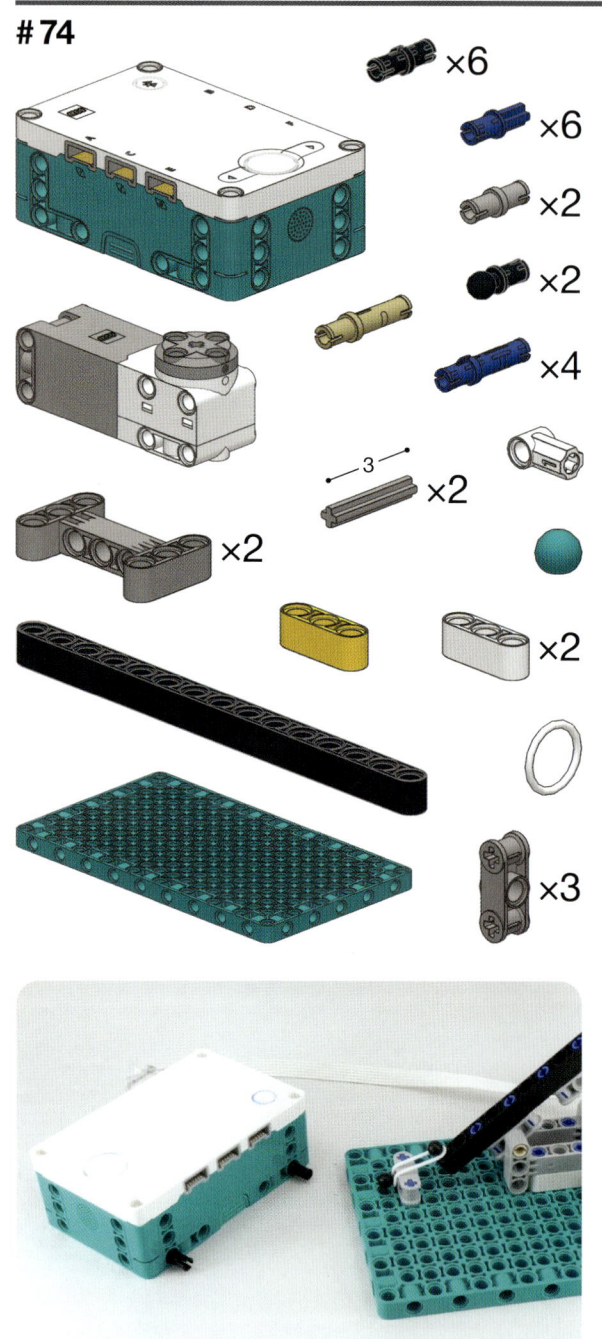

×6
×6
×2
×2
×4

3
×2
×2

×2

×3

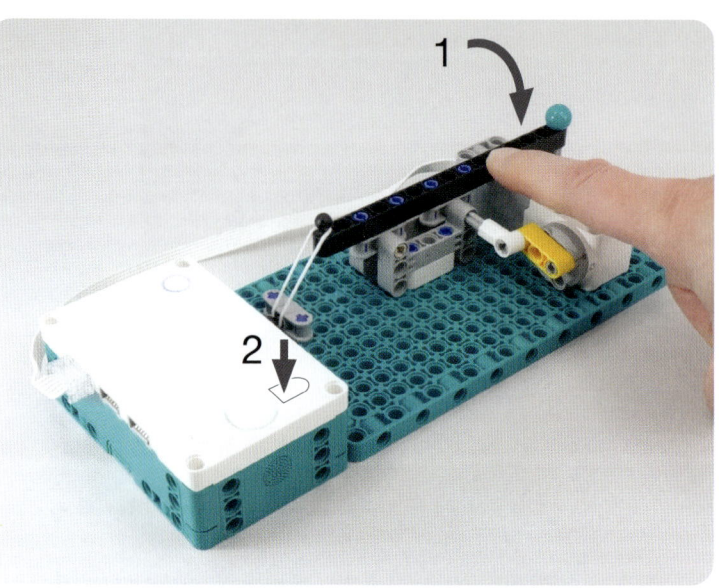

1

2

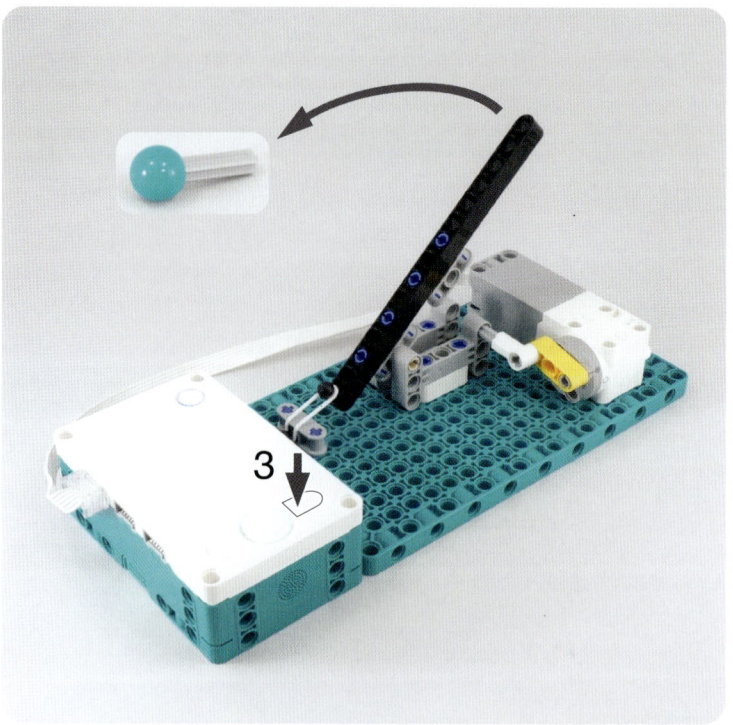

3

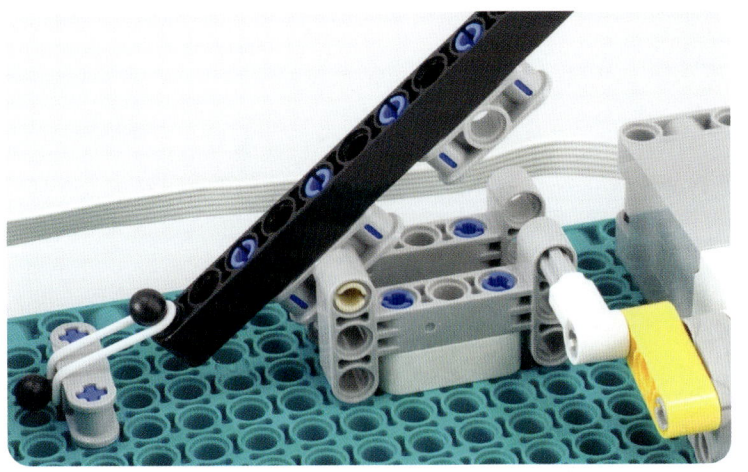

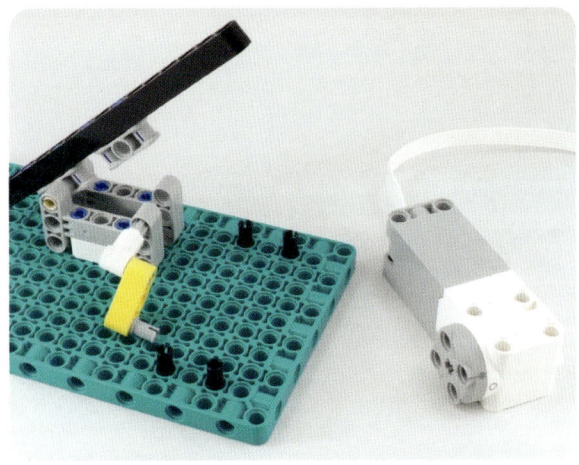

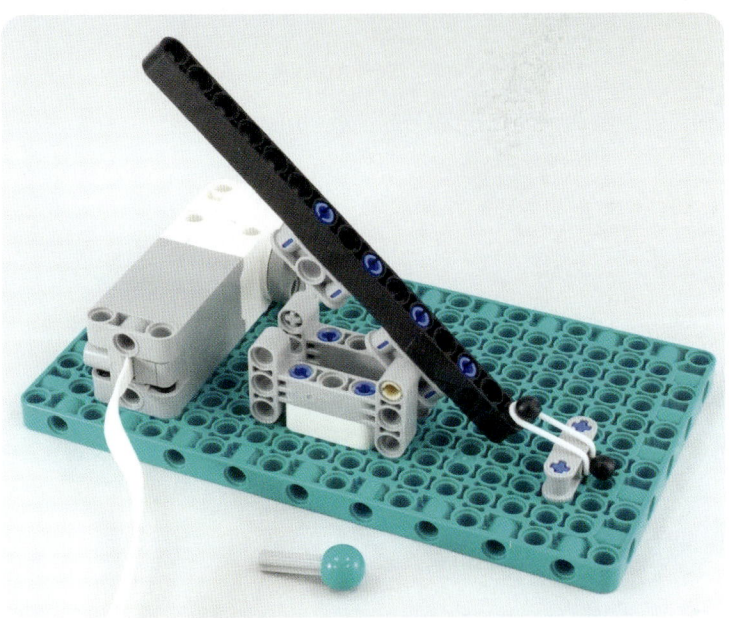

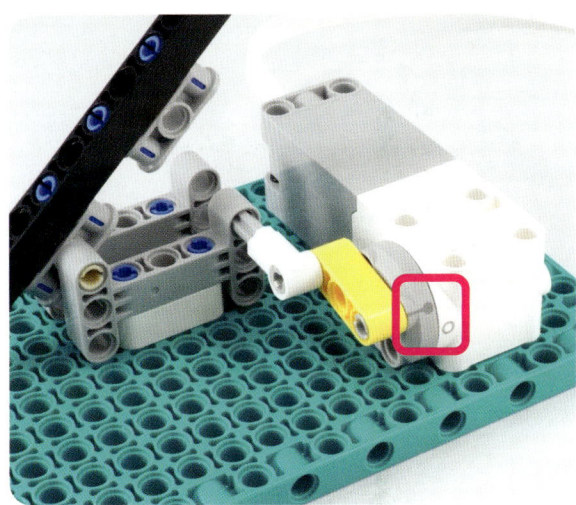

```
    ▶   when program starts

    forever

    ⬡   A ▼  go  shortest path ▼  to position  340

    wait until  ⬚  is  Right ▼  button  pressed ▼  ?

    ⬡   A ▼  go  shortest path ▼  to position  270

    wait until  ⬚  is  Right ▼  button  pressed ▼  ?

    ⬡   A ▼  go  shortest path ▼  to position  340

                                              ↵
```

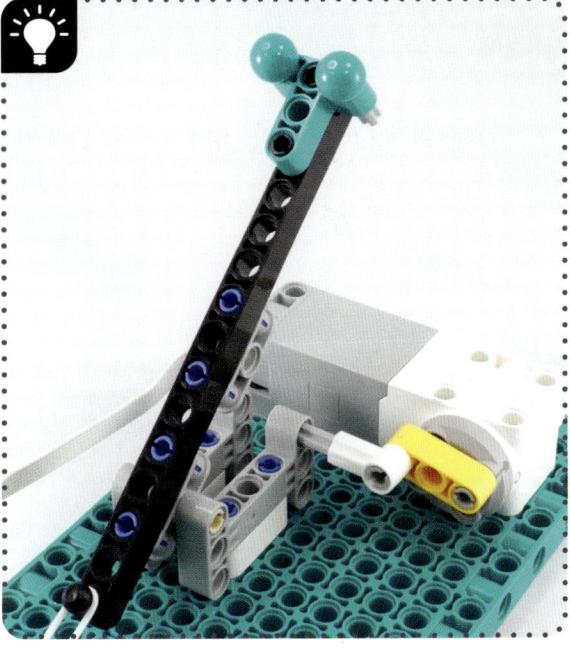

#75

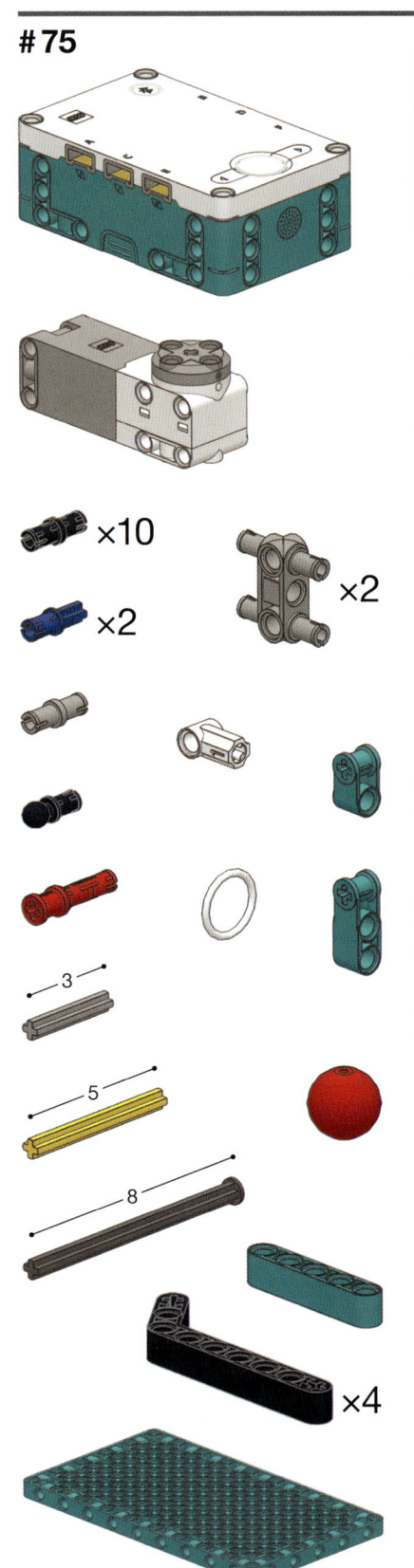

×10

×2

×2

—3—

—5—

—8—

×4

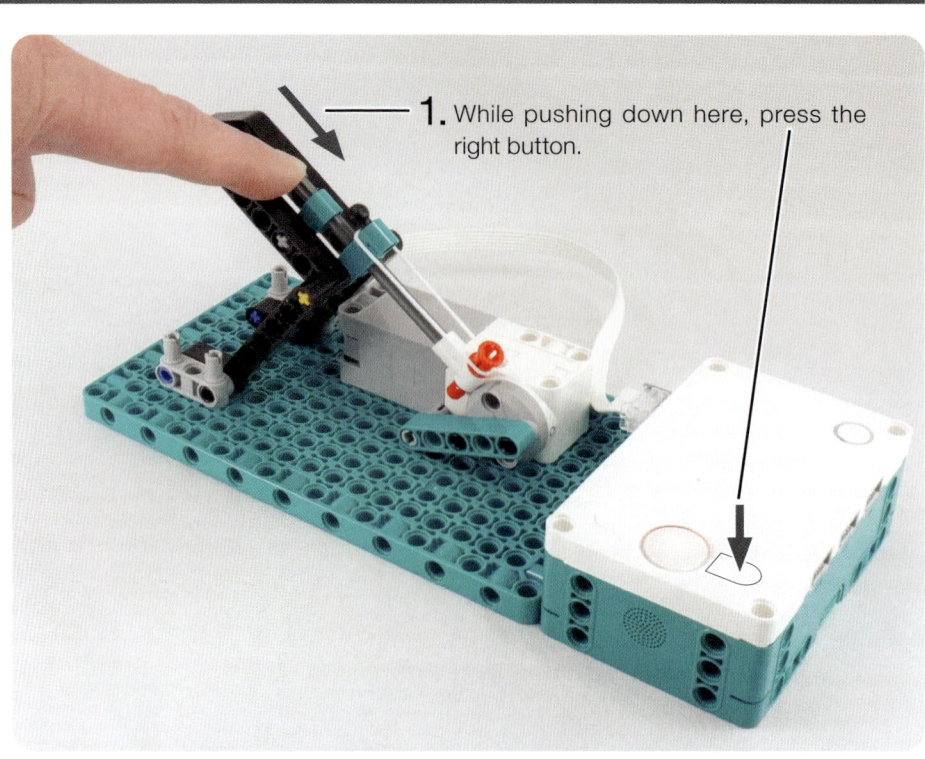

1. While pushing down here, press the right button.

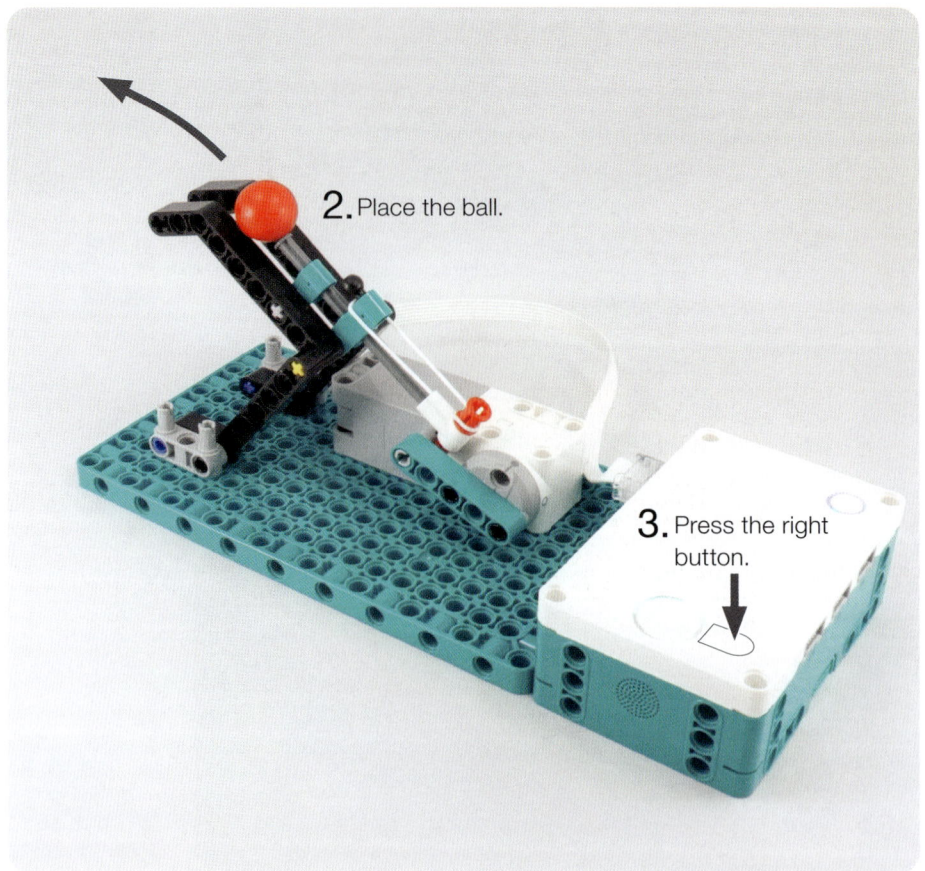

2. Place the ball.

3. Press the right button.

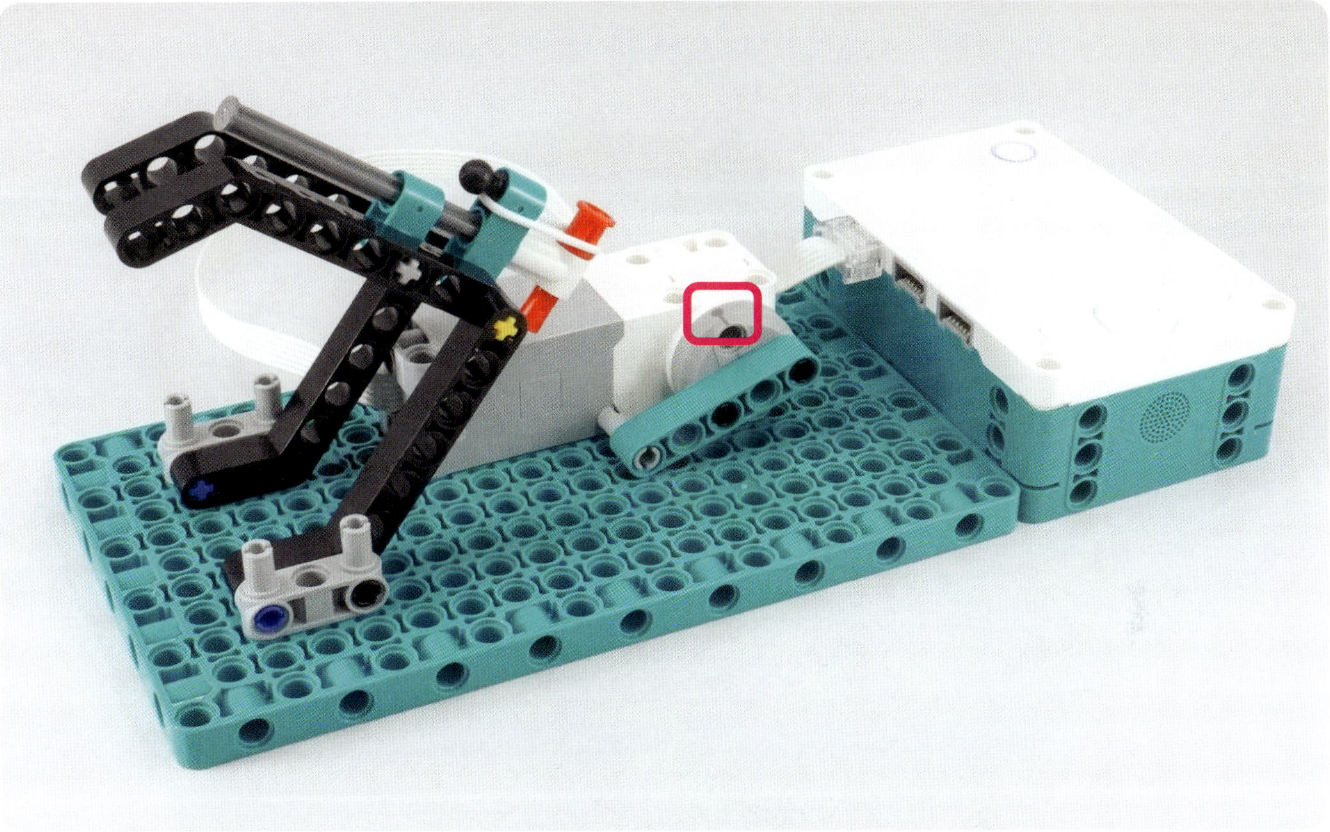

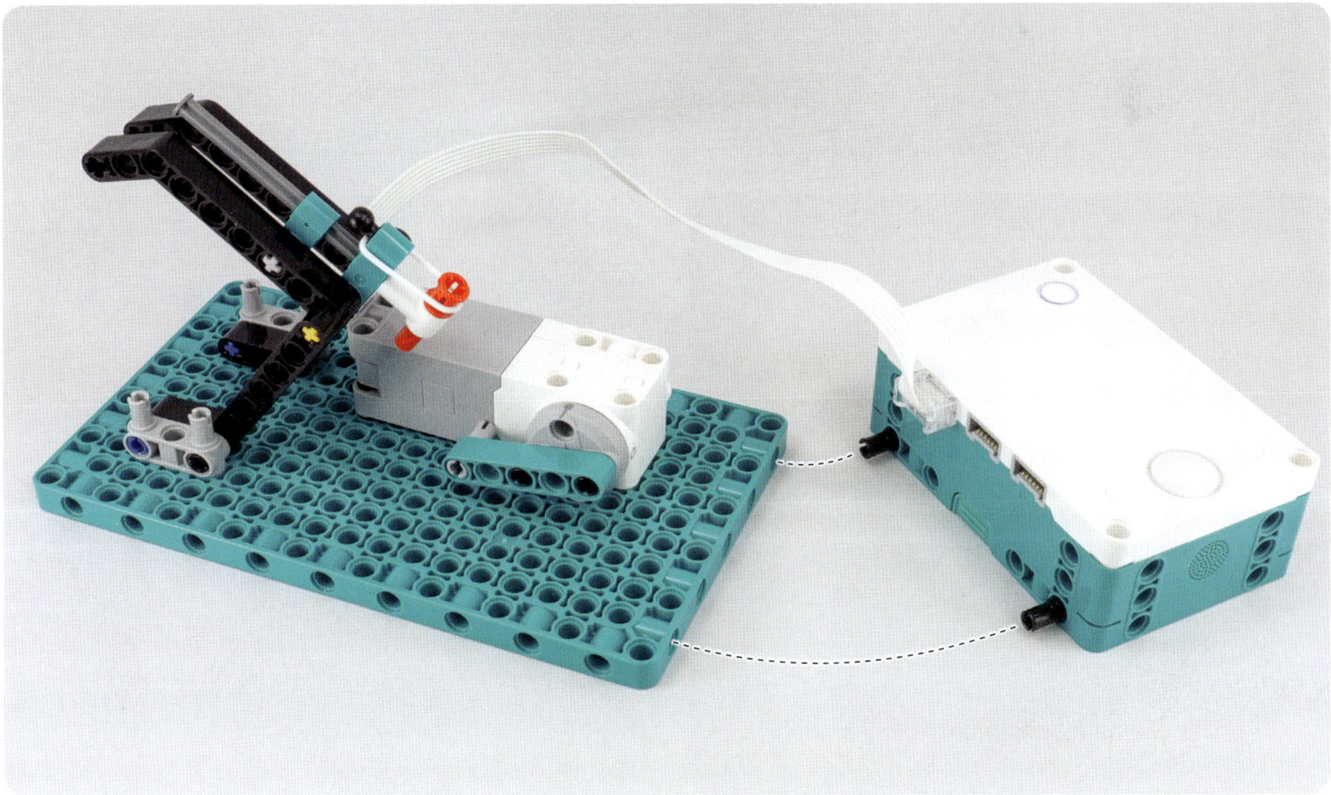

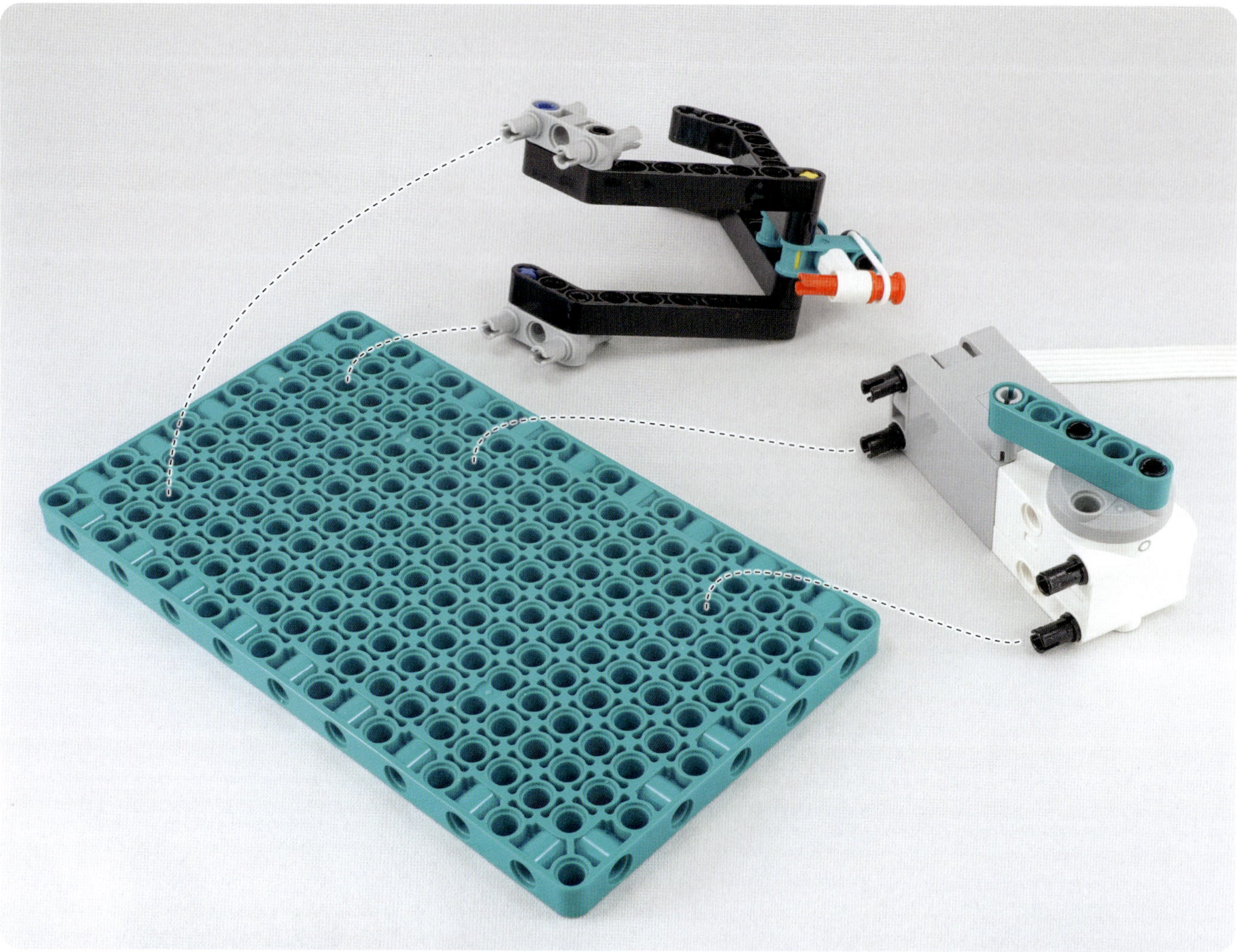

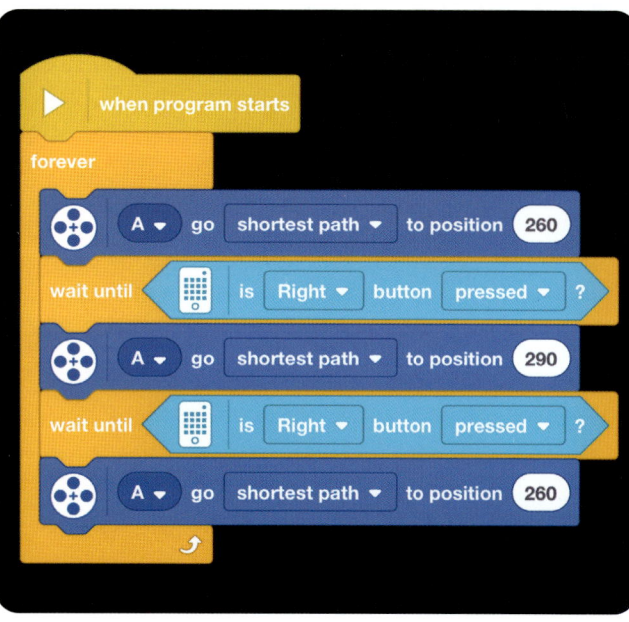

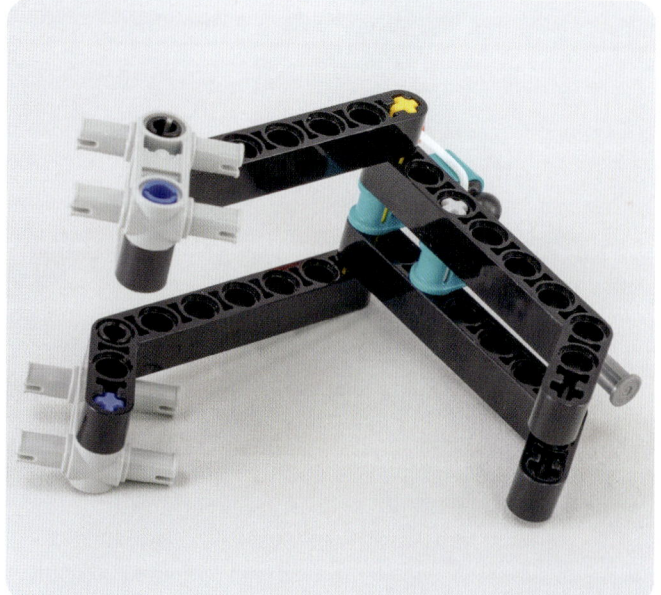

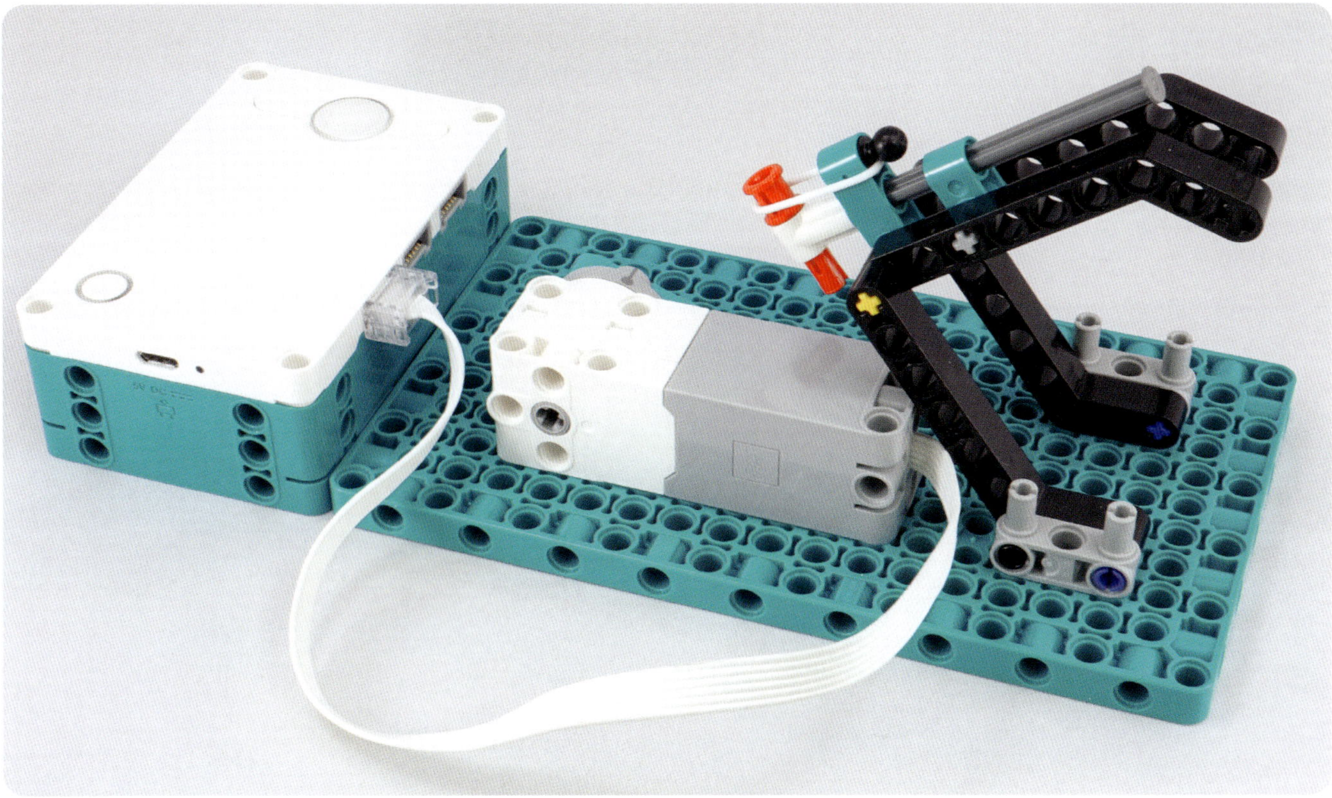

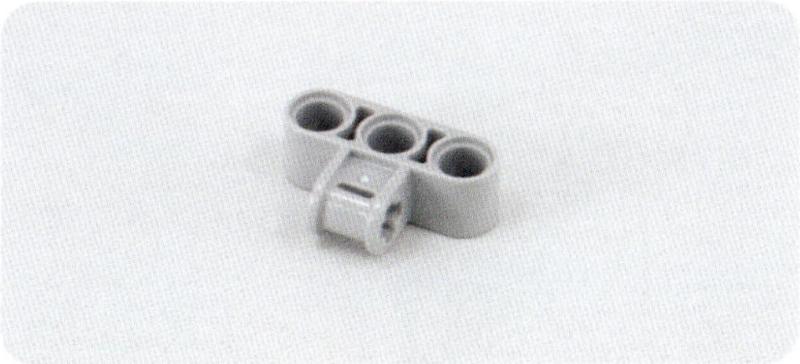

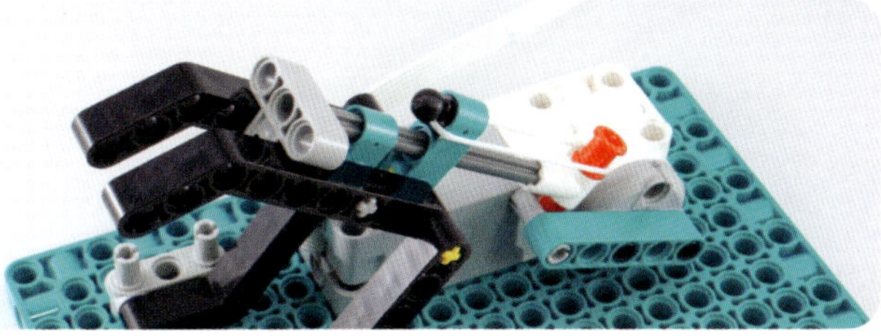

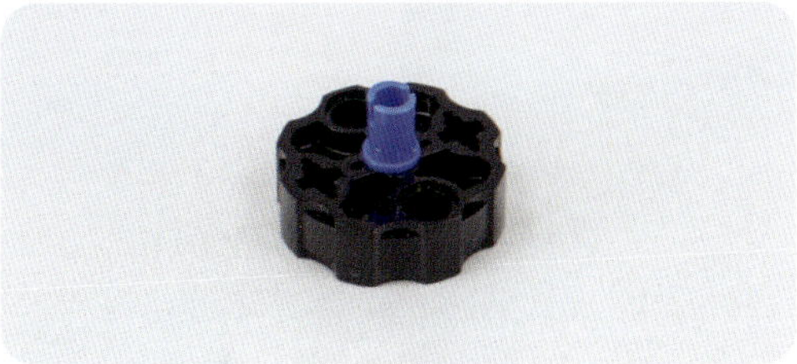

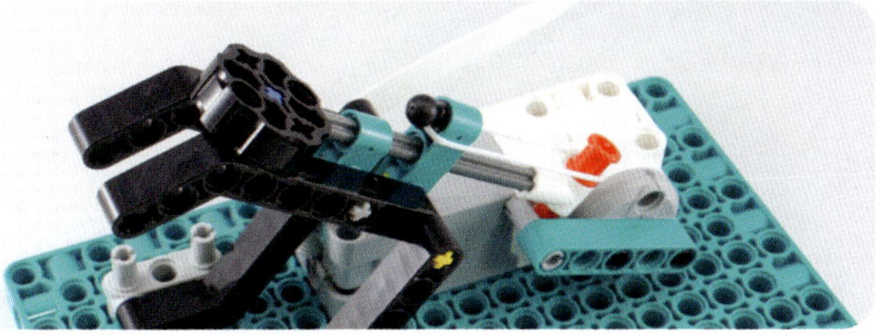

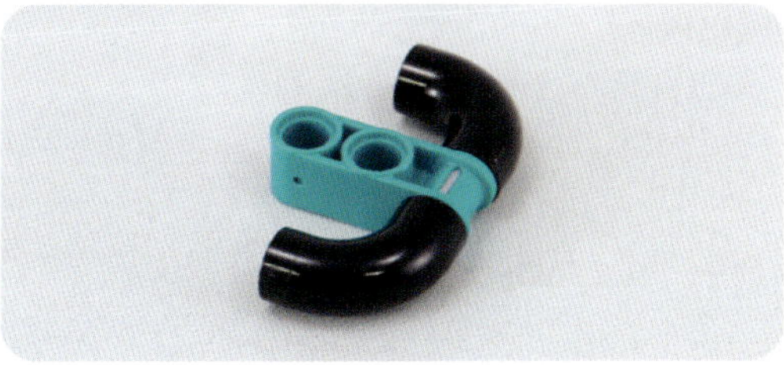

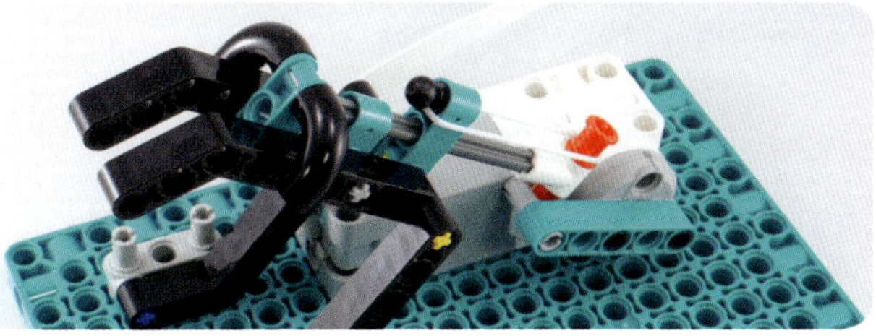

Wind devices

 # 76

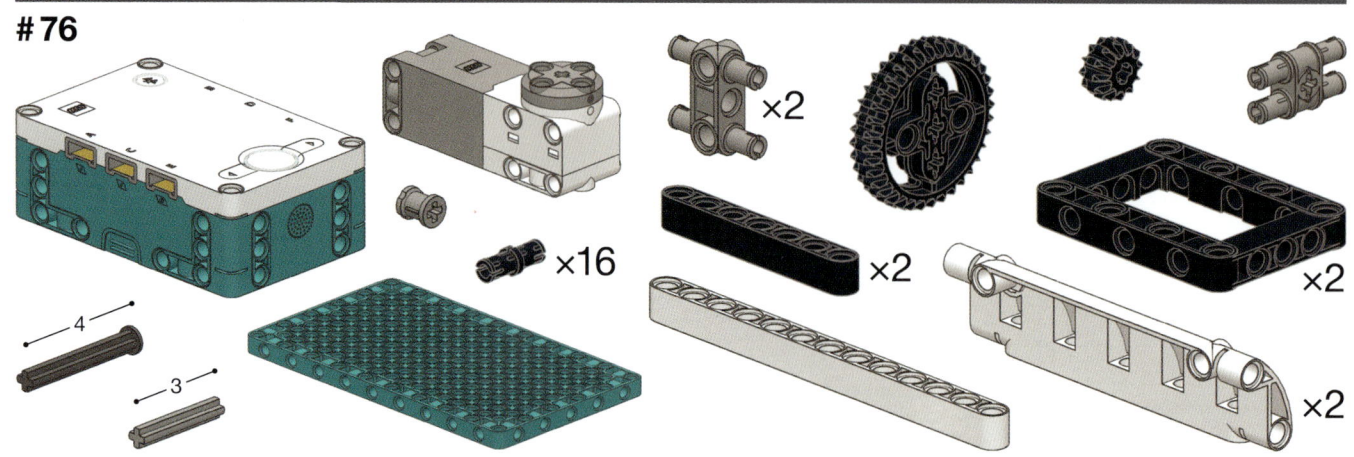

×2

×16

4

3

×2

×2

×2

×2

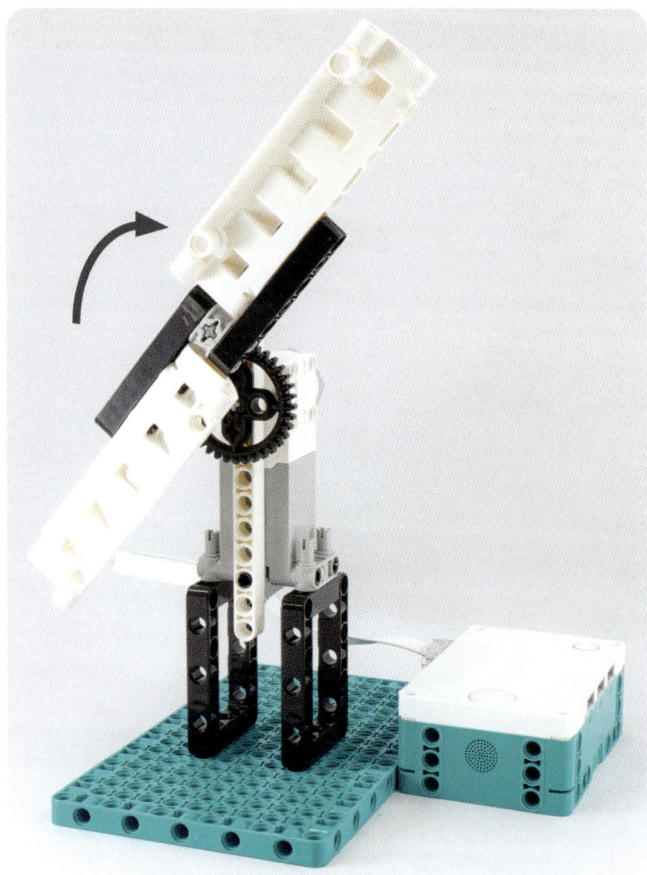

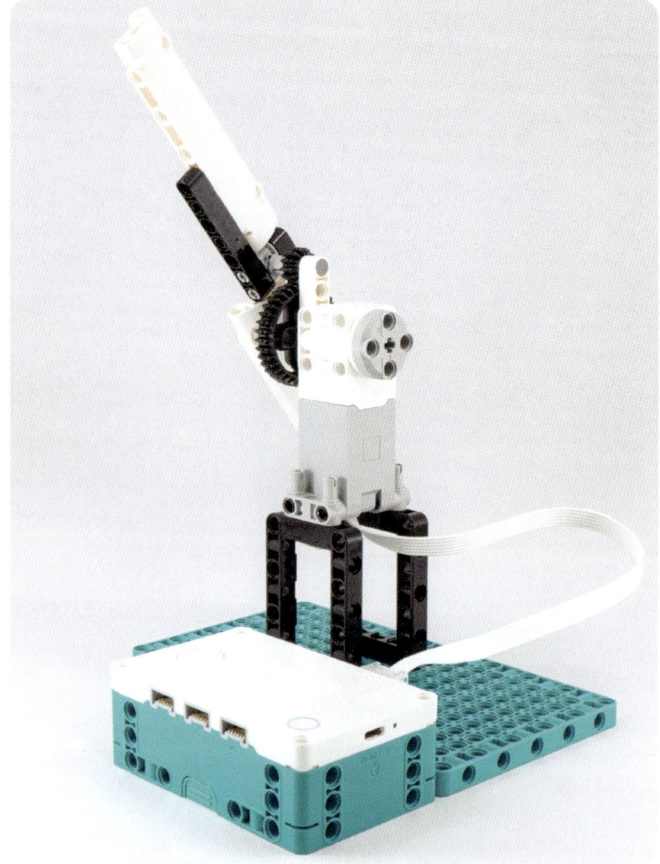

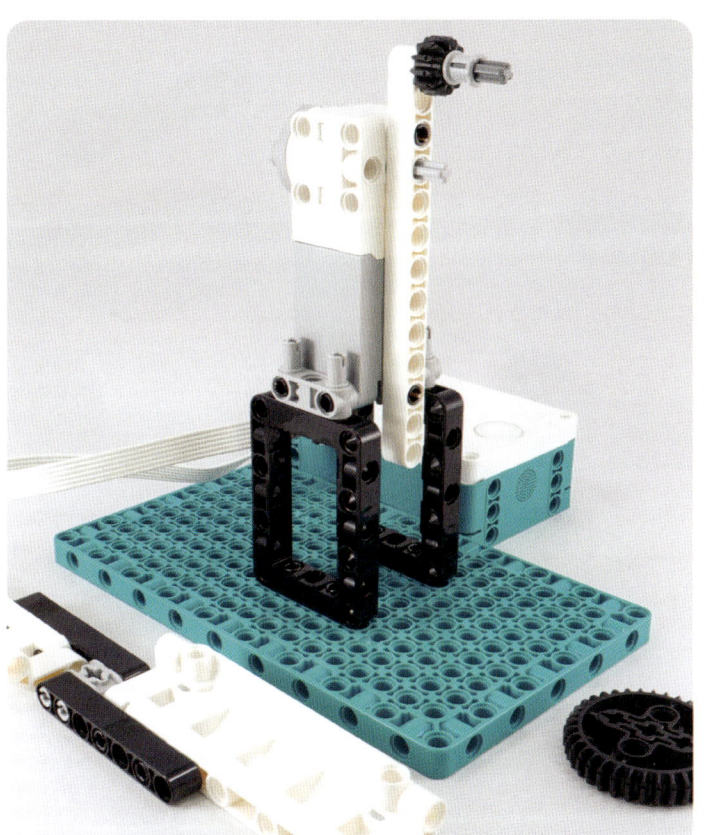

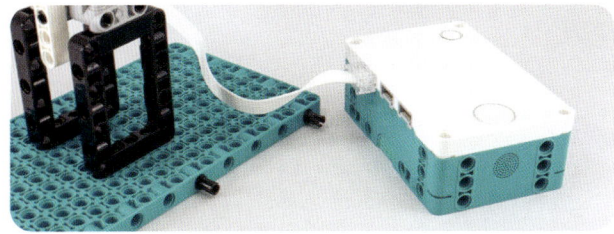

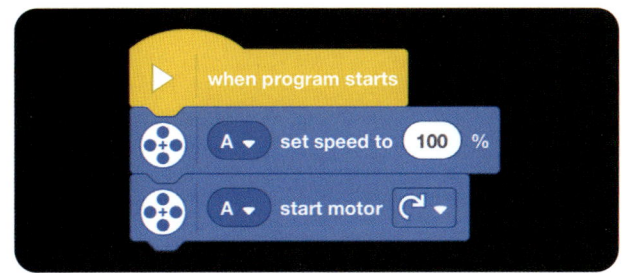

when program starts

A ▾ set speed to 100 %

A ▾ start motor ↻ ▾

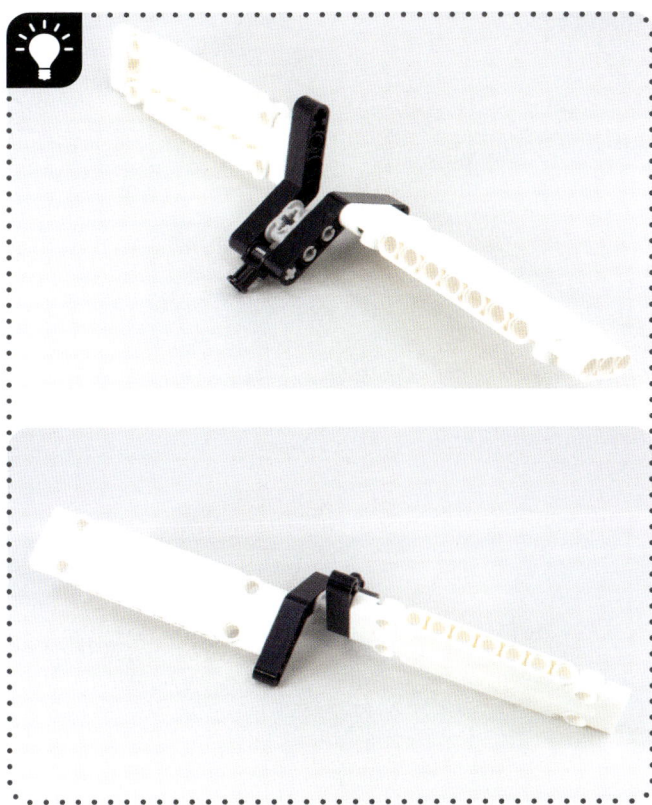

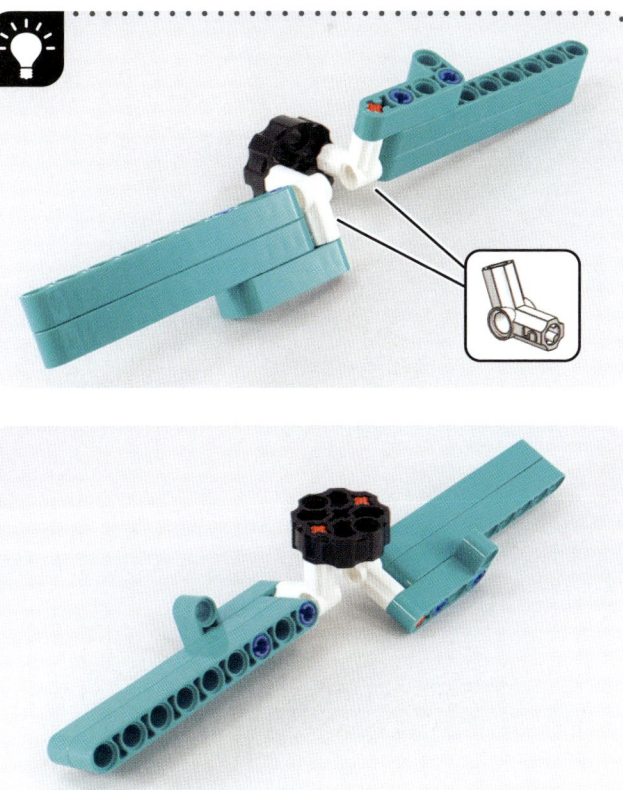

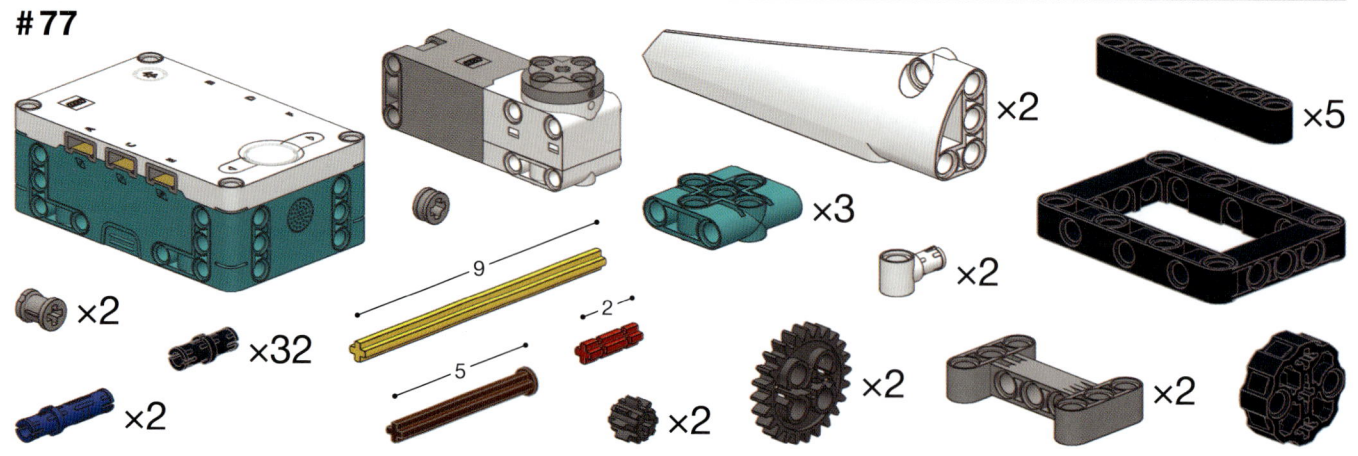

77

×2 ×32 ×2 9 5 2 ×2 ×2 ×3 ×2 ×5 ×2 ×2

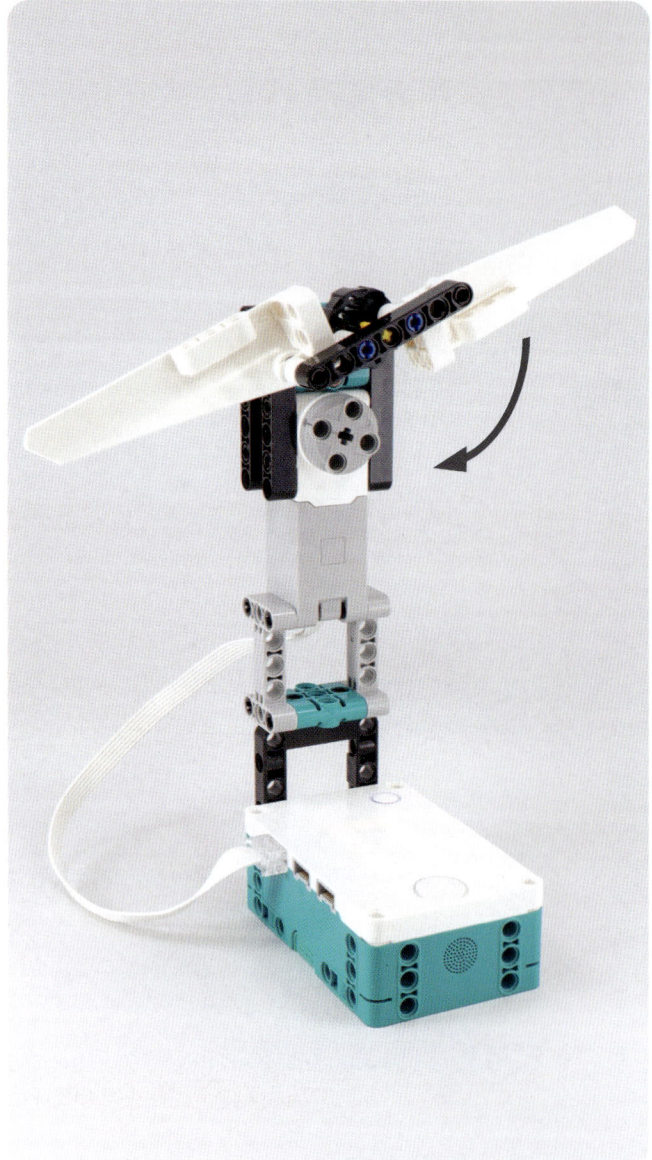

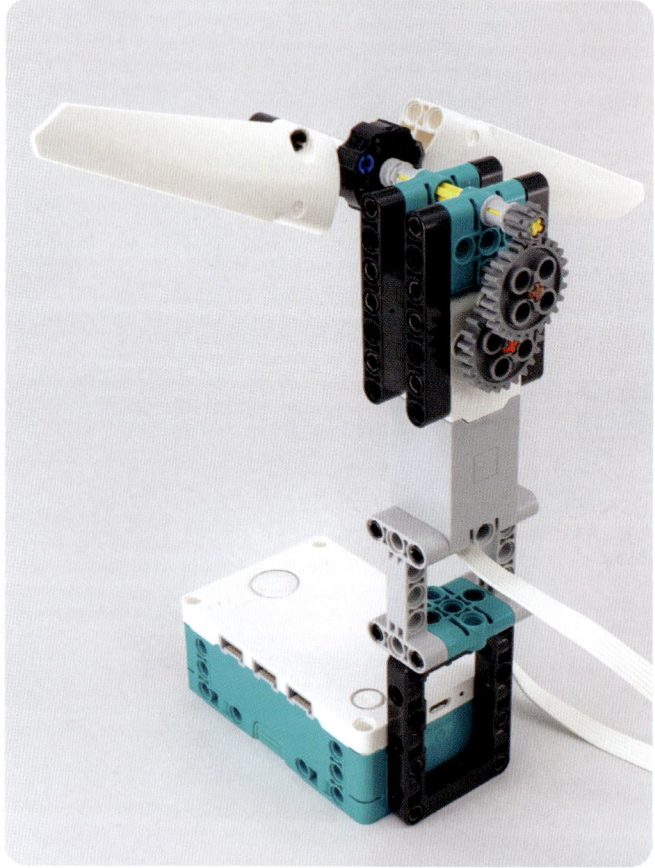

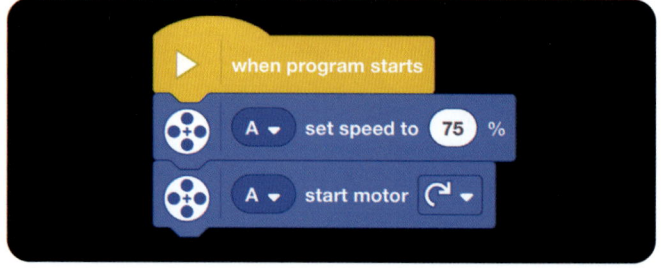

```
when program starts
A ▾  set speed to  75  %
A ▾  start motor  ↻ ▾
```

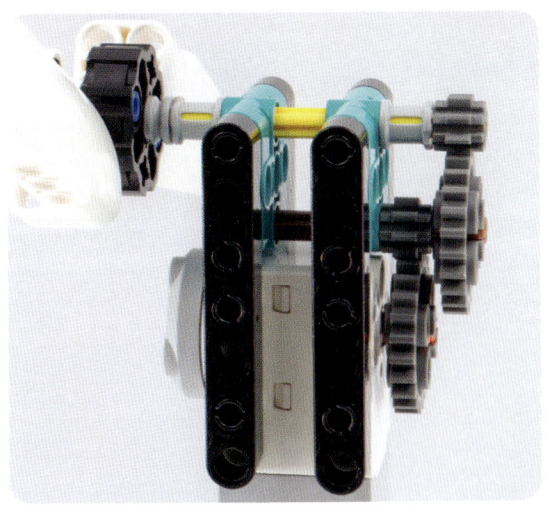

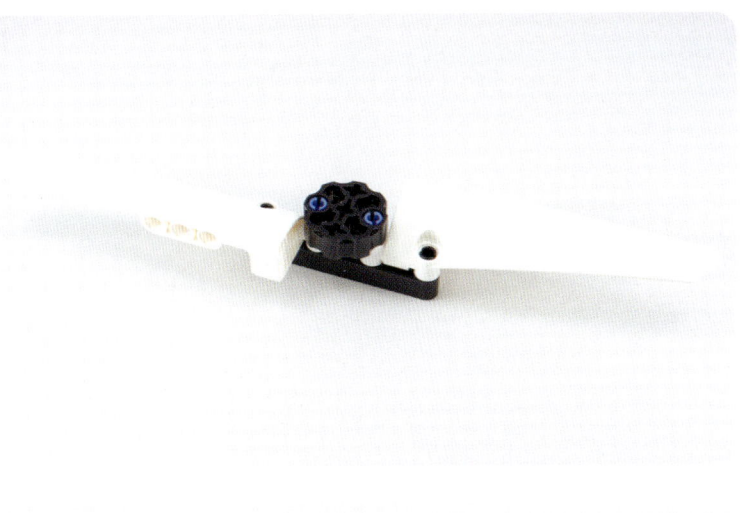

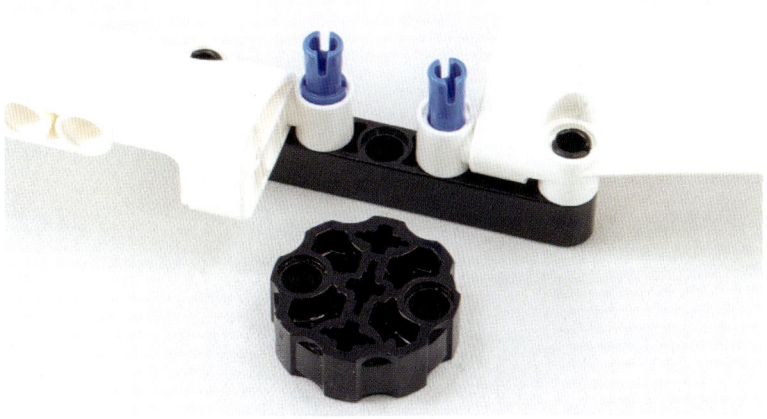

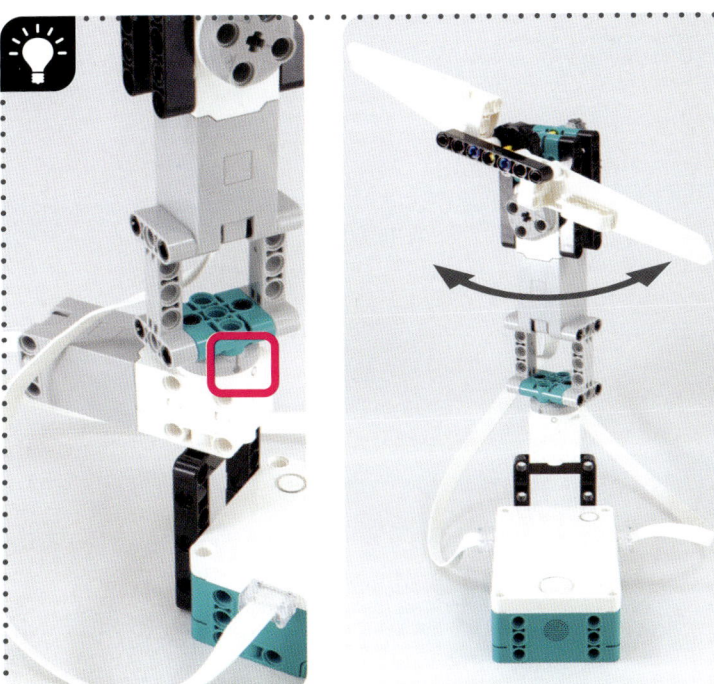

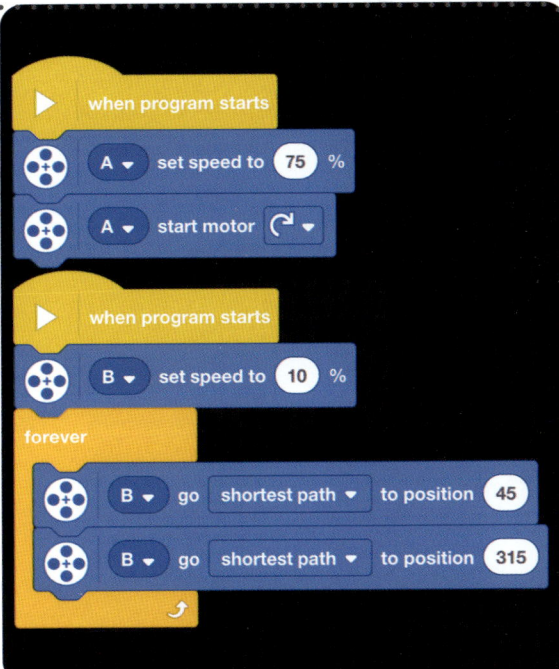

PART 4
Using Sensors

Using the Distance Sensor

#78

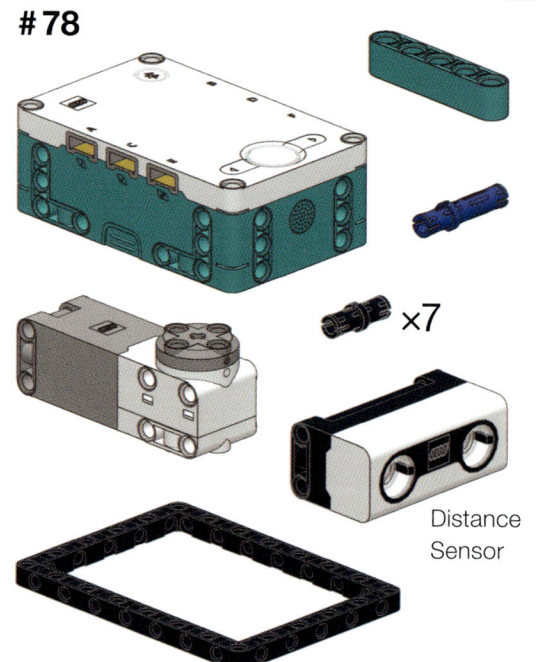

×7

Distance
Sensor

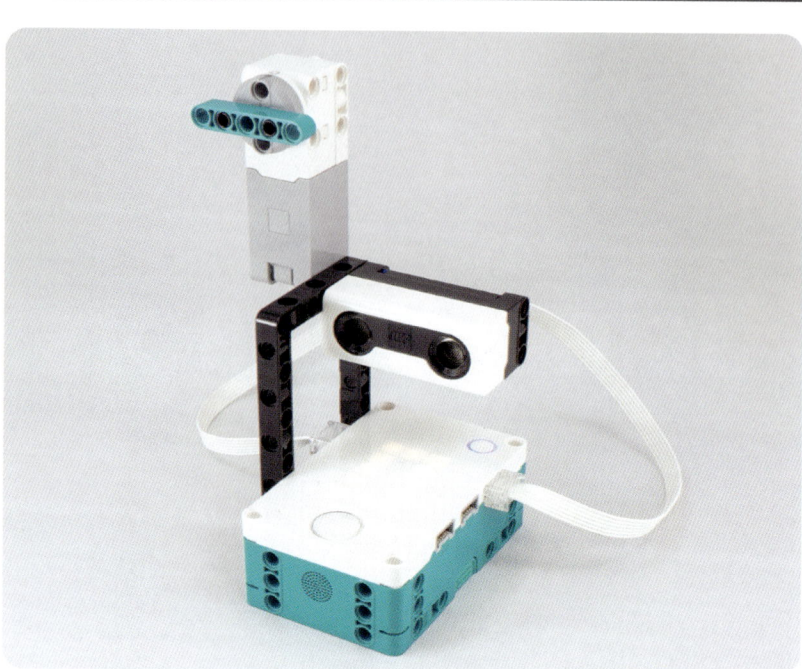

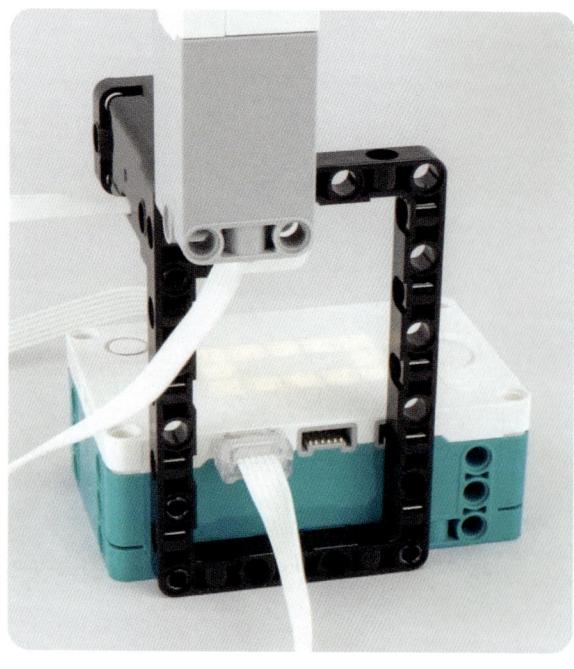

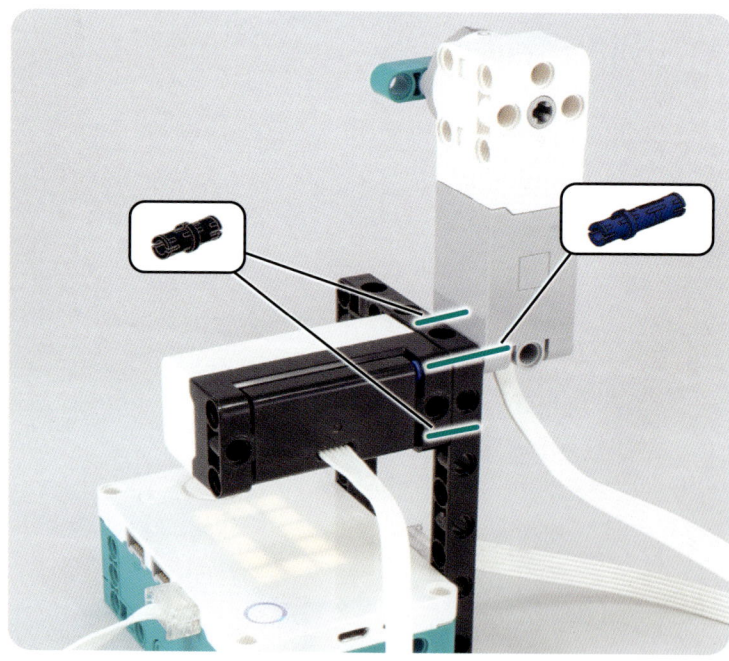

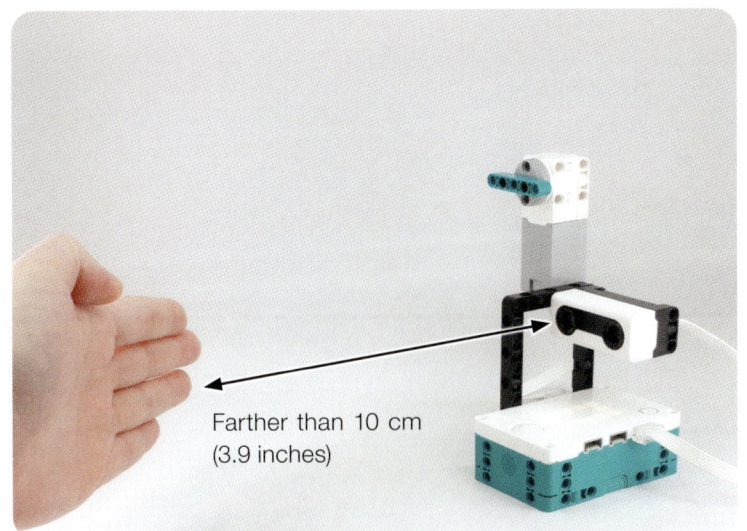

Farther than 10 cm
(3.9 inches)

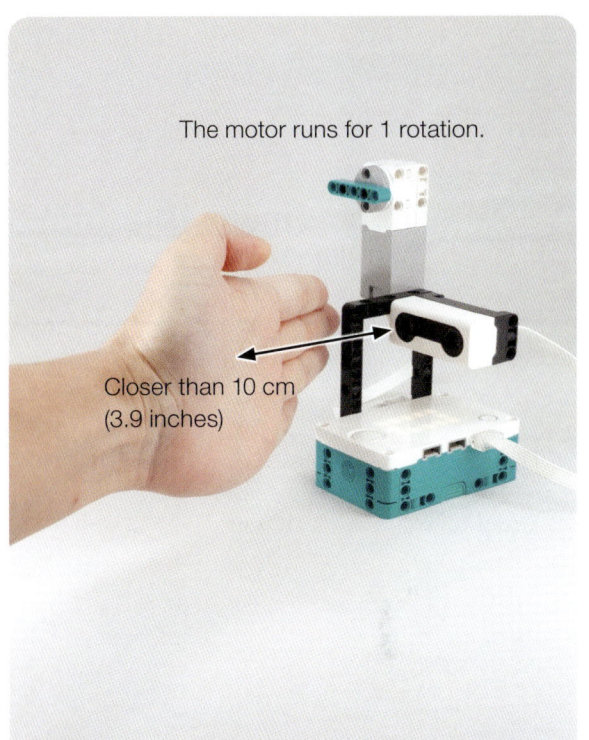

The motor runs for 1 rotation.

Closer than 10 cm
(3.9 inches)

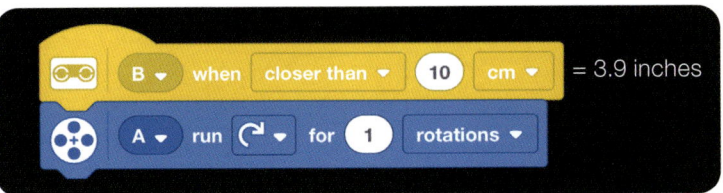

B ▾ when closer than ▾ 10 cm ▾ = 3.9 inches

A ▾ run ↻ ▾ for 1 rotations ▾

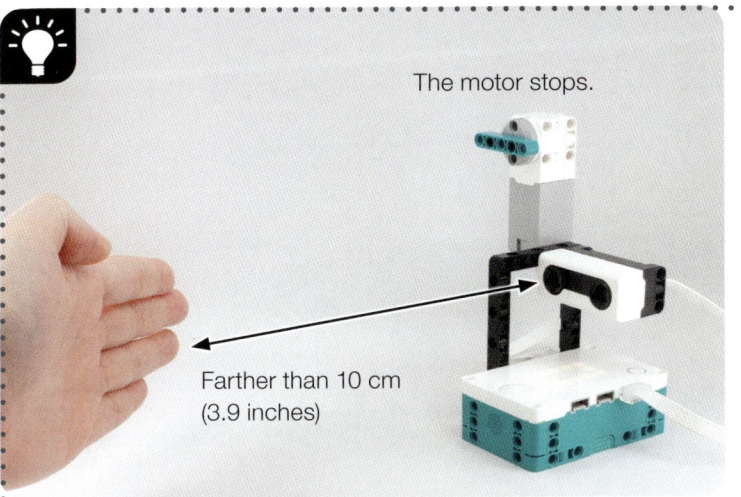

The motor stops.

Farther than 10 cm
(3.9 inches)

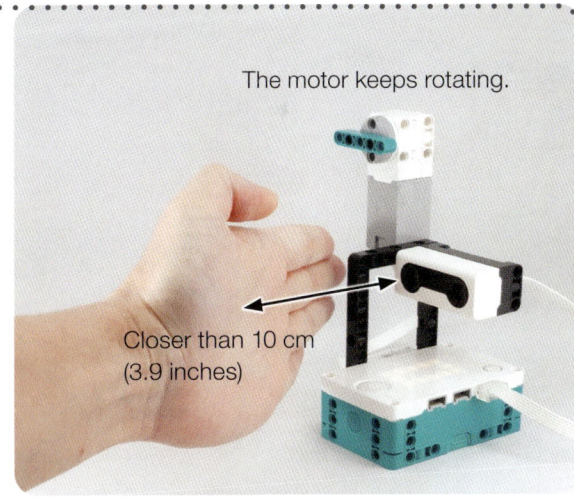

The motor keeps rotating.

Closer than 10 cm
(3.9 inches)

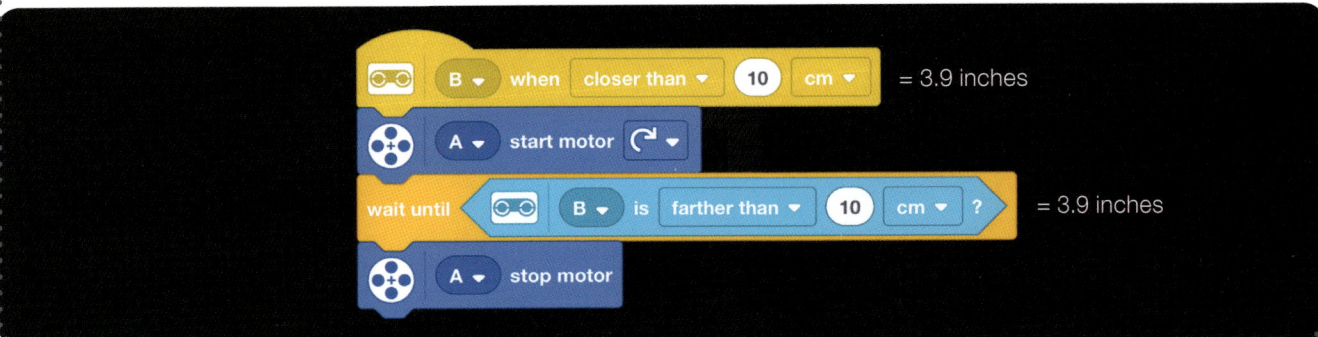

B ▾ when closer than ▾ 10 cm ▾ = 3.9 inches

A ▾ start motor ↻ ▾

wait until B ▾ is farther than ▾ 10 cm ▾ ? = 3.9 inches

A ▾ stop motor

133

Using the Color Sensor

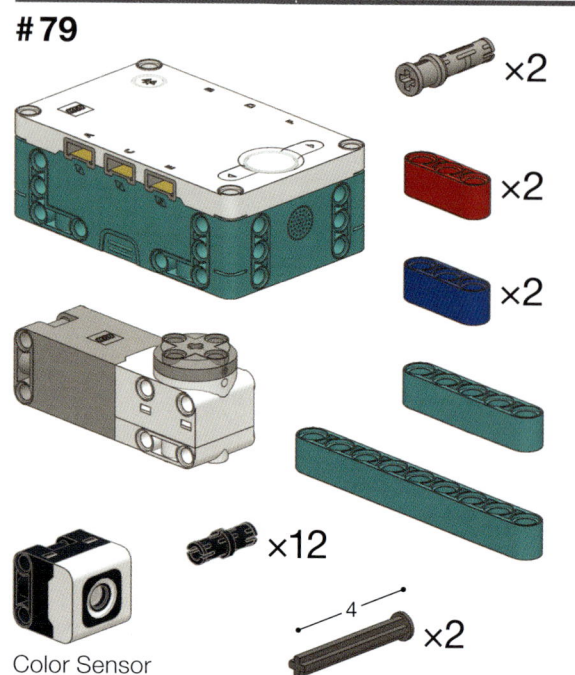

79

×2

×2

×2

×12

4

×2

Color Sensor

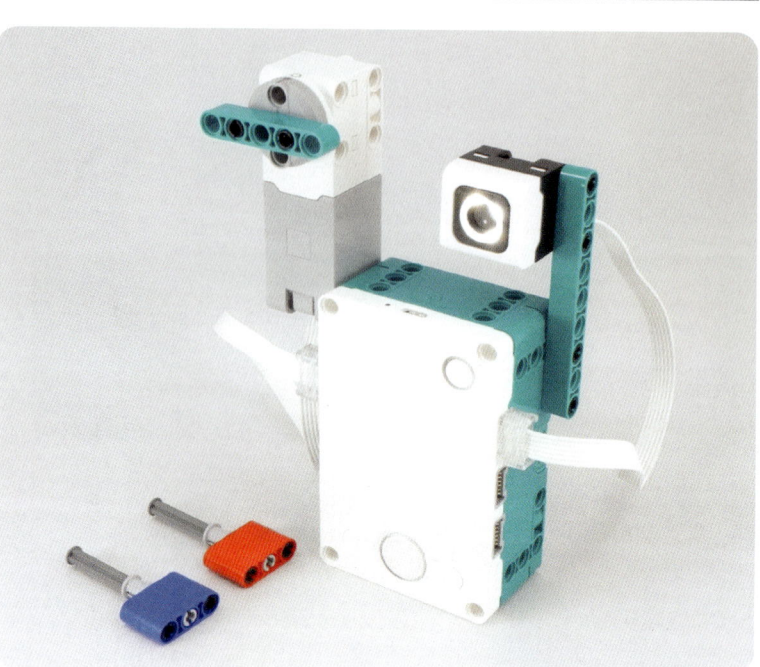

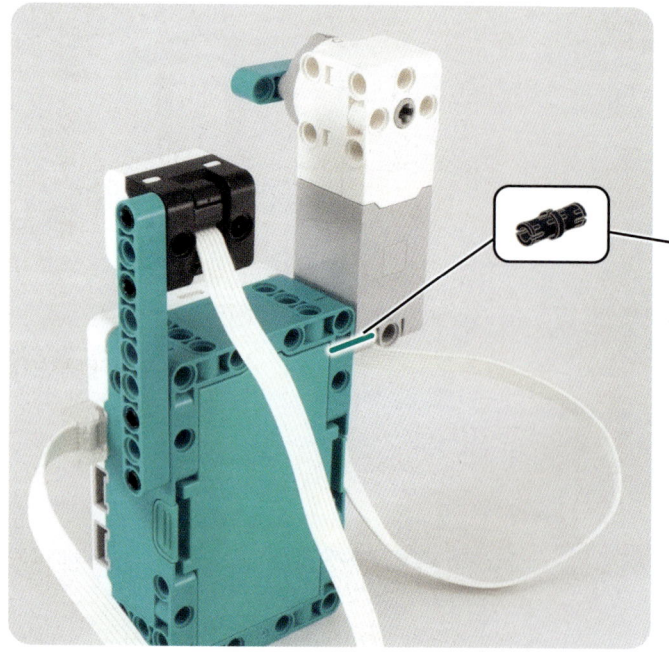

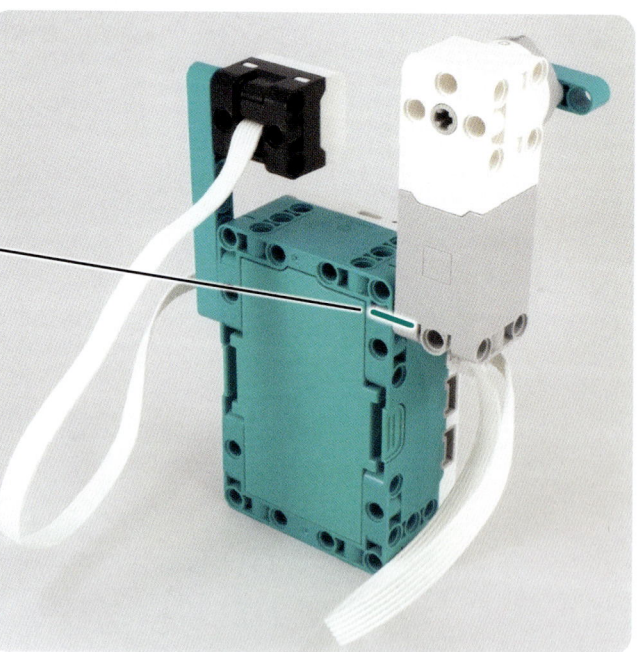

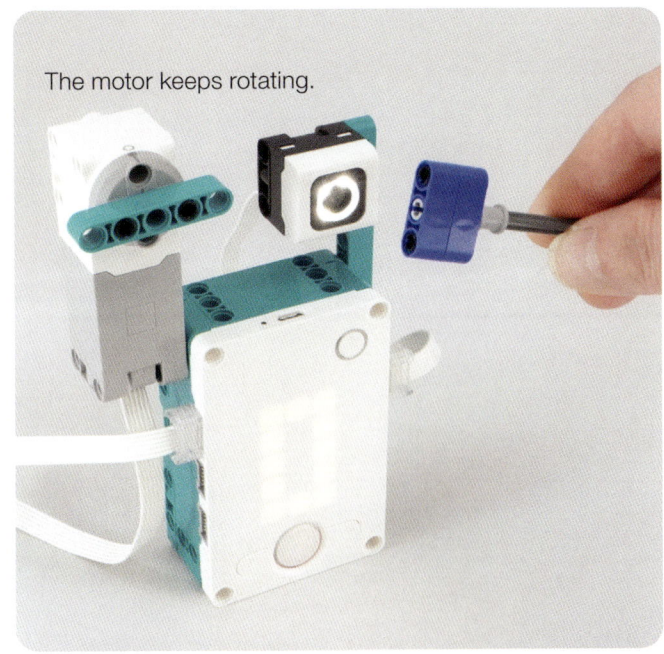

The motor keeps rotating.

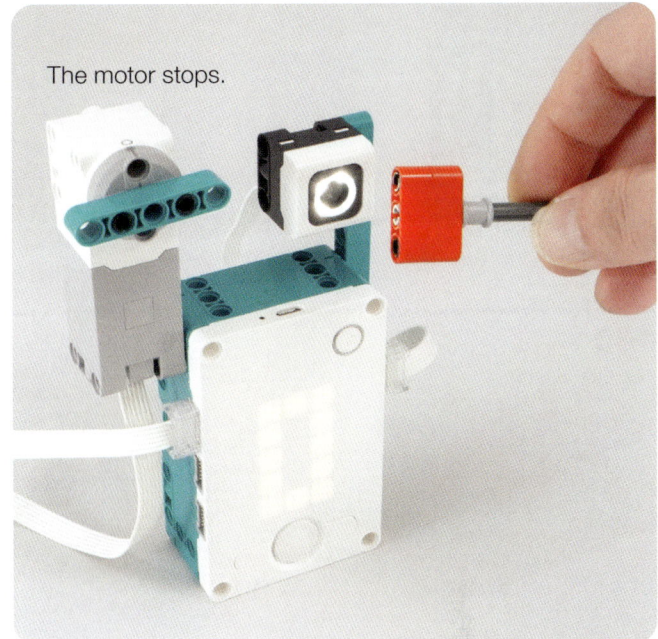

The motor stops.

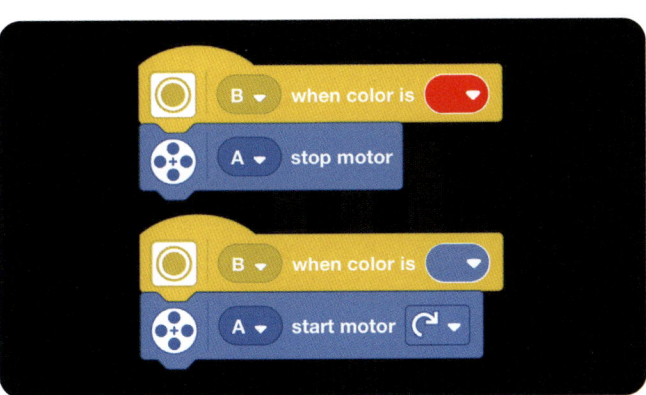

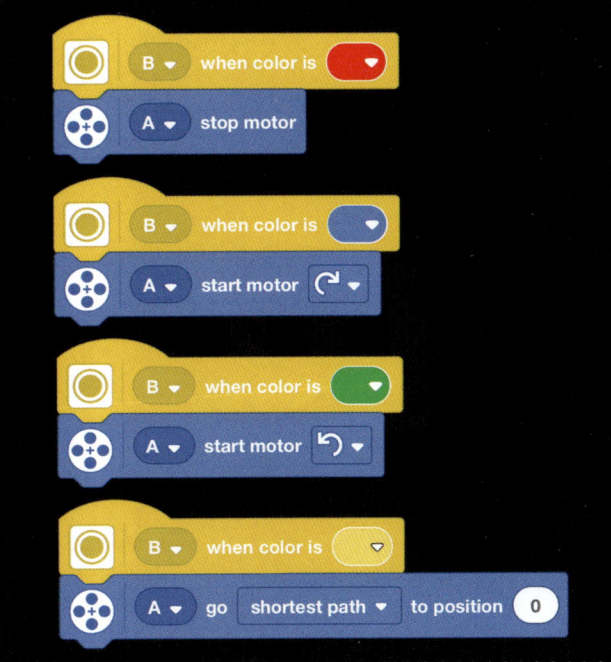

Using the Hub's built-in sensor

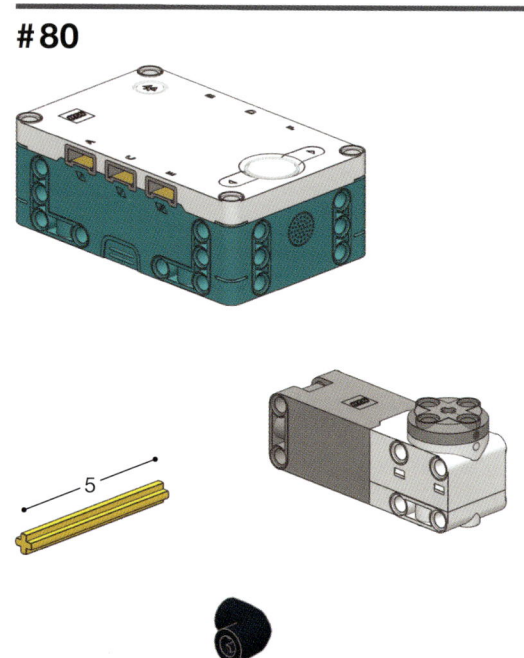

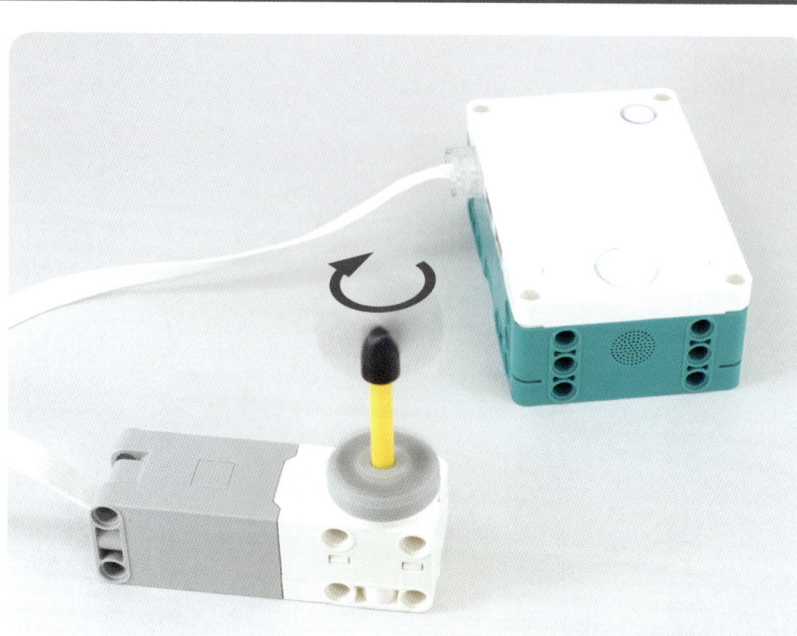

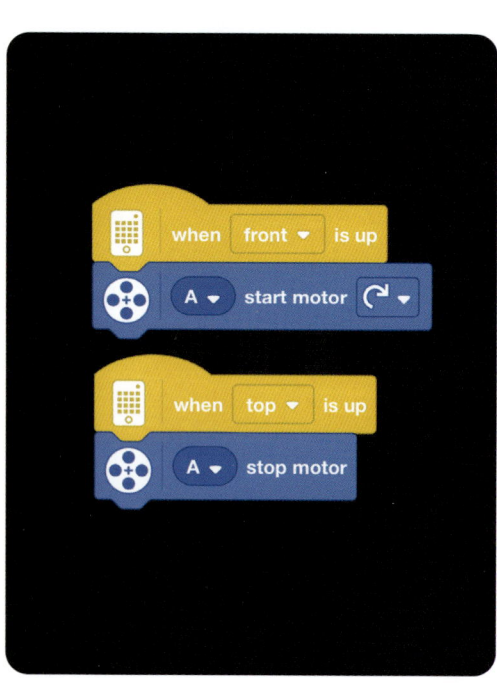

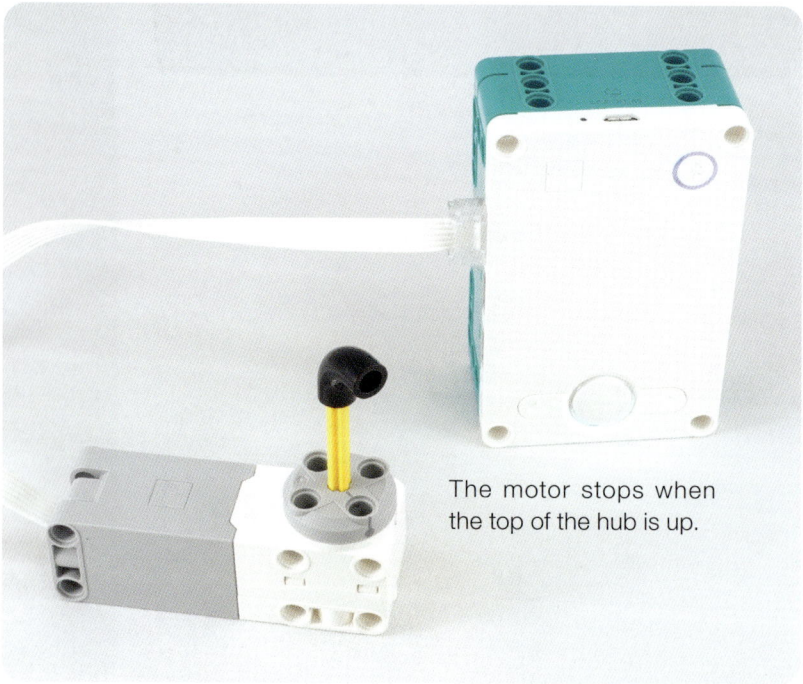

The motor stops when the top of the hub is up.

#81

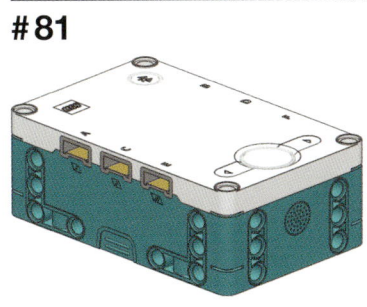

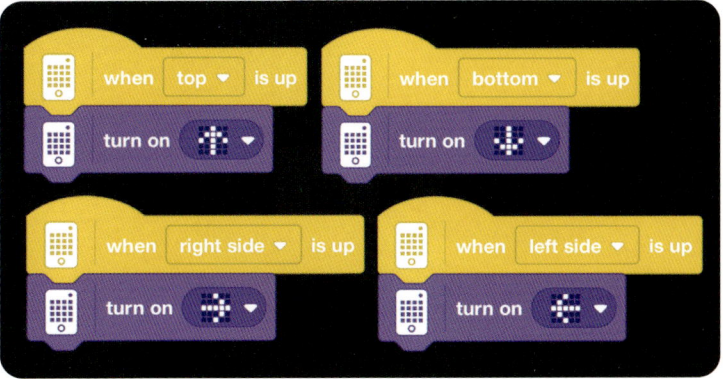

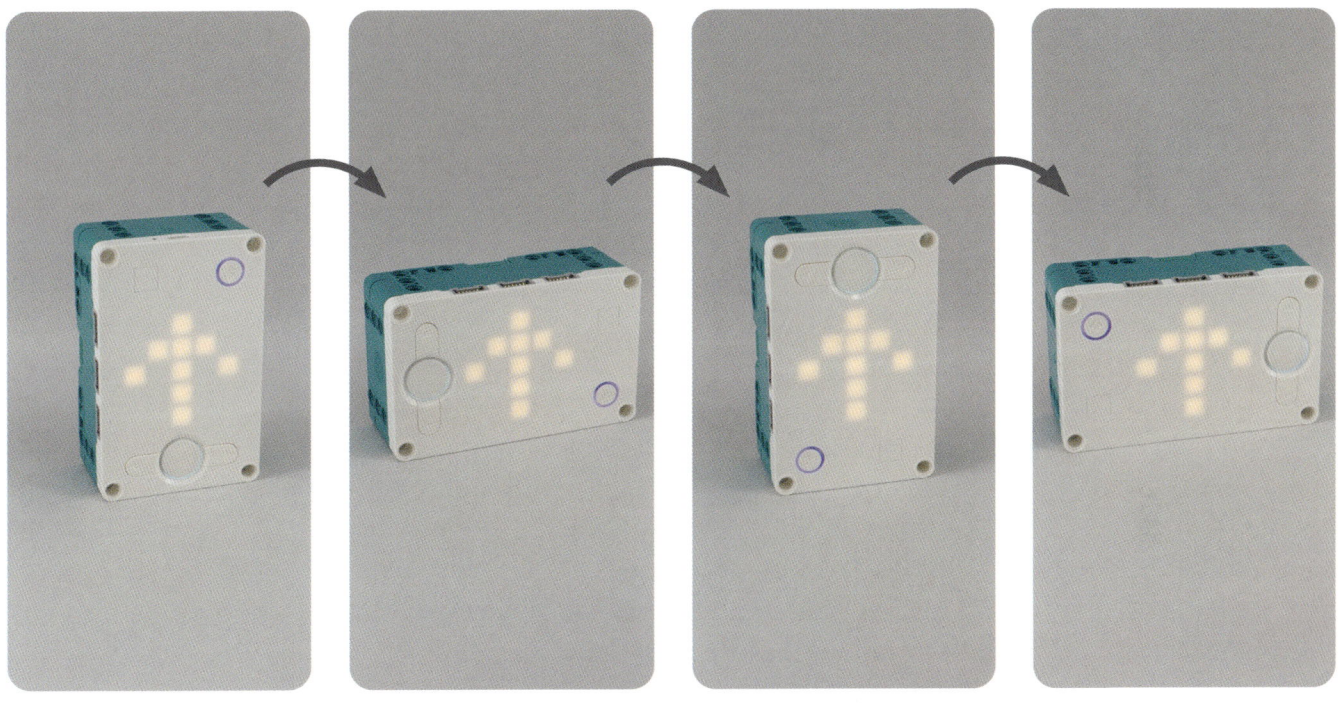

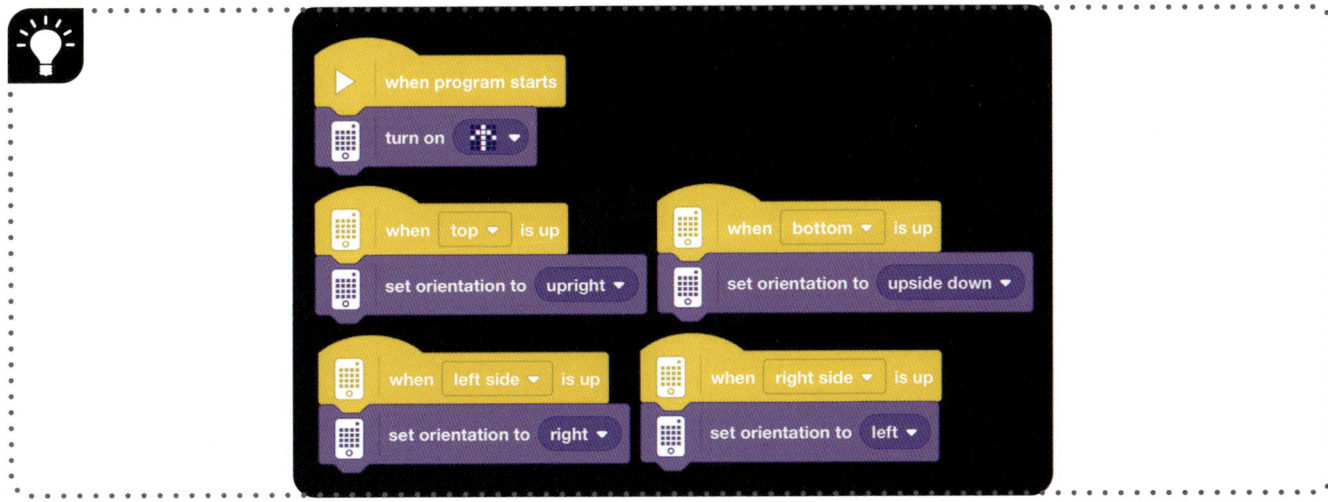

#82

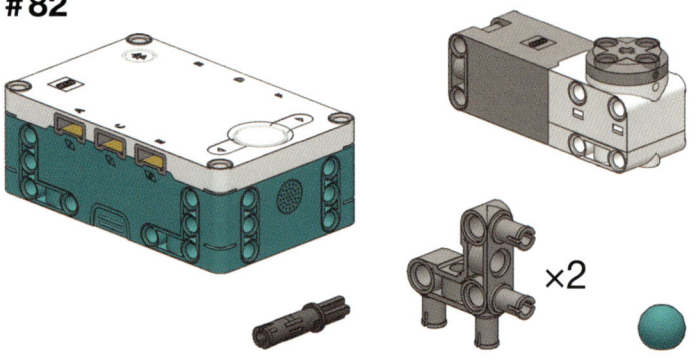

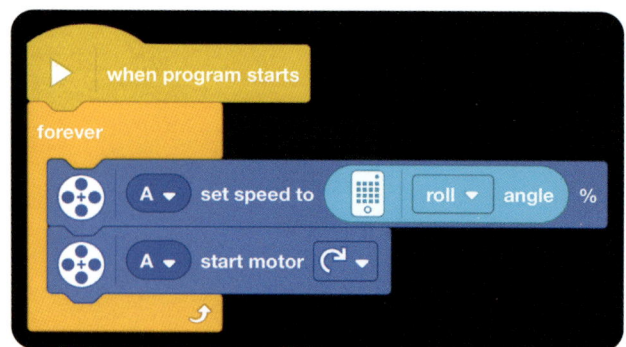

×2

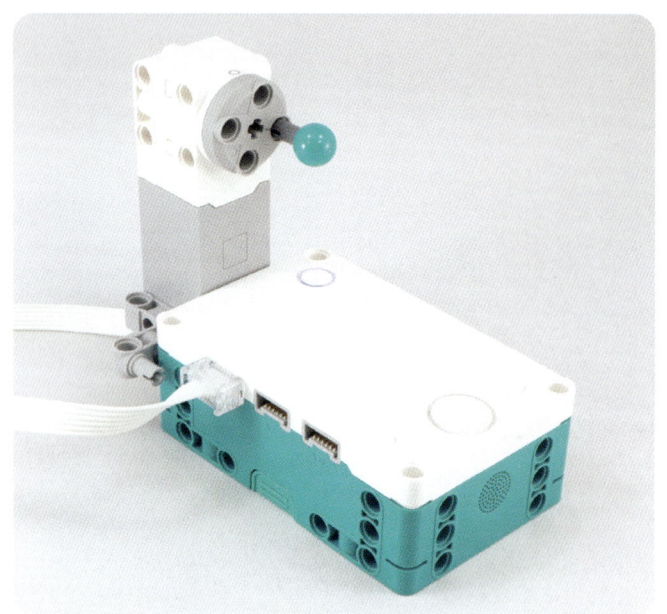

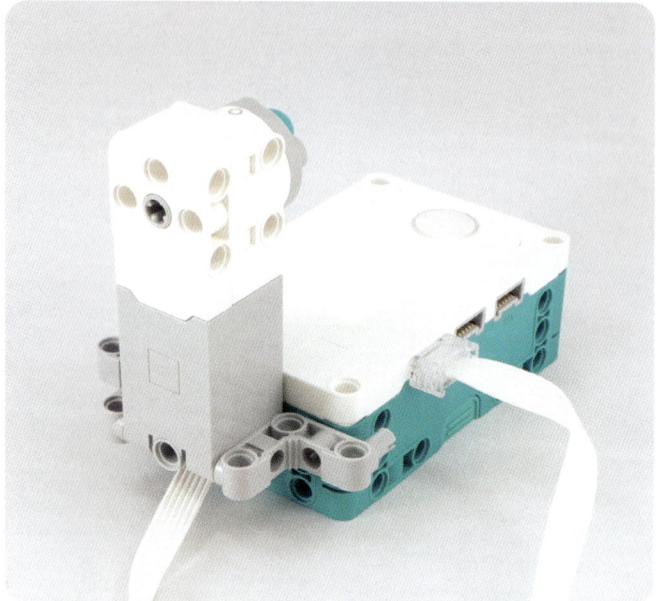

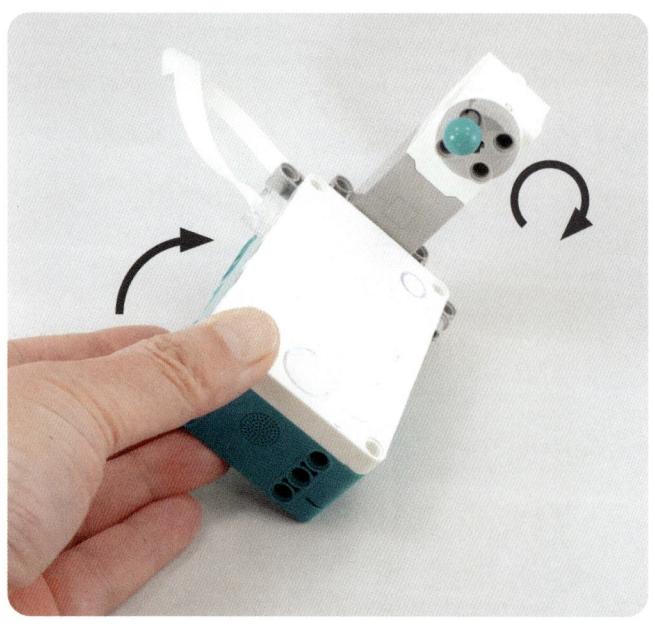

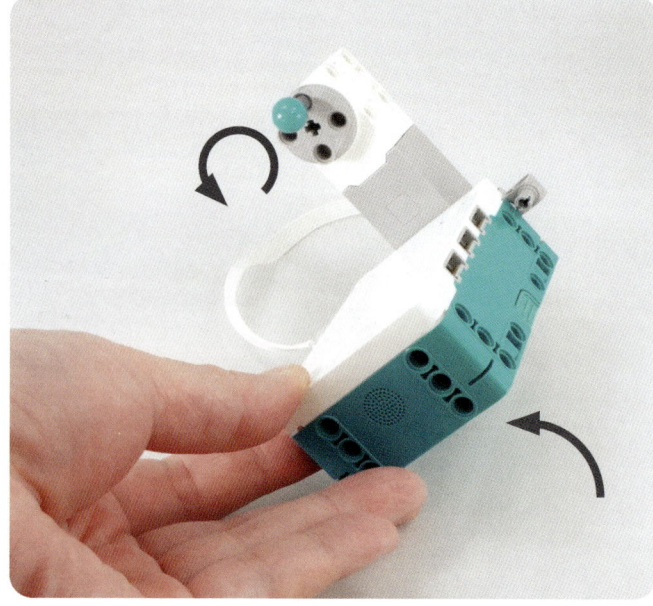

#83

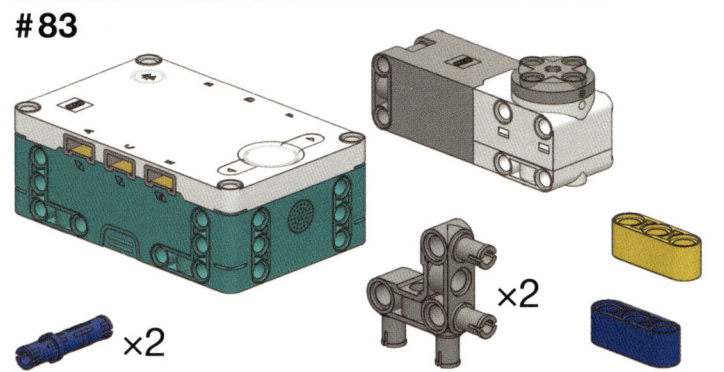

×2

×2

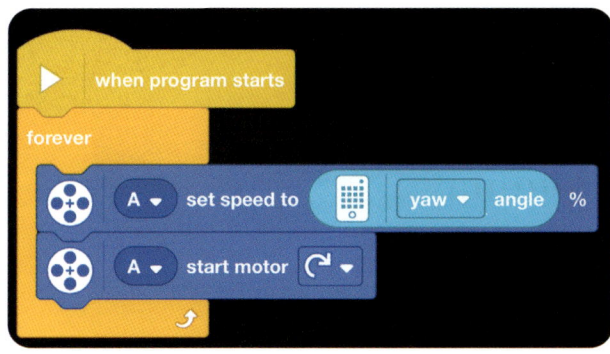

when program starts

forever

A ▼ set speed to [ ] yaw ▼ angle %

A ▼ start motor ↻ ▼

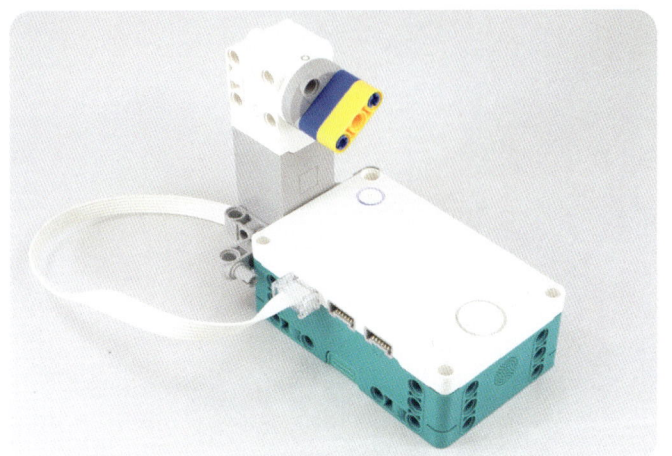

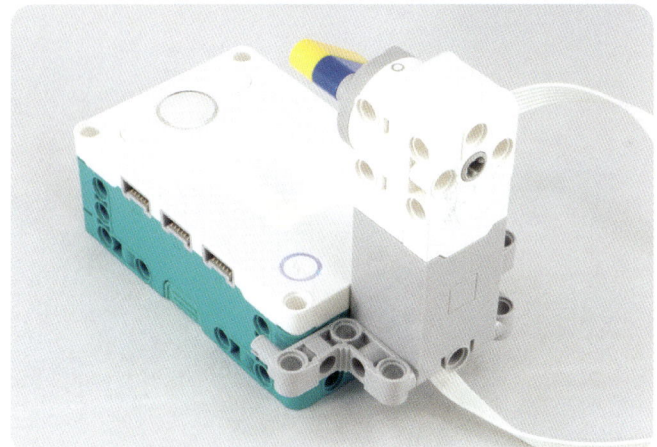

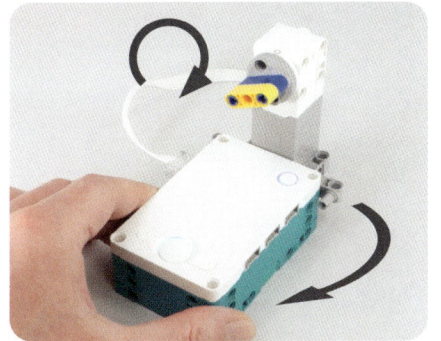

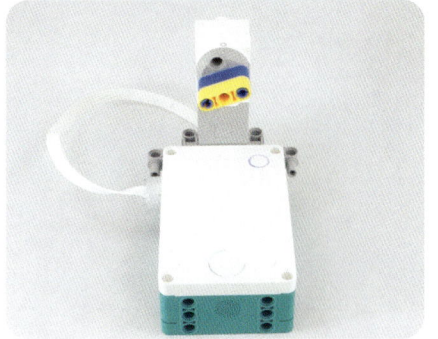

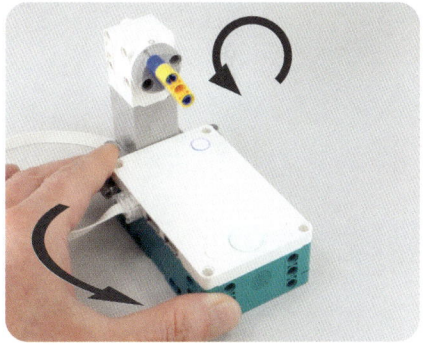

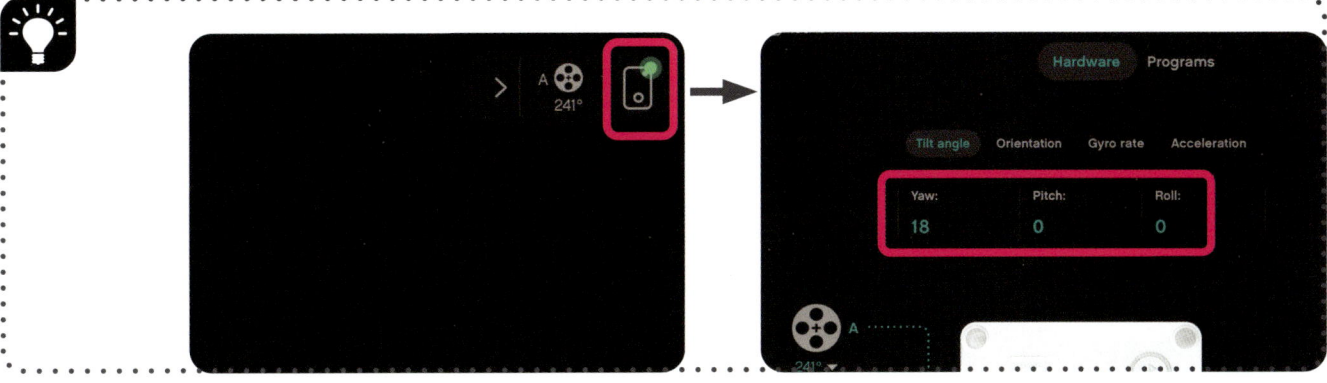

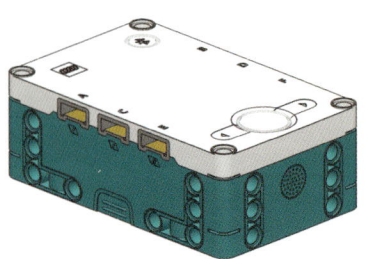

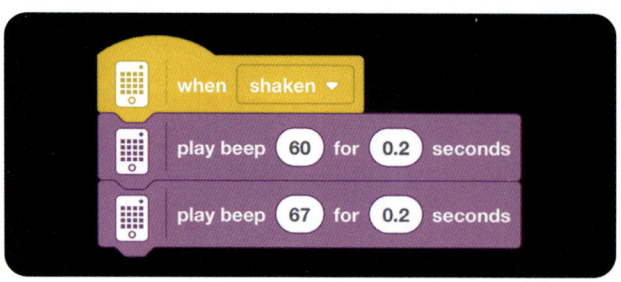

Shake!

Shake!

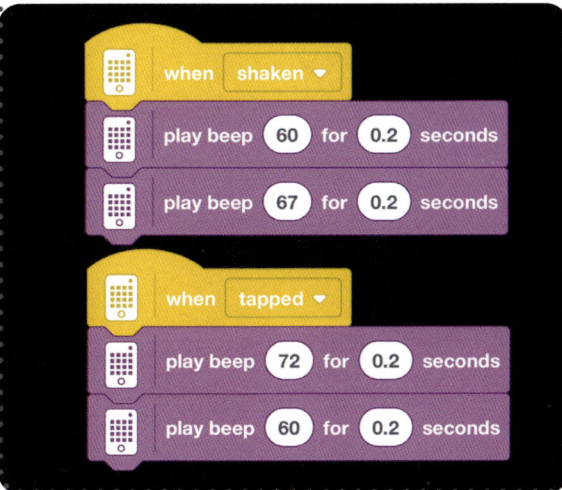

```
when shaken ▾
play beep 60 for 0.2 seconds
play beep 67 for 0.2 seconds

when tapped ▾
play beep 72 for 0.2 seconds
play beep 60 for 0.2 seconds
```

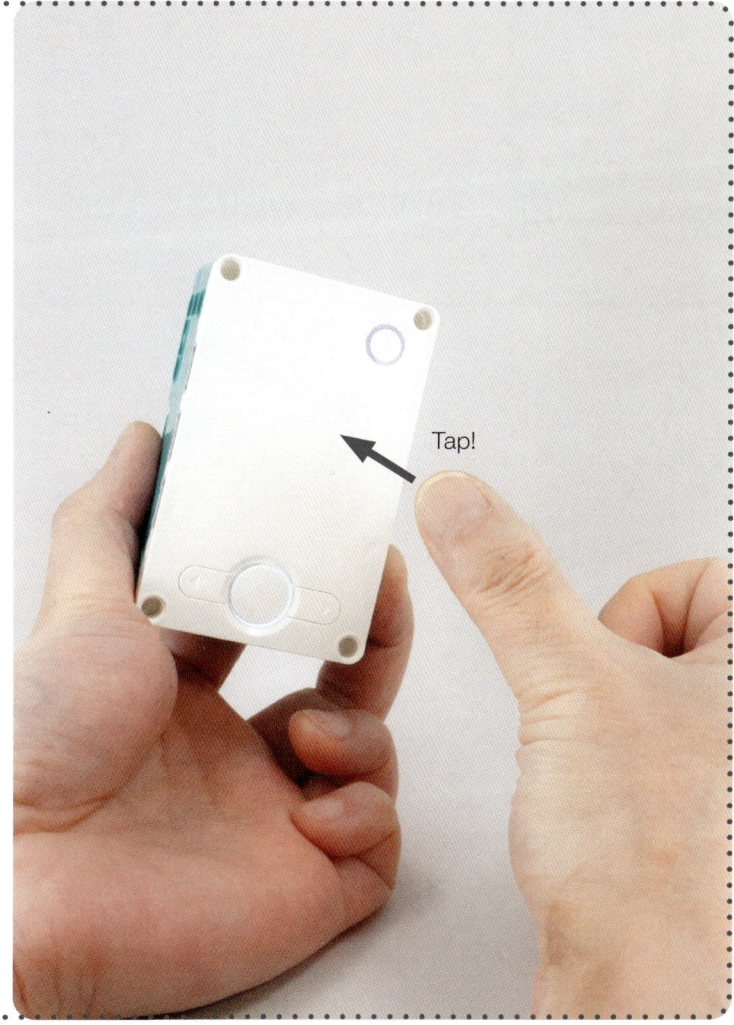

Tap!

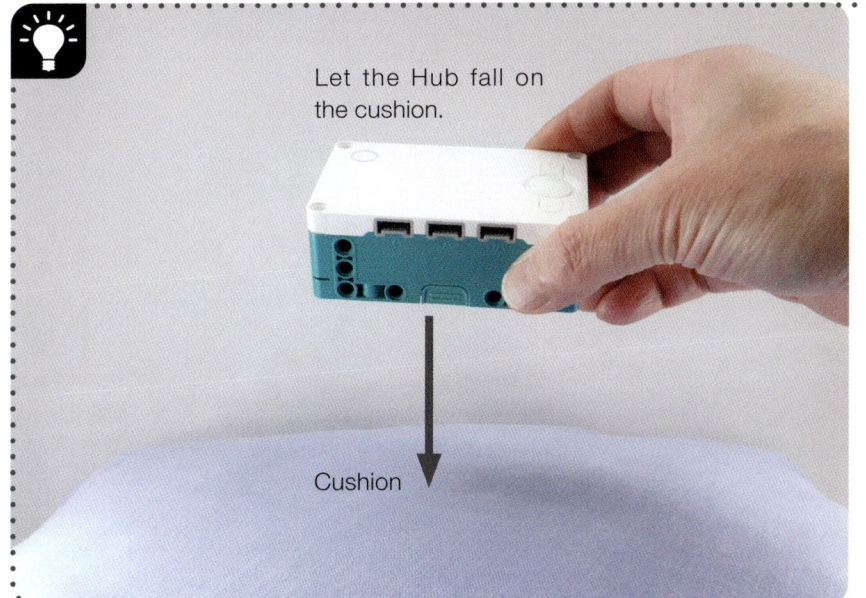

Let the Hub fall on the cushion.

Cushion

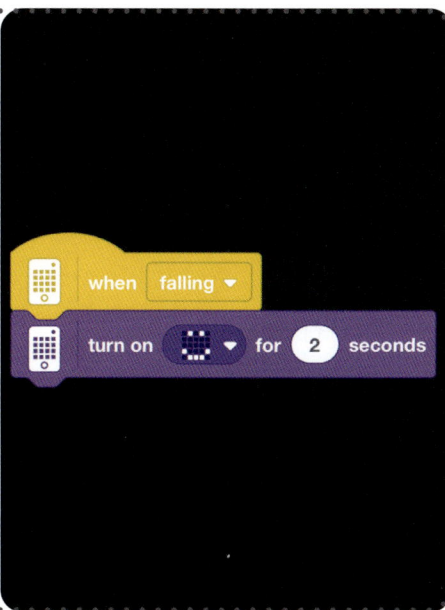

```
when falling ▾
turn on 🔲 ▾ for 2 seconds
```

Vehicles with sensors

#85

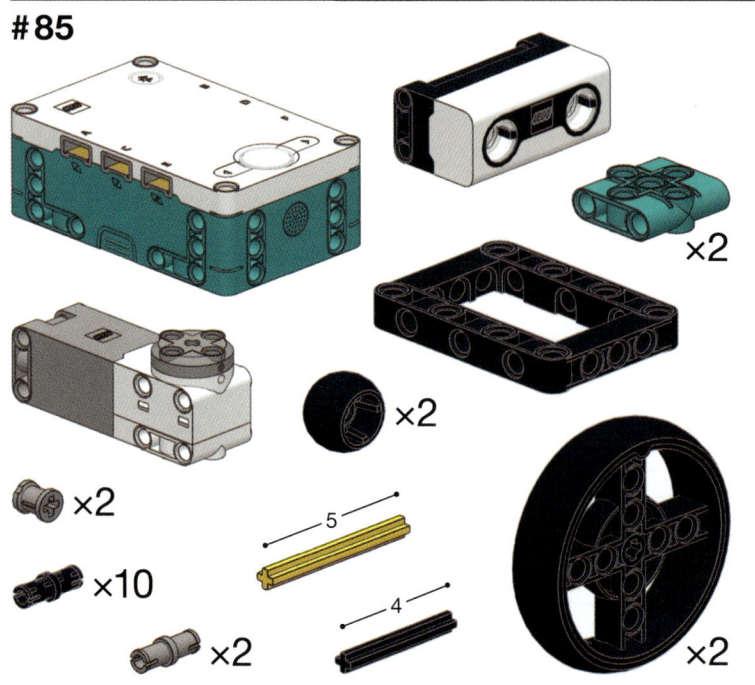

×2

5

×2

×10

4

×2

×2

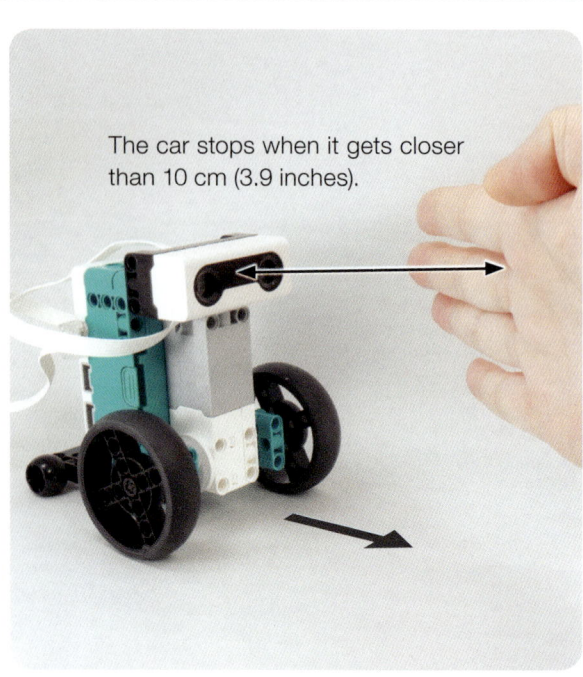

The car stops when it gets closer than 10 cm (3.9 inches).

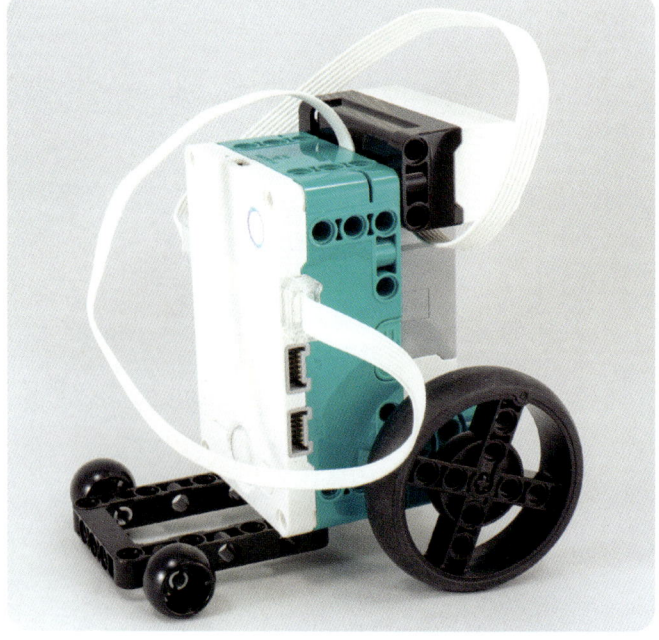

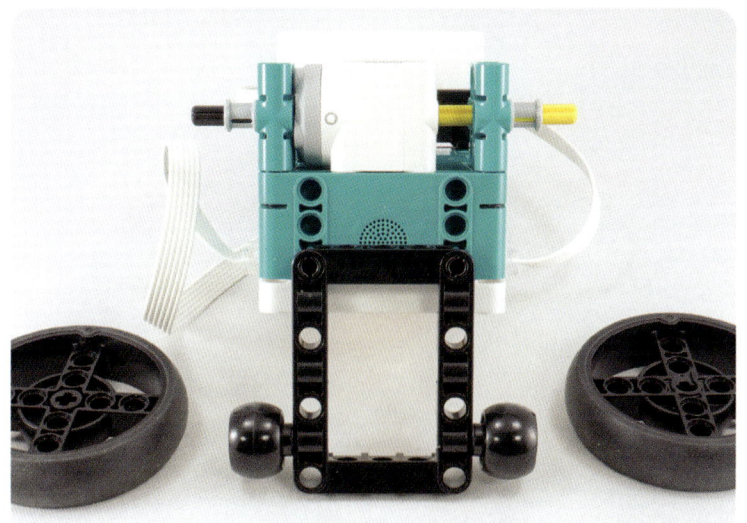

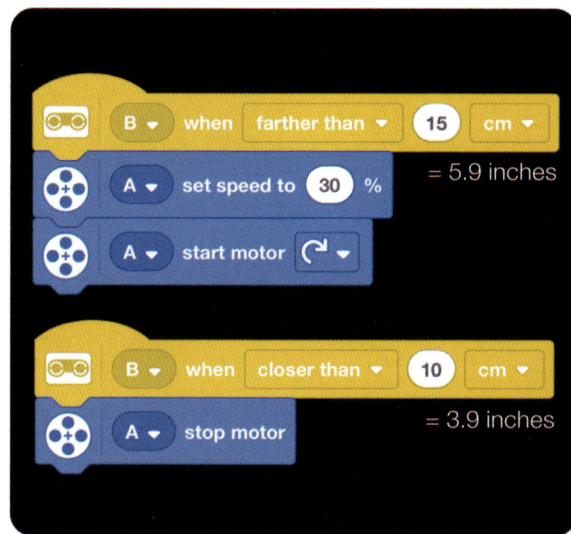

= 5.9 inches

= 3.9 inches

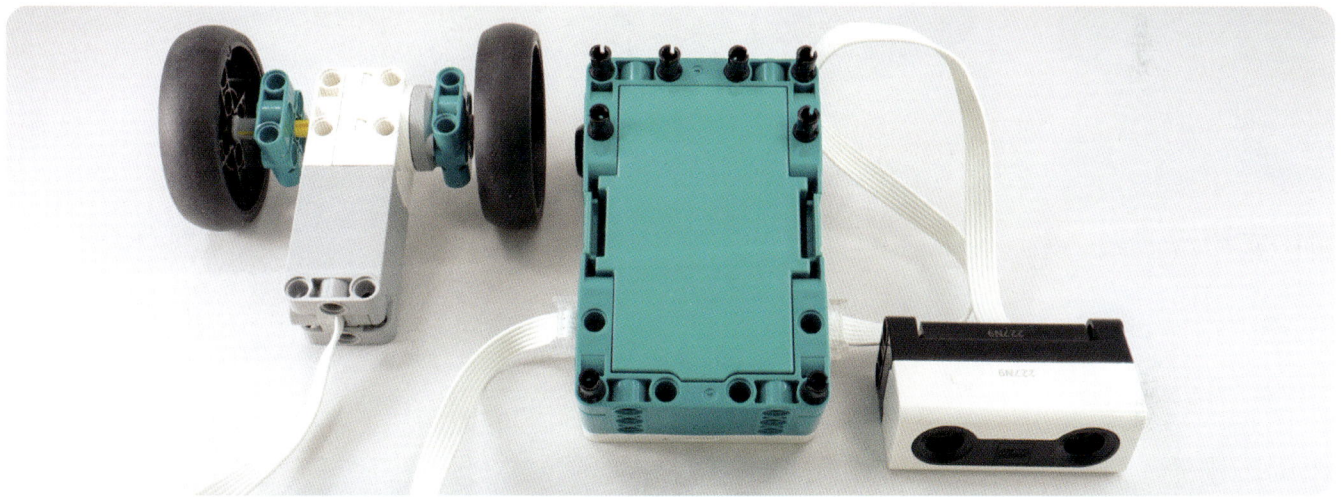

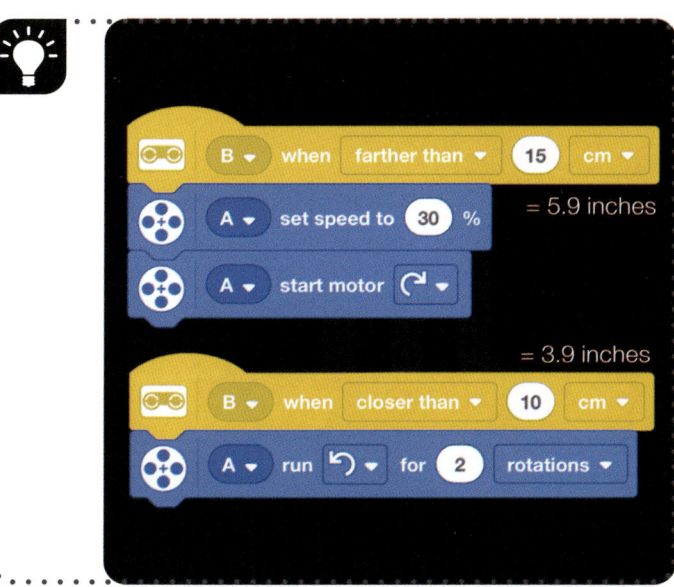

= 5.9 inches

= 3.9 inches

#86

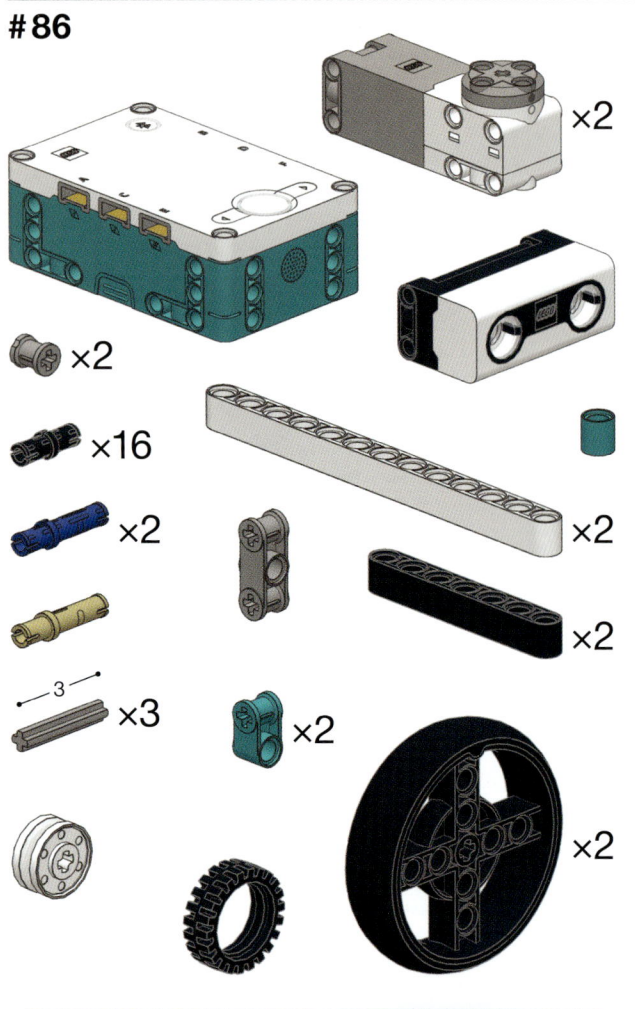

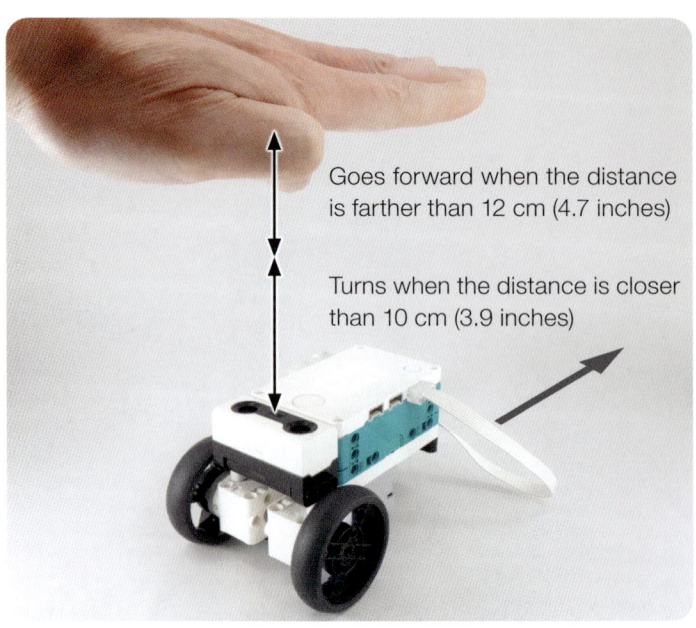

Goes forward when the distance is farther than 12 cm (4.7 inches)

Turns when the distance is closer than 10 cm (3.9 inches)

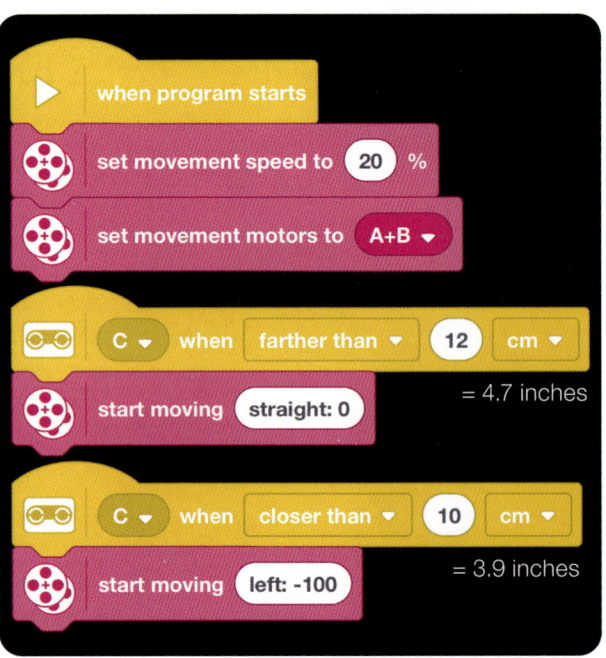

```
when program starts
set movement speed to 20 %
set movement motors to A+B ▾
C ▾ when farther than ▾ 12 cm ▾      = 4.7 inches
start moving straight: 0
C ▾ when closer than ▾ 10 cm ▾       = 3.9 inches
start moving left: -100
```

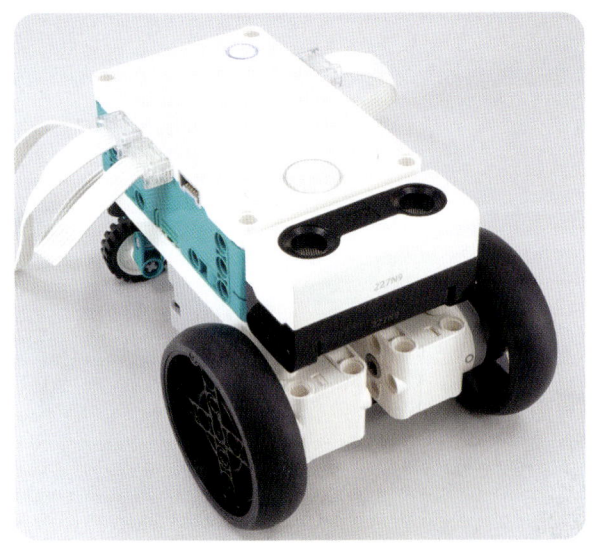

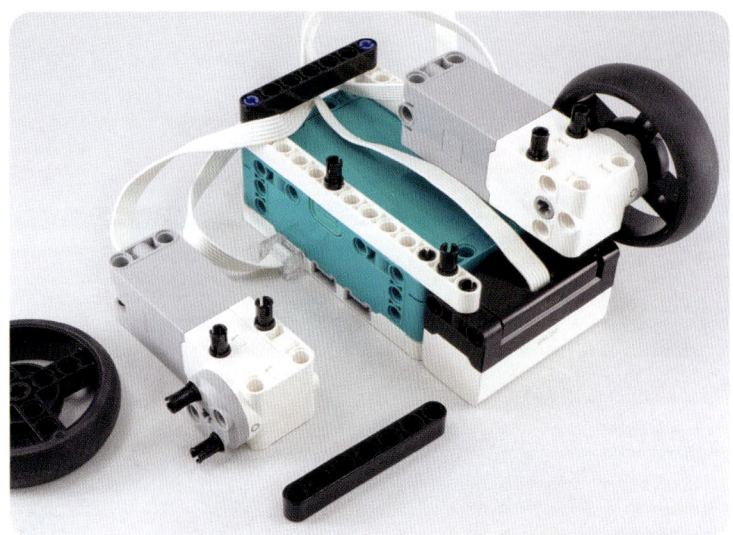

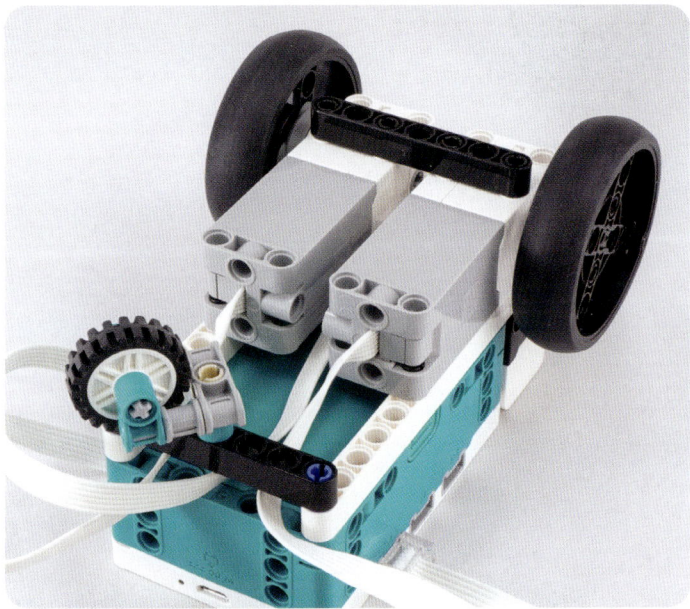

The car stops when the distance is farther than 20 cm (7.9 inches). It goes forward when the distance is 10–20 cm (3.9–7.9 inches). It turns when the distance is closer than 10 cm (3.9 inches).

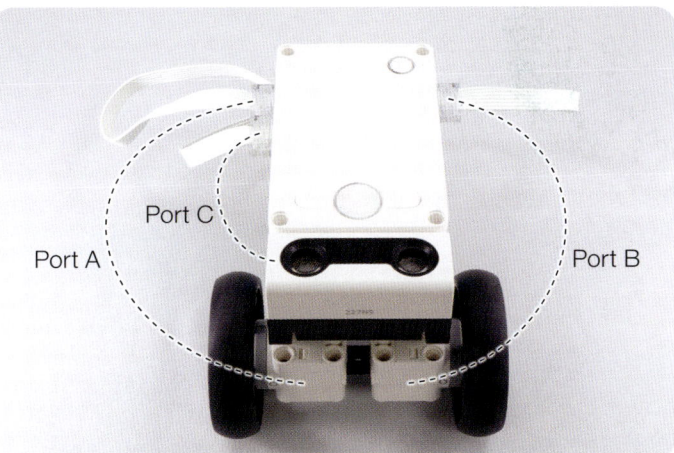

Port C

Port A

Port B

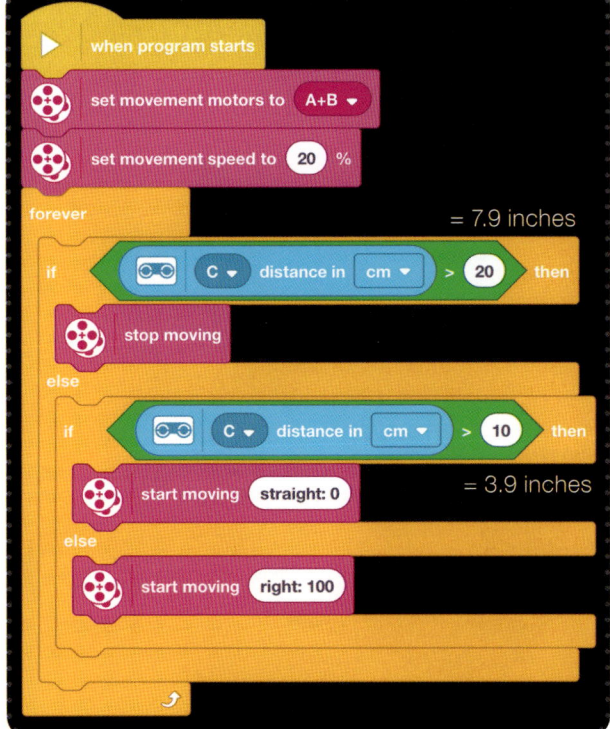

```
when program starts

set movement motors to  A+B ▾

set movement speed to  20 %

forever
                                                          = 7.9 inches
    if    [C ▾] distance in [cm ▾]  > 20    then
        stop moving
    else
        if    [C ▾] distance in [cm ▾]  > 10    then
                                                          = 3.9 inches
            start moving  straight: 0
        else
            start moving  right: 100
```

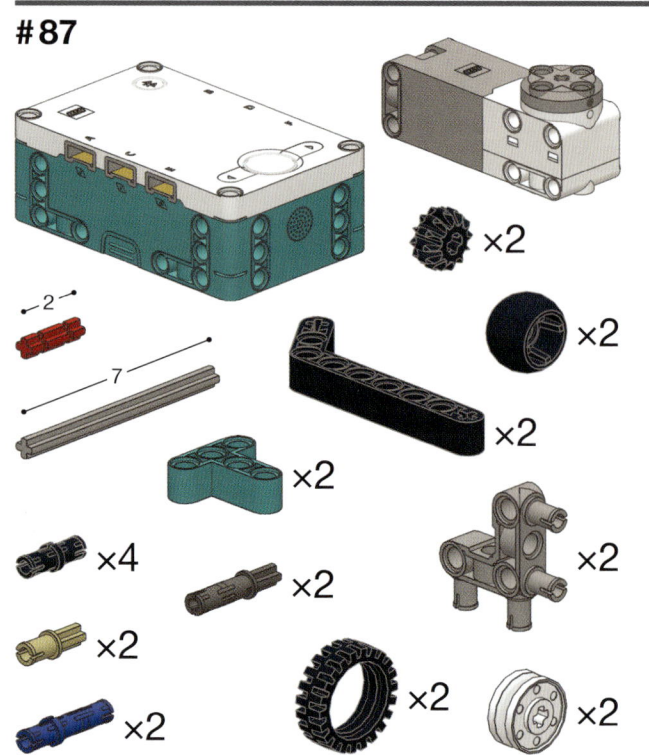

87

×2
×2
−2−
×2
7
×2
×4
×2
×2
×2
×2
×2
×2

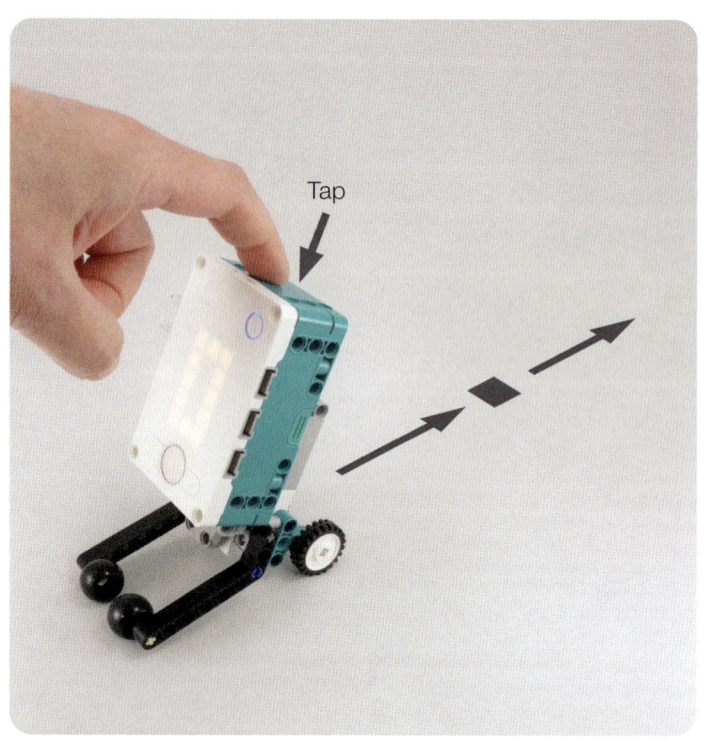

Tap

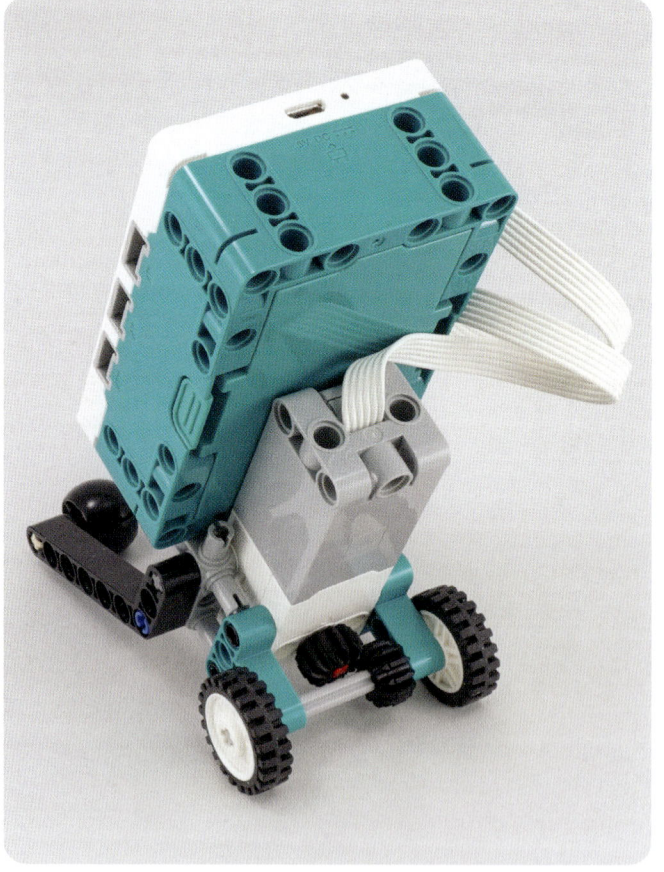

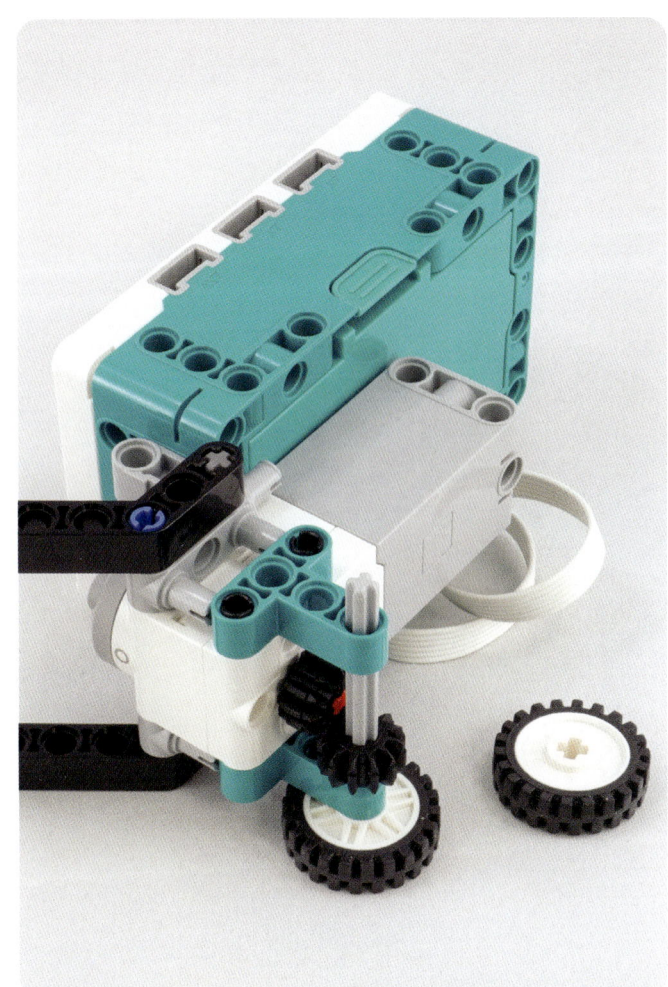

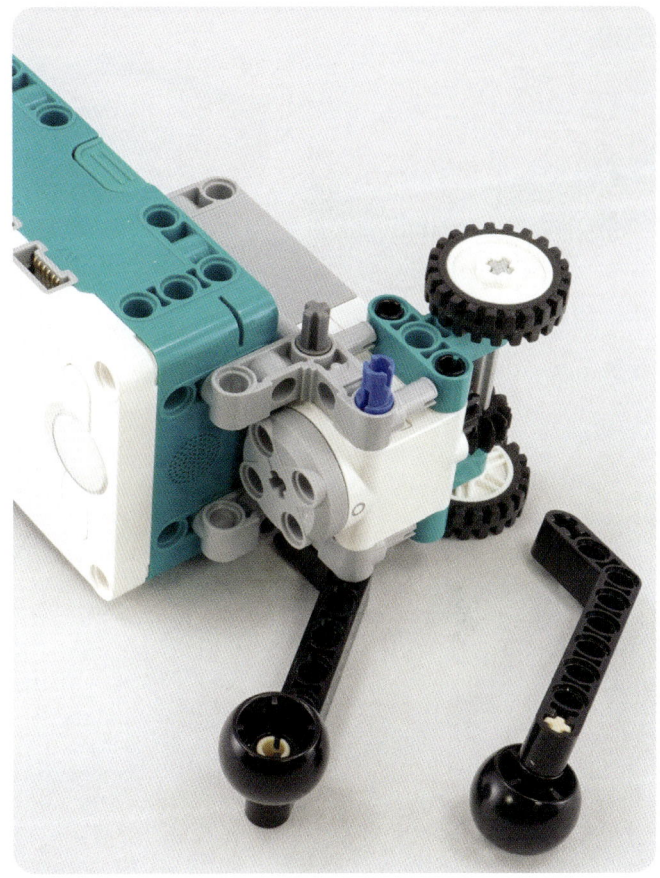

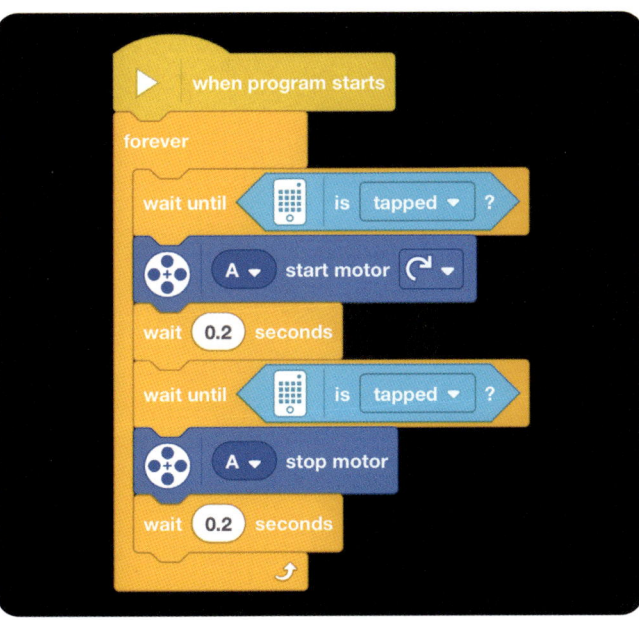

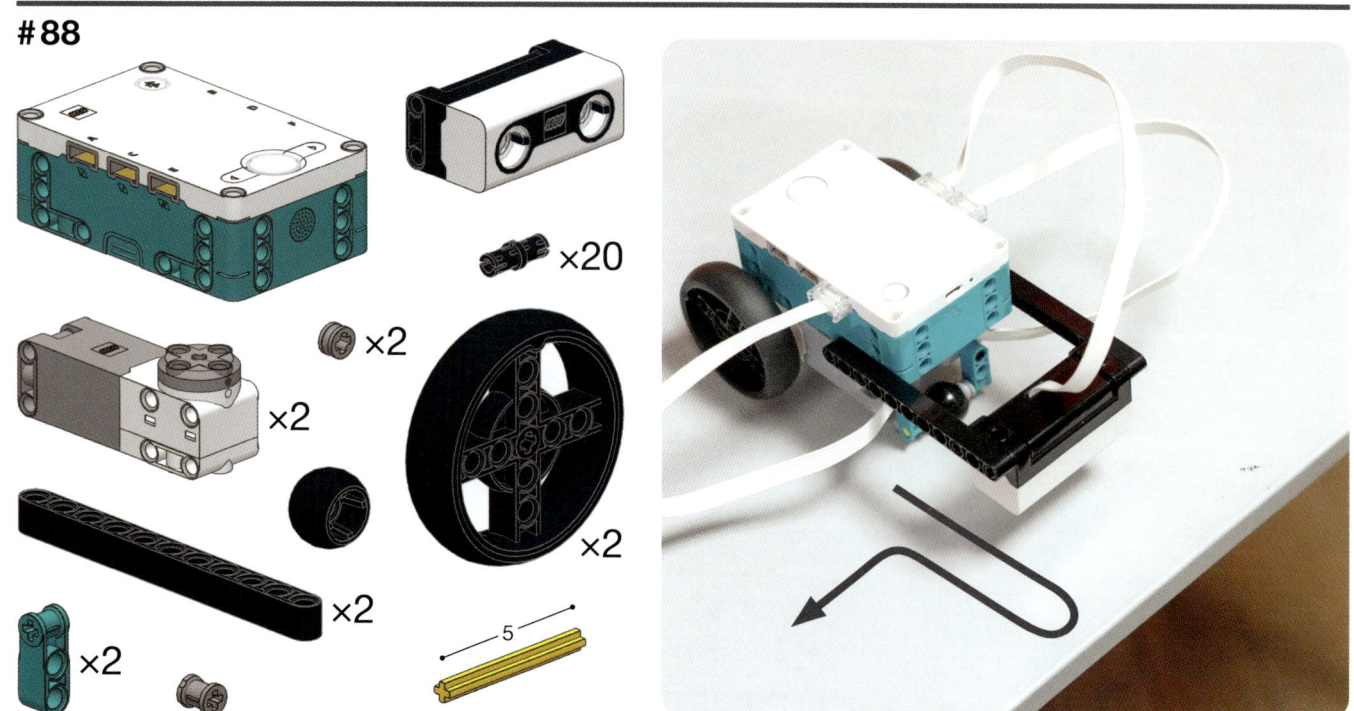

88

×2

×2

×2

×20

×2

×2

5

×2

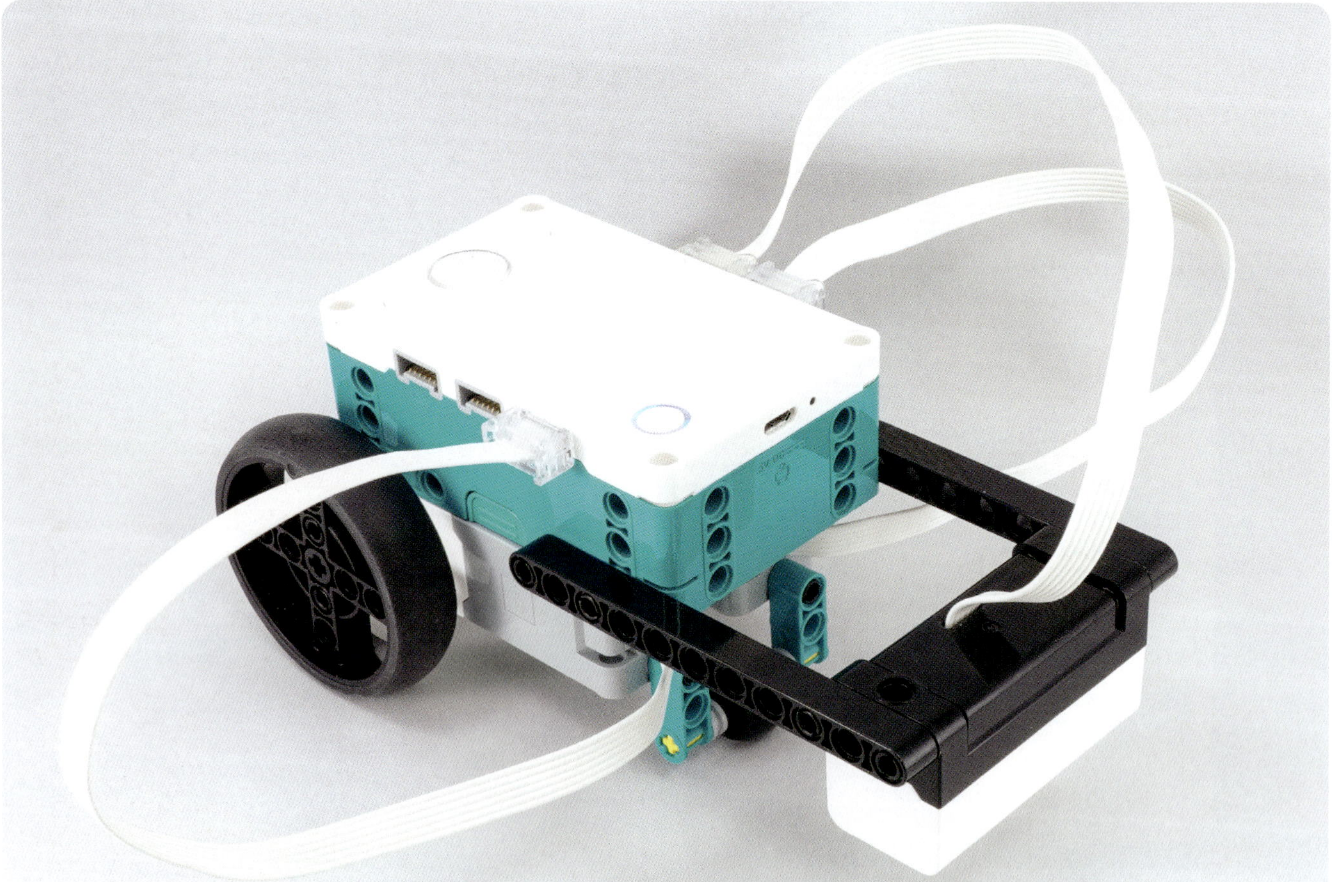

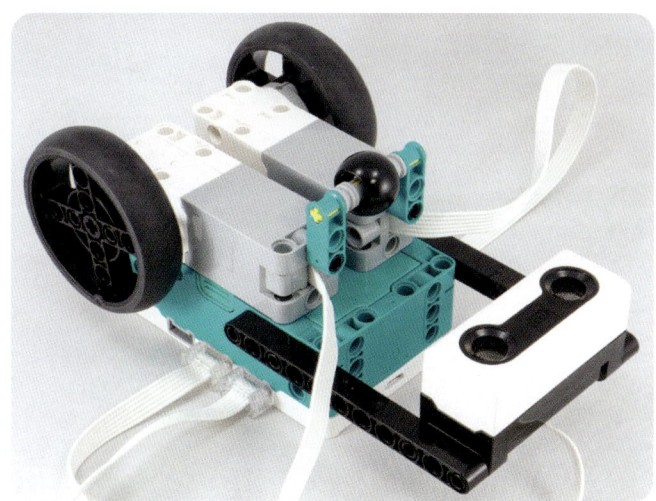

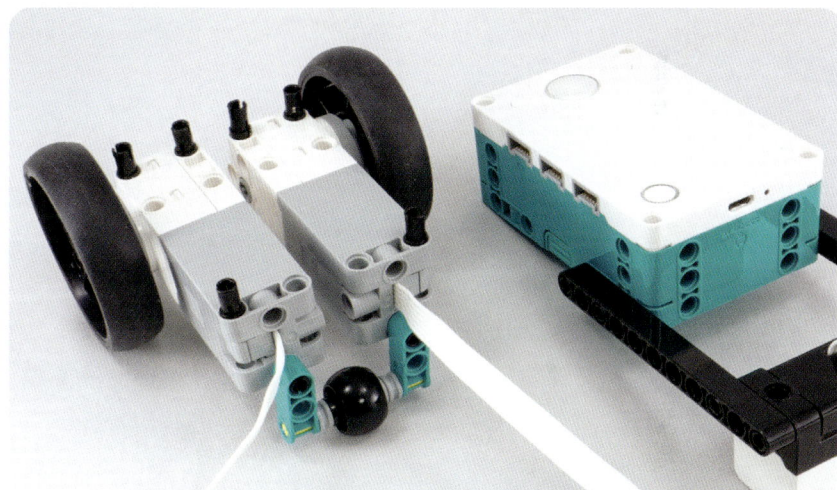

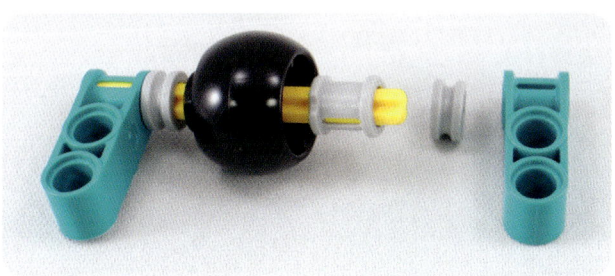

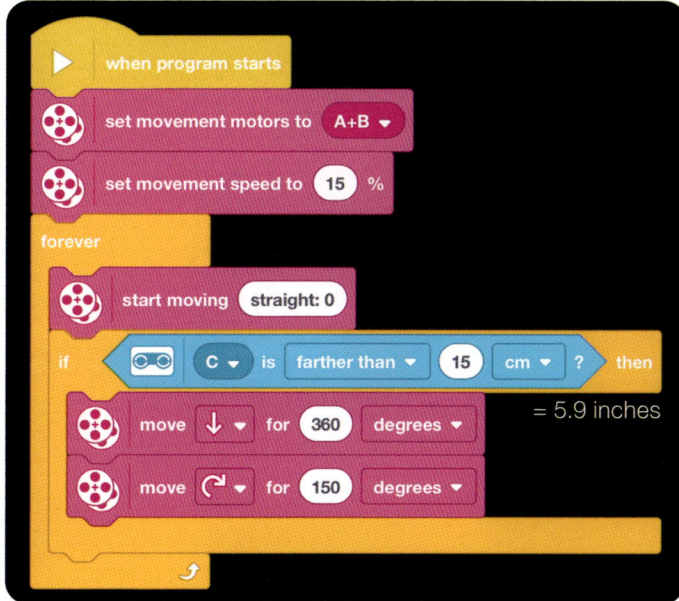

= 5.9 inches

#89

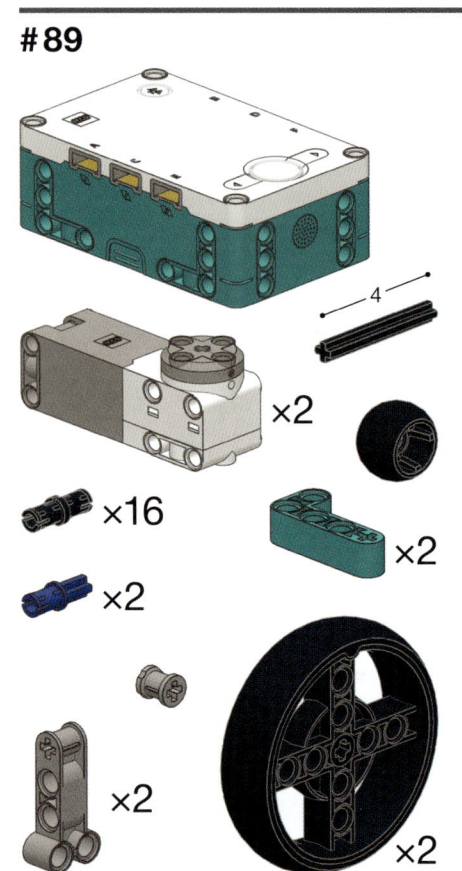

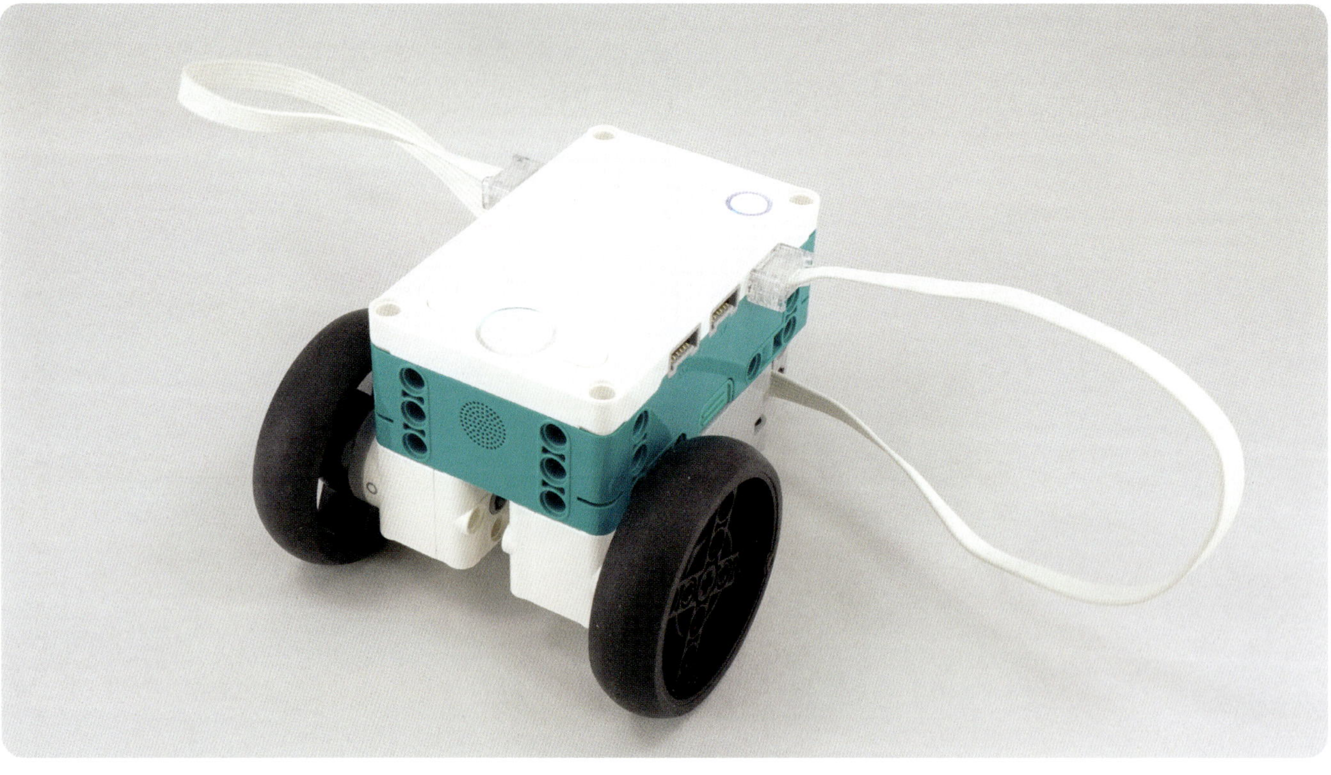

×2
×16
×2
×2
4
×2
×2

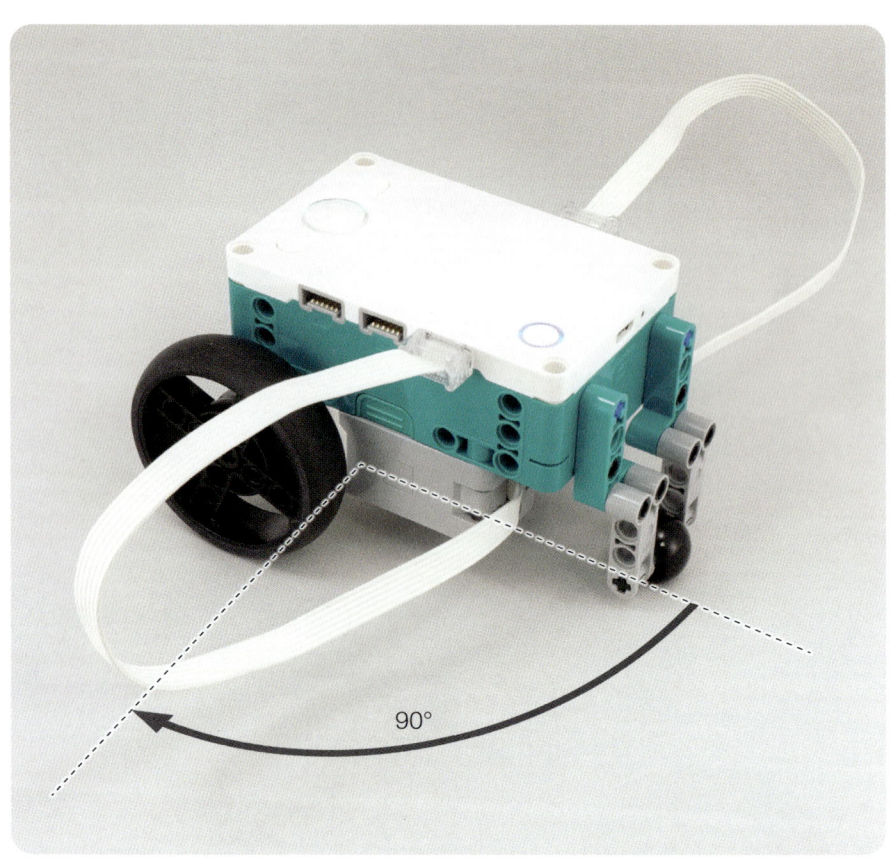

90°

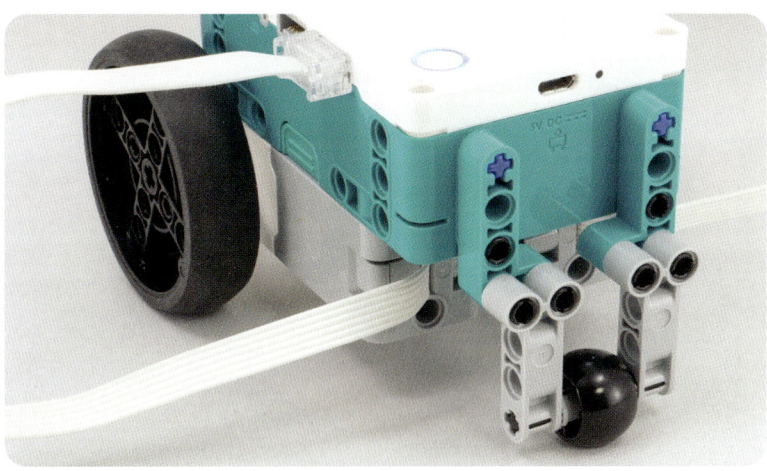

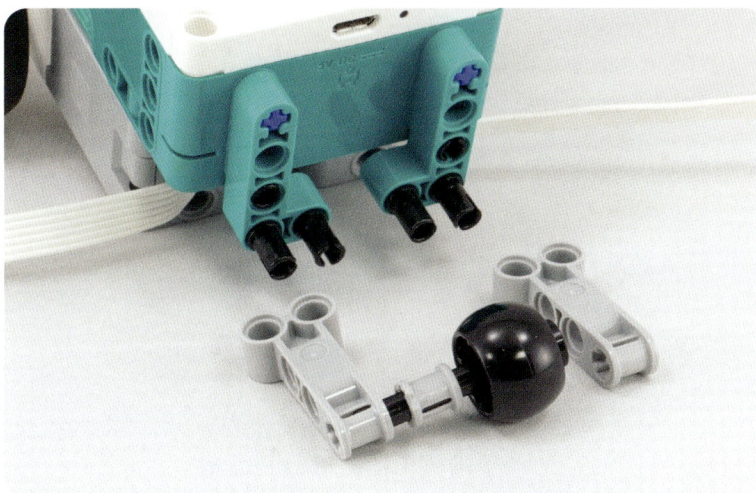

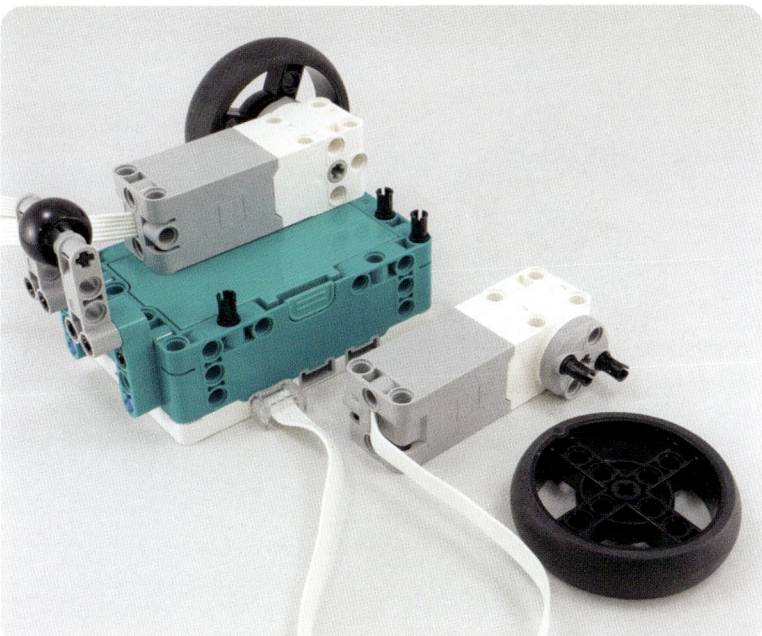

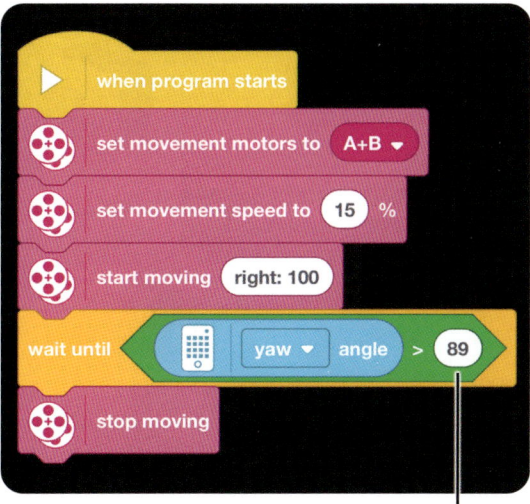

The car stops when the turning angle reaches 90 degrees or more.

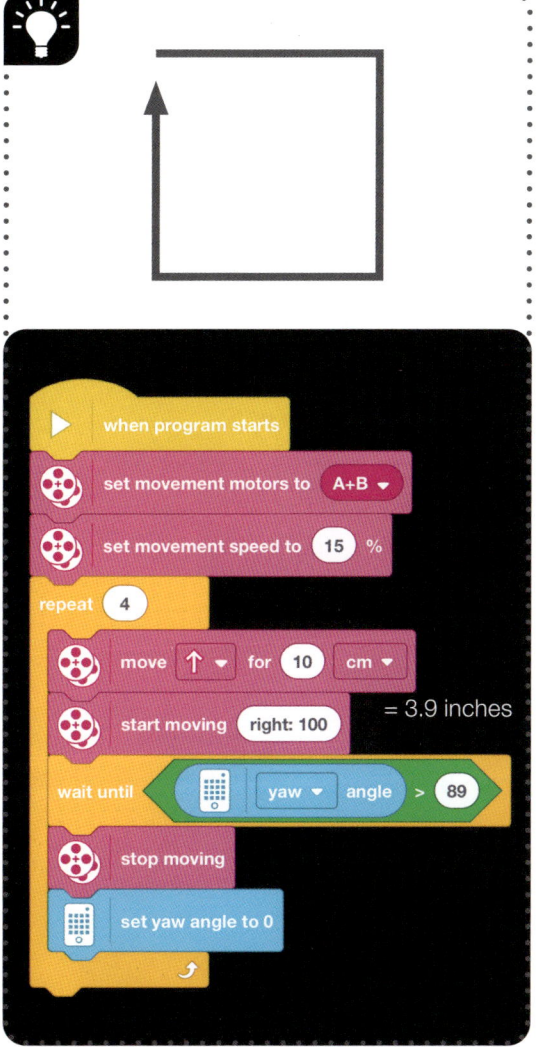

90

×2

×2

×2

×2 ×12 ×4 ×3 ×4 ×2

5 ×3

4

8 ×2

9

×2

×2 ×2

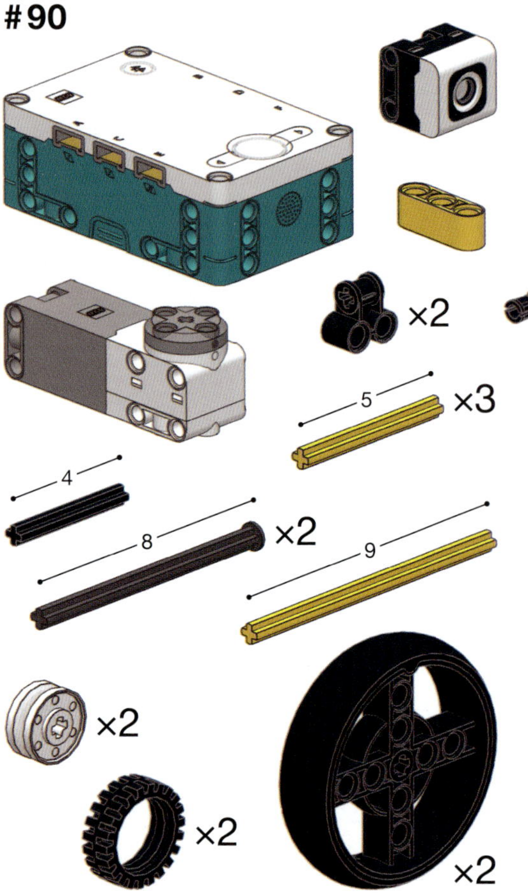

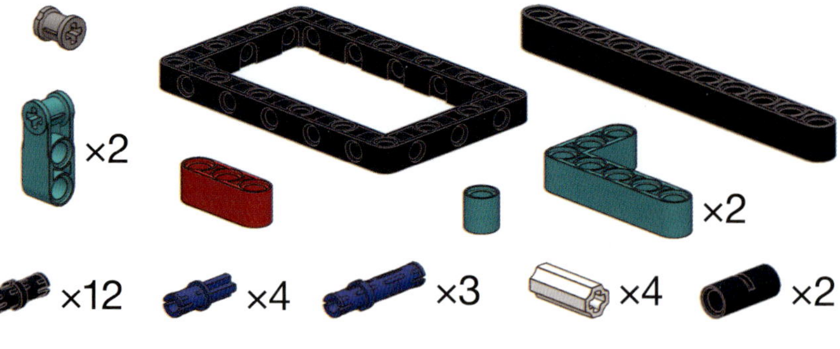

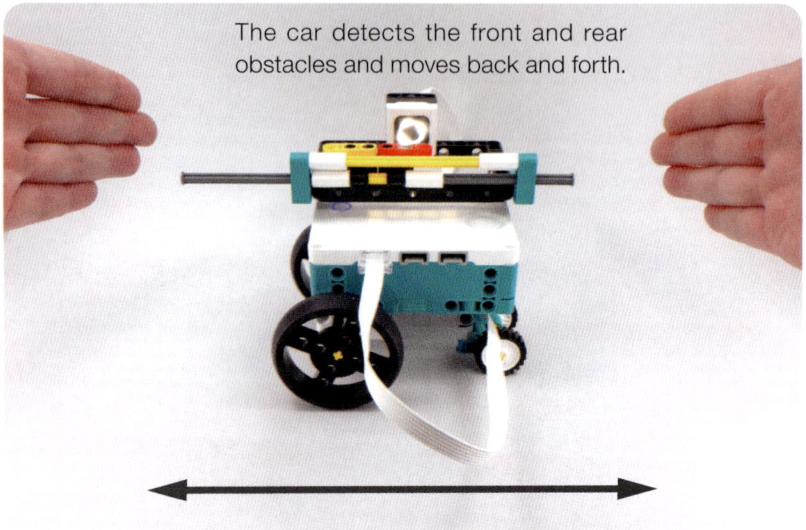

The car detects the front and rear obstacles and moves back and forth.

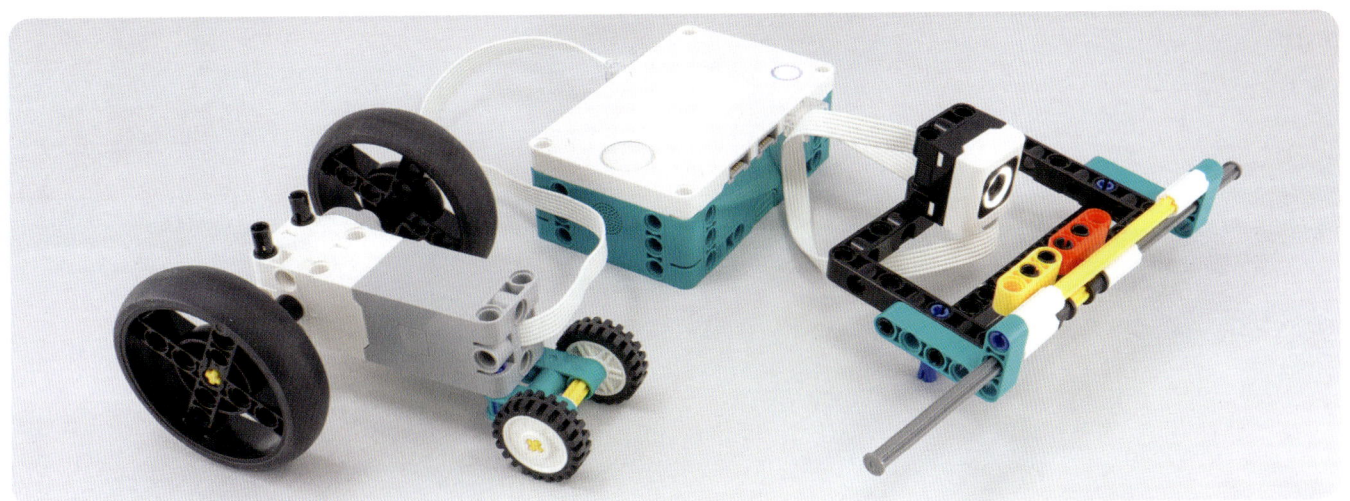

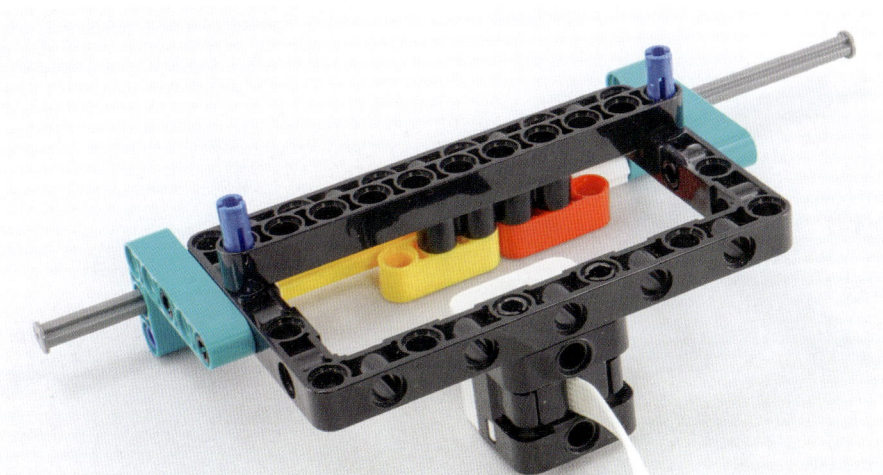

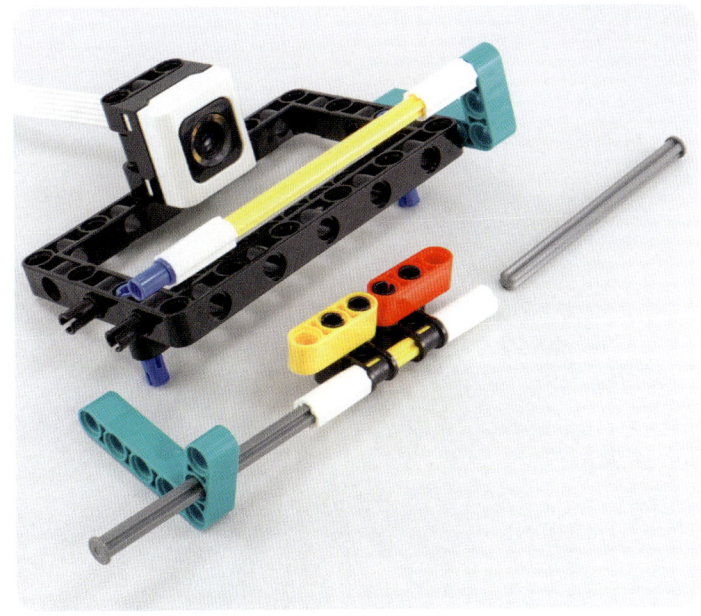

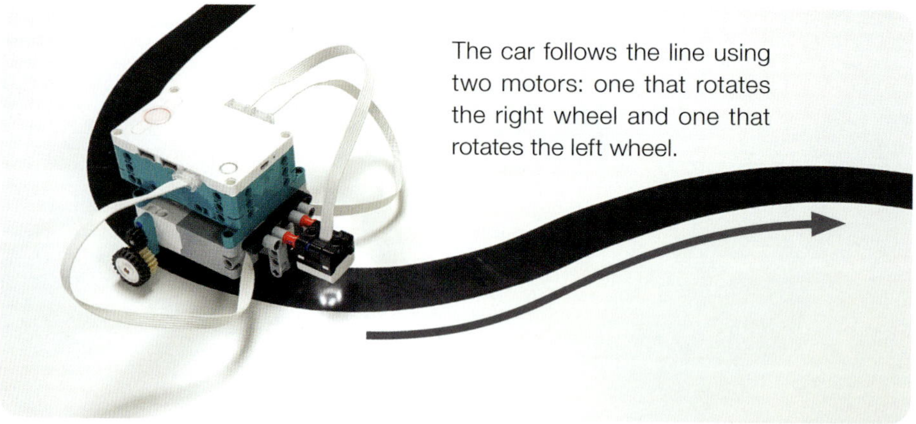

91

The car follows the line using two motors: one that rotates the right wheel and one that rotates the left wheel.

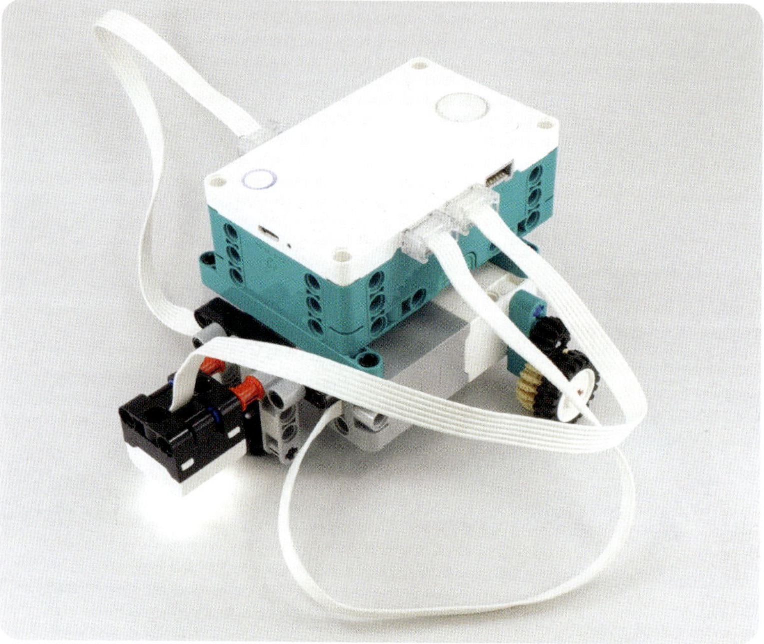

×2

×4

×2

×18

×2

×2

3 — ×2

3 — ×2

4 — ×2

×2

×2

×2

×2

×2

×2

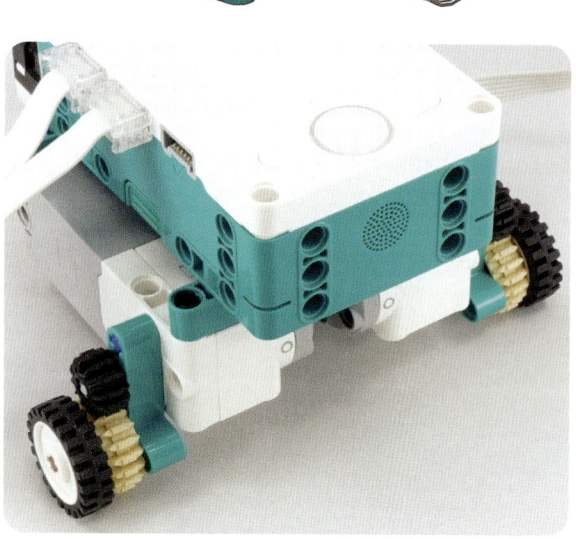

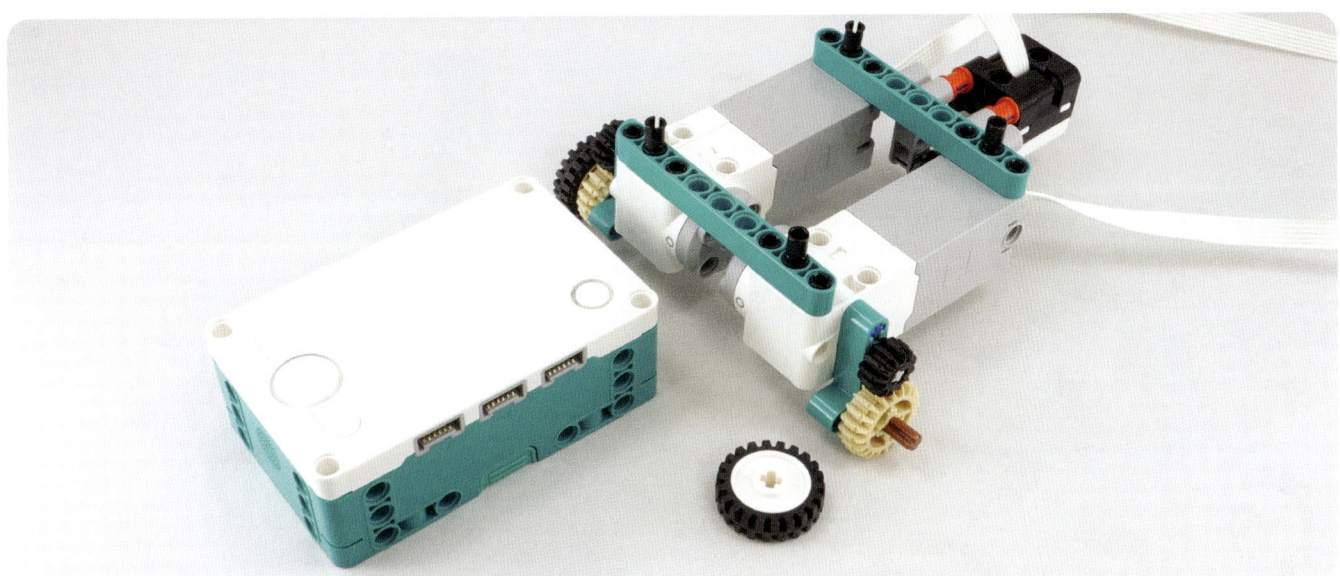

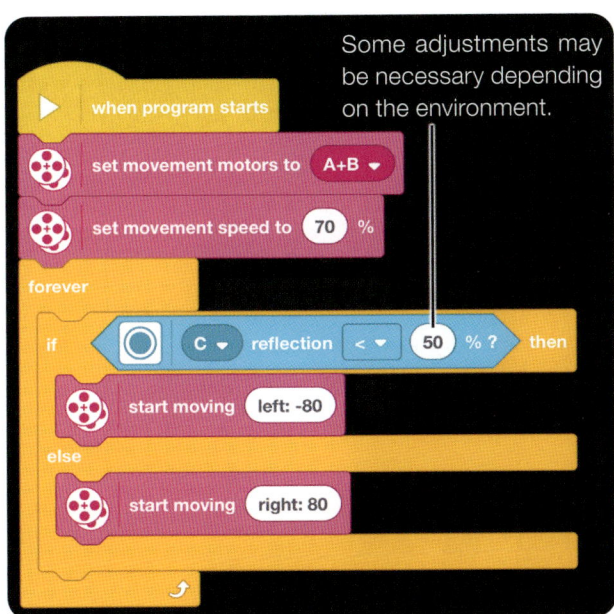

Some adjustments may be necessary depending on the environment.

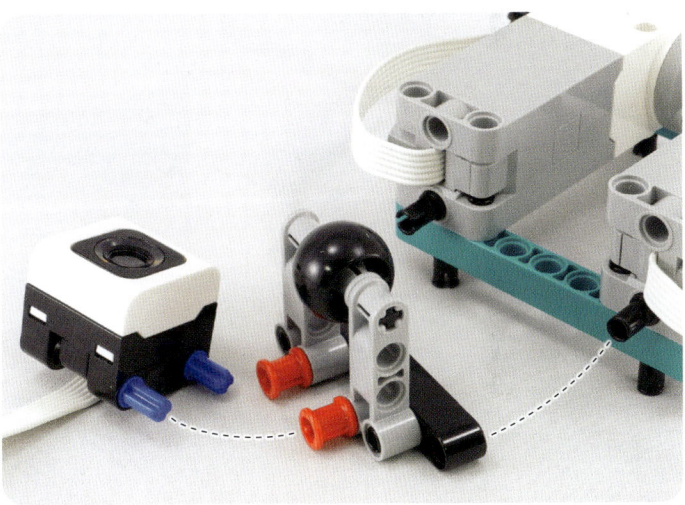

This program uses the difference between the threshold value and the sensor value as the movement direction. The car moves slightly more smoothly.

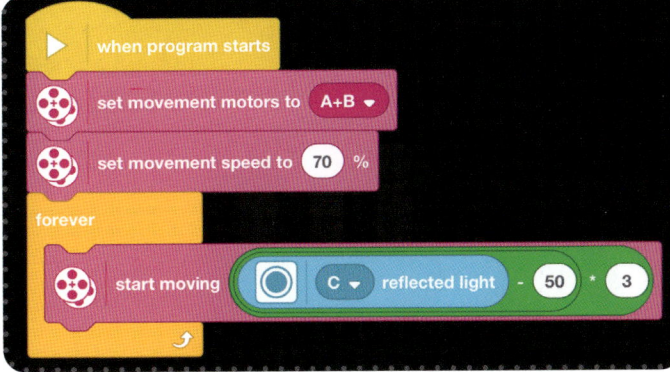

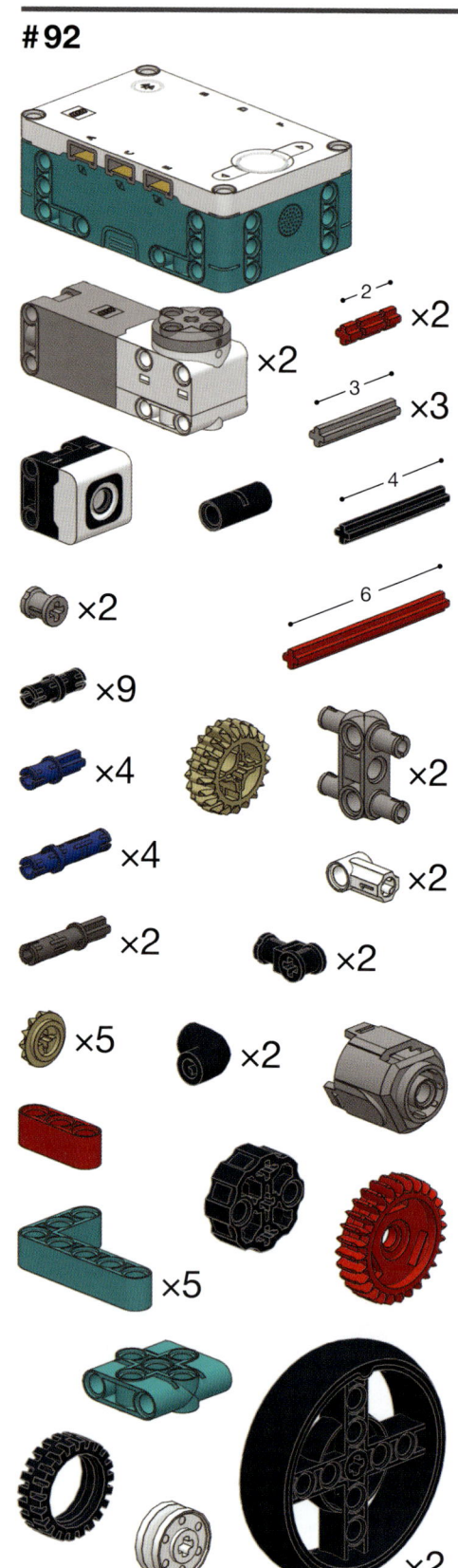

92

×2

→ 2 ←
×2

3
×3

4

6

×2

×9

×4

×2

×4

×2

×2

×2

×5

×2

×5

×2

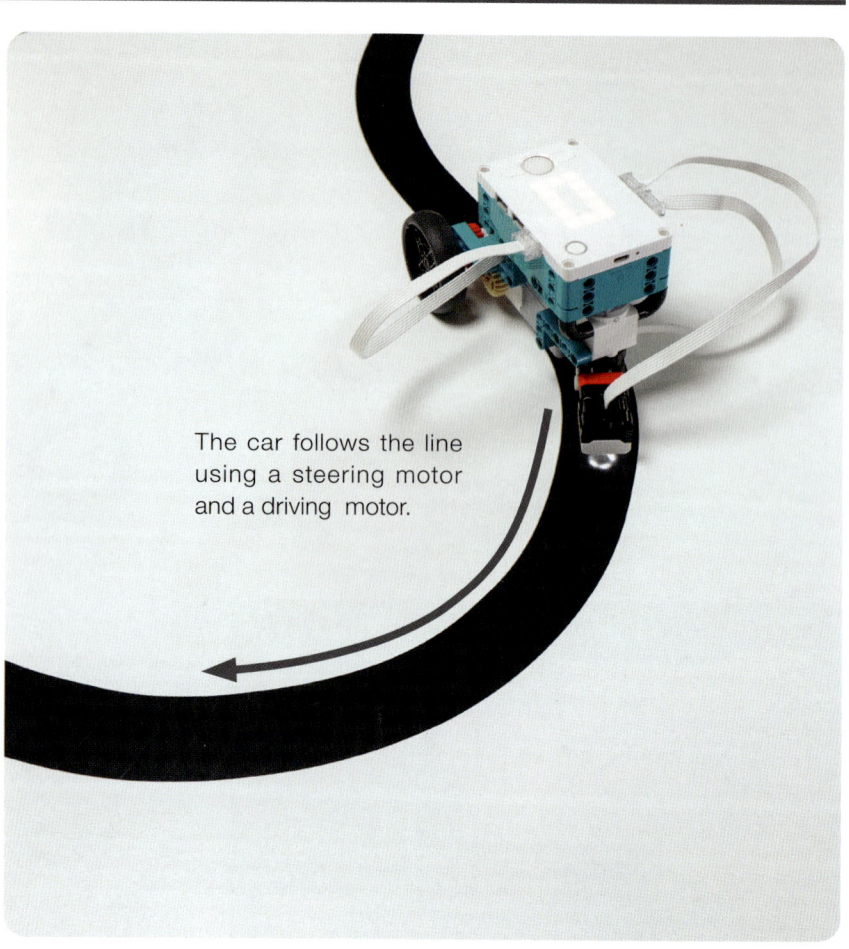

The car follows the line
using a steering motor
and a driving motor.

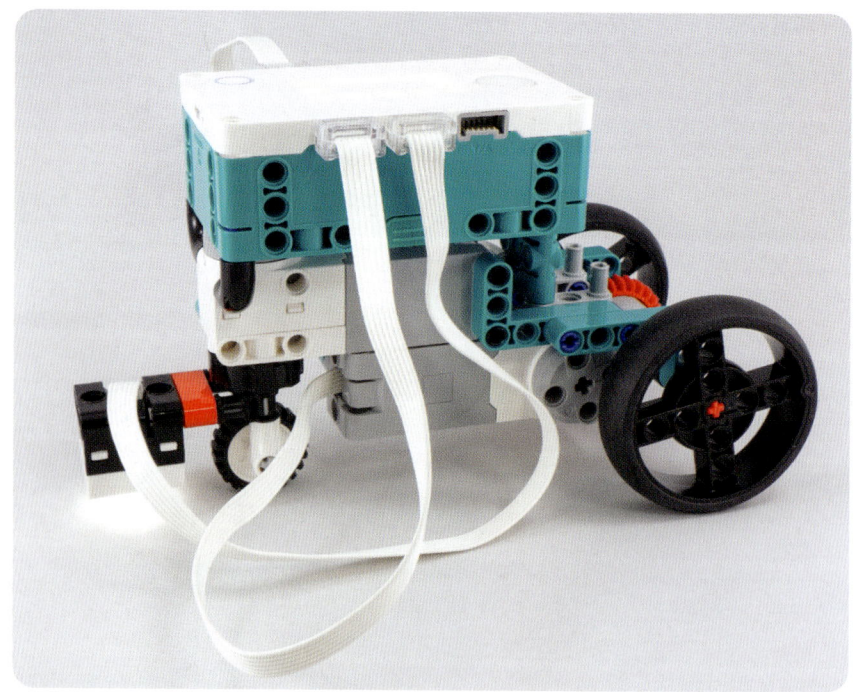

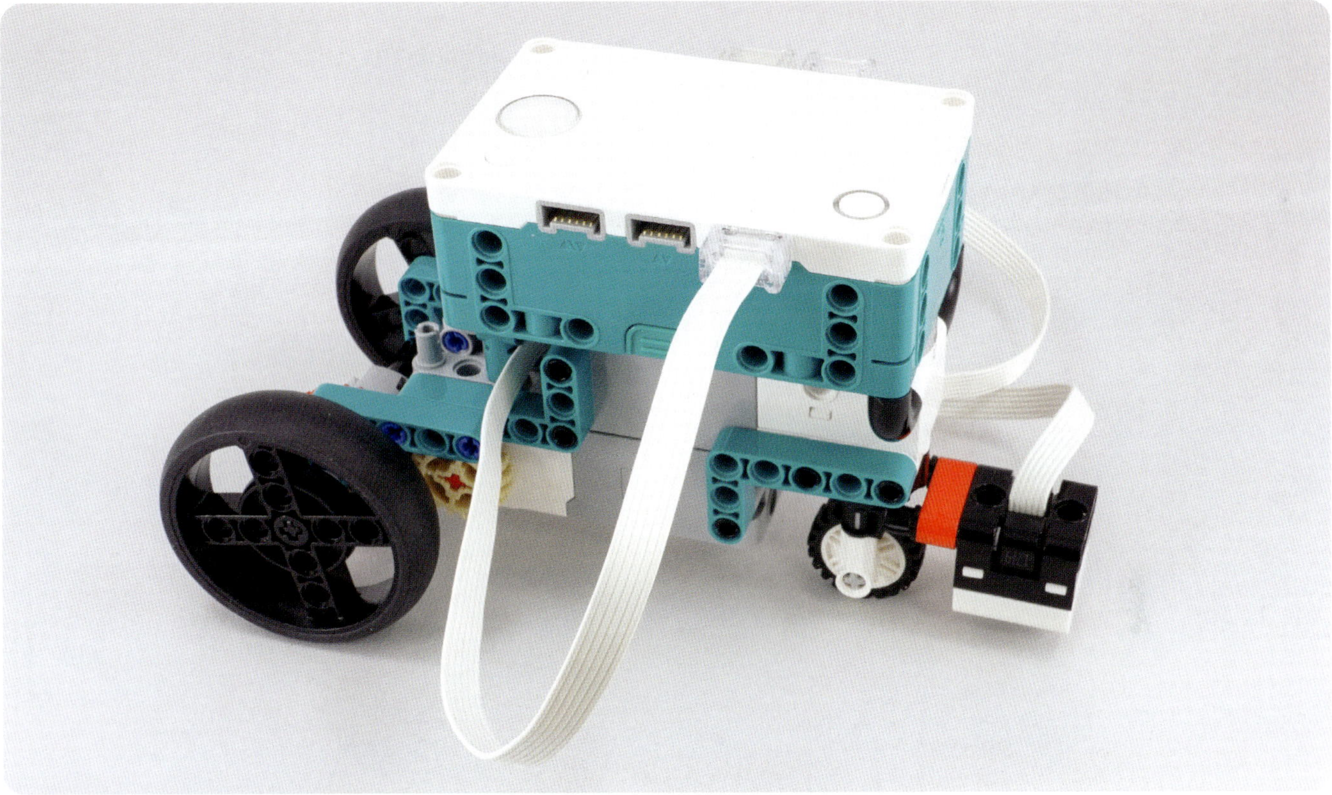

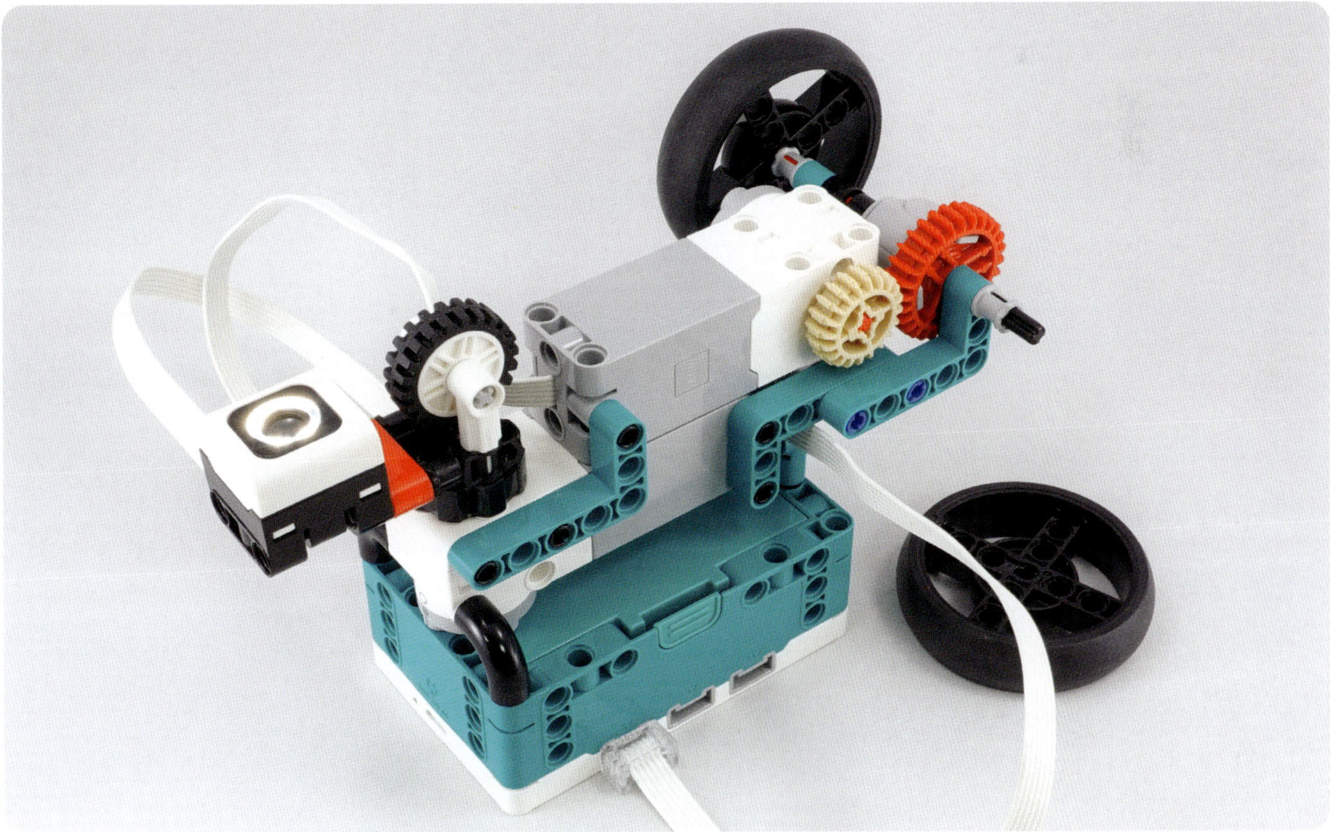

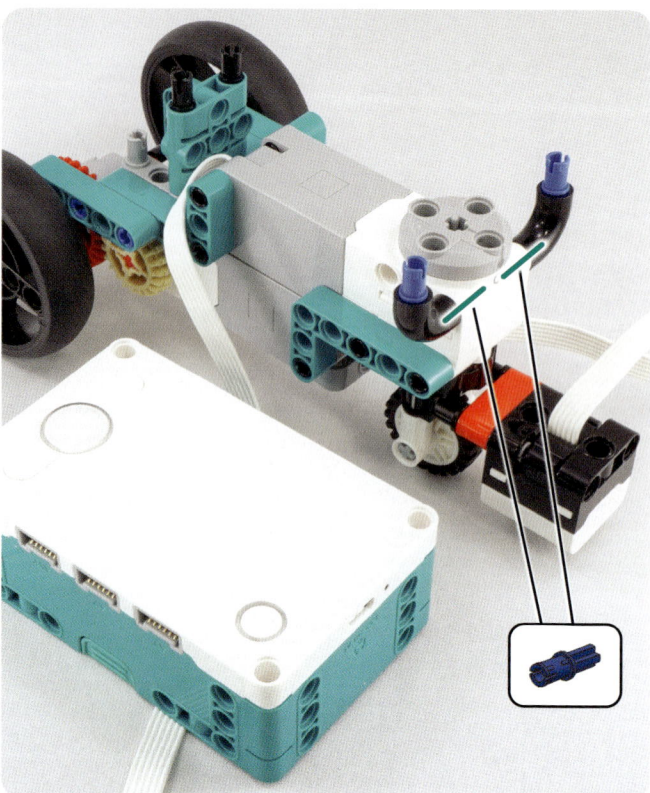

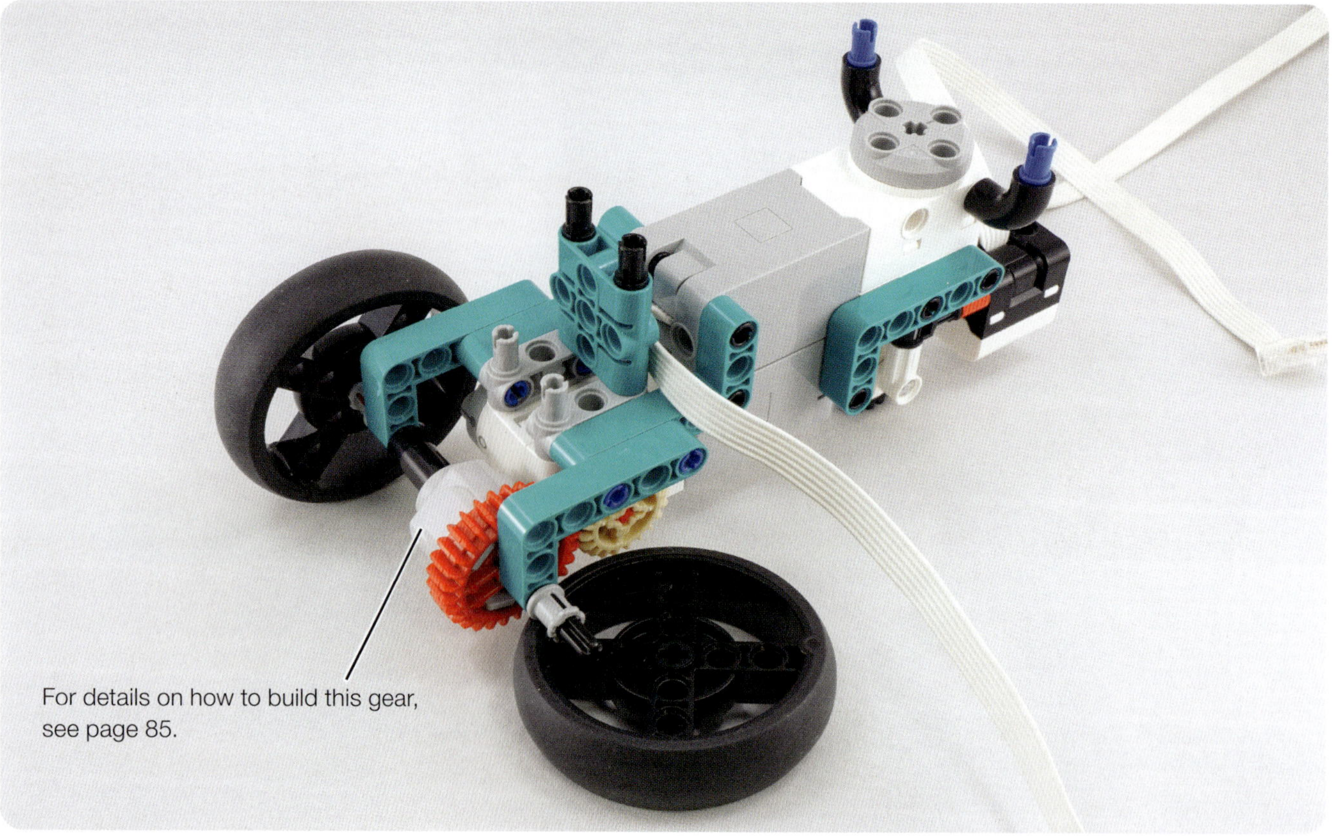

For details on how to build this gear, see page 85.

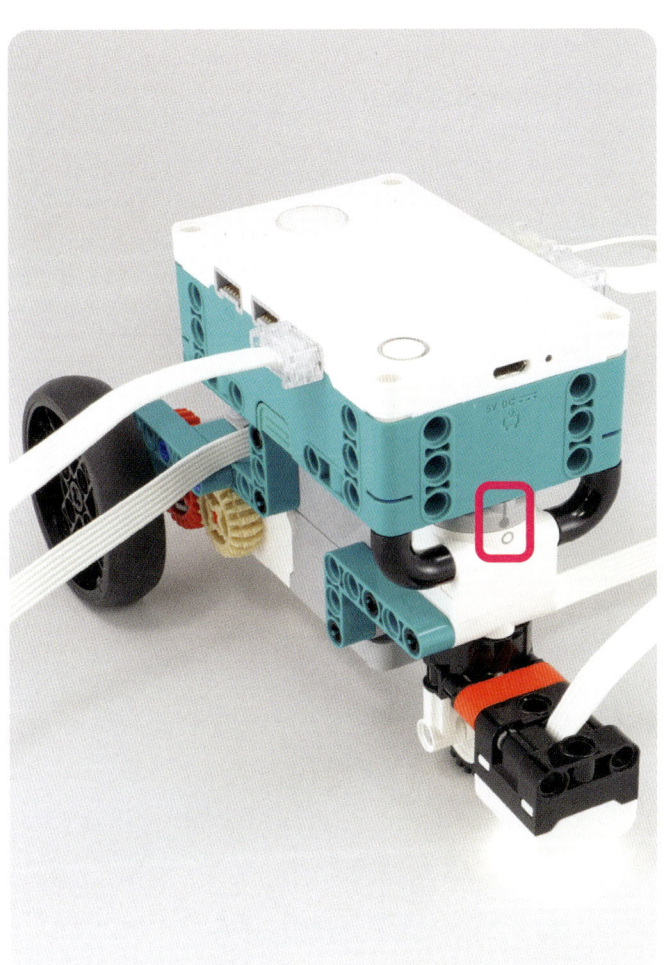

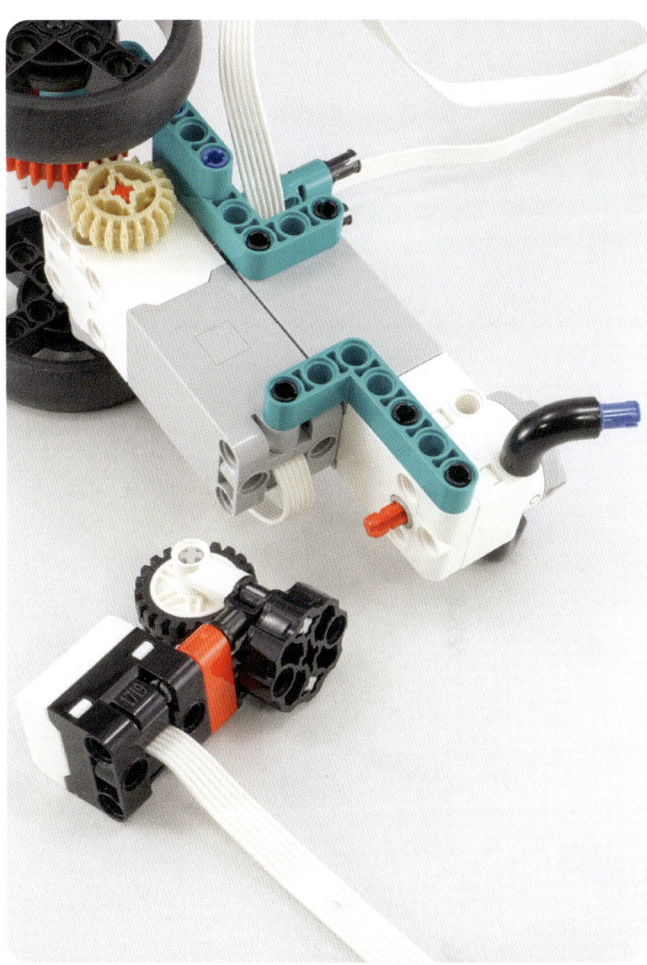

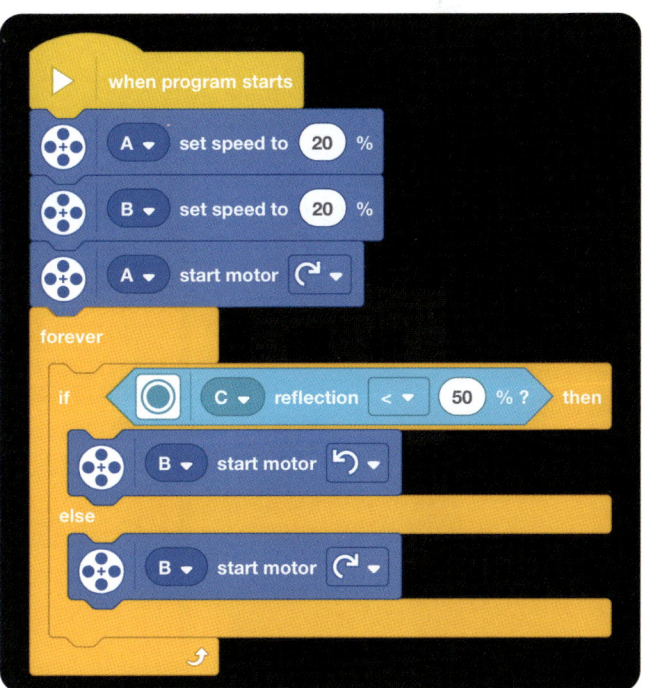

More ways to use sensors

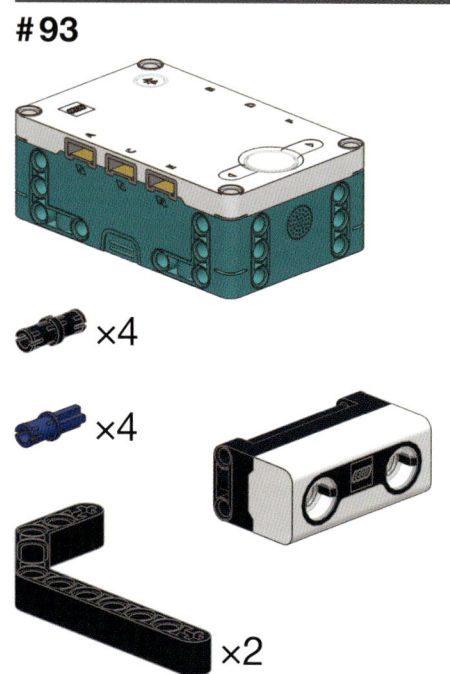

×4

×4

×2

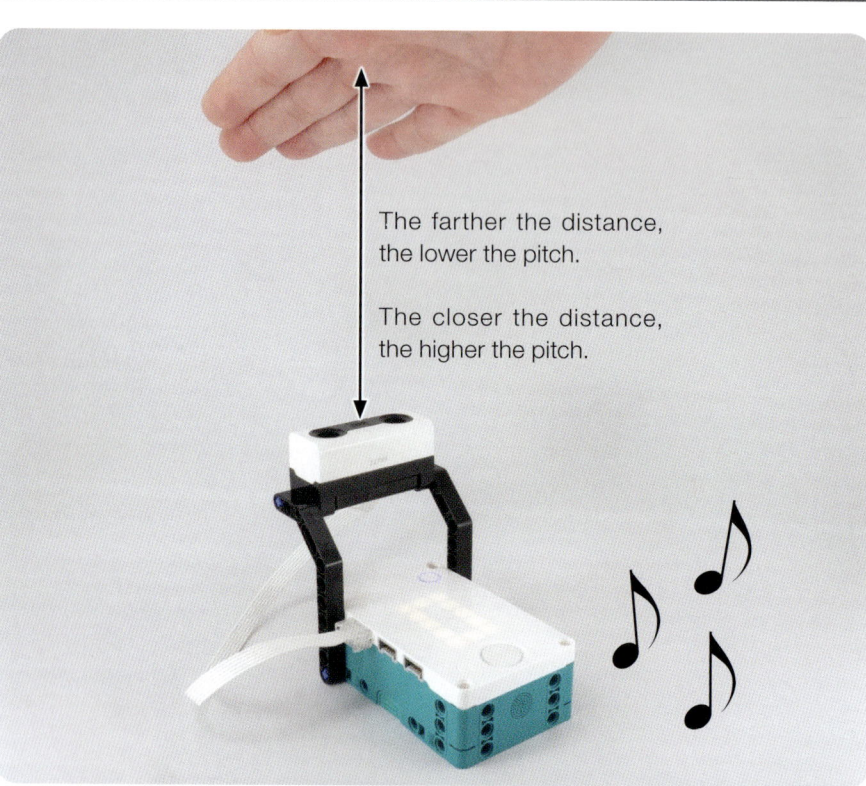

The farther the distance,
the lower the pitch.

The closer the distance,
the higher the pitch.

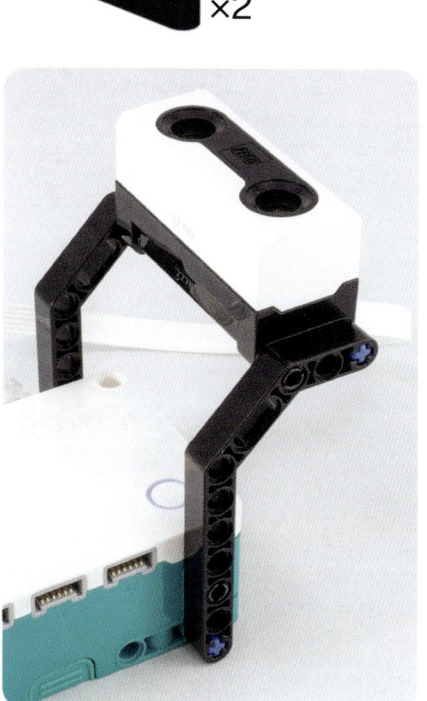

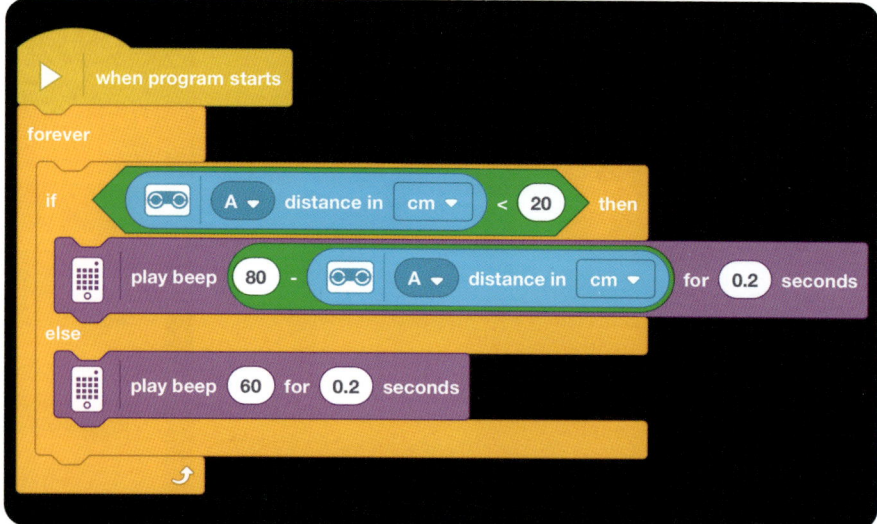

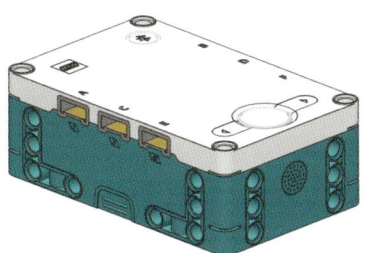

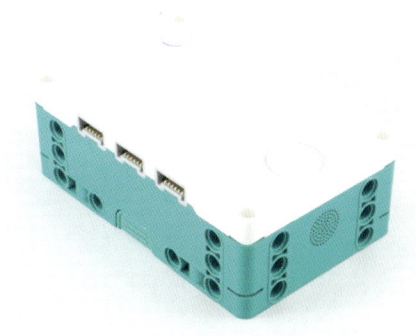

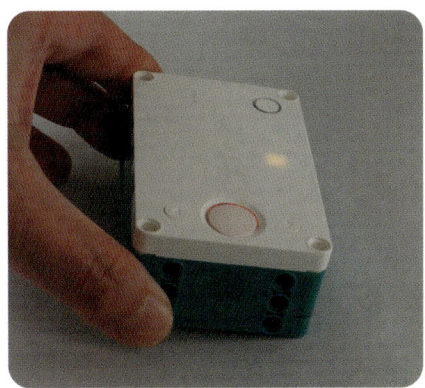

Examine the tilt from side to side and front to back.

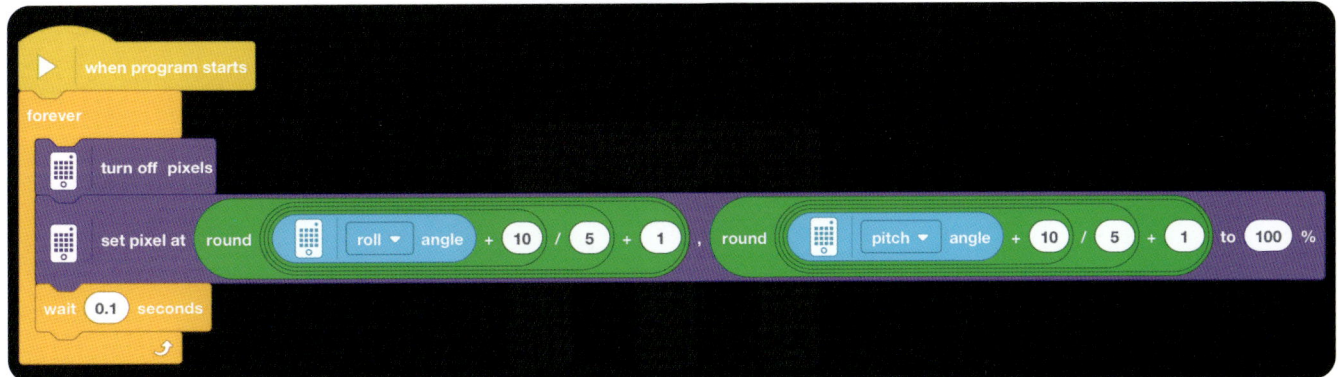

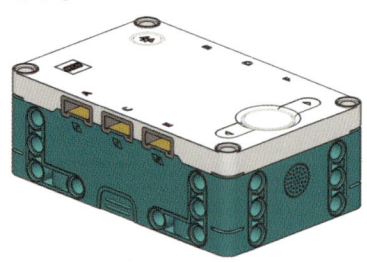

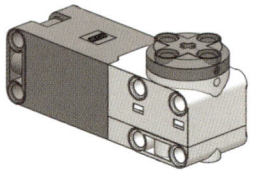

 ×2

 ×26

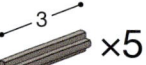 ×2

×3 ×5

 ×2

 ×4

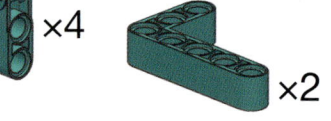 ×2

 ×2

×2

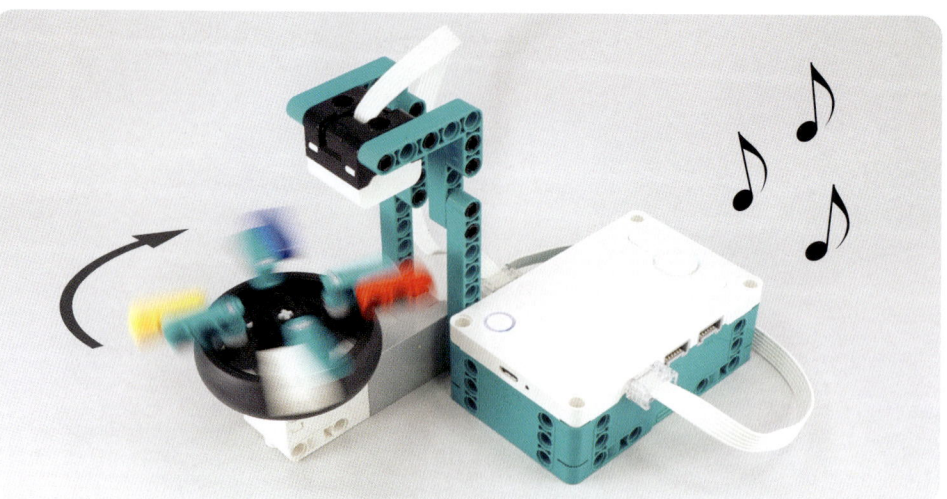

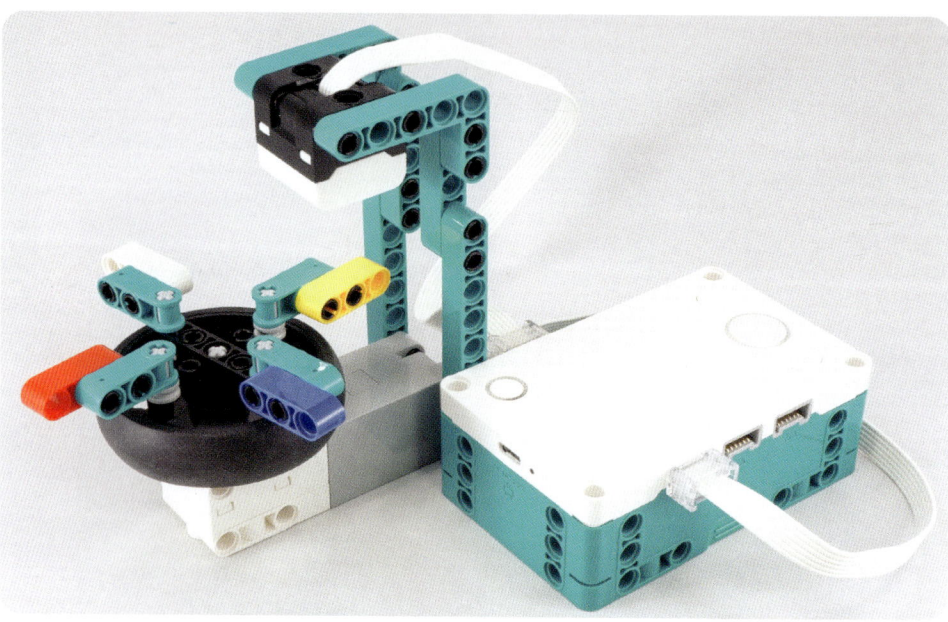

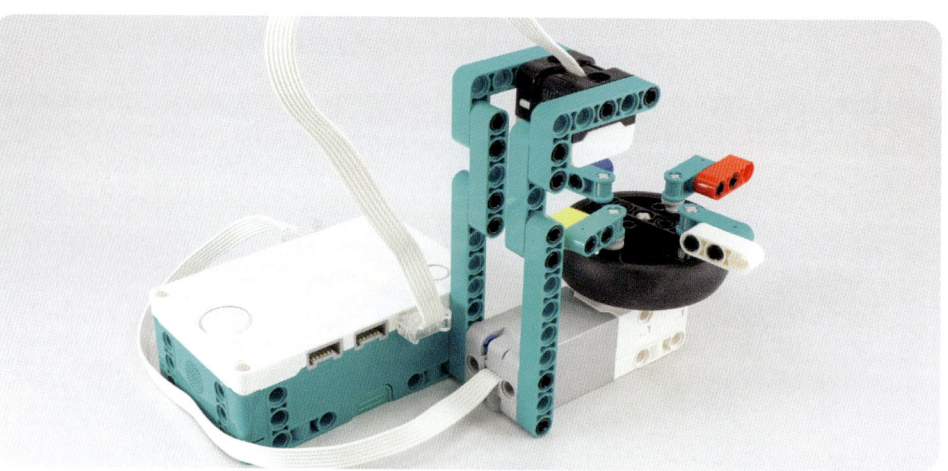

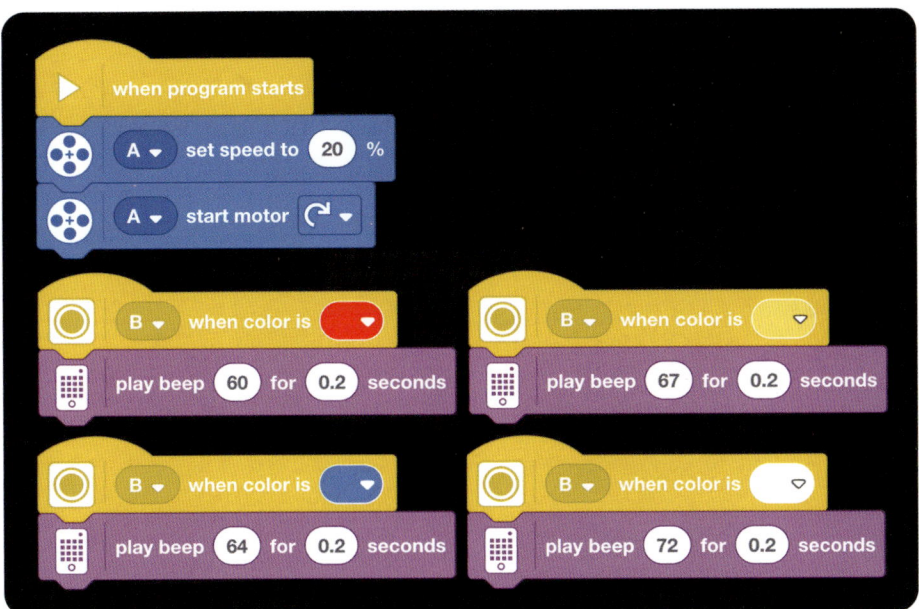

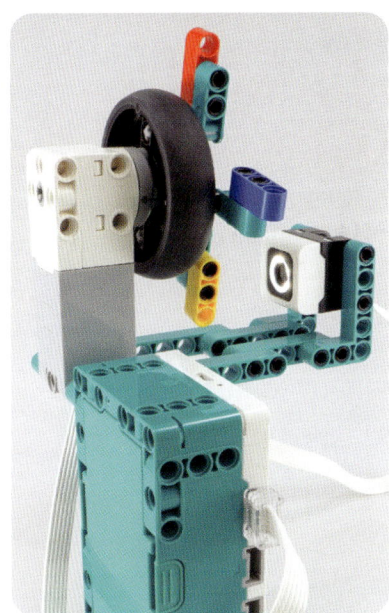

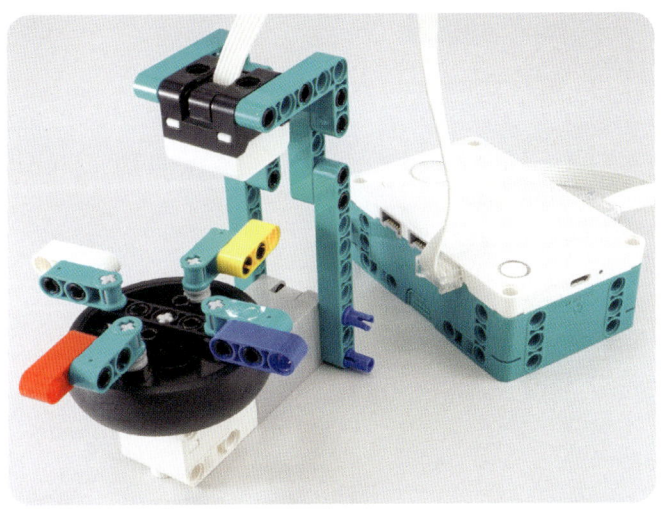

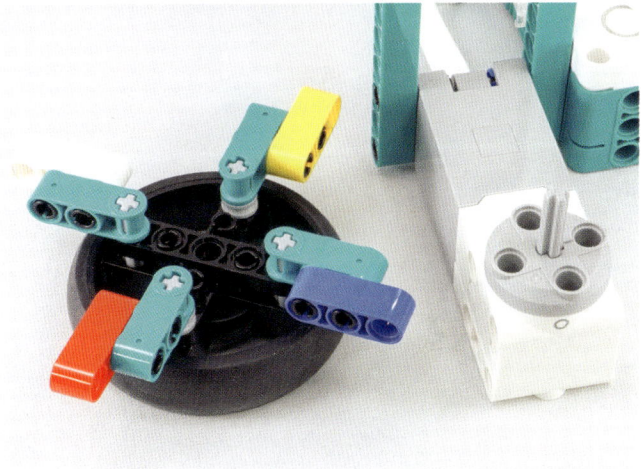

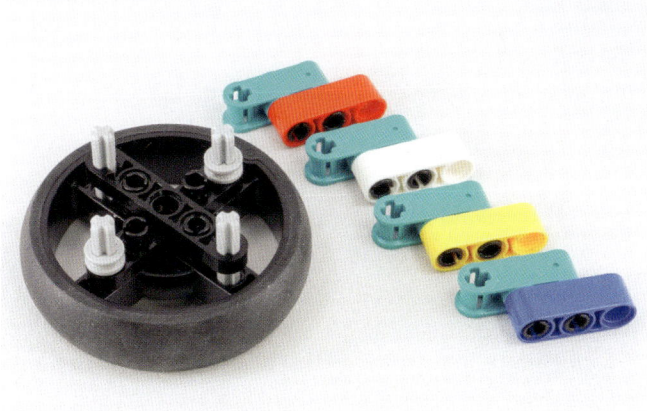

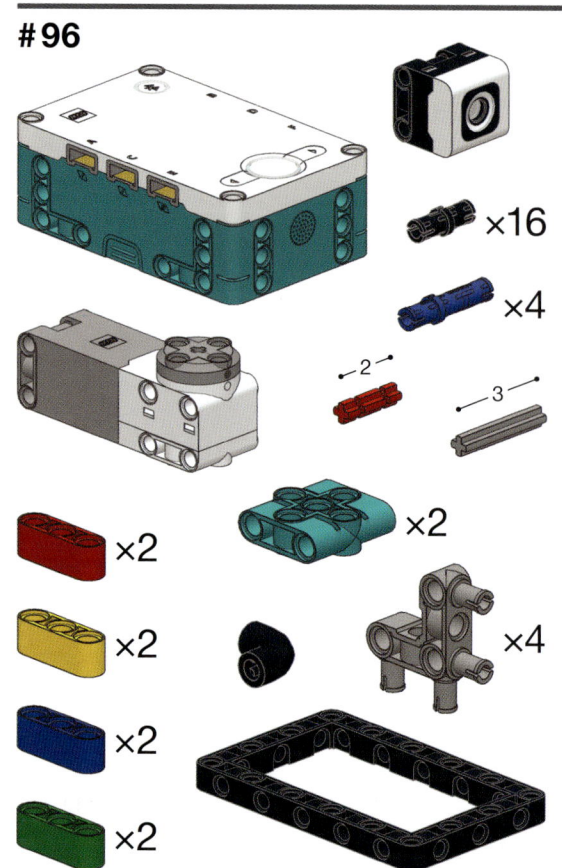

×16

×4

‹–2–›

‹––3––›

×2

×2

×2

×2

×2

×4

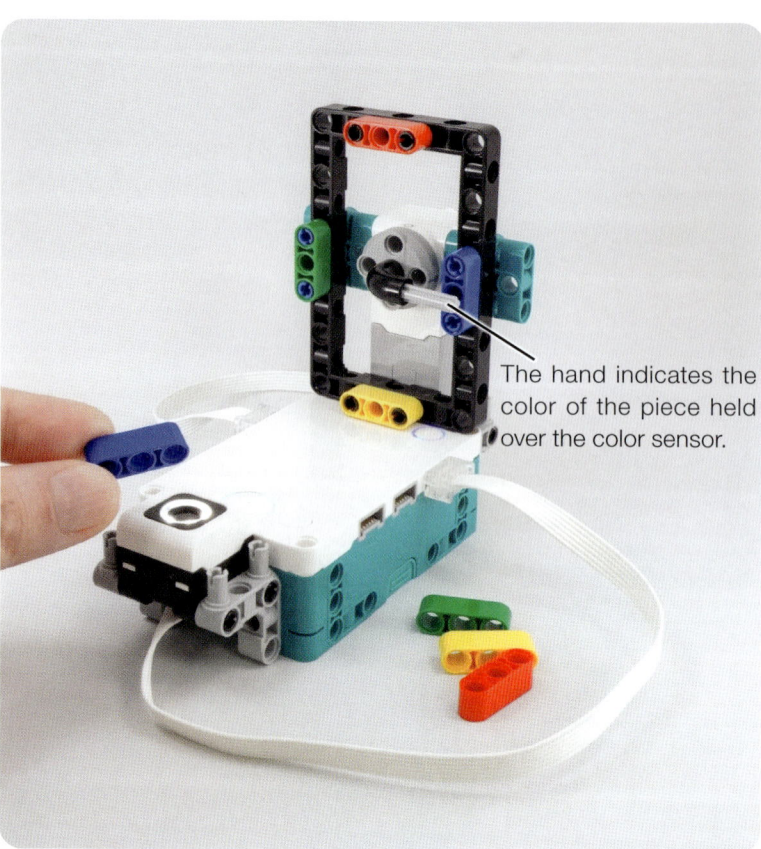

The hand indicates the color of the piece held over the color sensor.

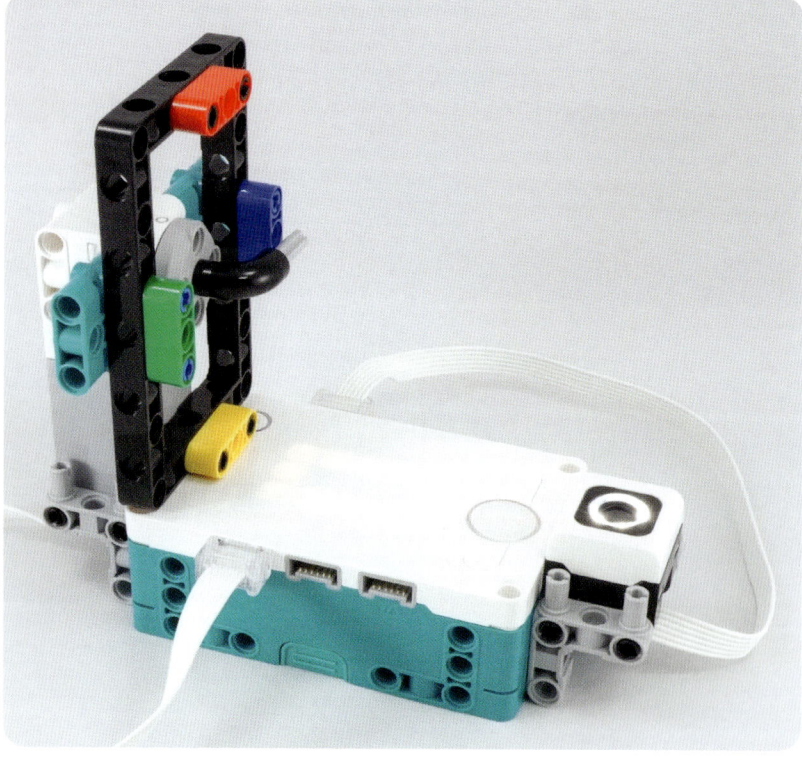

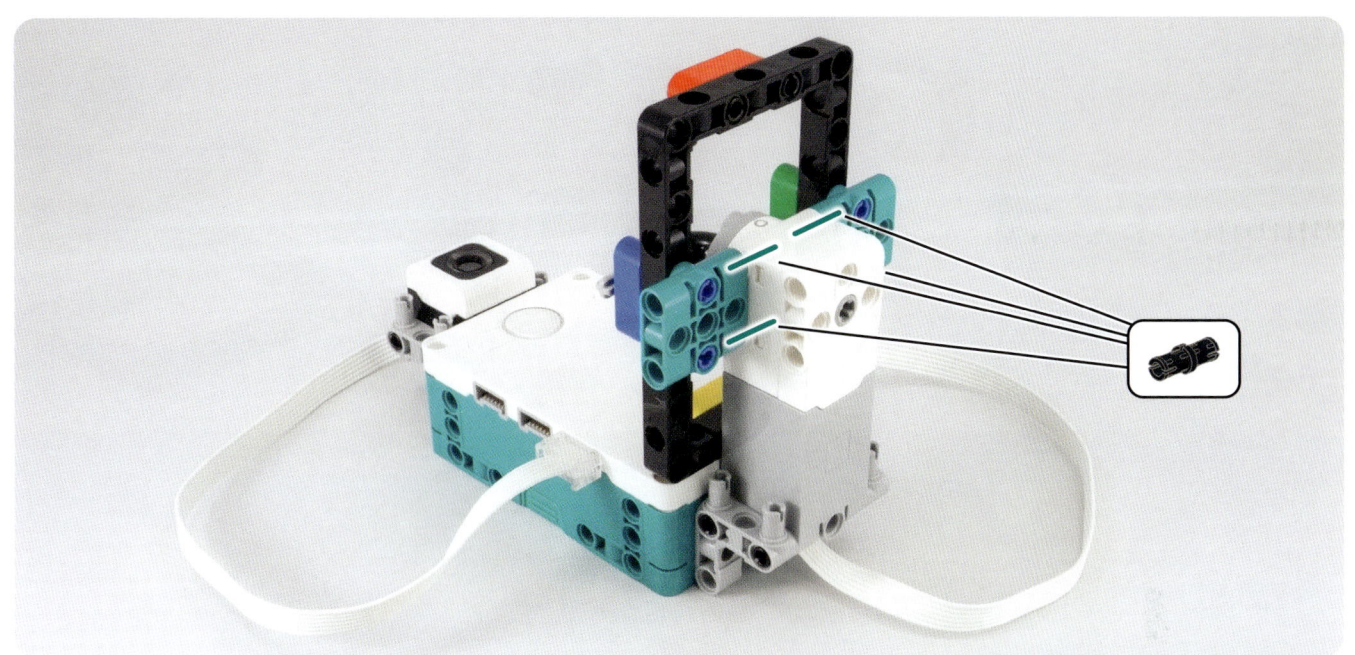

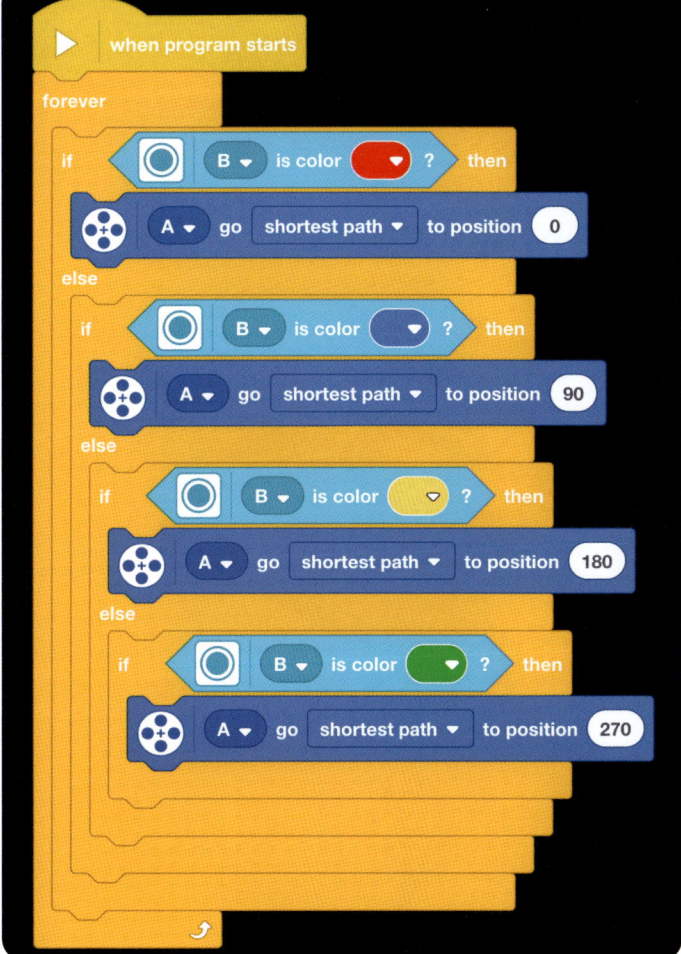

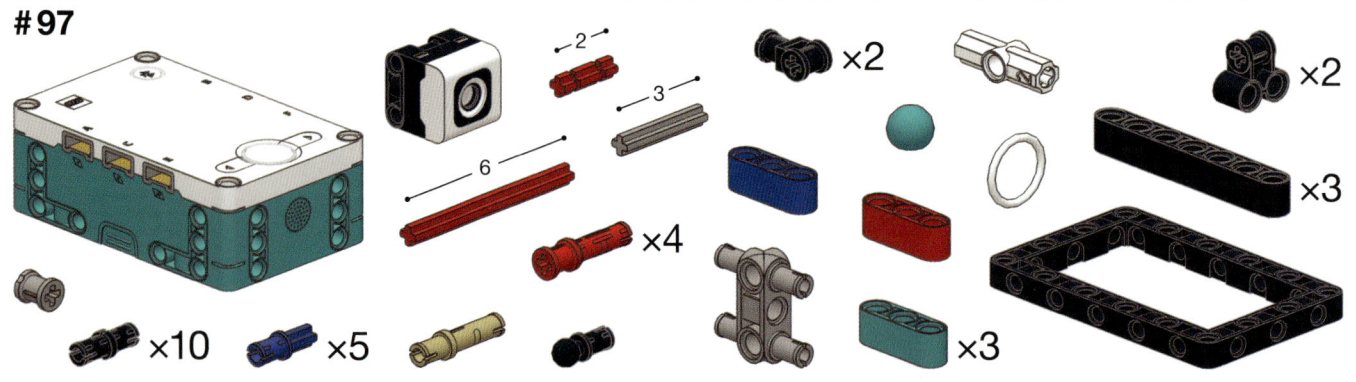

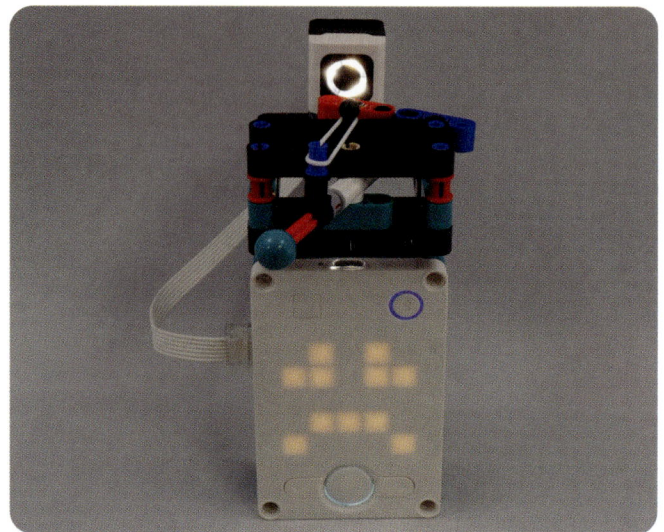

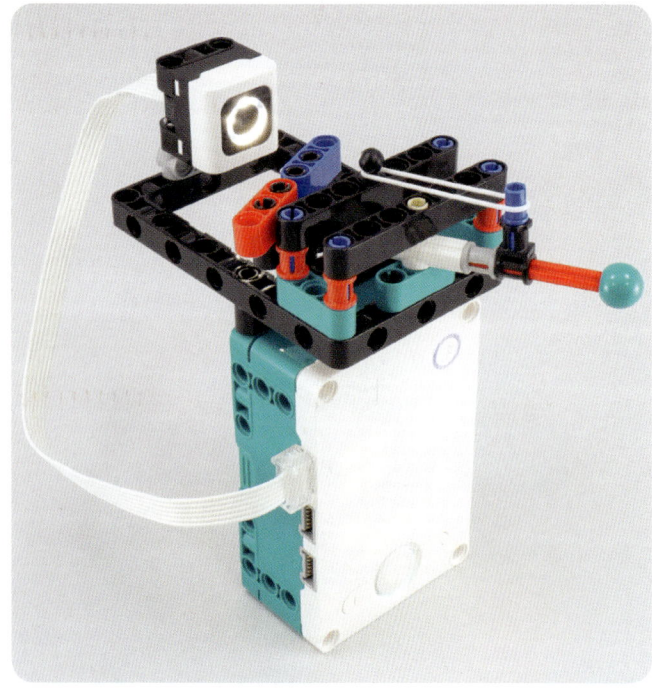

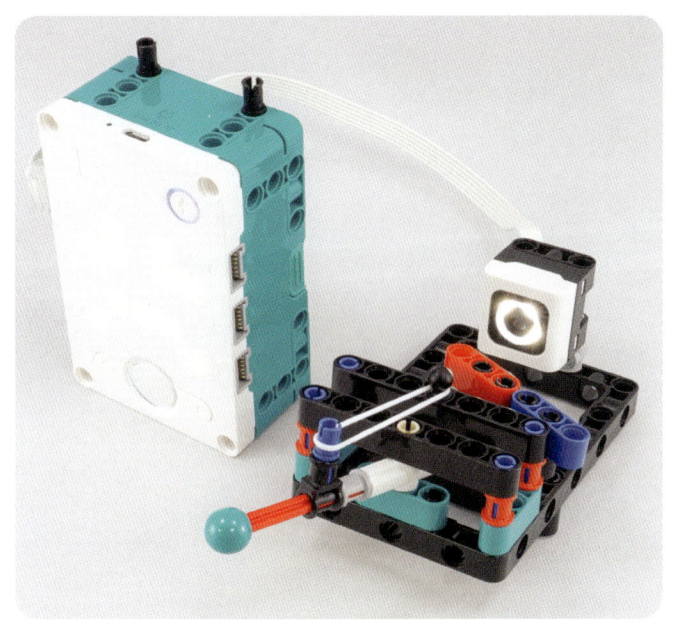

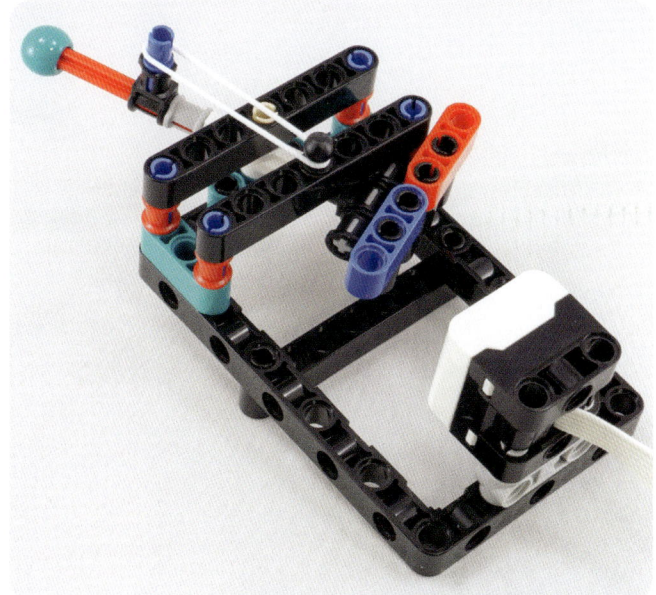

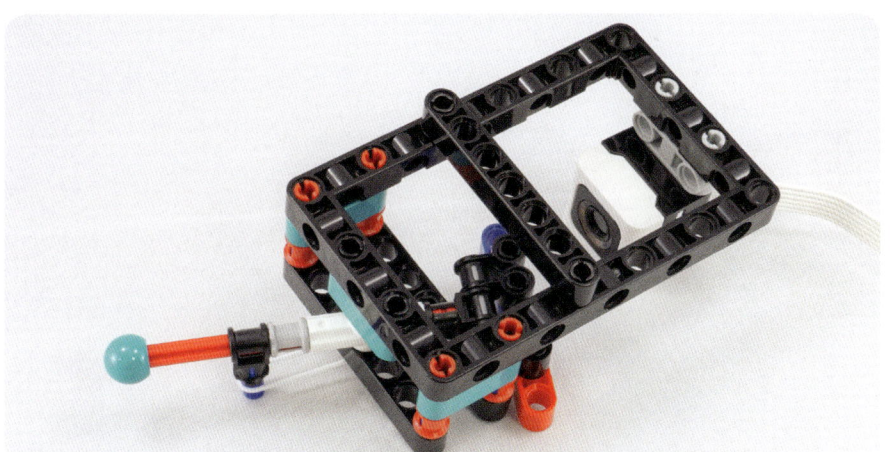

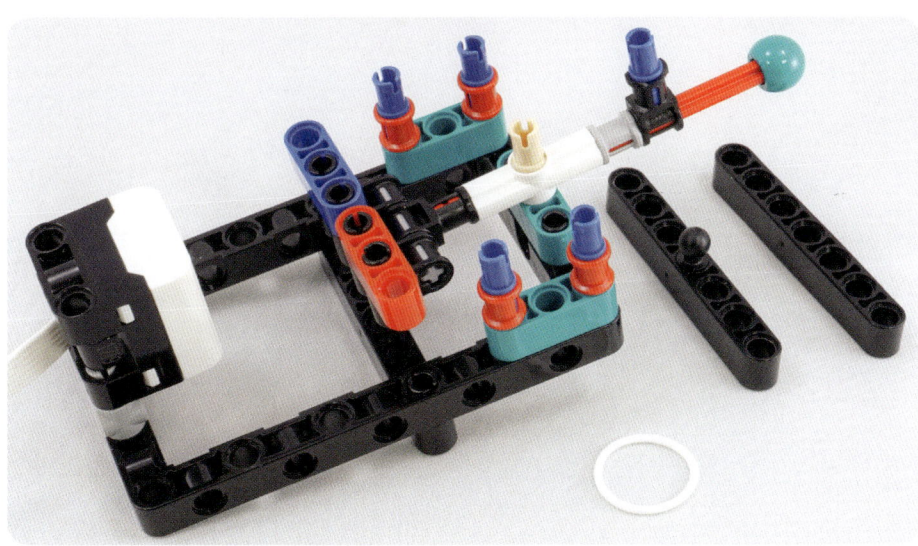

PART 5
Other Enjoyable Mechanisms

Various moving mechanisms

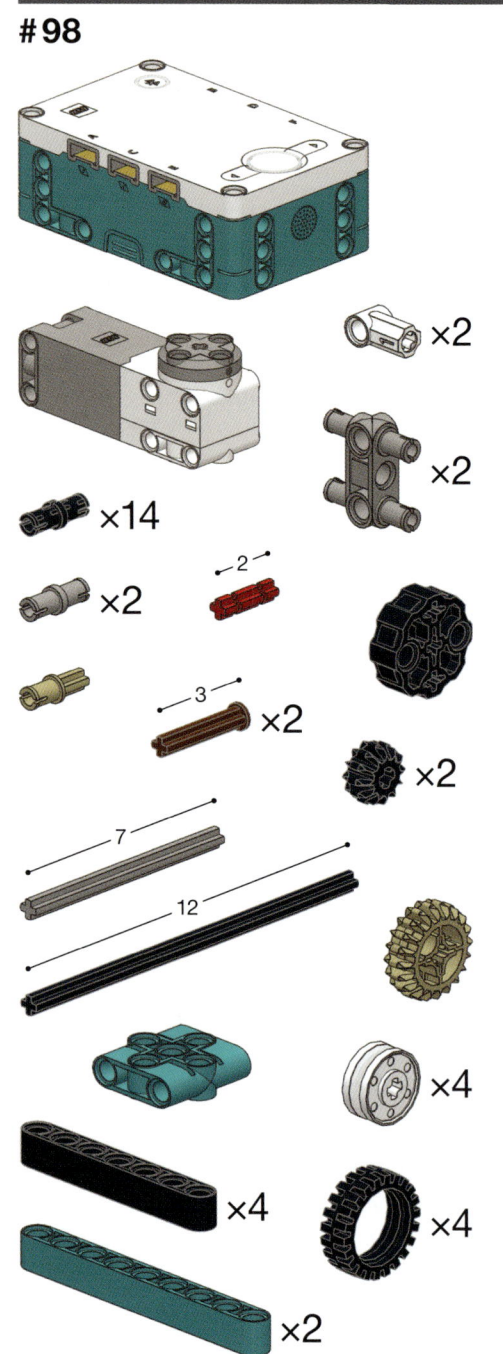

×2

×2

×14

×2   — 2 —

— 3 —   ×2

×2

— 7 —

— 12 —

×4

×4   ×4

×2

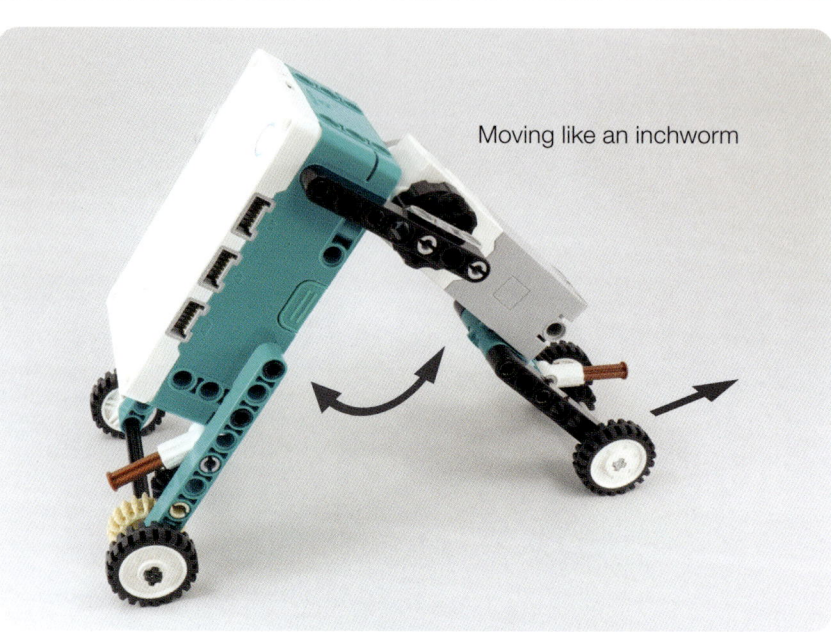

Moving like an inchworm

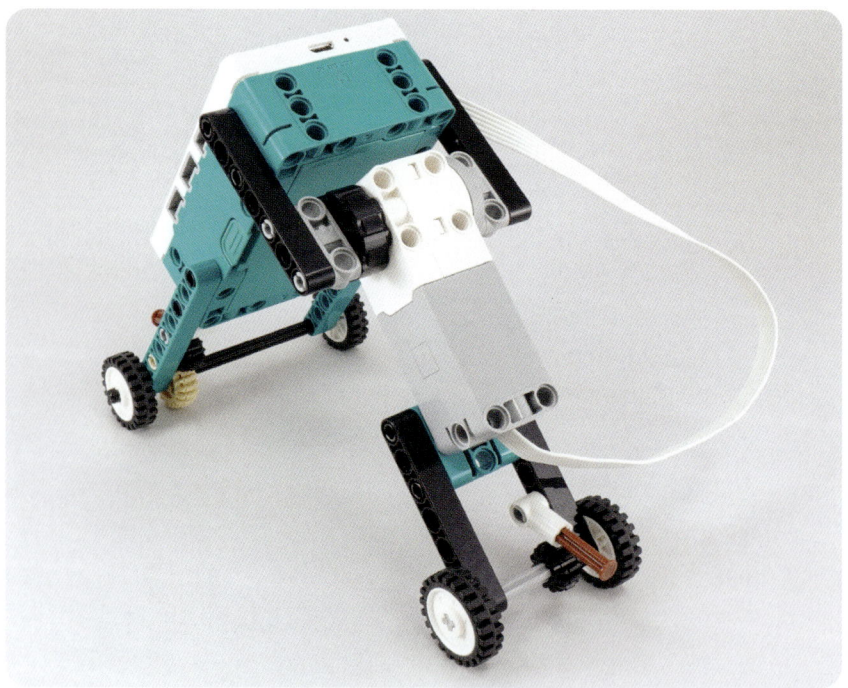

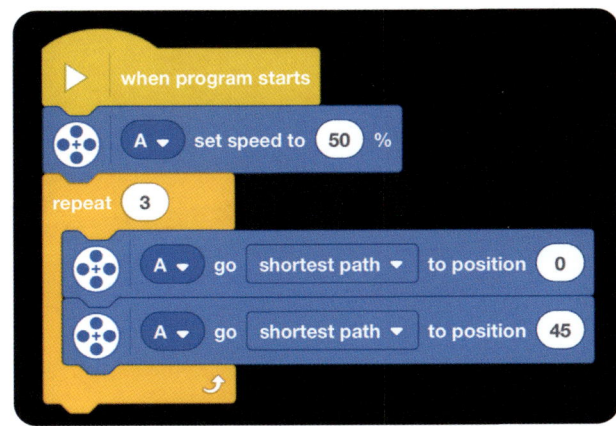

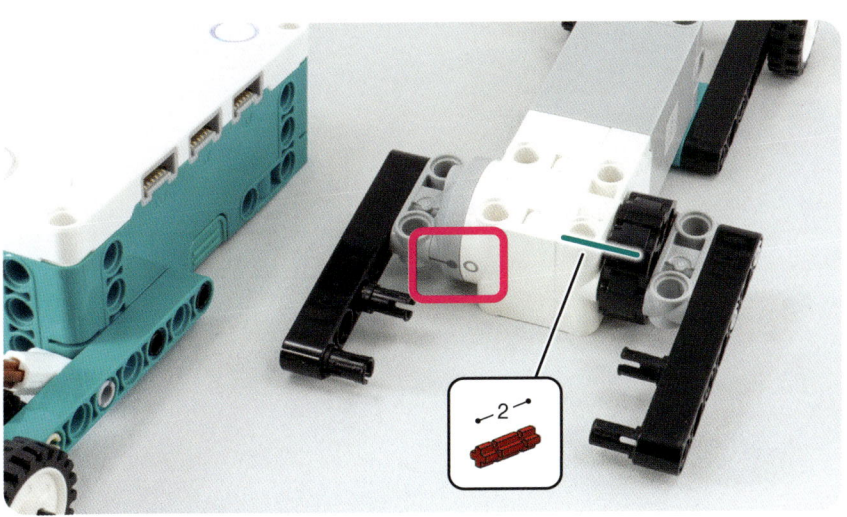

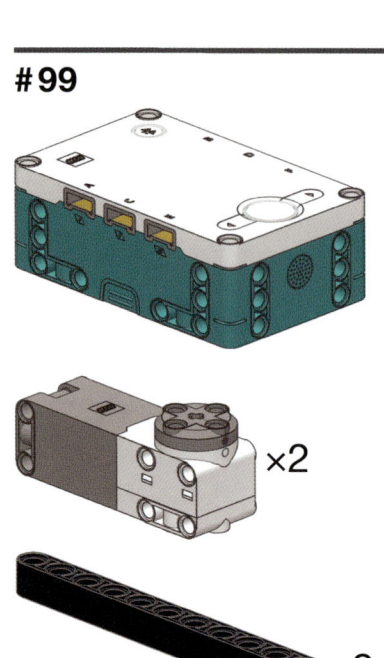

 ×2

 ×4

 ×4

 ×4

×28

×4

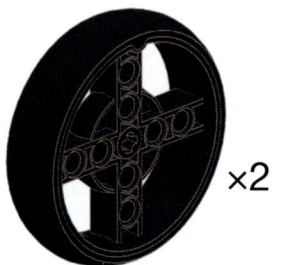 ×2

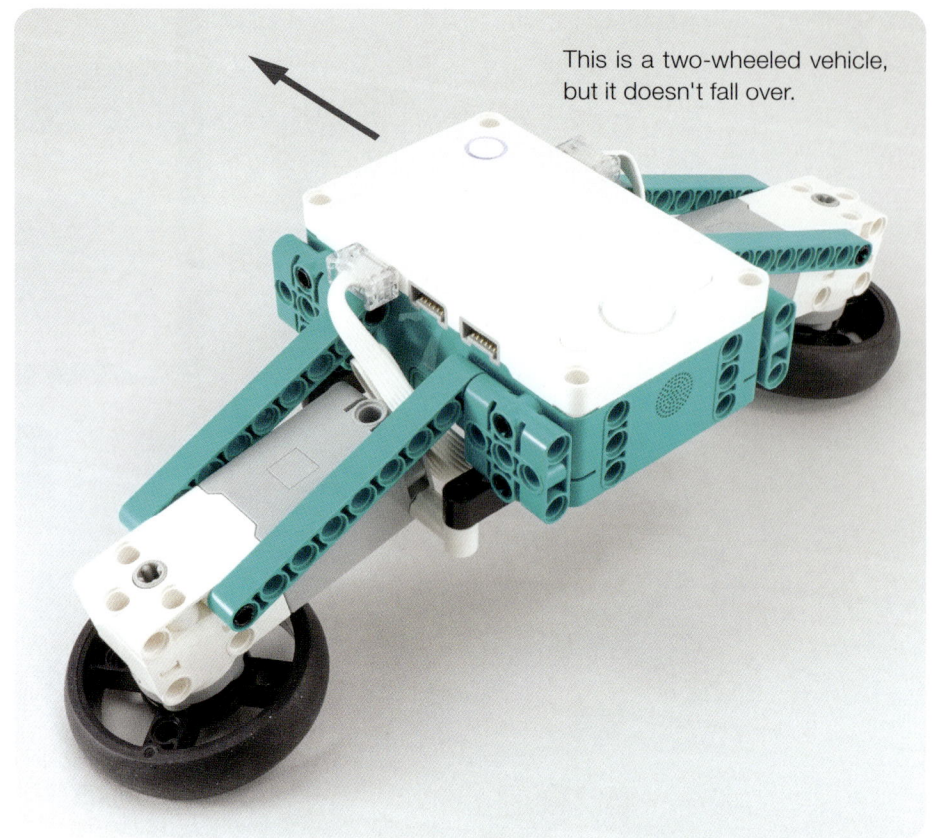

This is a two-wheeled vehicle, but it doesn't fall over.

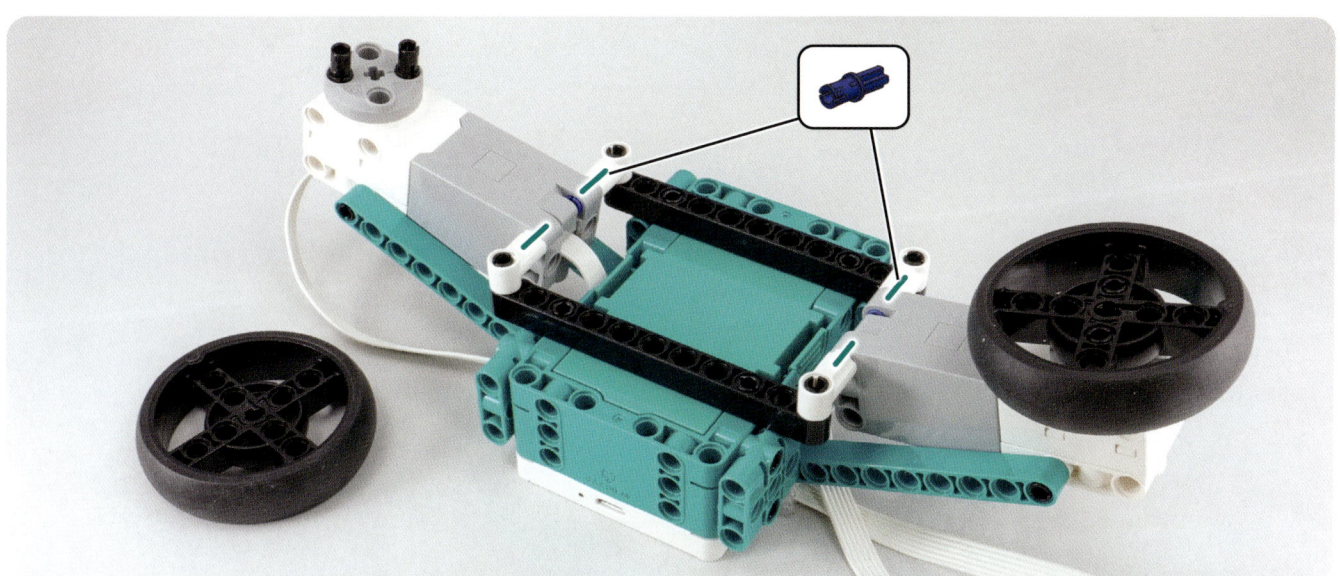

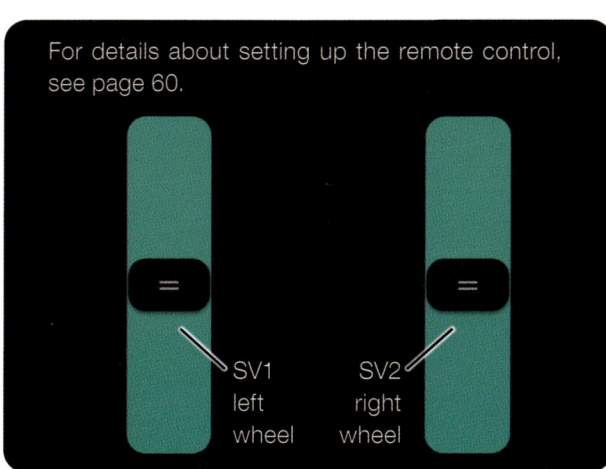

For details about setting up the remote control, see page 60.

SV1
left
wheel

SV2
right
wheel

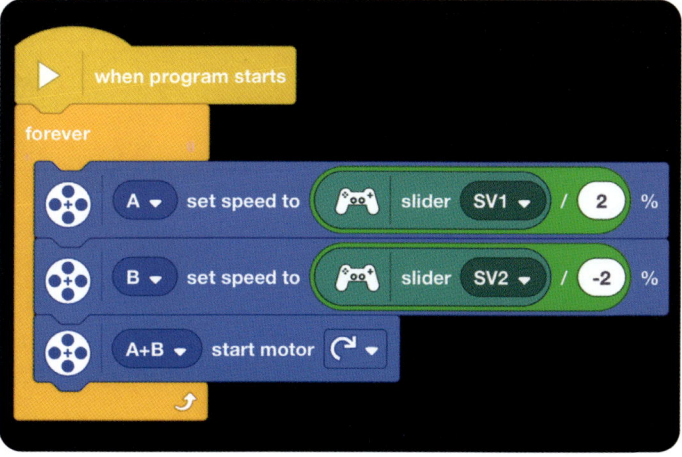

```
when program starts
forever
    A ▼  set speed to   slider  SV1 ▼  /  2   %
    B ▼  set speed to   slider  SV2 ▼  /  -2  %
    A+B ▼  start motor  ↻ ▼
```

#100

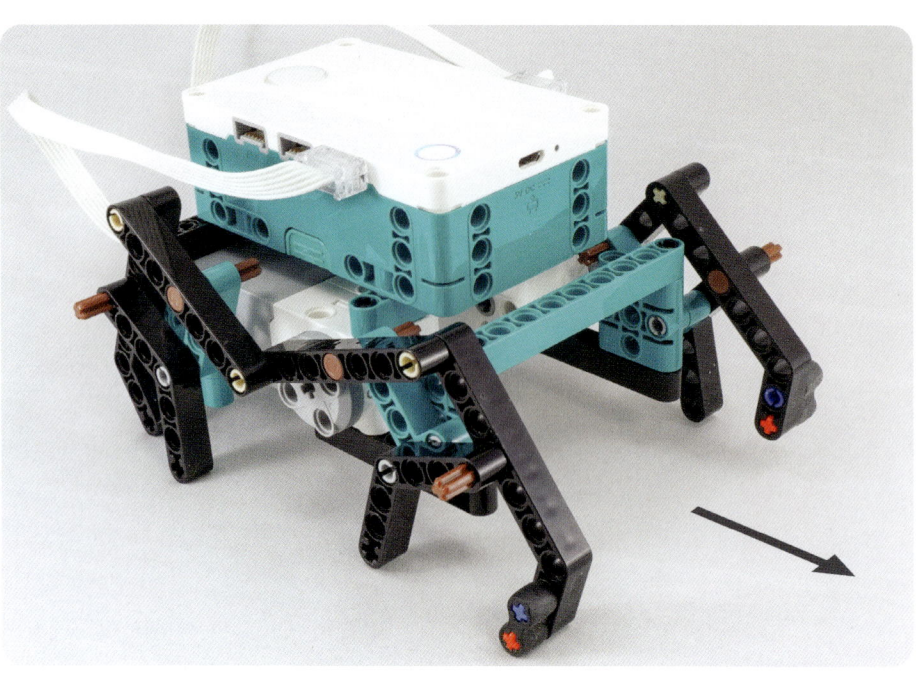

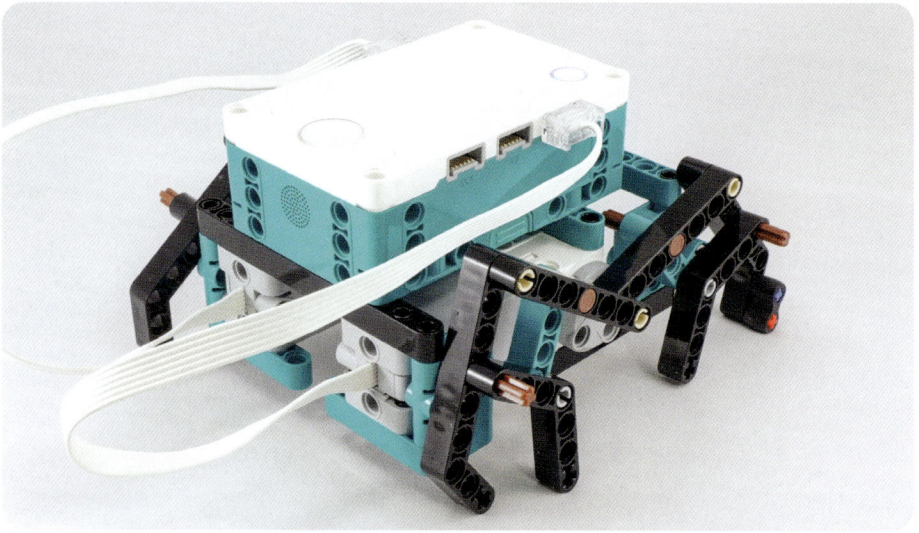

×2

×32 ×2

×8

×2 ×4

×4 ×2

×2

←2→ ×2 ×4

←3→ ×8

×4

×6

×4

×2

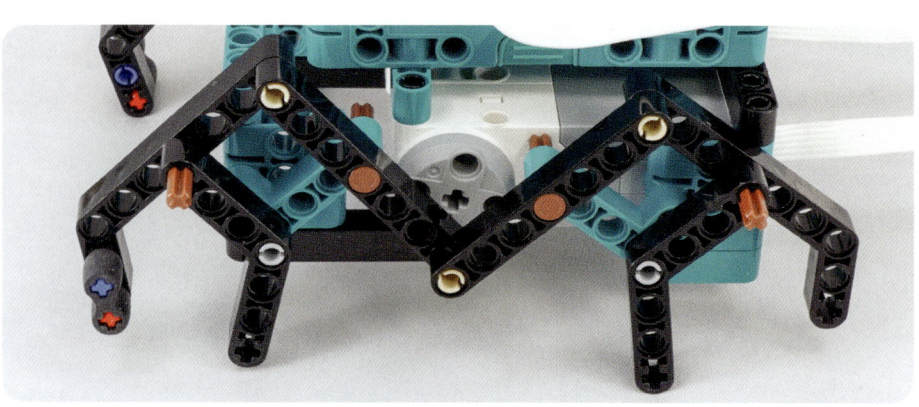

●●●●●

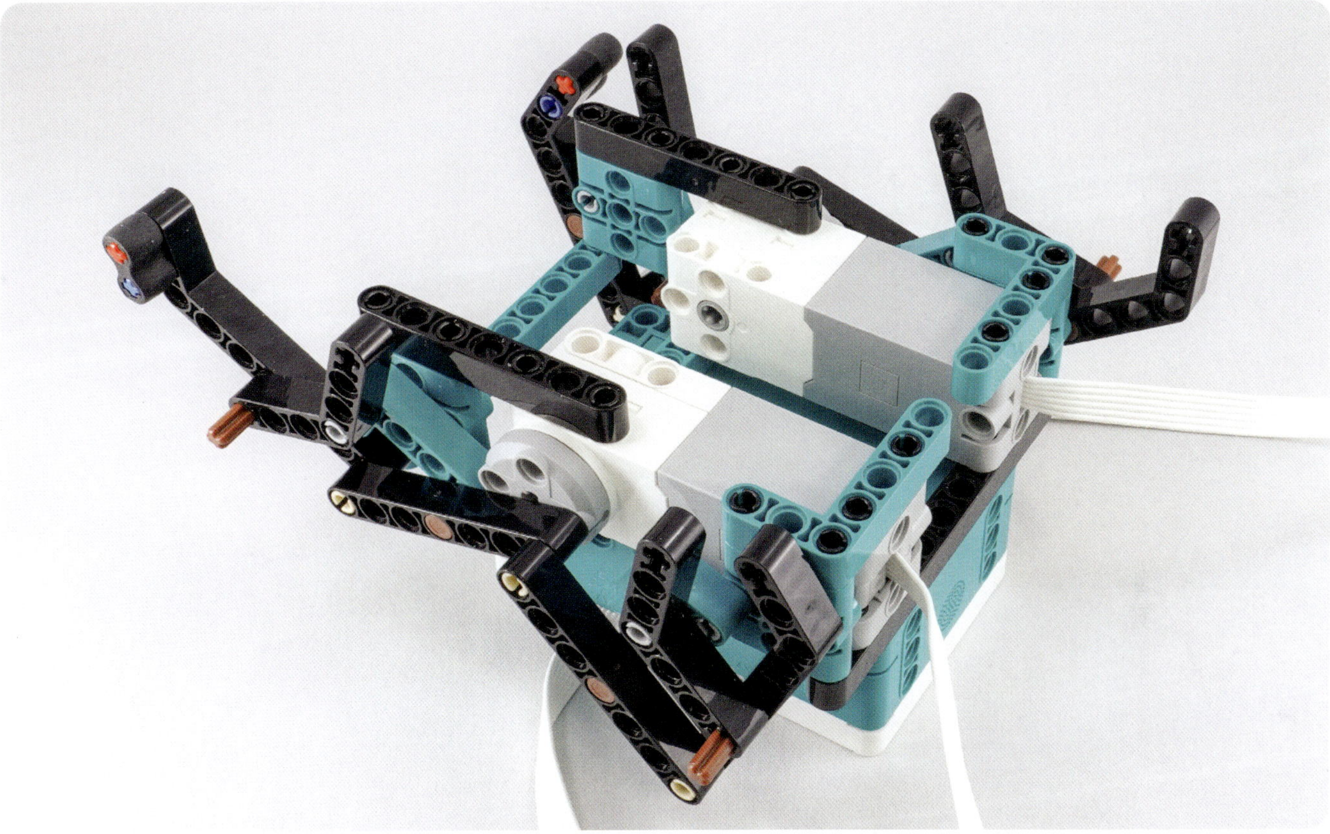

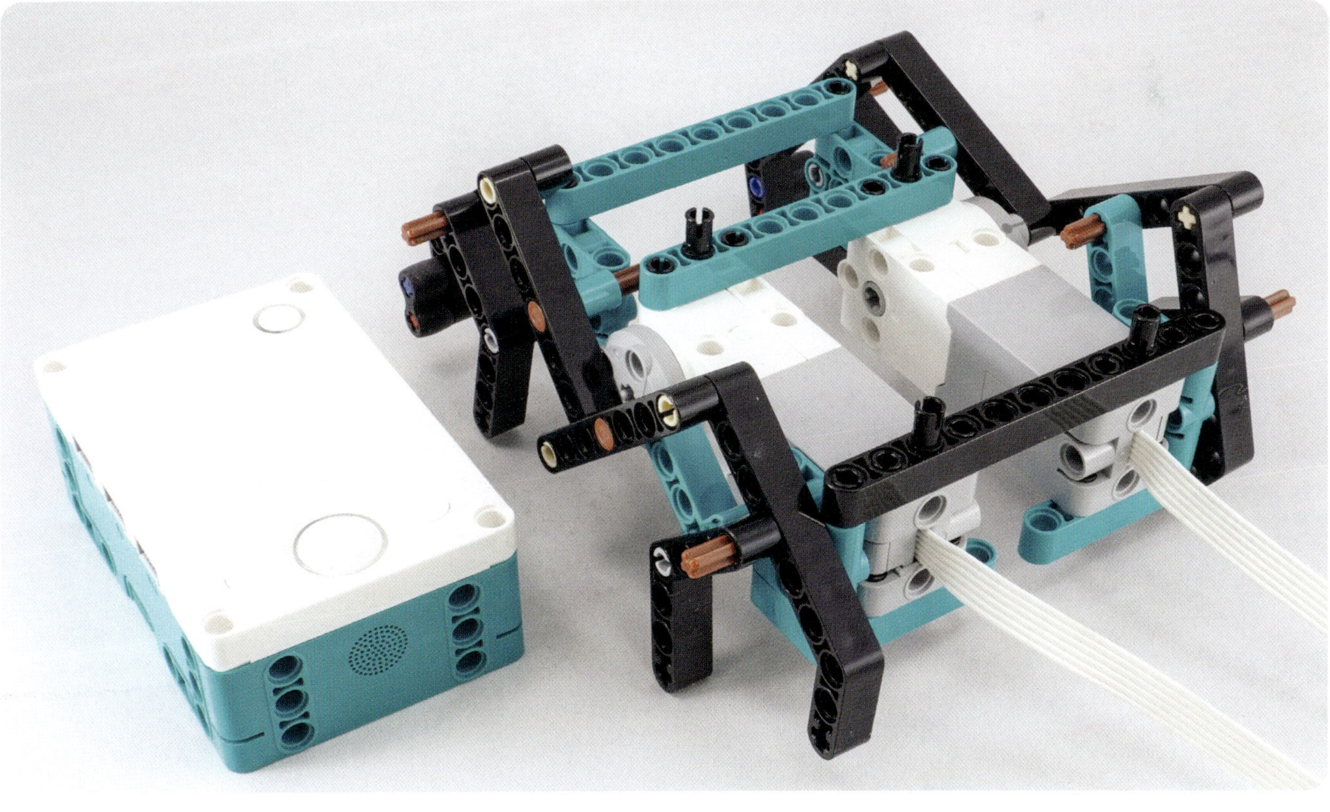

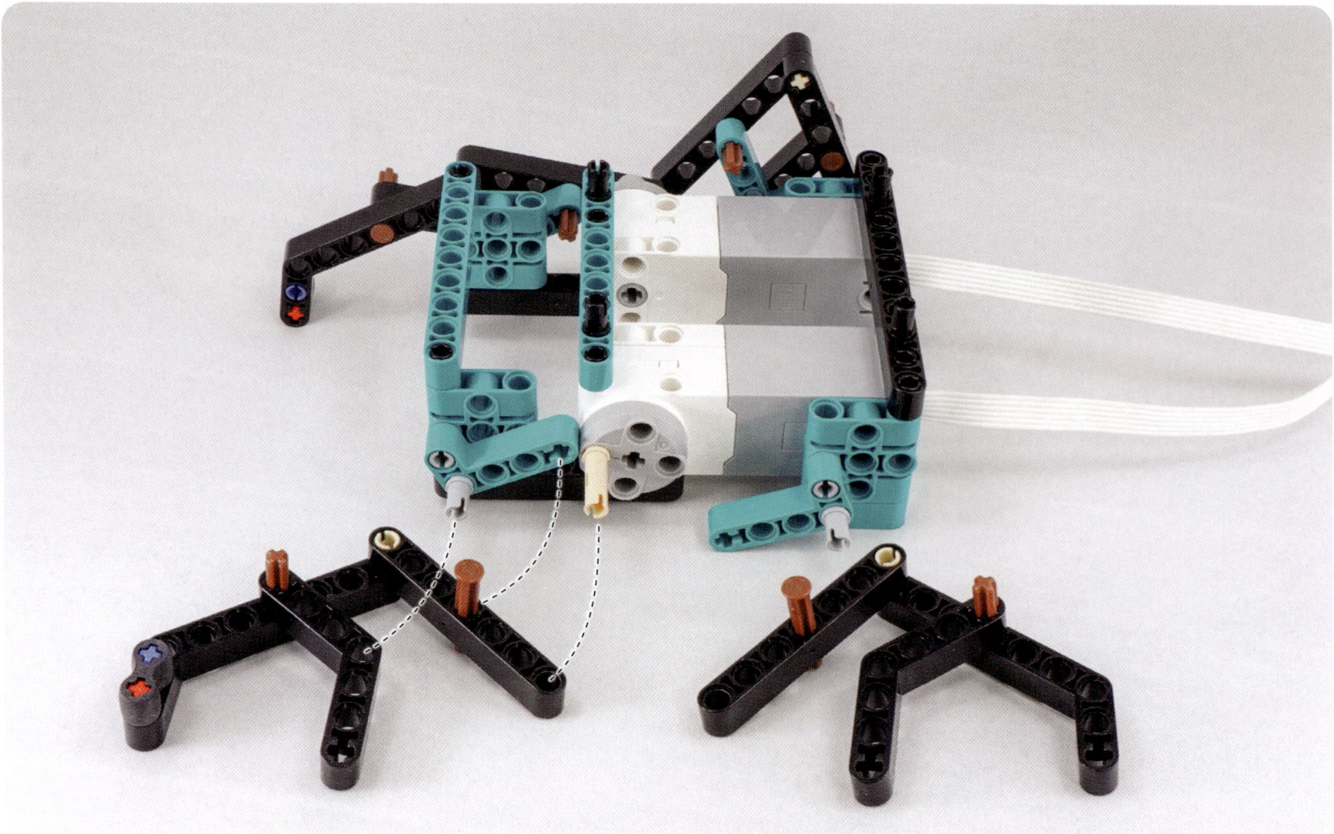

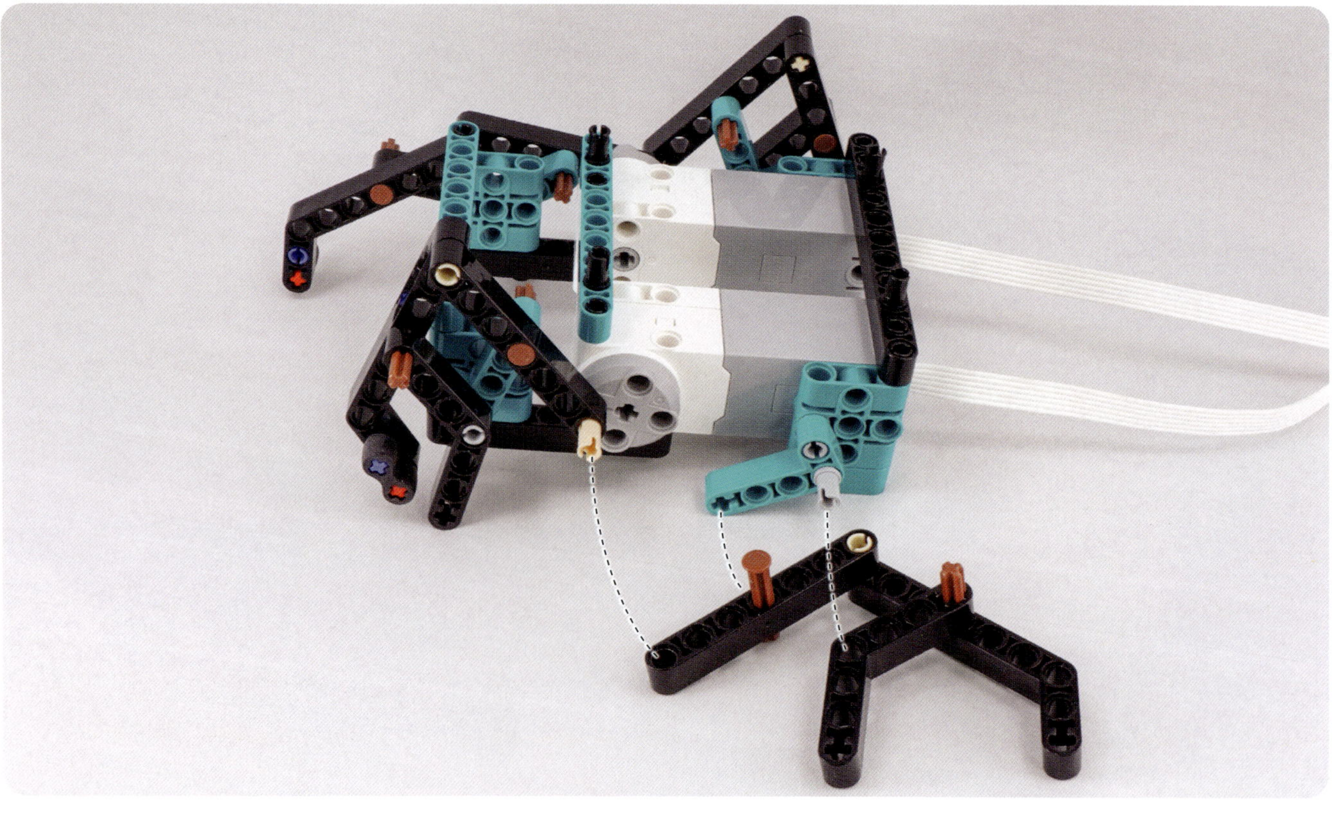

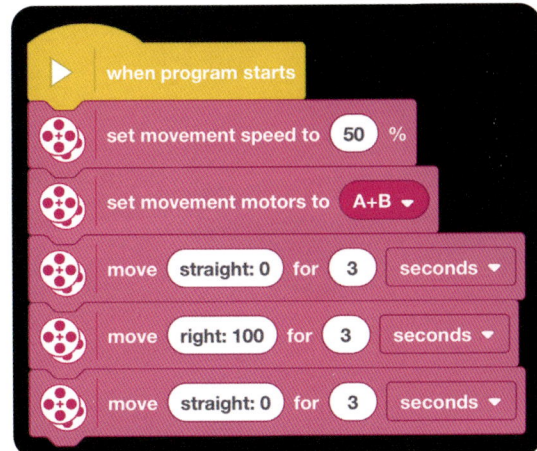

> when program starts
> set movement speed to 50 %
> set movement motors to A+B ▼
> move straight: 0 for 3 seconds ▼
> move right: 100 for 3 seconds ▼
> move straight: 0 for 3 seconds ▼

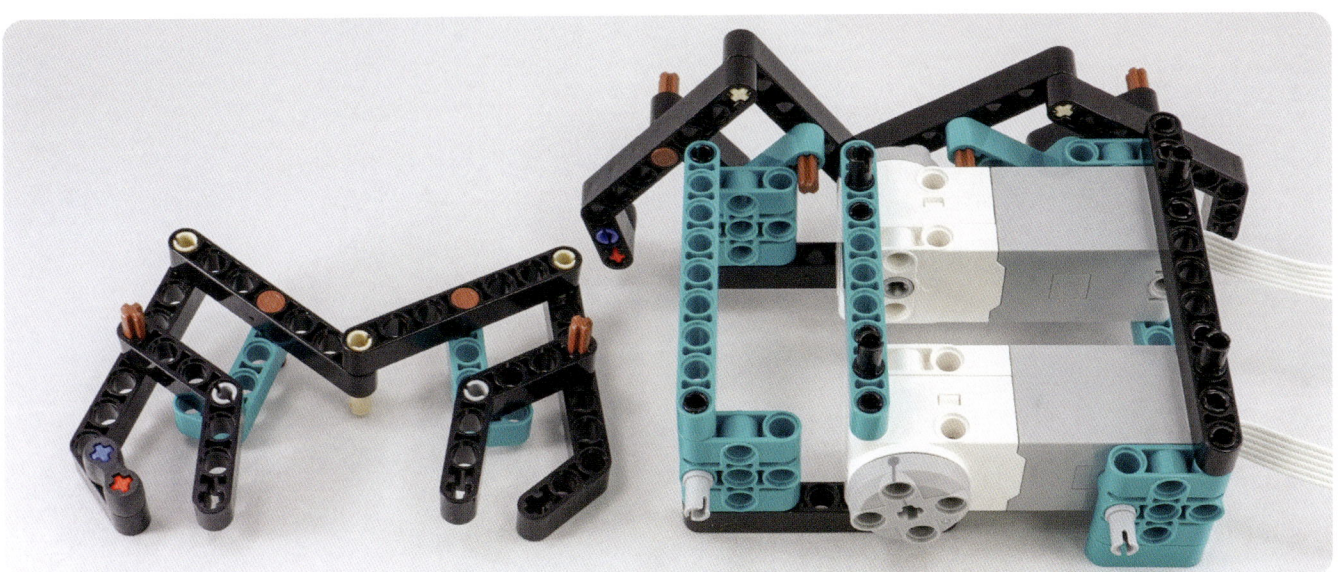

 For details about setting up the remote control, see page 60.

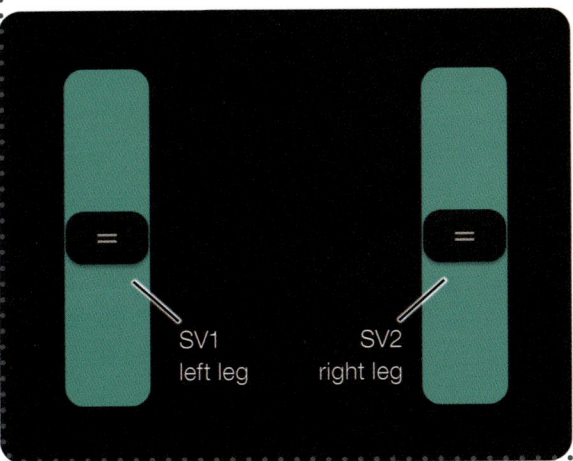

SV1
left leg

SV2
right leg

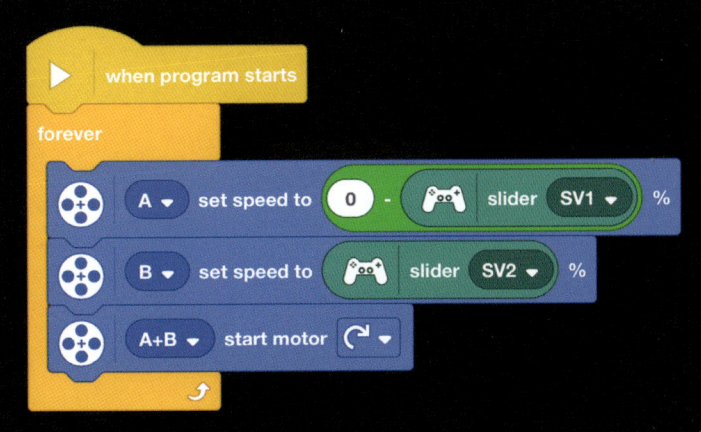

> when program starts
> forever
>   A ▼ set speed to 0 - slider SV1 ▼ %
>   B ▼ set speed to slider SV2 ▼ %
>   A+B ▼ start motor ↻ ▼

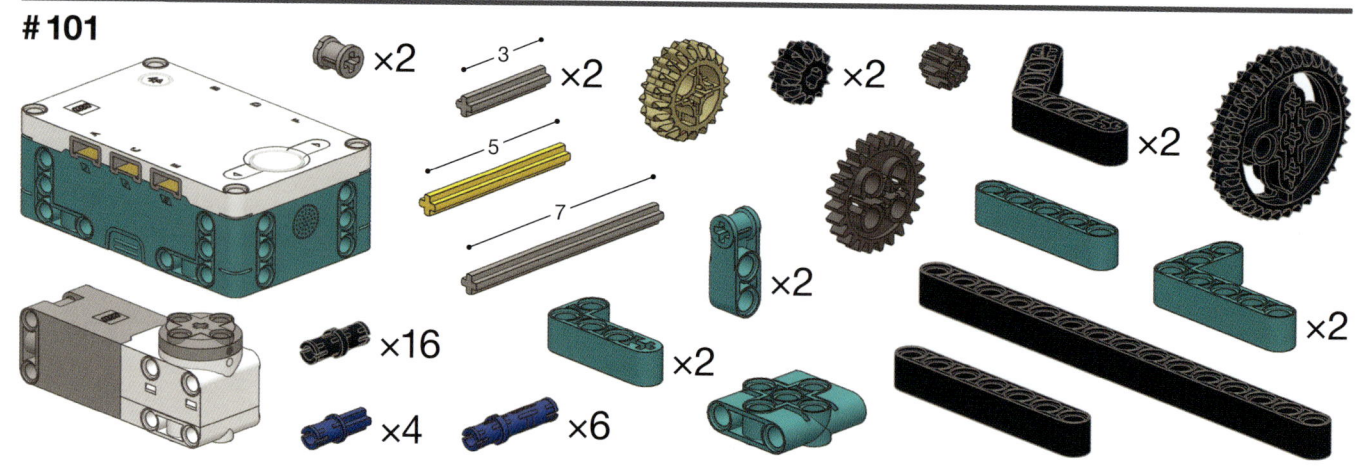

Moving through vibration

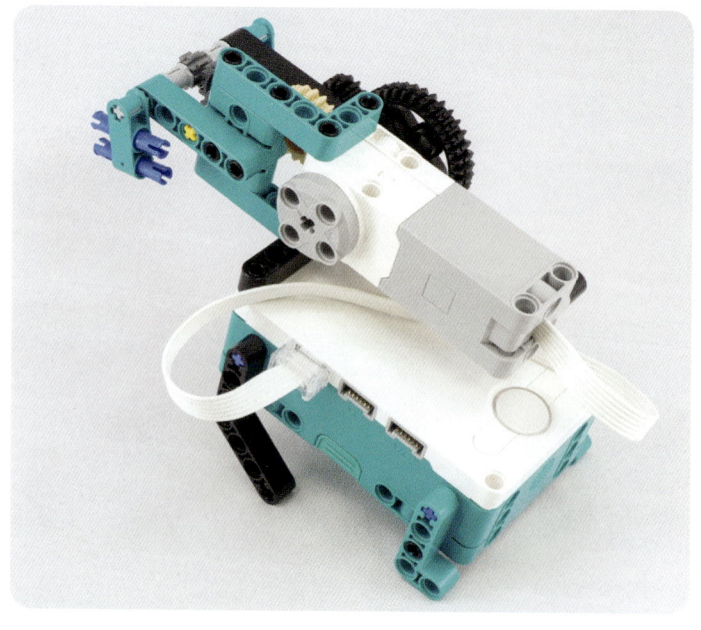

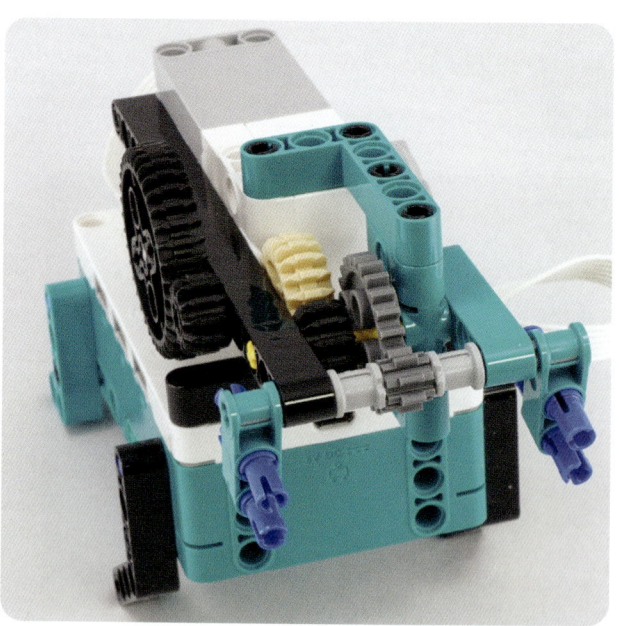

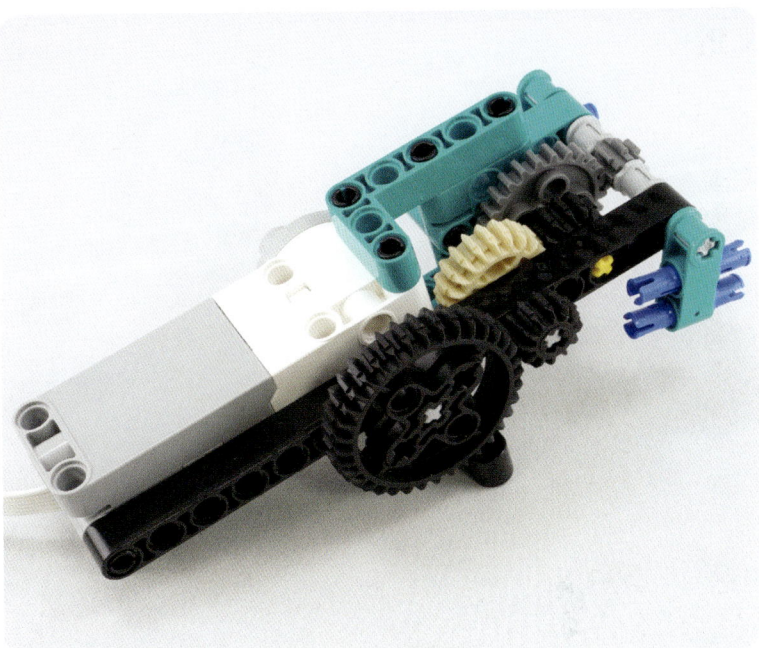

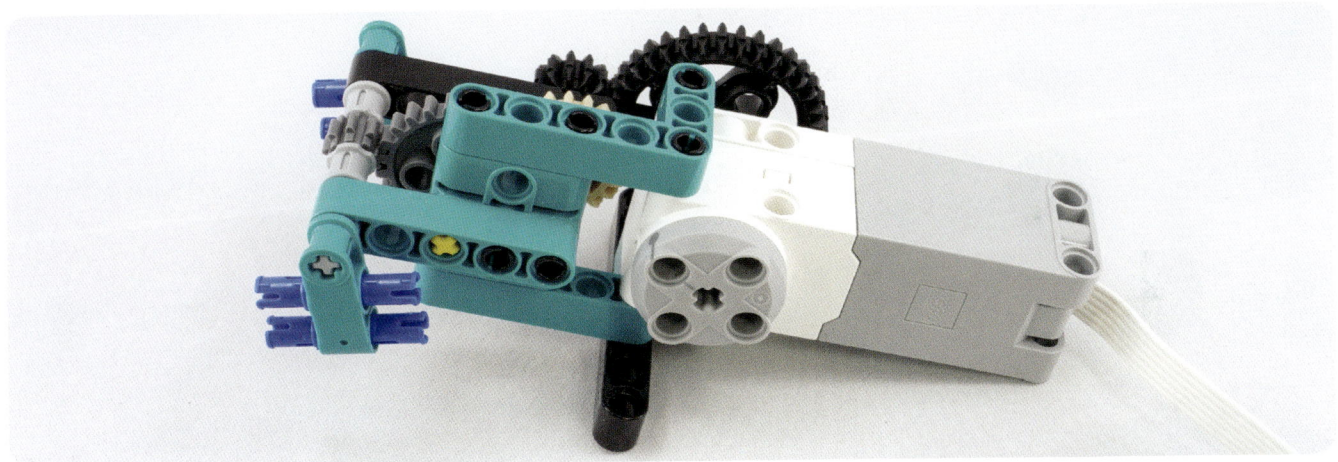

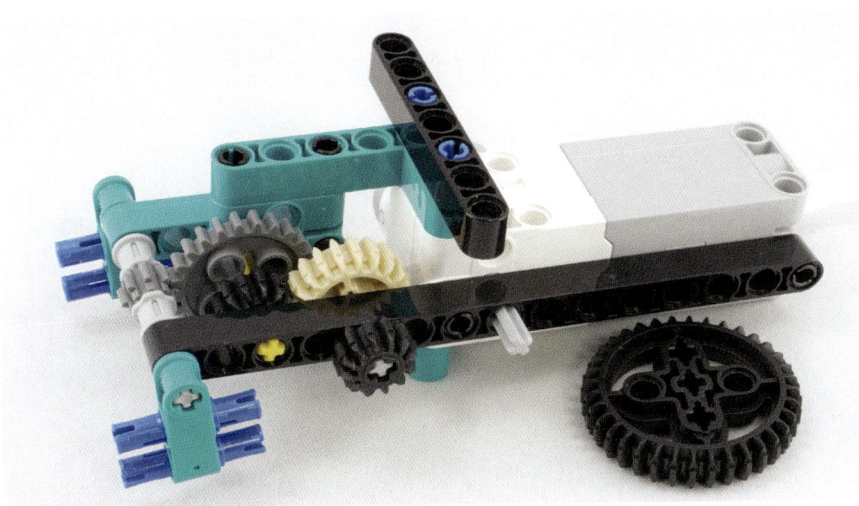

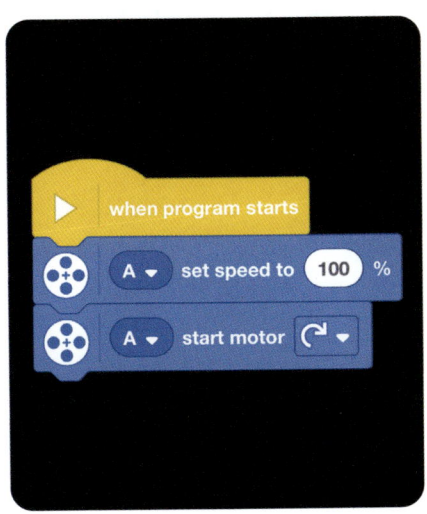

×14

×6

×4

×2

×2

×2

×2

×3

×2

×2

×2

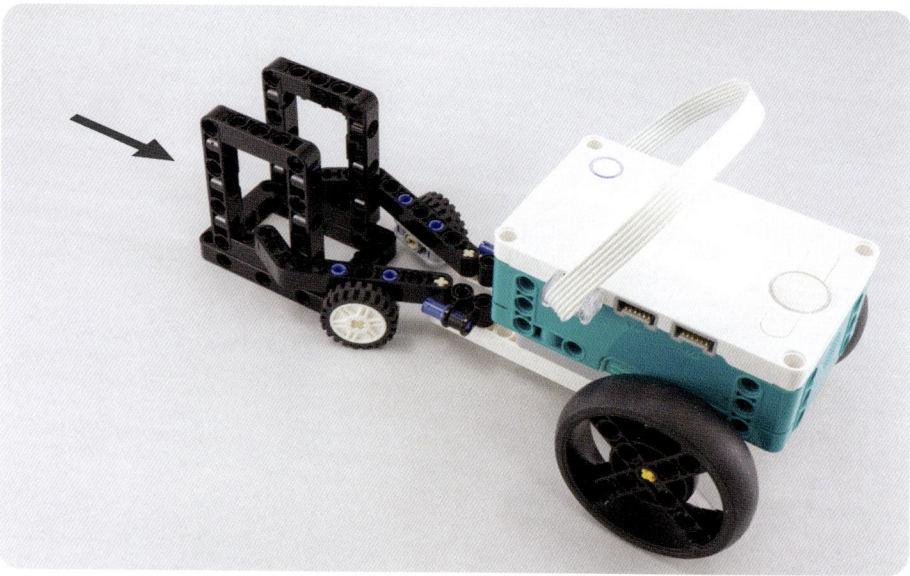

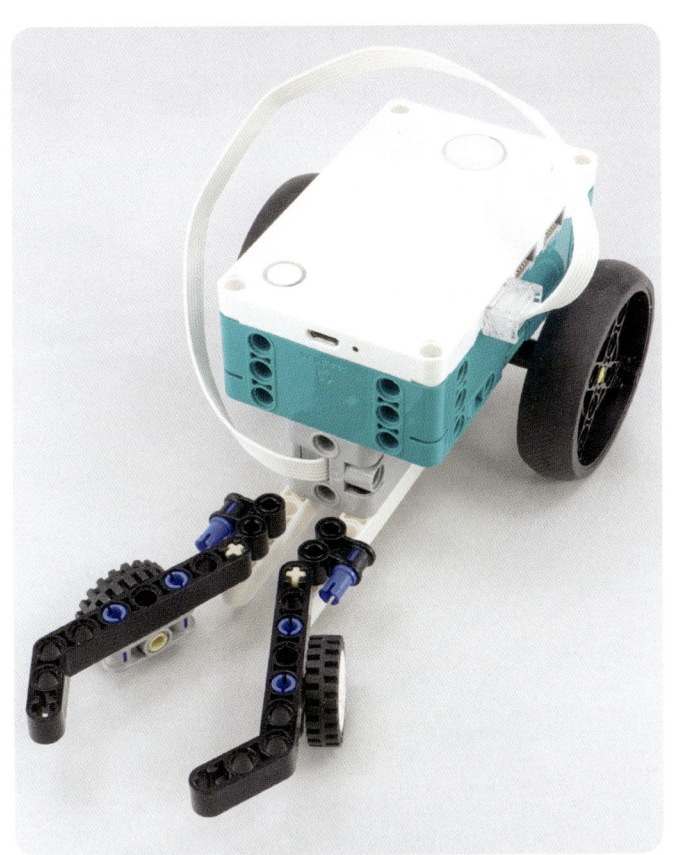

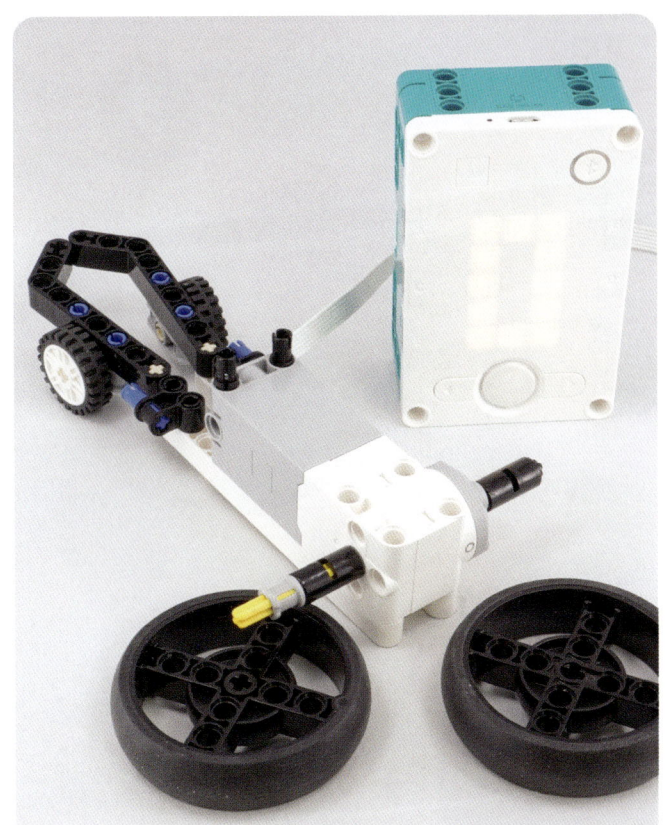

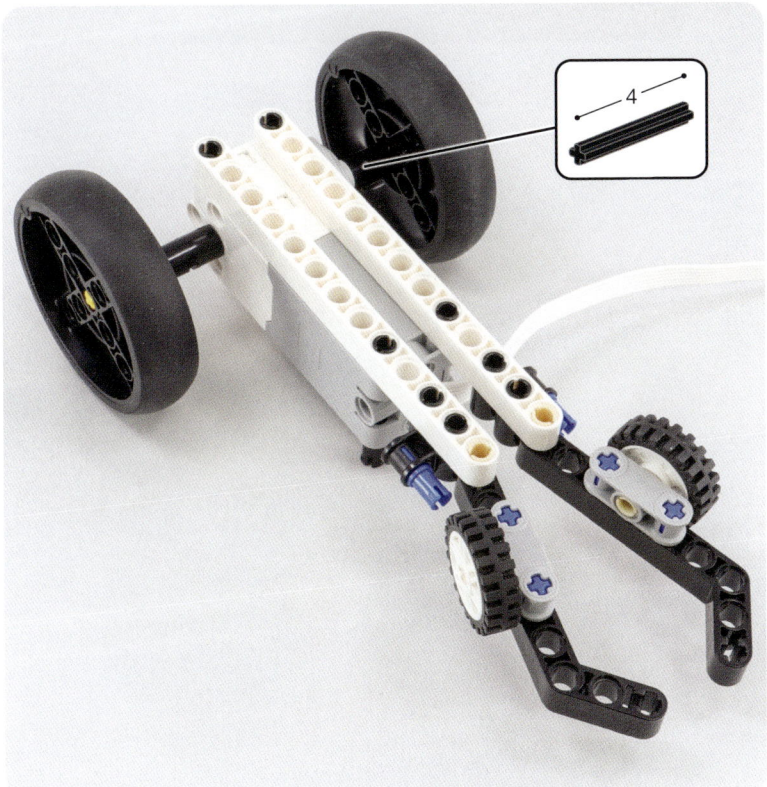

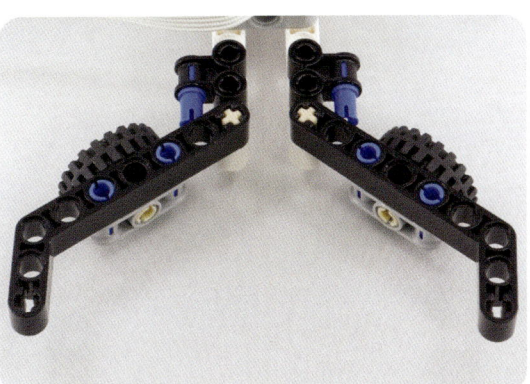

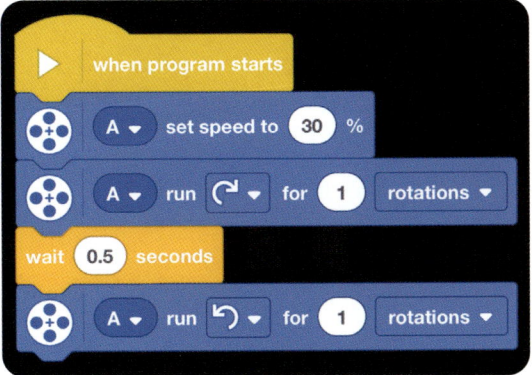

4

#103

×23
×3
×2
×4
×5
2
3
×2
×3
5
×2
7
×2

×2
×2
×2
×2
×2
×2

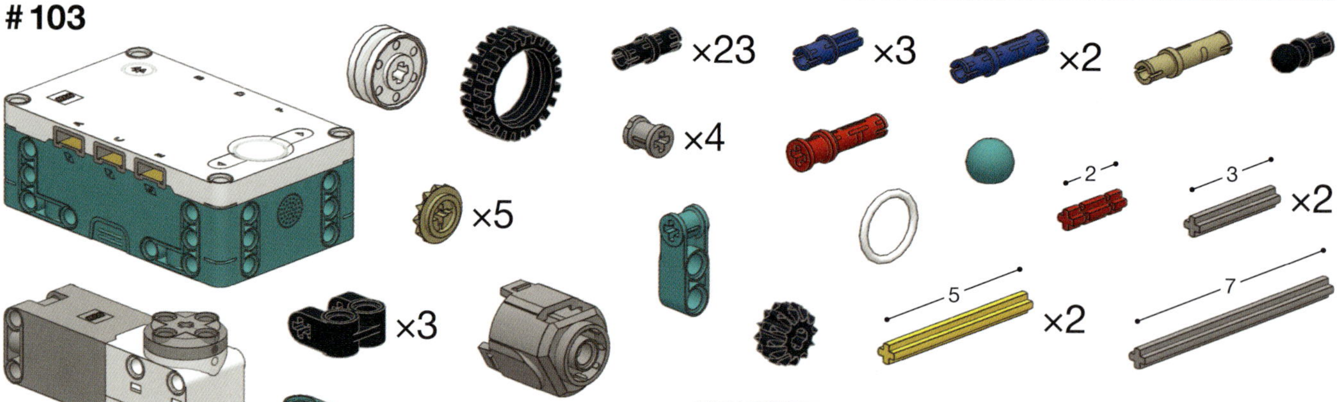

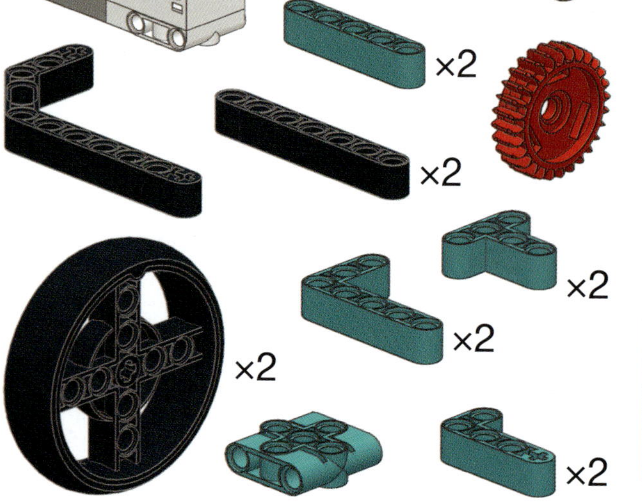

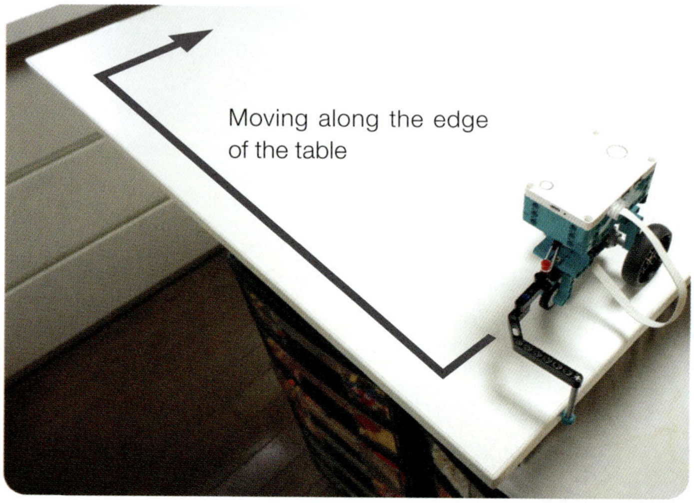

Moving along the edge
of the table

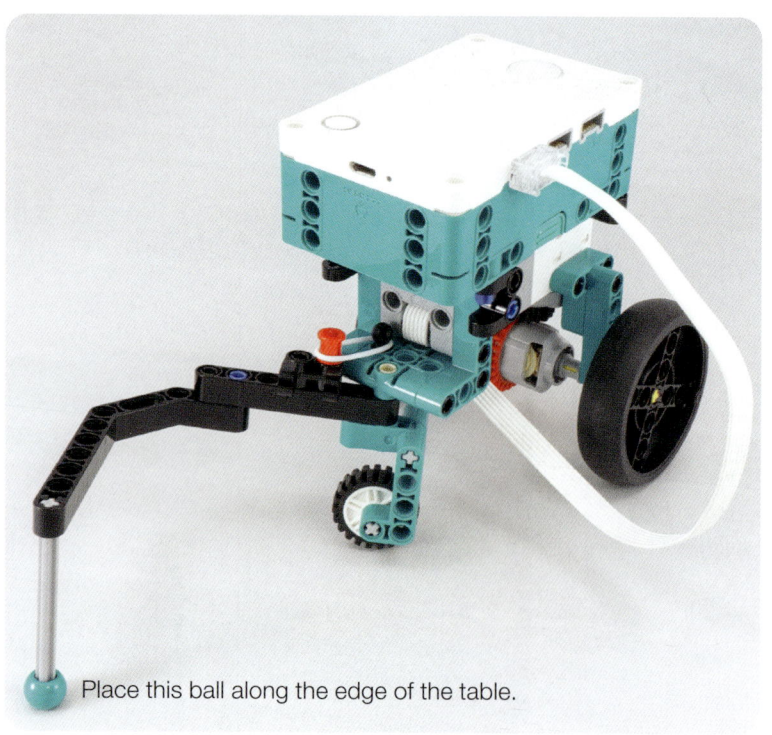

Place this ball along the edge of the table.

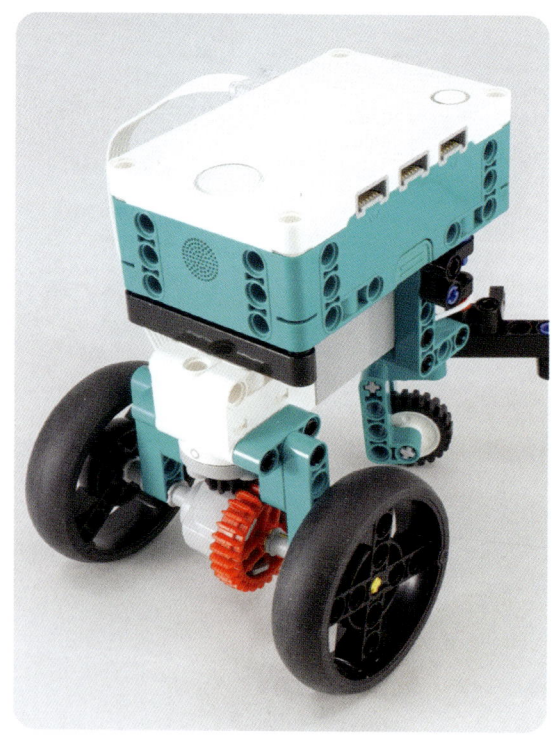

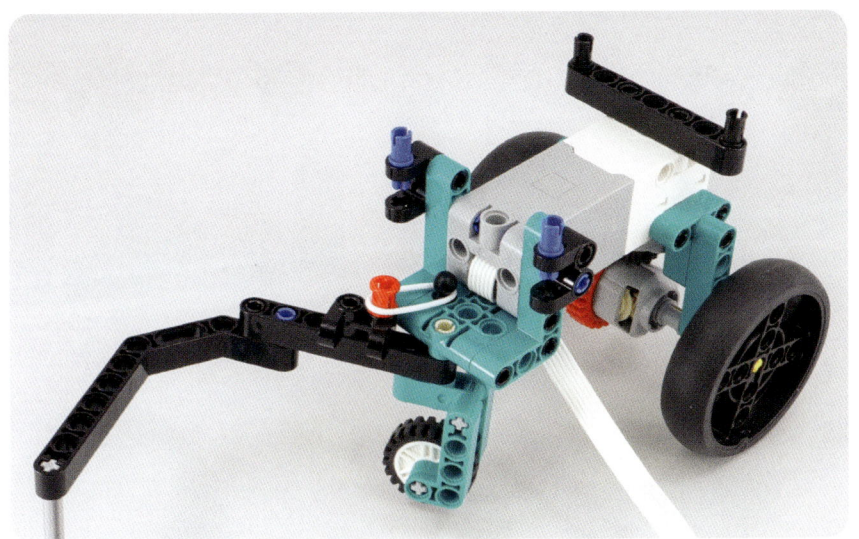

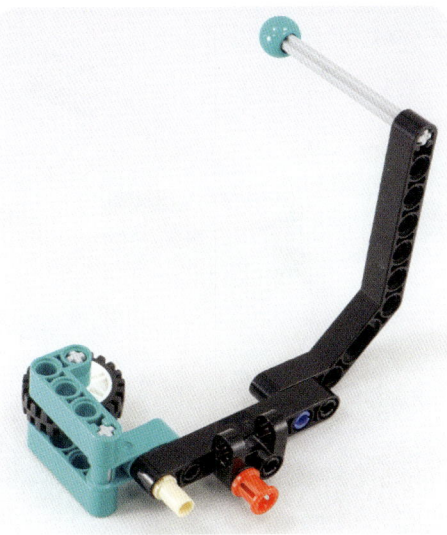

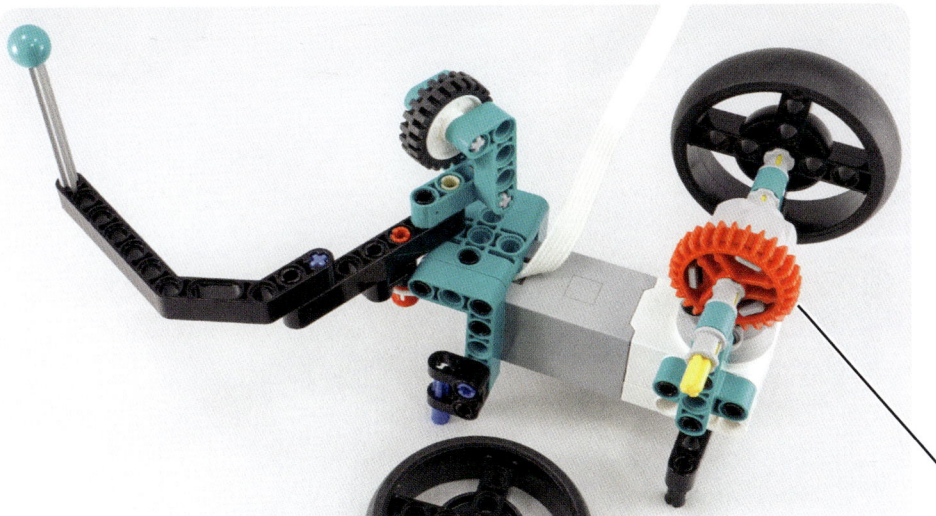

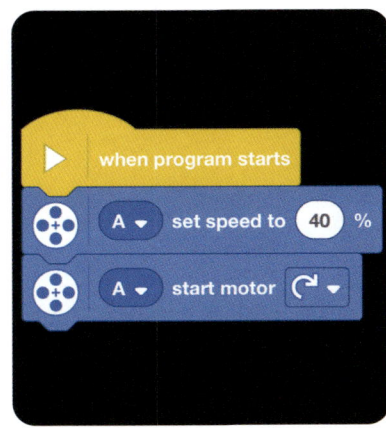

For details on how to build this gear, see page 85.

#104

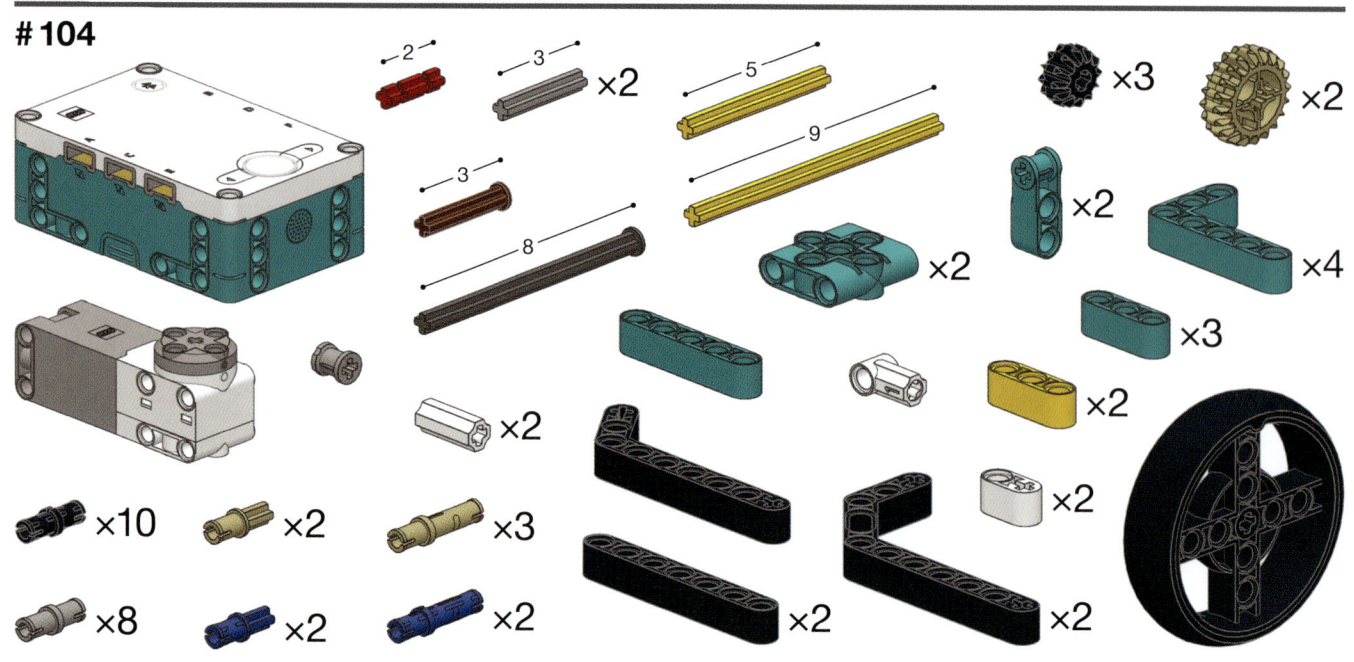

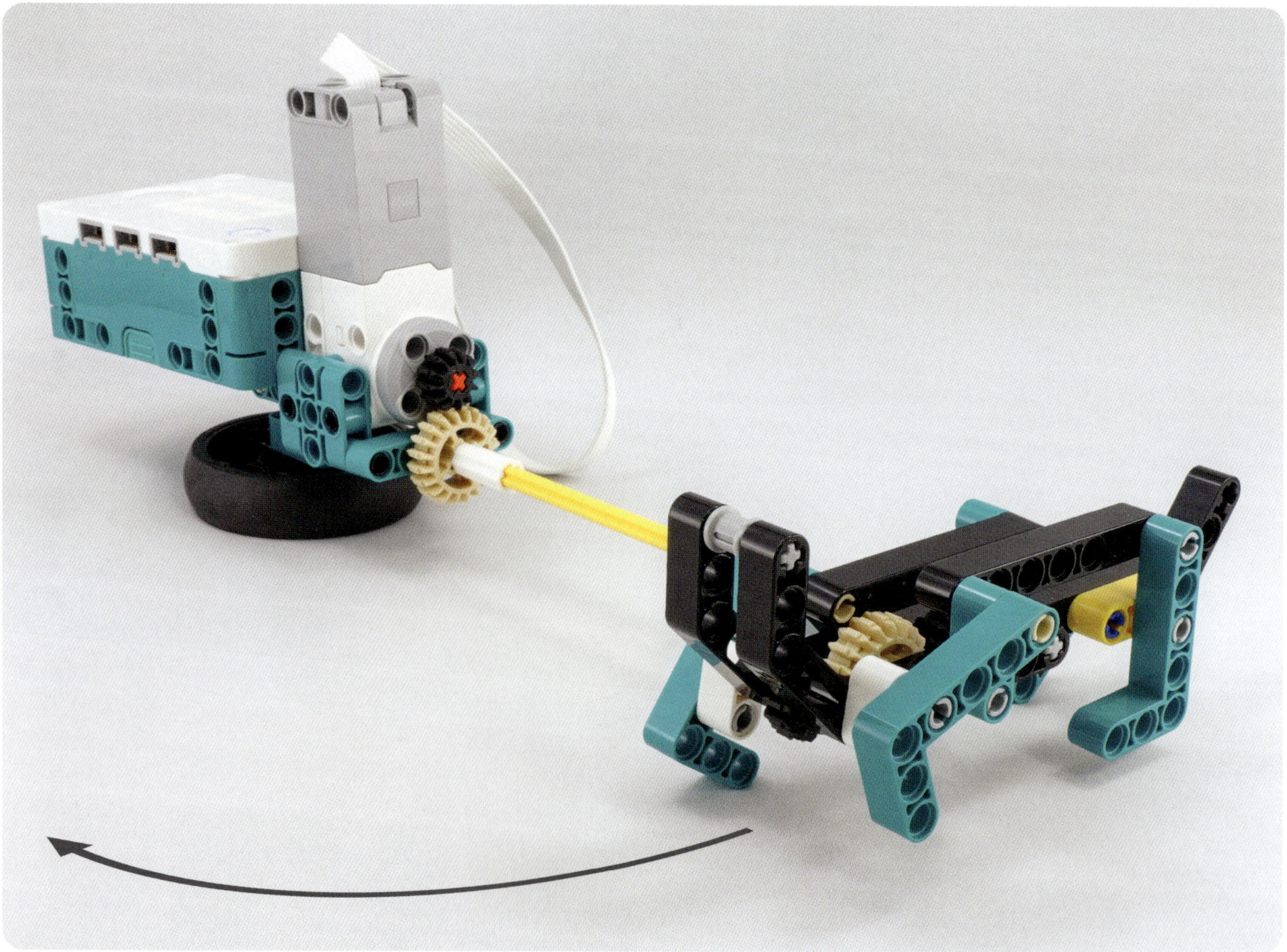

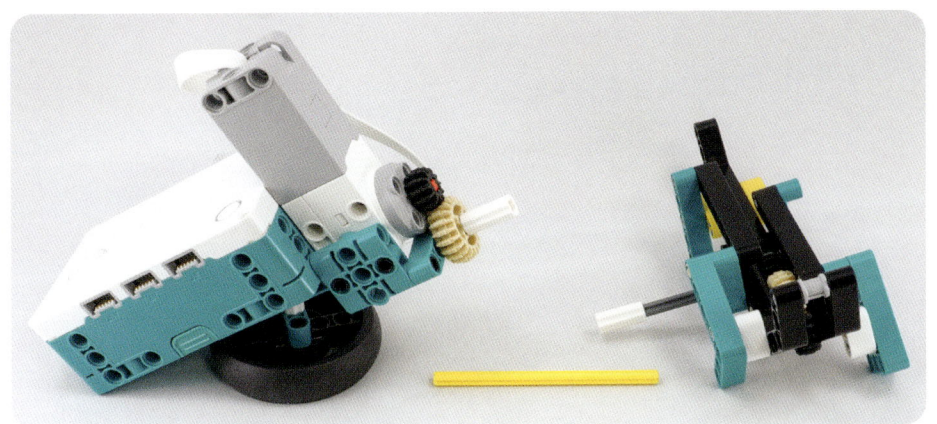
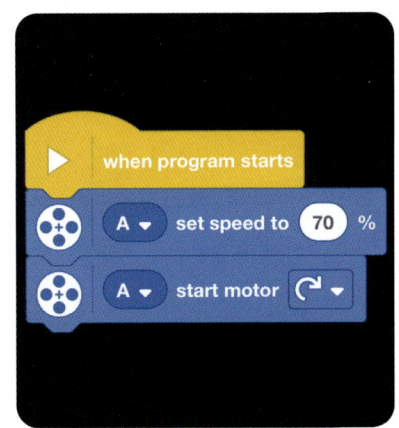
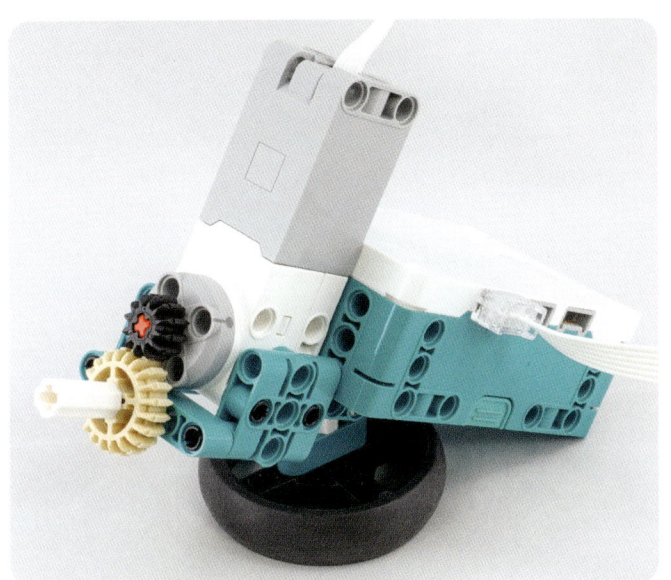
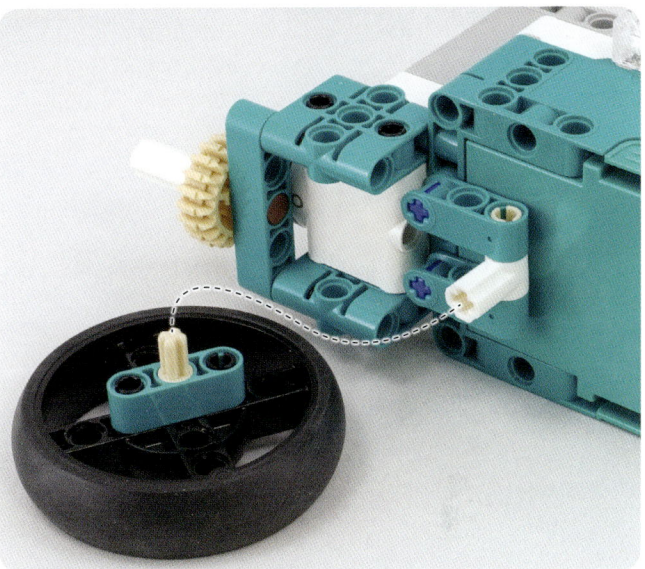
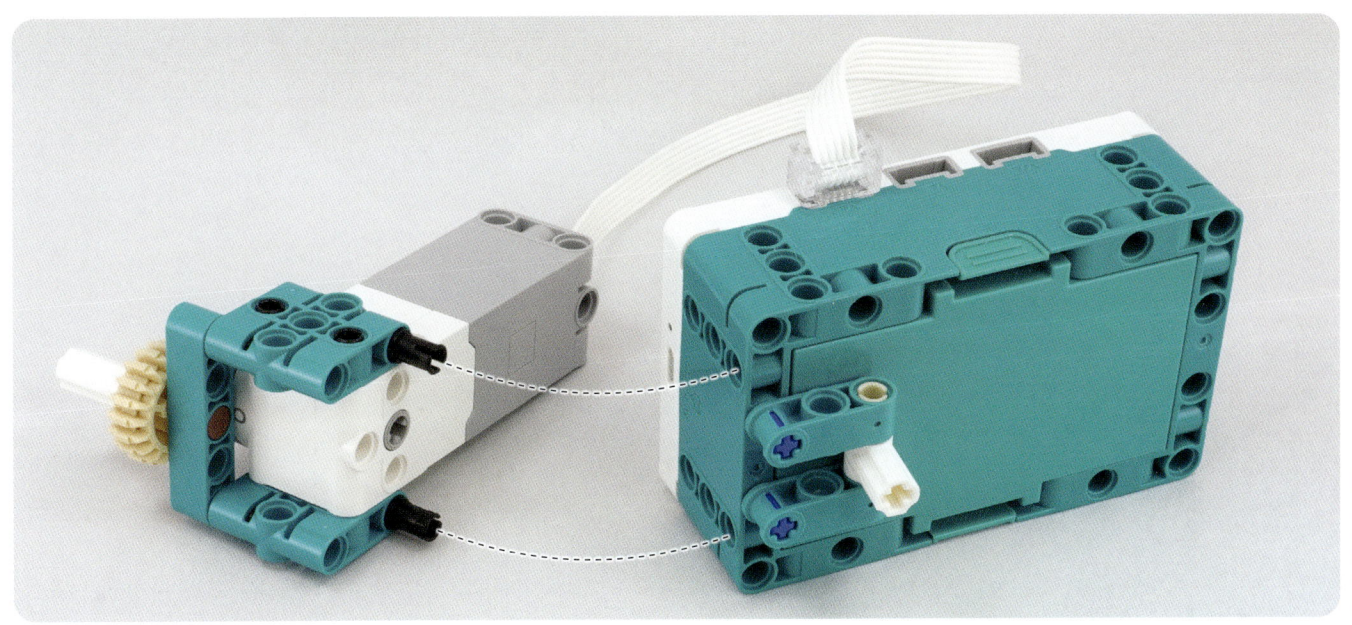

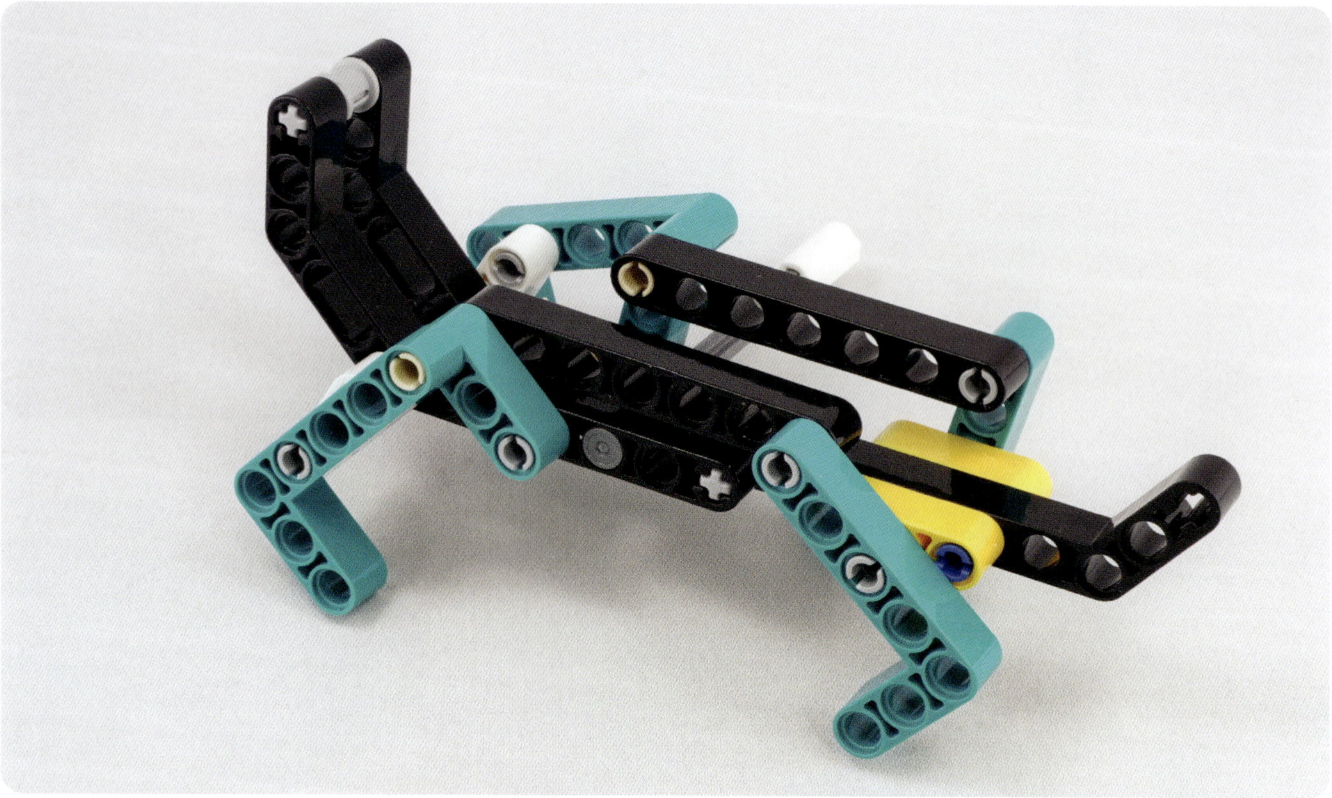

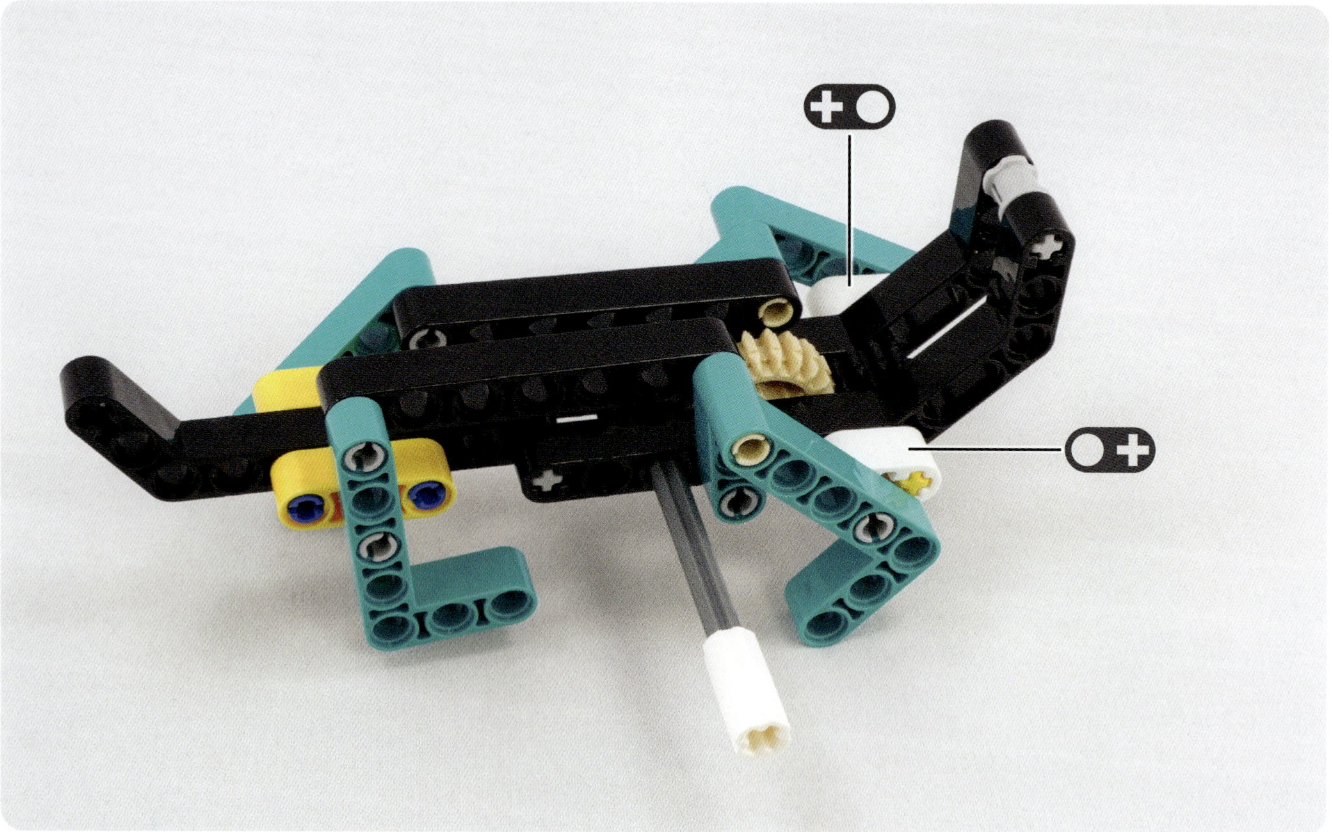

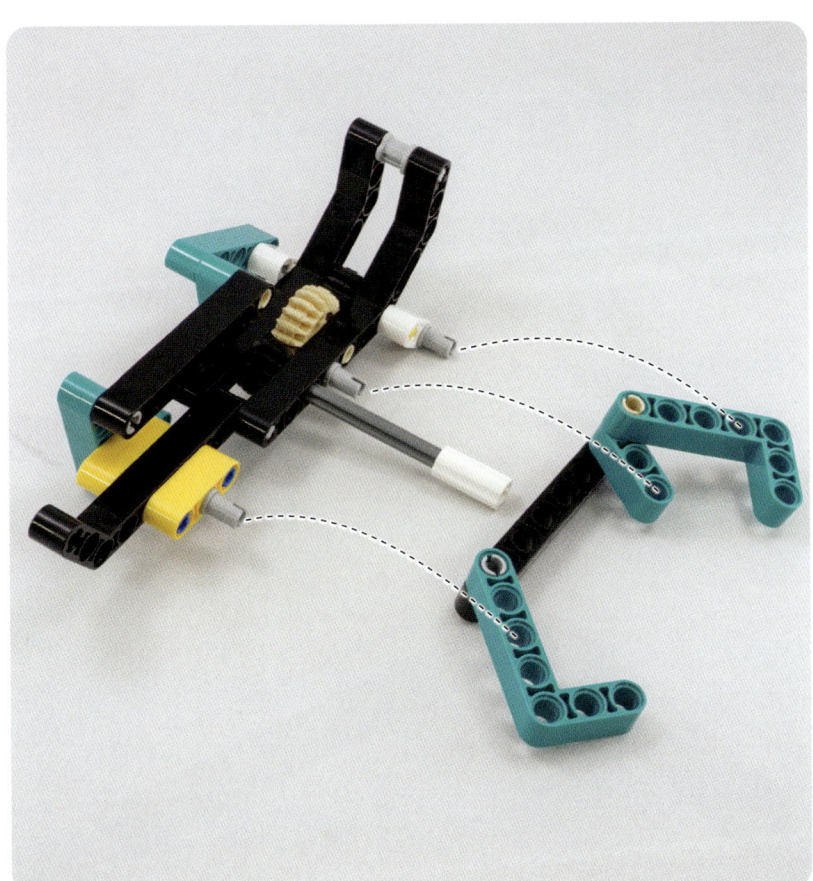

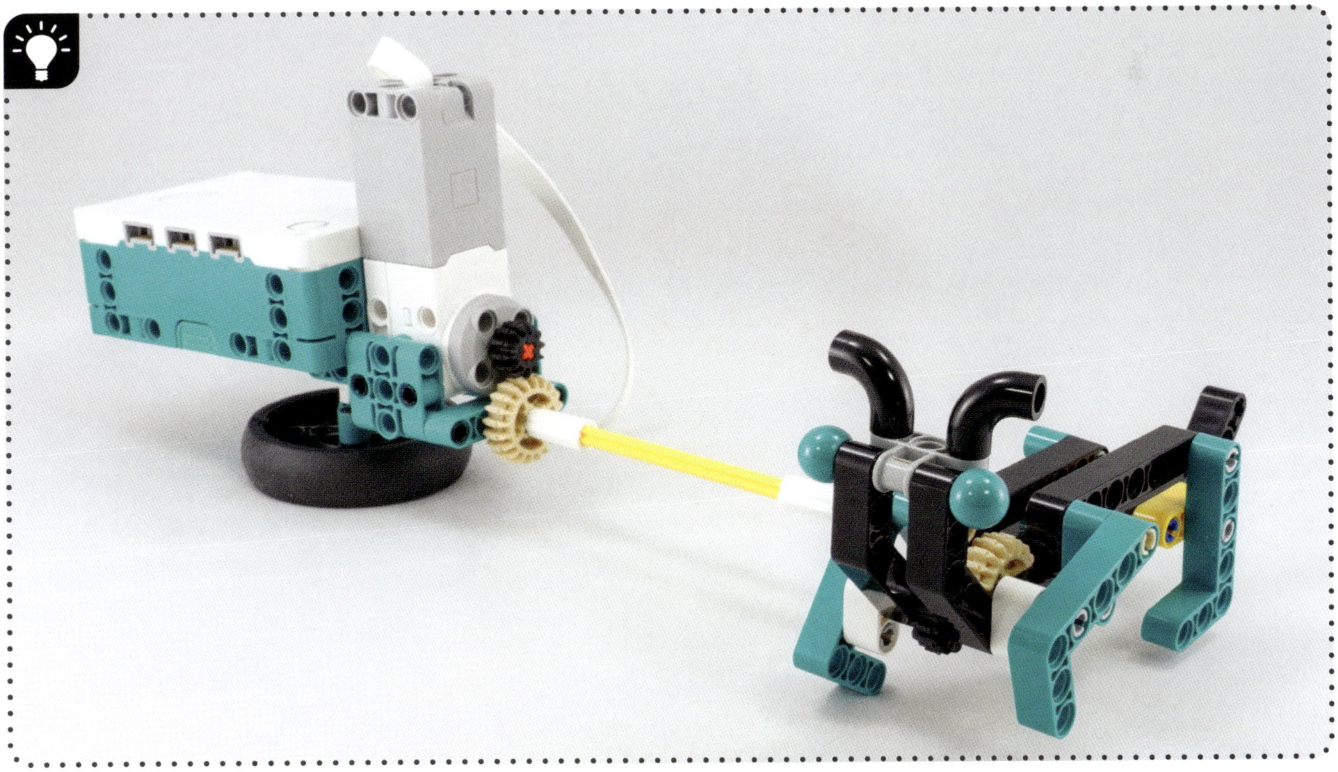

#105

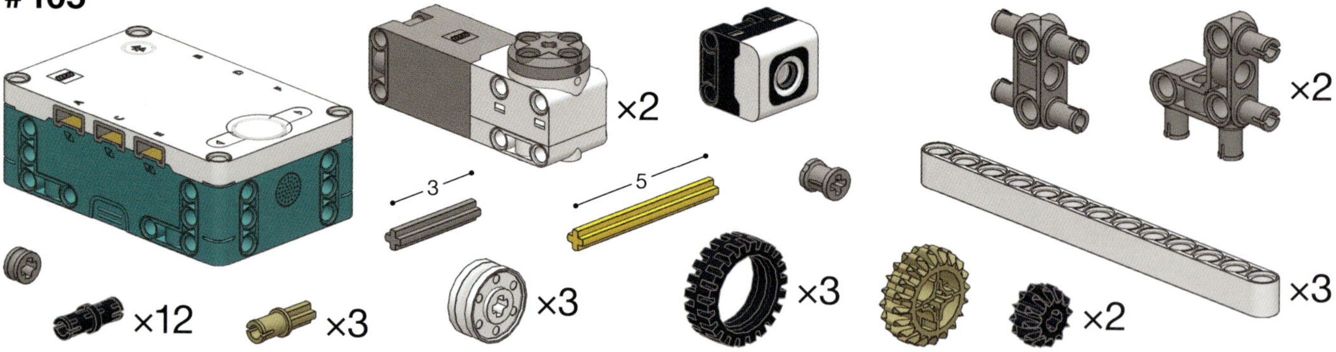

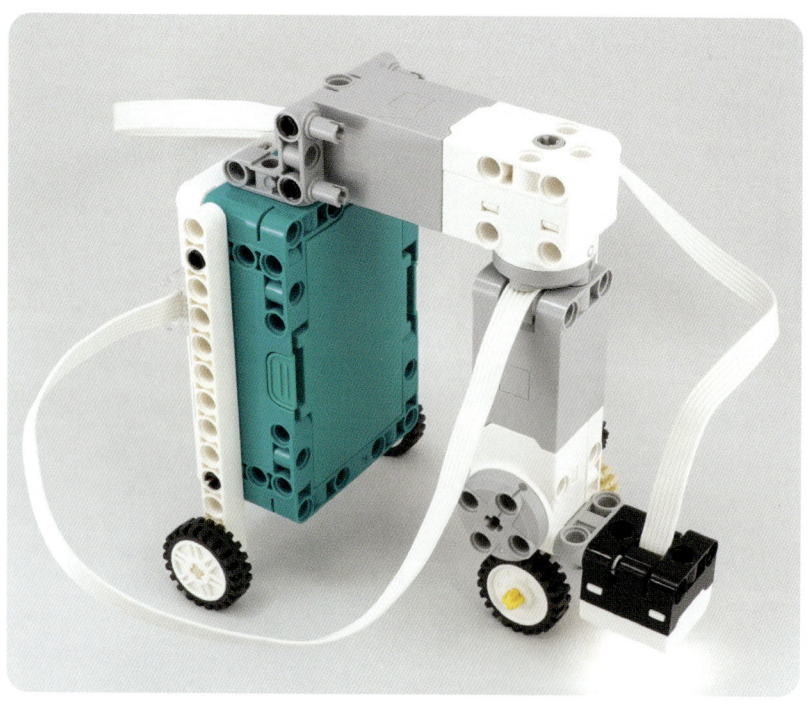

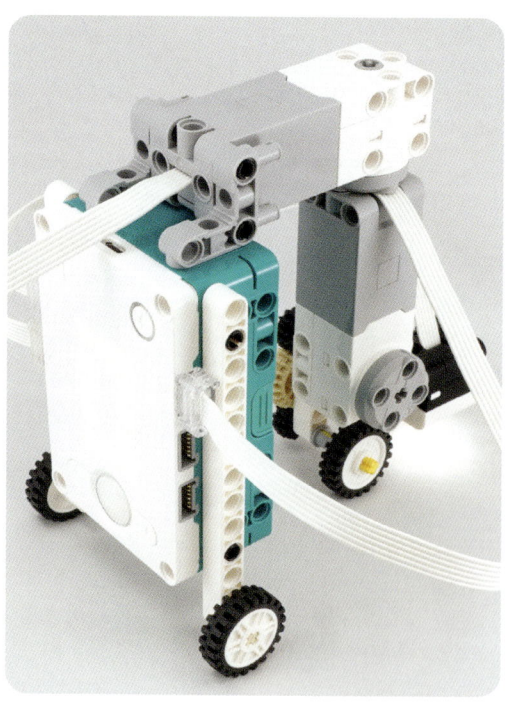

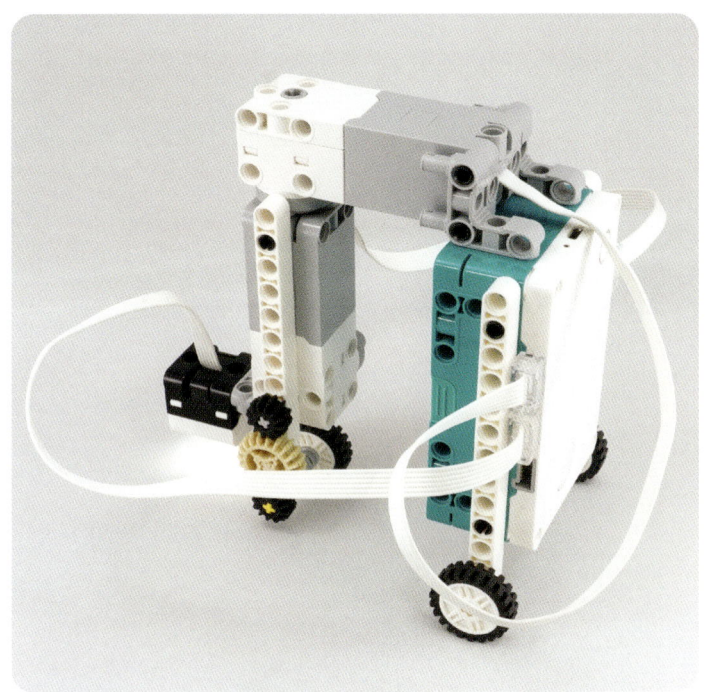

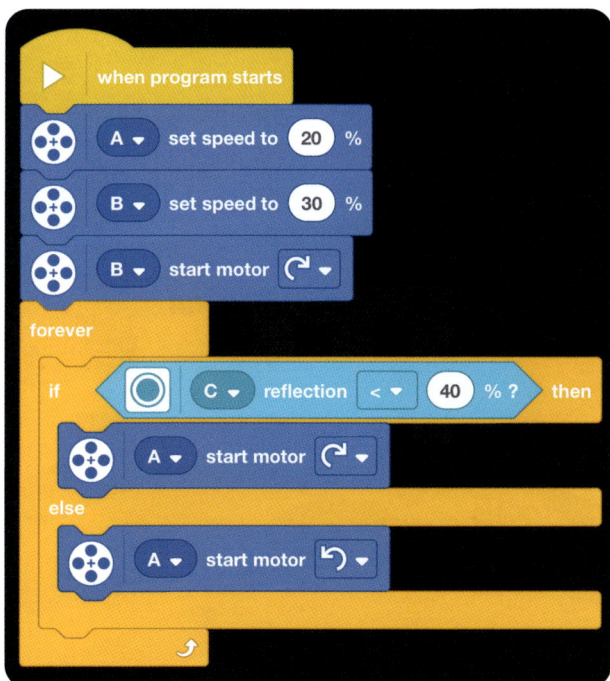

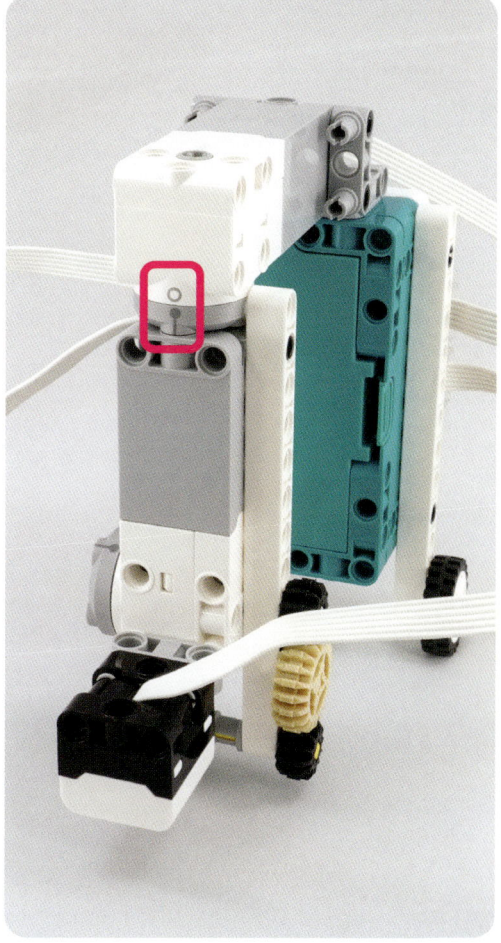

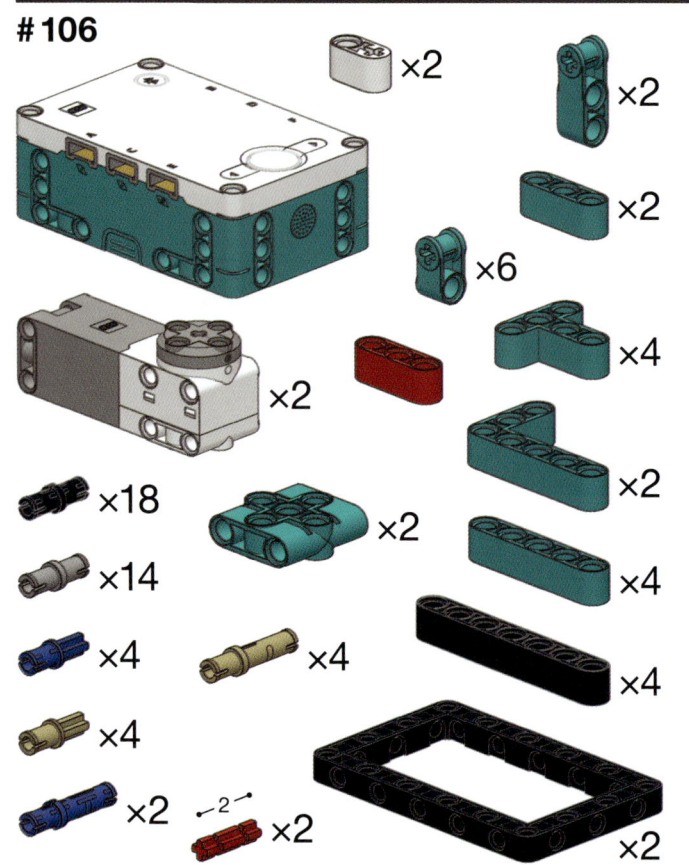

106

×2
×2
×2
×6
×2
×4
×2
×18
×2
×14
×4
×4
×4
×4
×4
×2
←2→ ×2
×2

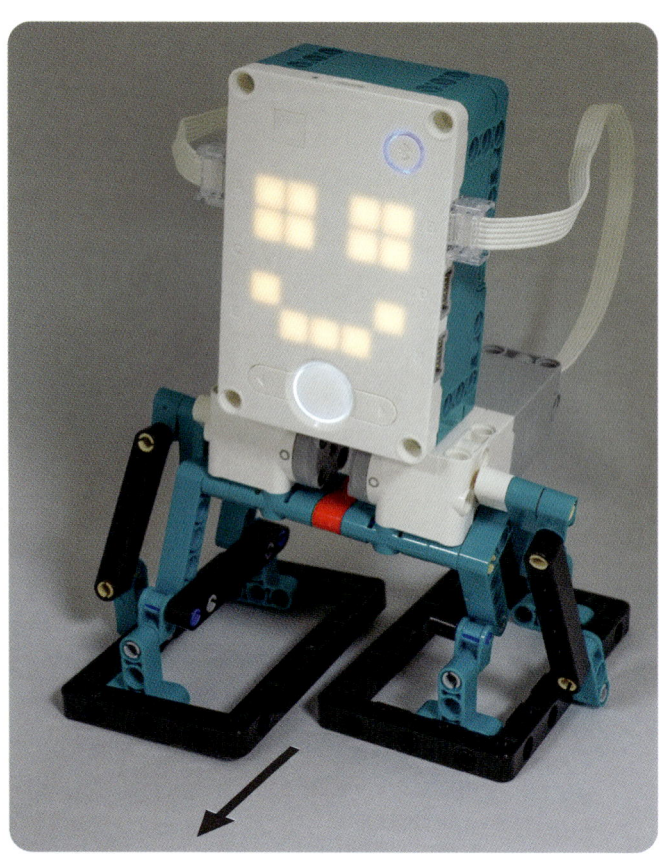

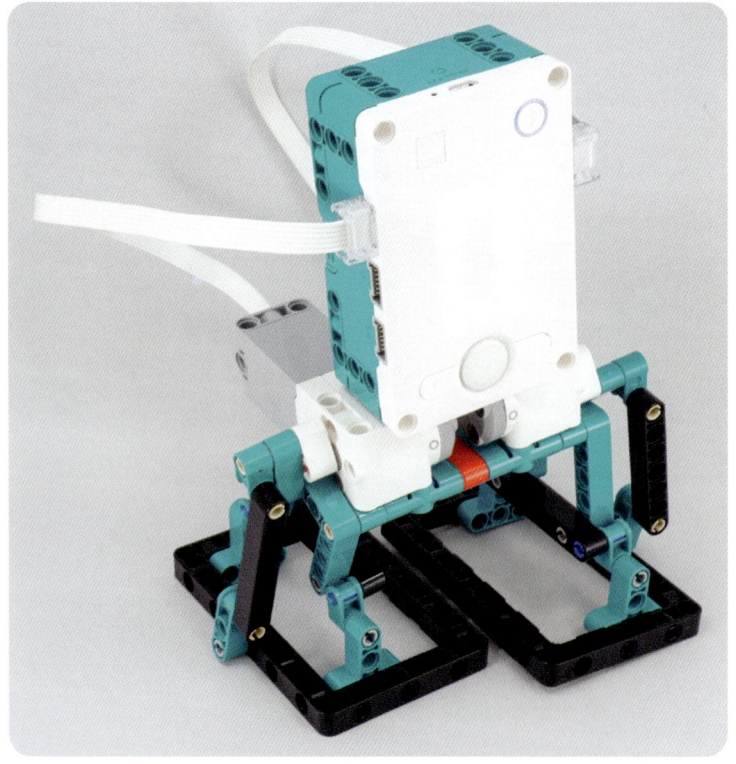

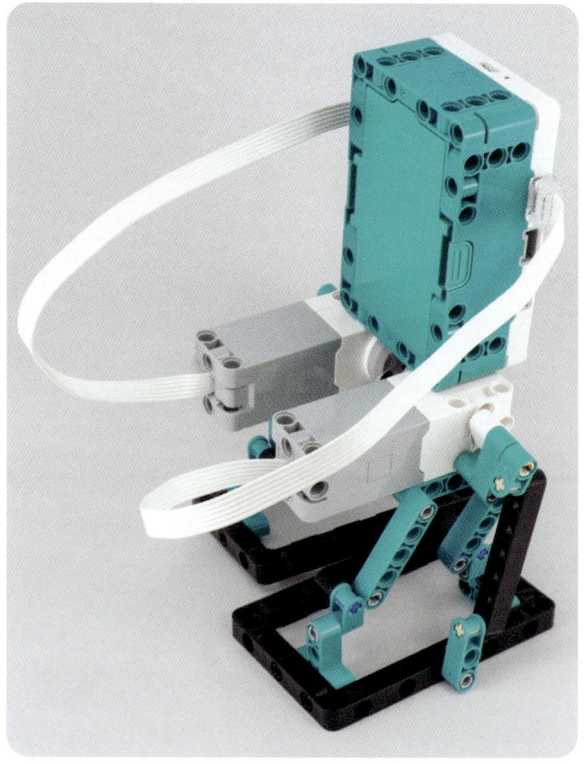

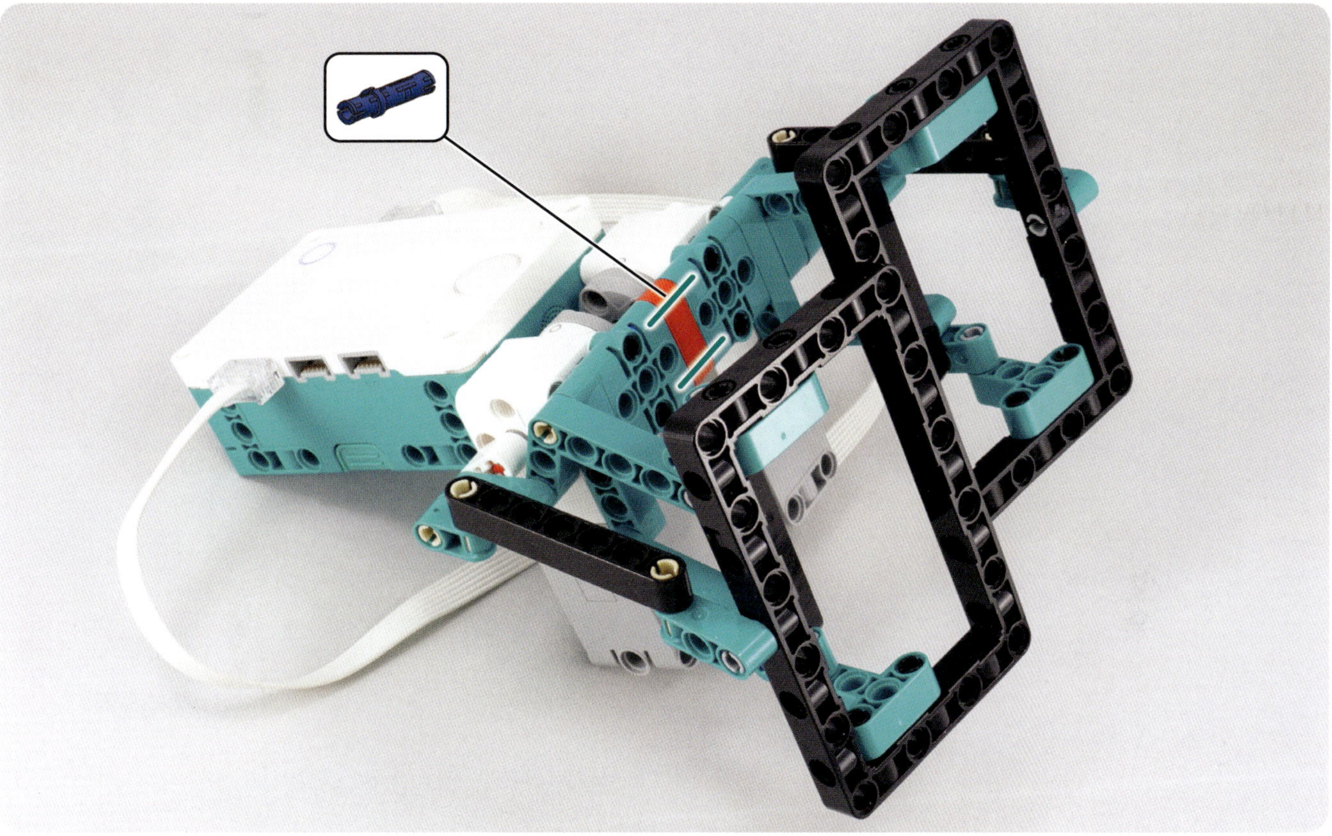

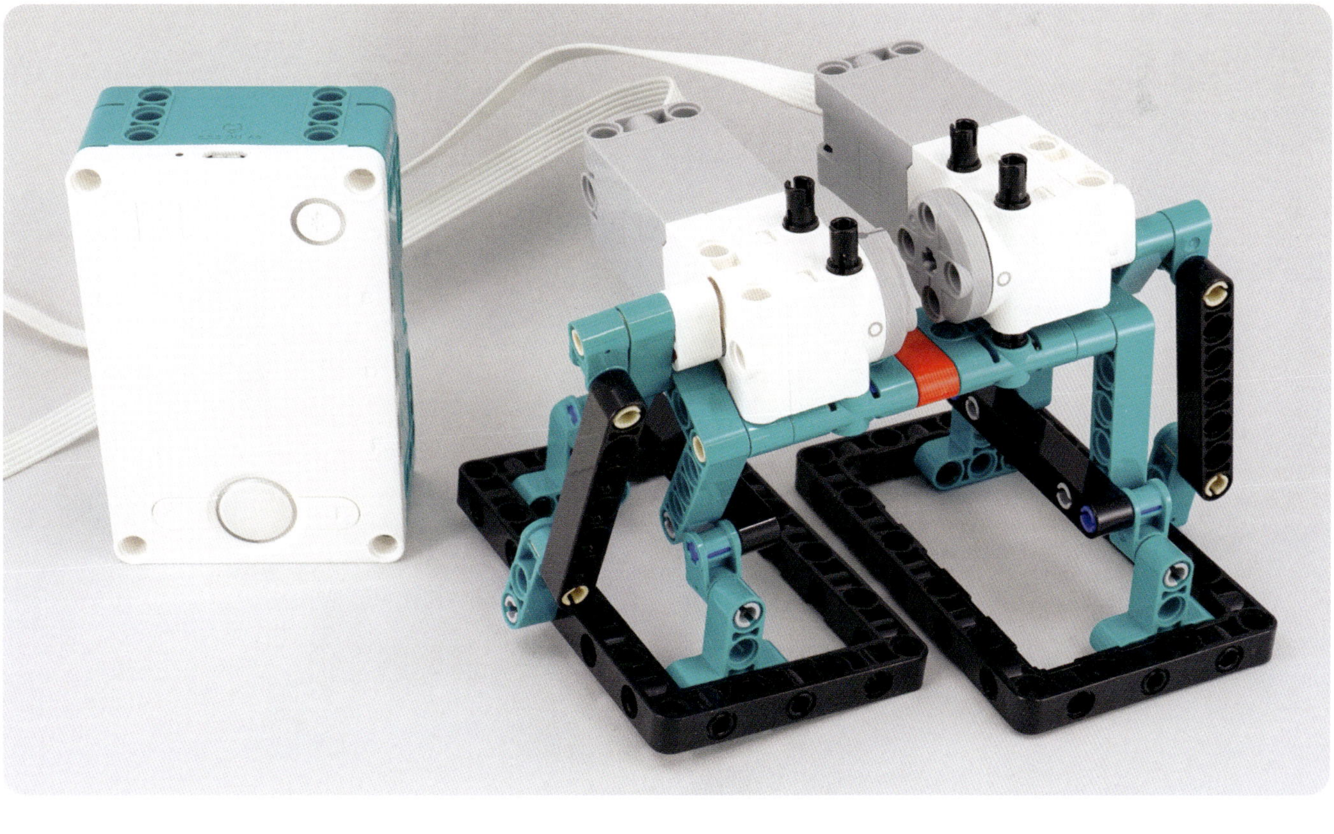

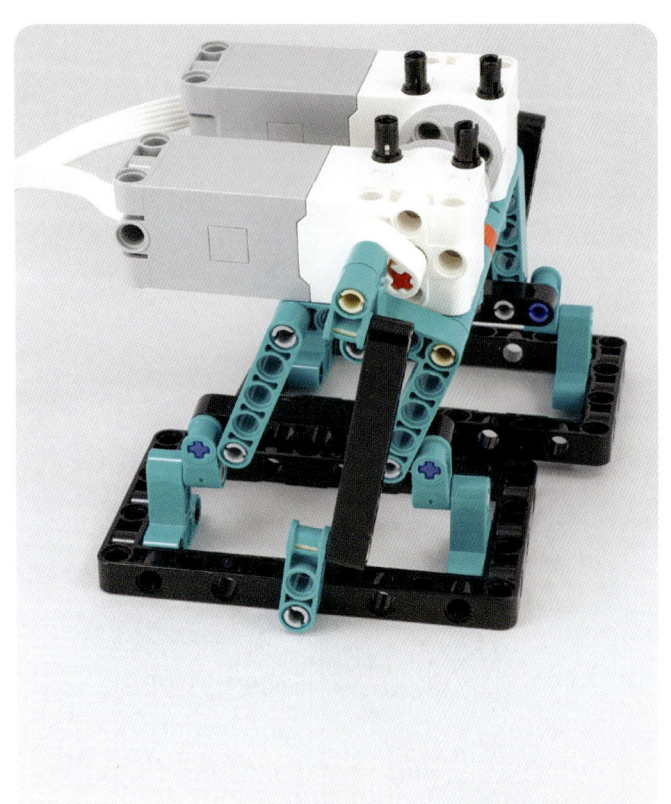

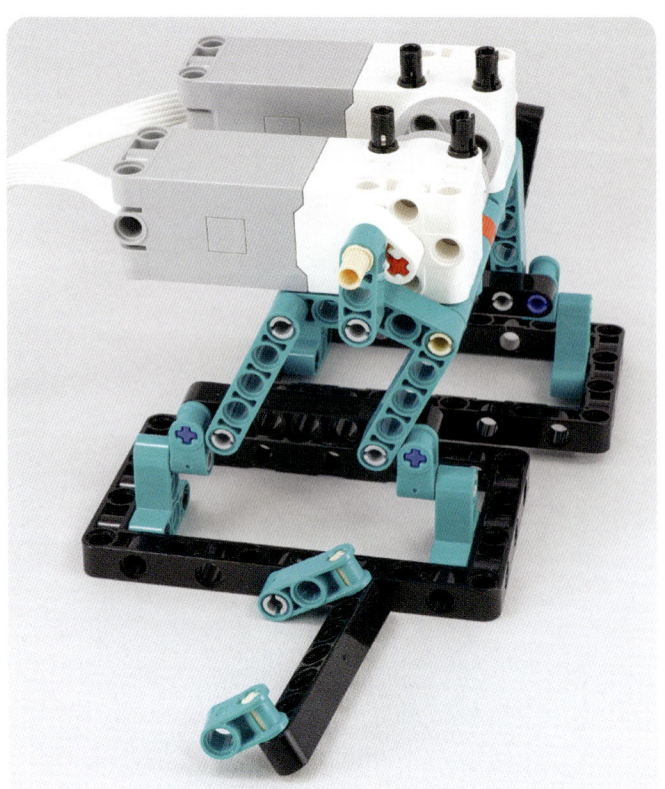

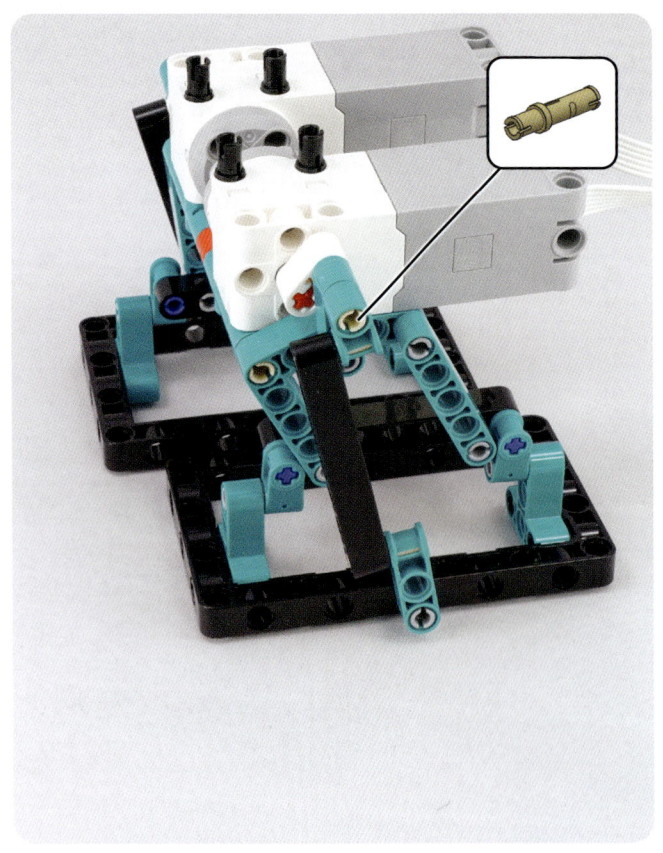

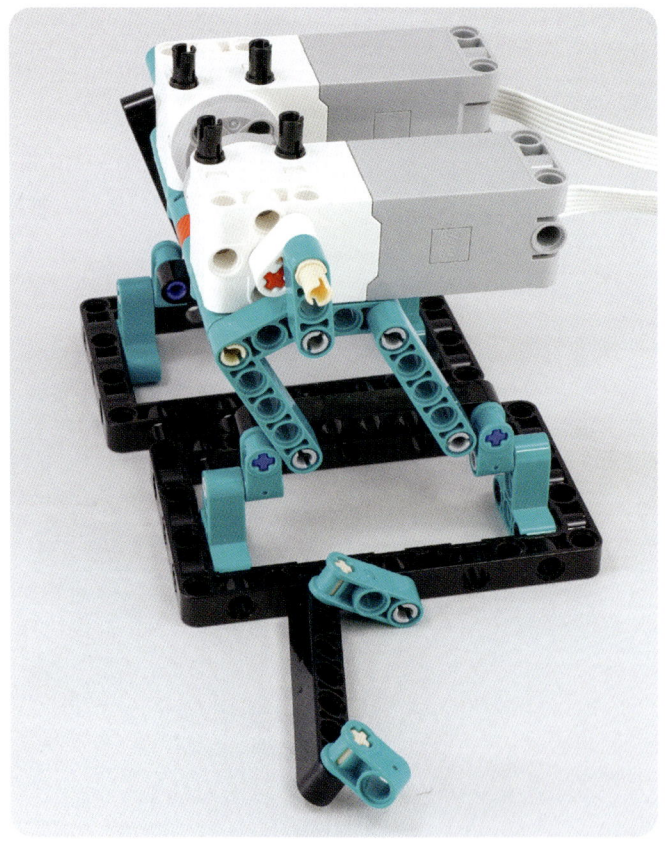

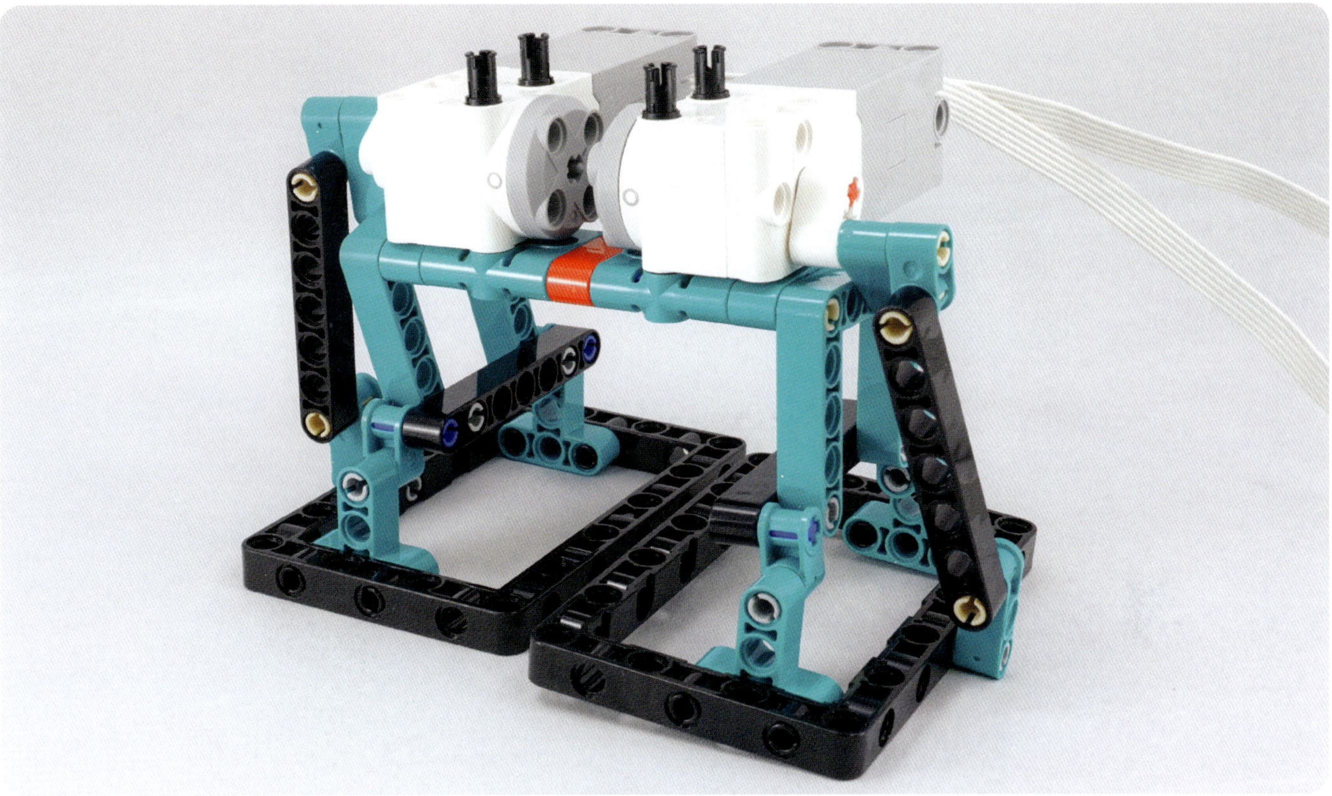

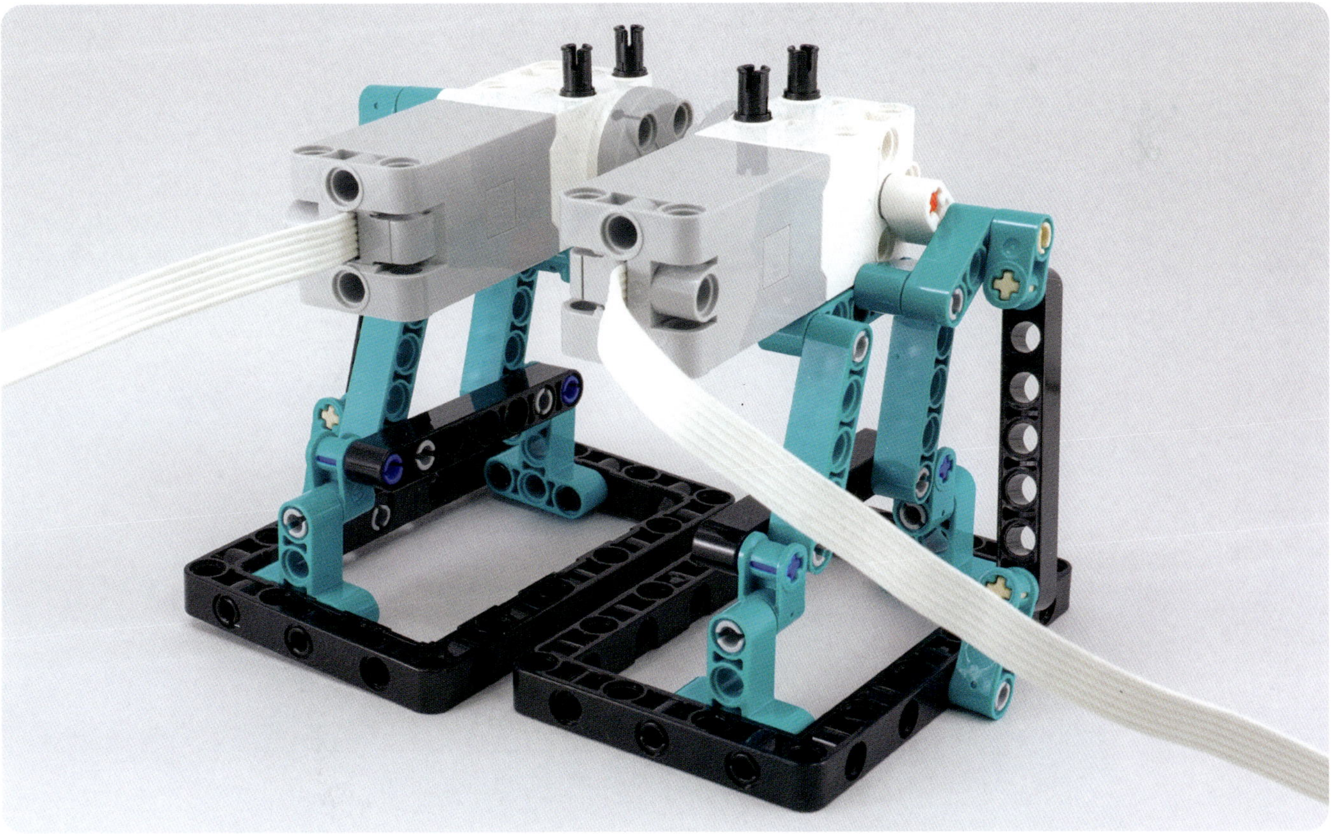

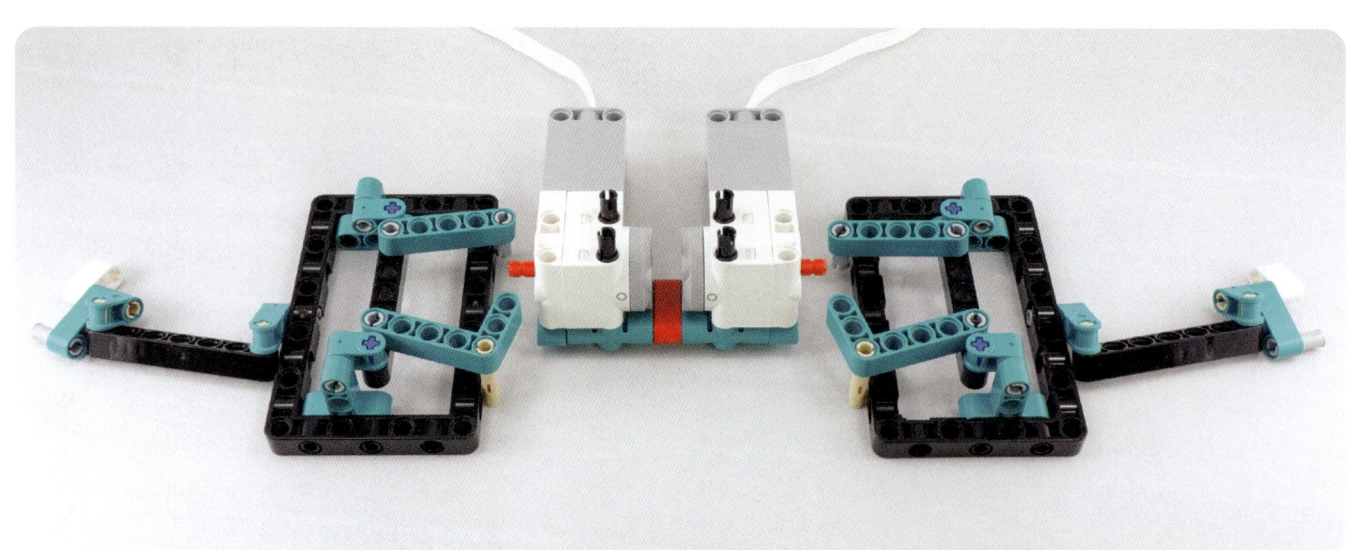

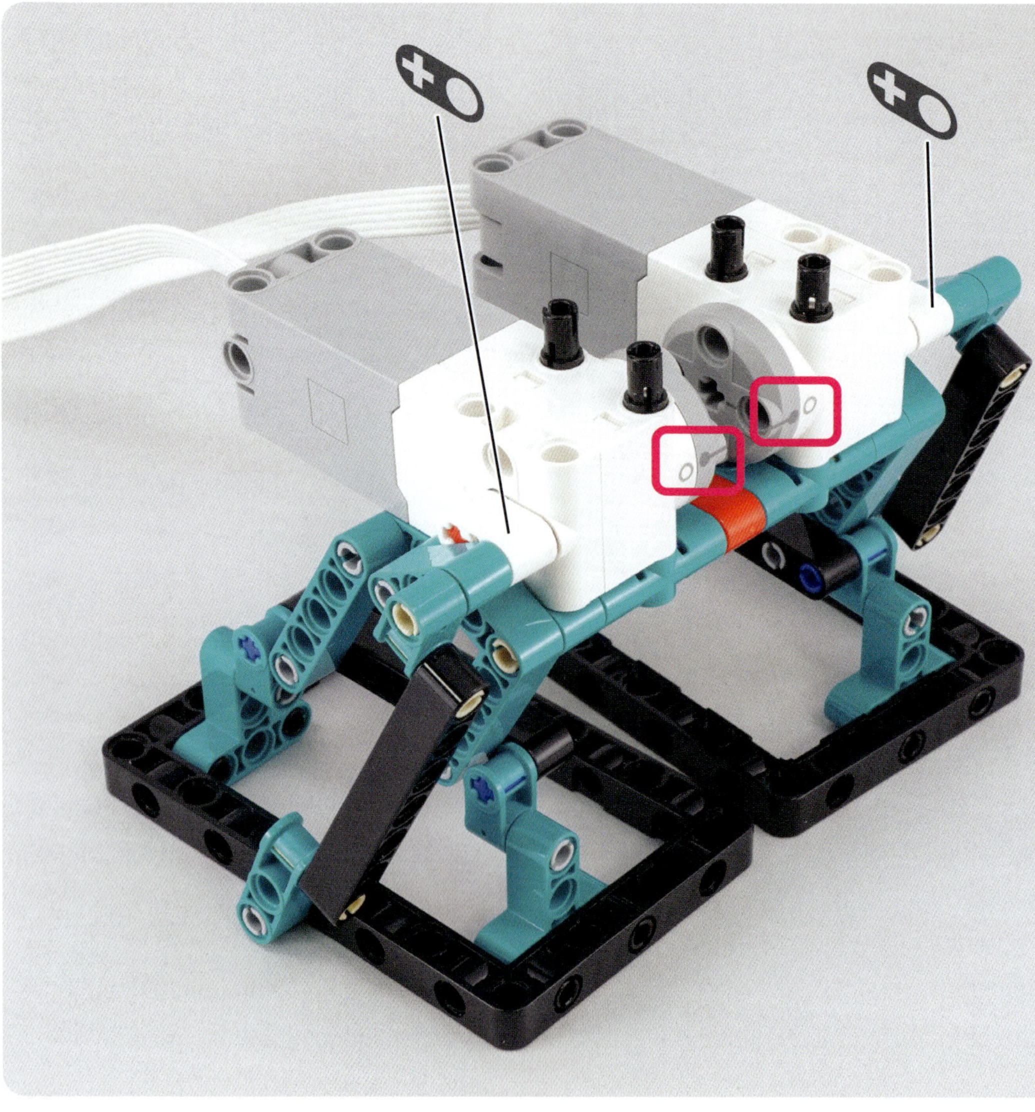

Line up the marks on each motor. Then place the white parts on the left and right in the same direction as shown.

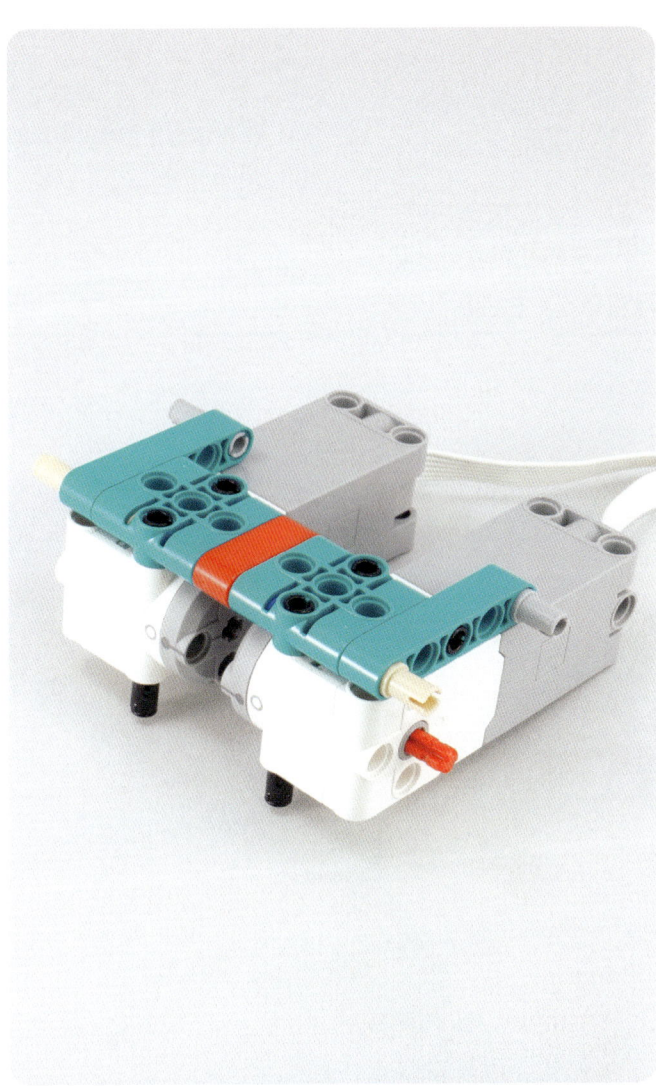

Go straight
ahead

Turn to the
left

Go straight
ahead

Turn to the
right

Go straight
ahead

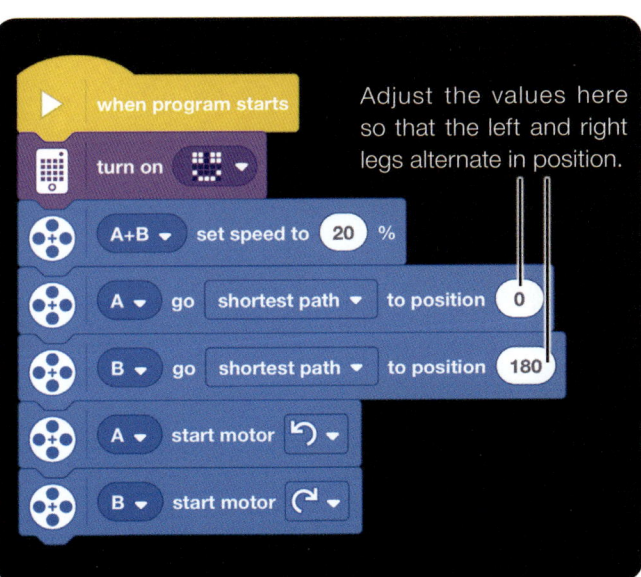

Adjust the values here
so that the left and right
legs alternate in position.

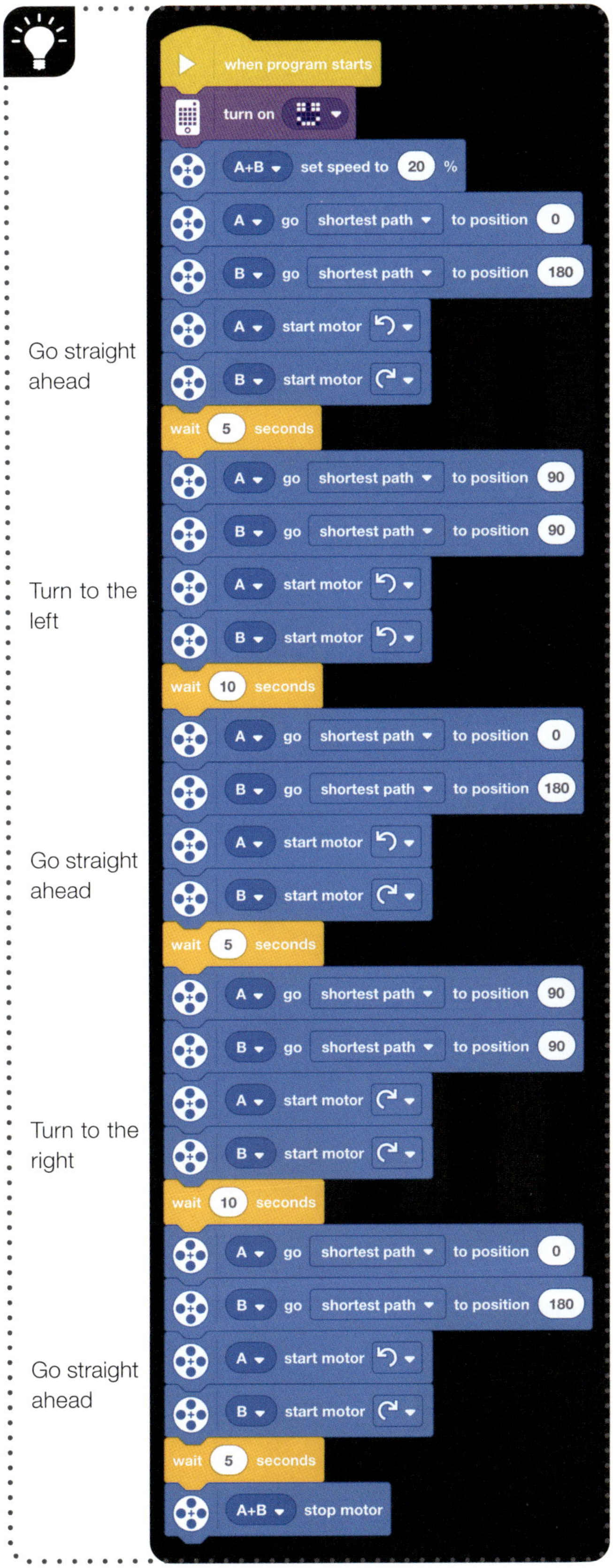

Spinning tops

#107

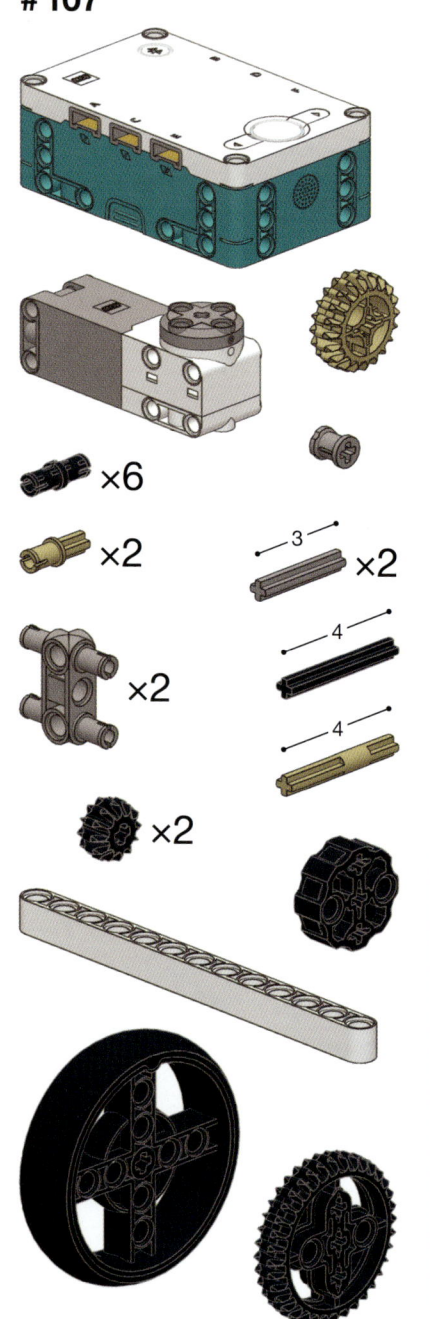

×6

×2

3 ×2

4

4

×2

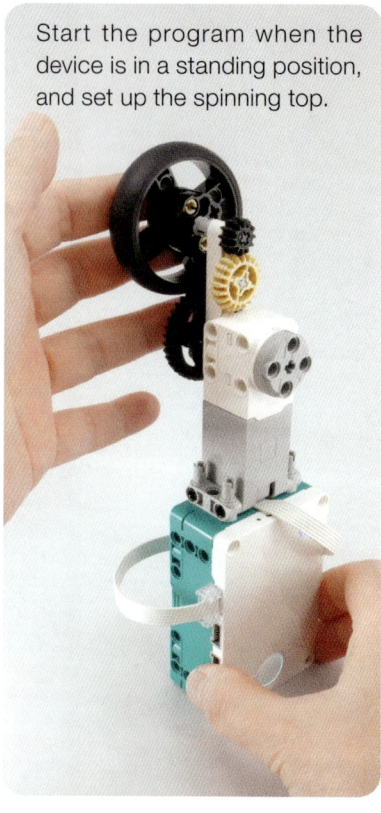

Start the program when the device is in a standing position, and set up the spinning top.

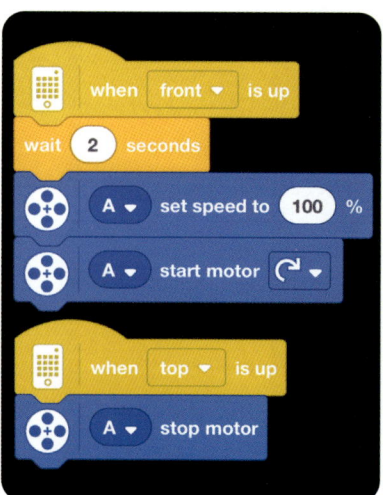

```
when  front ▼  is up

wait  2  seconds

A ▼  set speed to  100  %

A ▼  start motor ↻ ▼

when  top ▼  is up

A ▼  stop motor
```

Support the spinning top with your hand and put it down.

After 2 seconds, it begins to spin.

Once the rotation is stable, release the spinning top. Then allow the device to stand.

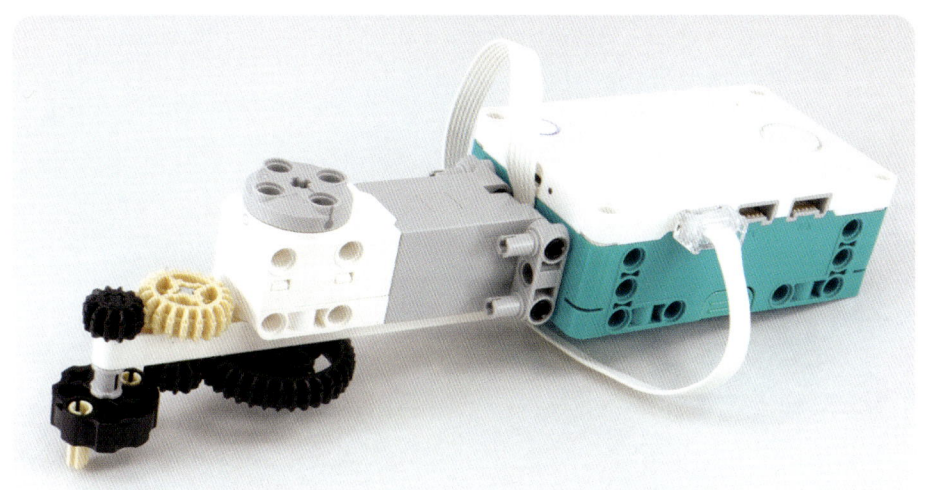

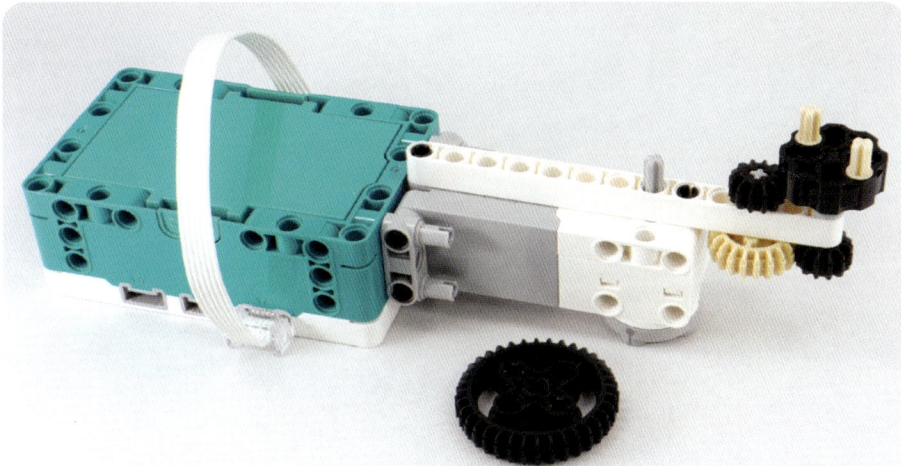

4

108

×16

×3

×2

×2

×2

×2

×2

×2

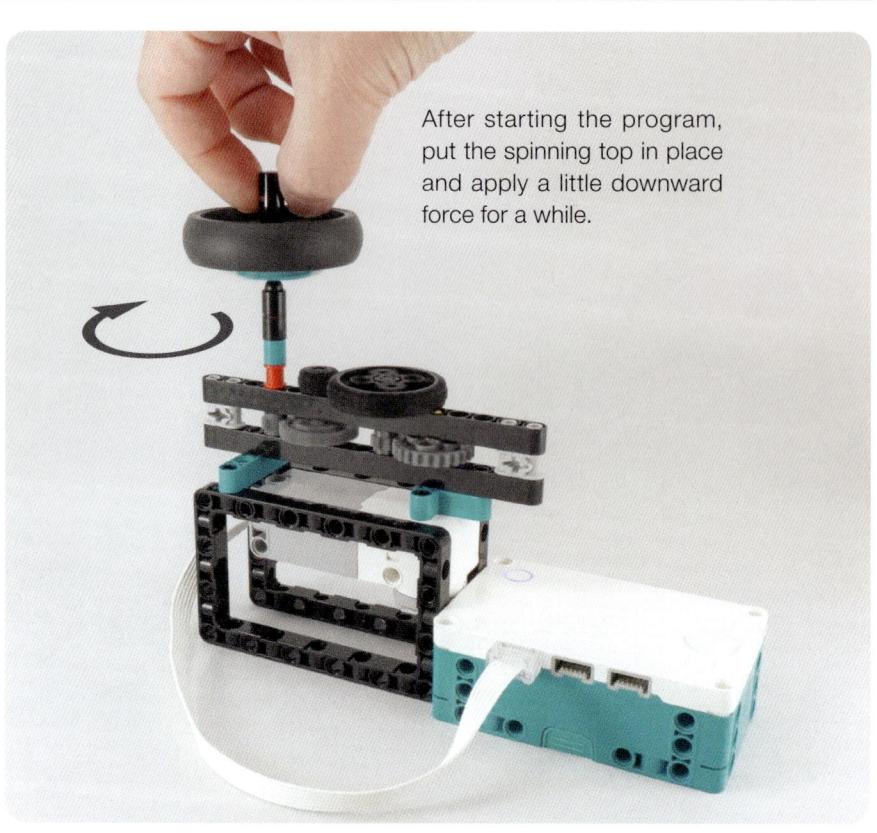

After starting the program, put the spinning top in place and apply a little downward force for a while.

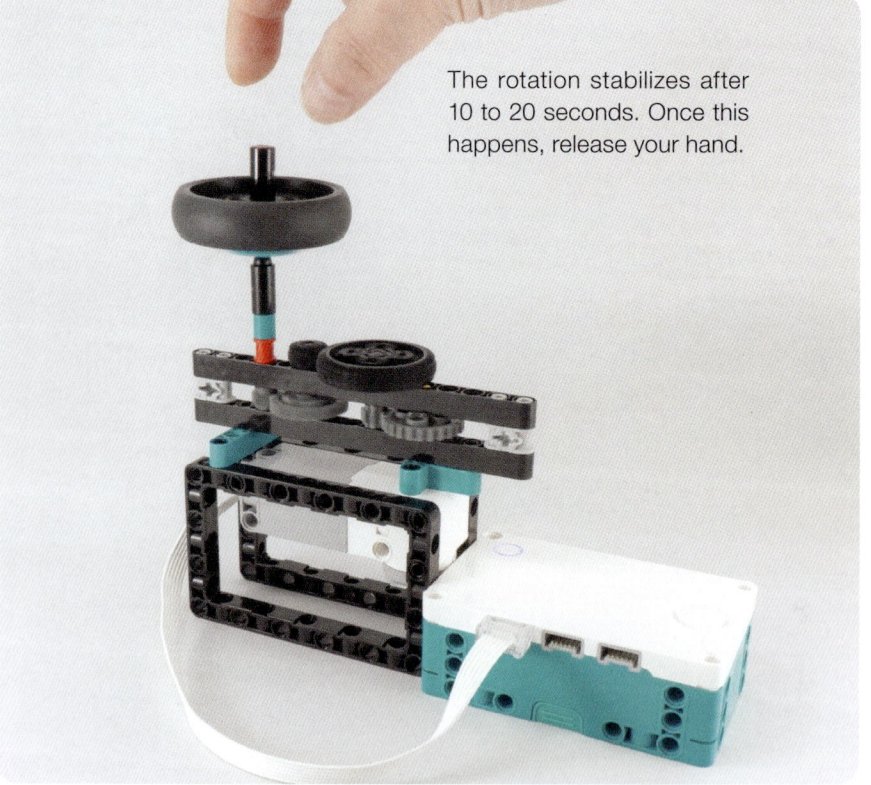

The rotation stabilizes after 10 to 20 seconds. Once this happens, release your hand.

The spinning top stays horizontal even if you tilt the base.

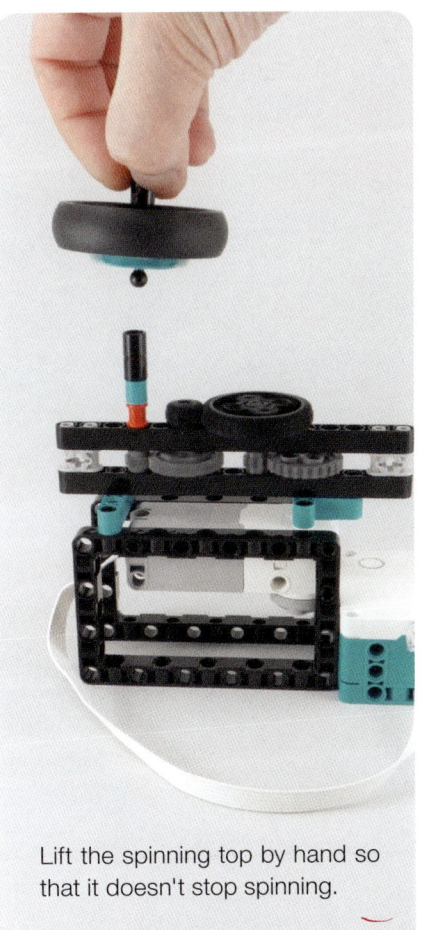

Lift the spinning top by hand so that it doesn't stop spinning.

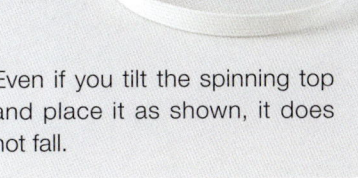

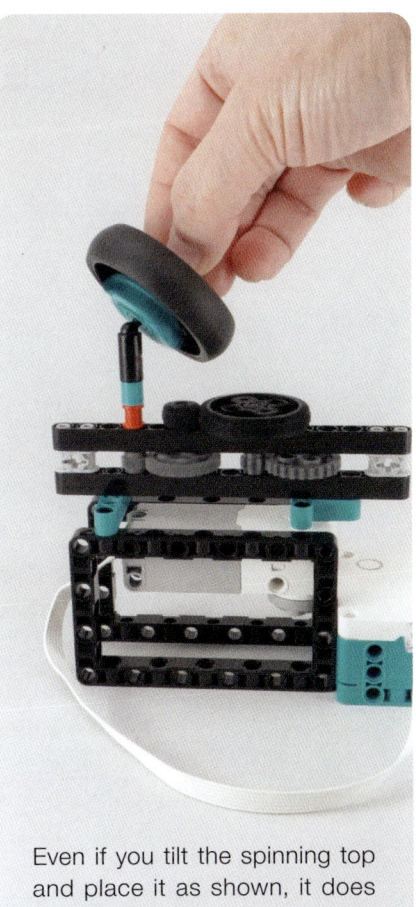

Even if you tilt the spinning top and place it as shown, it does not fall.

It gradually returns to its original position.

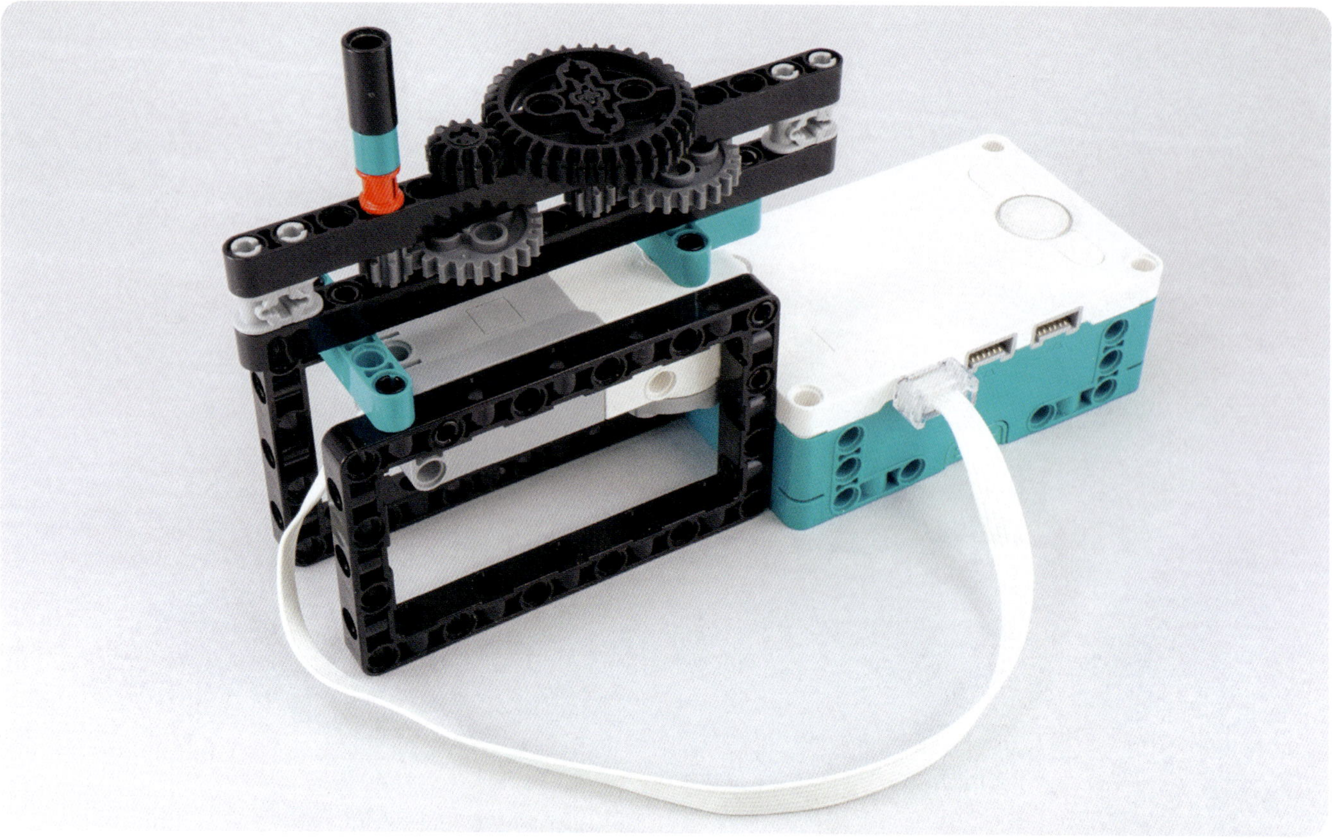

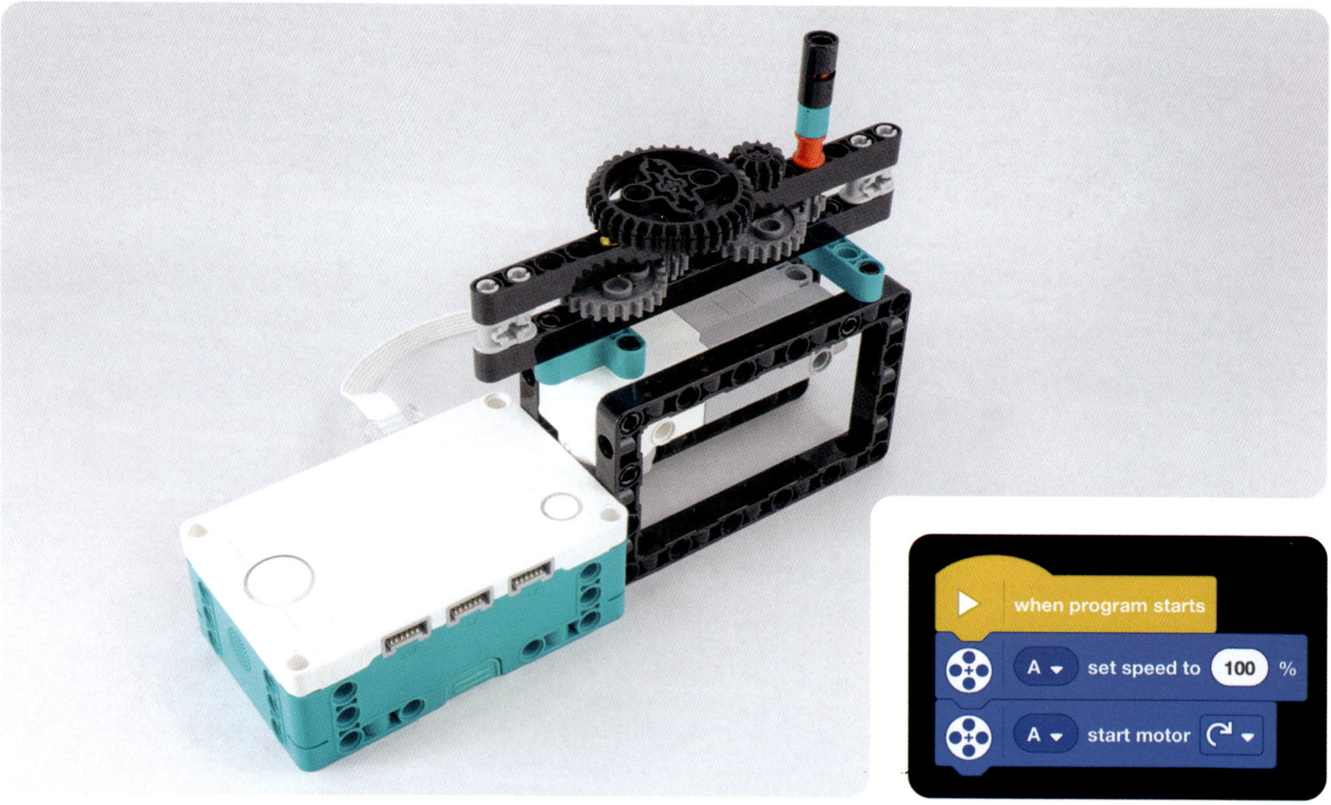

```
▶  when program starts

⚙  A ▾  set speed to  100  %

⚙  A ▾  start motor  ↻ ▾
```

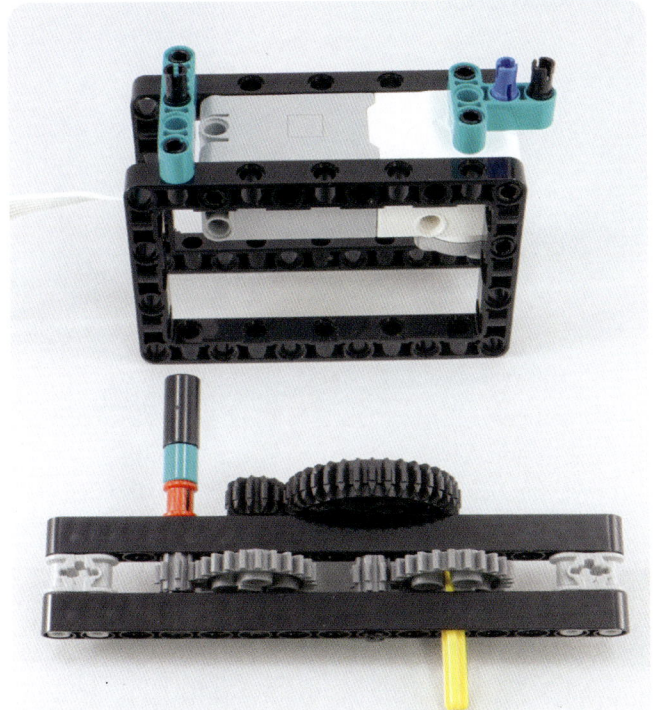

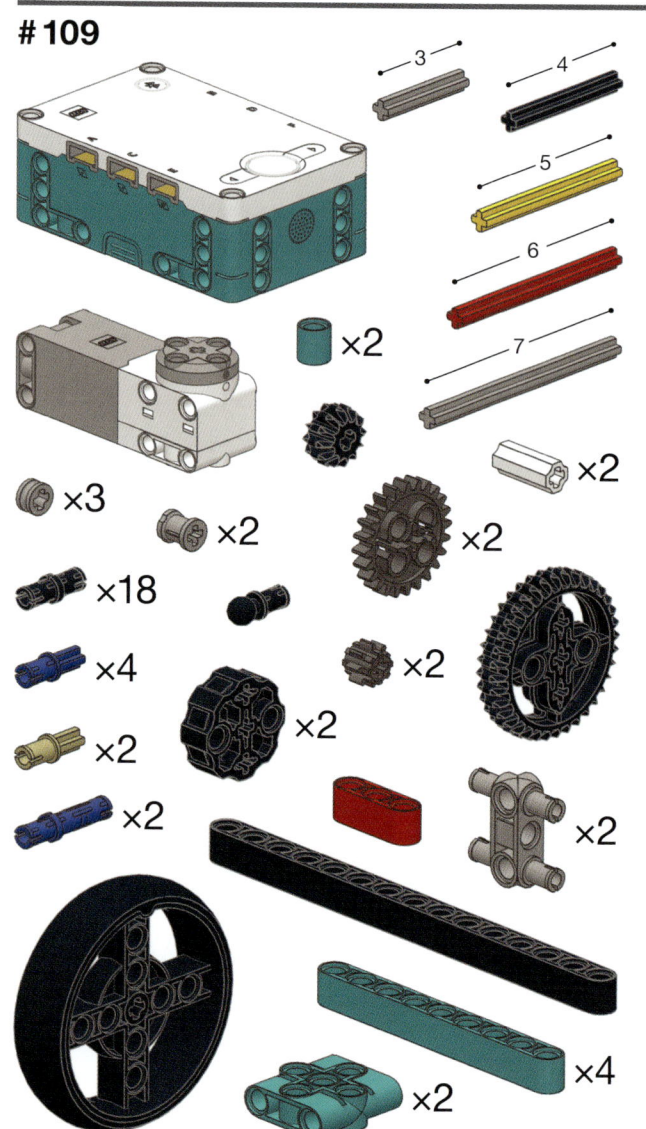

109

3
4
5
6
7

×2
×3
×2
×18
×4
×2
×2
×2
×2
×2
×2
×2
×4
×2

Start the program when the device is in a standing position, and set up the spinning top.

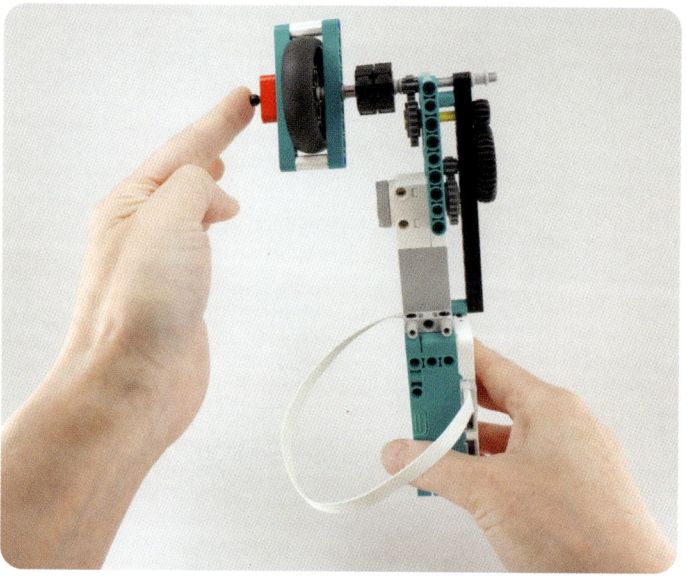

When the device is tilted 90 degrees, the spinning top starts to spin.

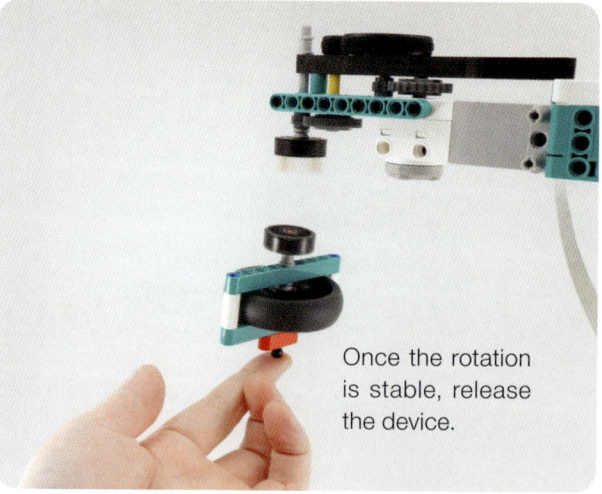

Once the rotation is stable, release the device.

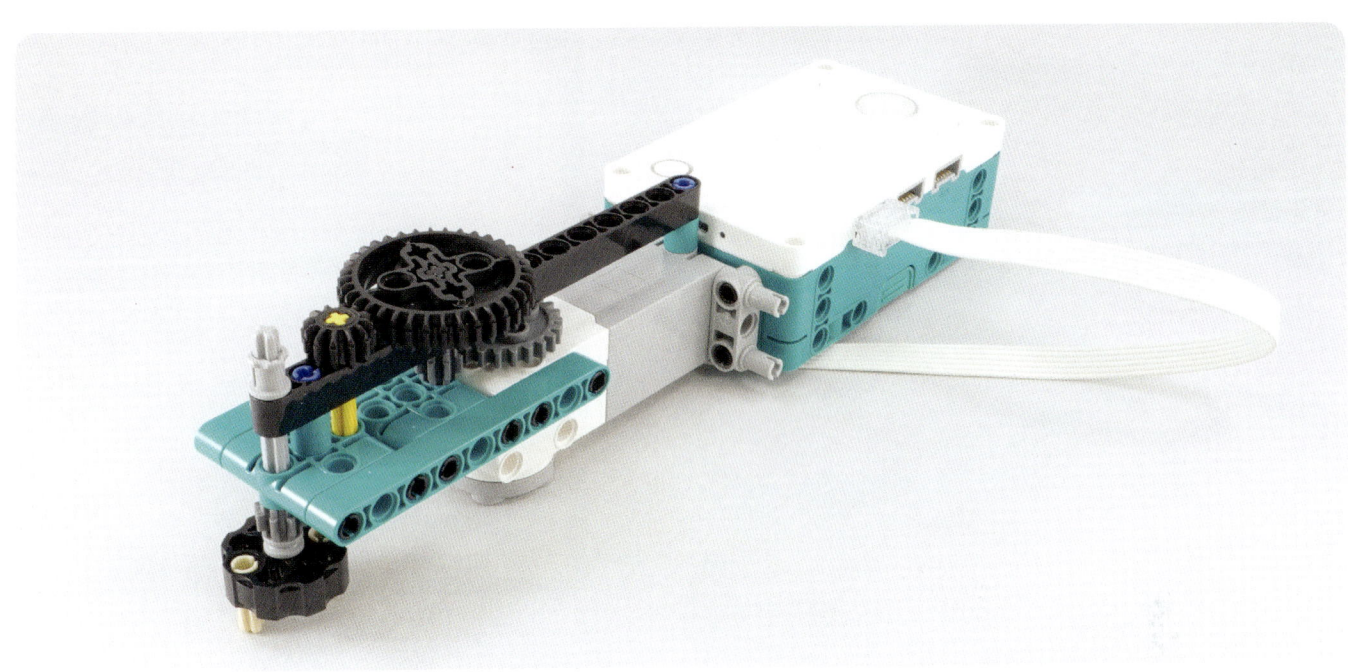

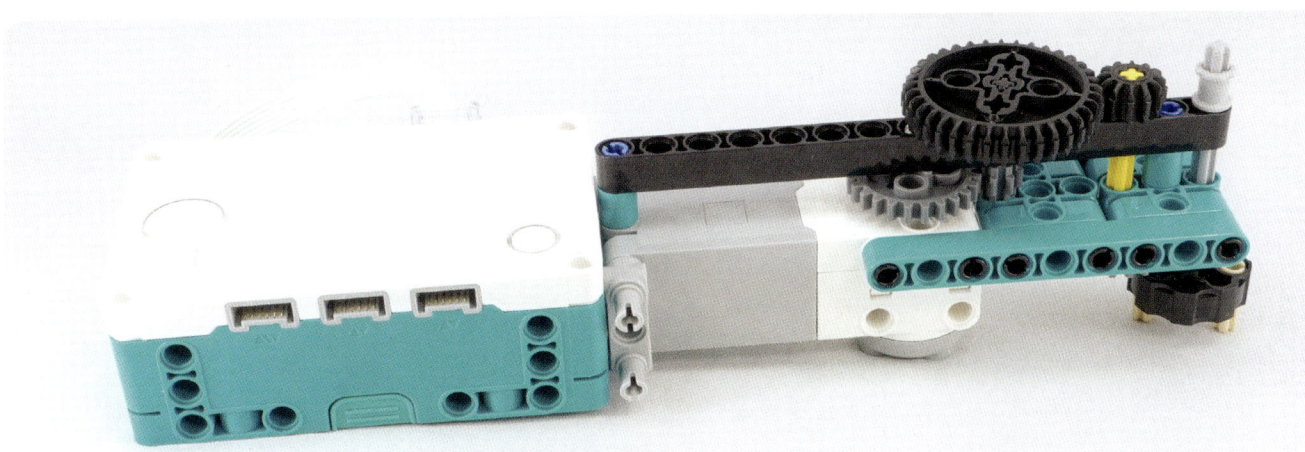

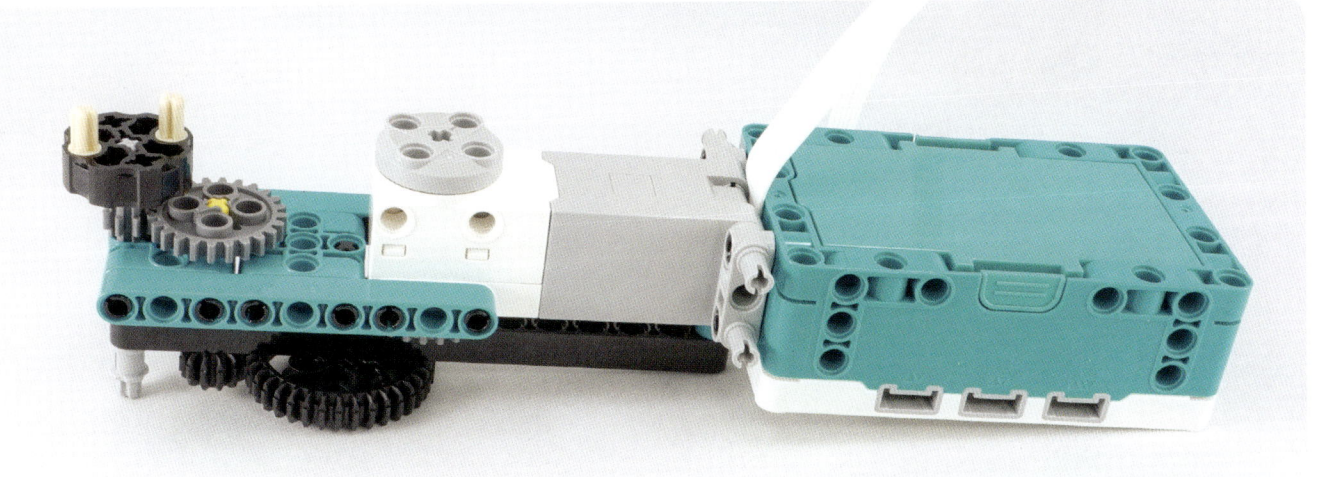

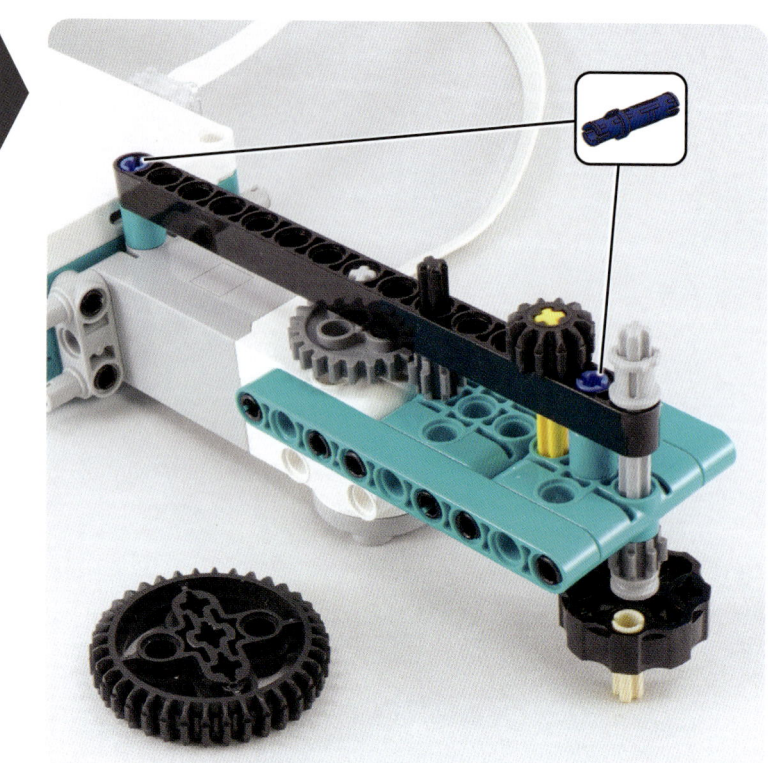

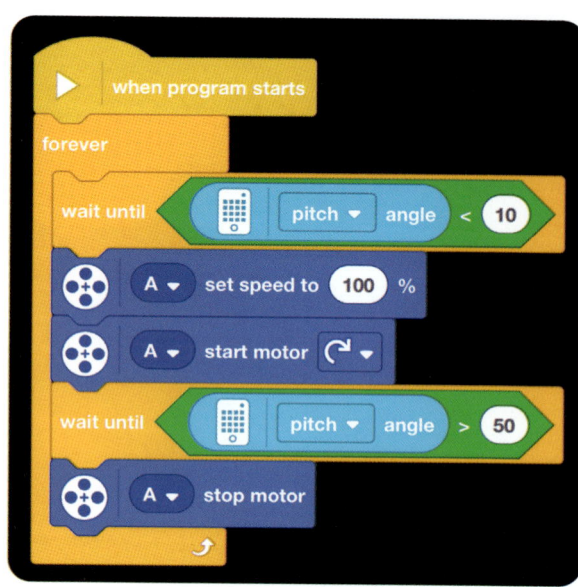

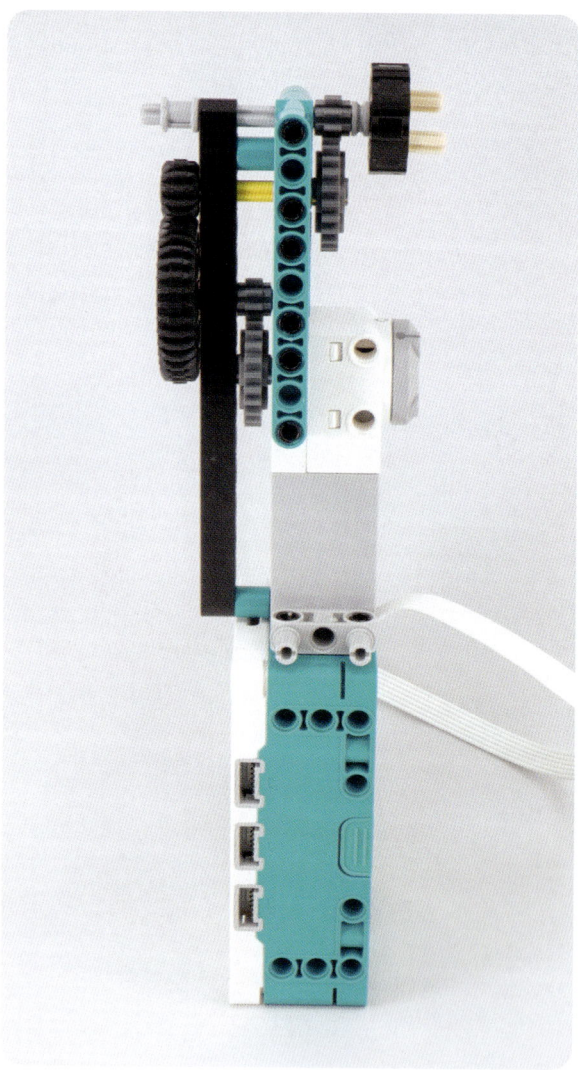

Drawing devices

#110

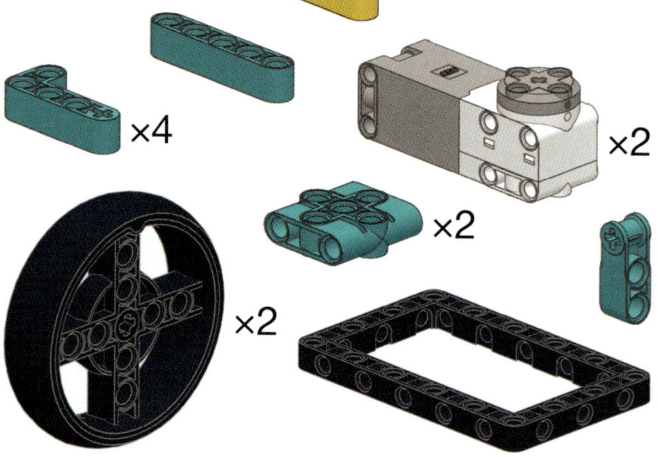

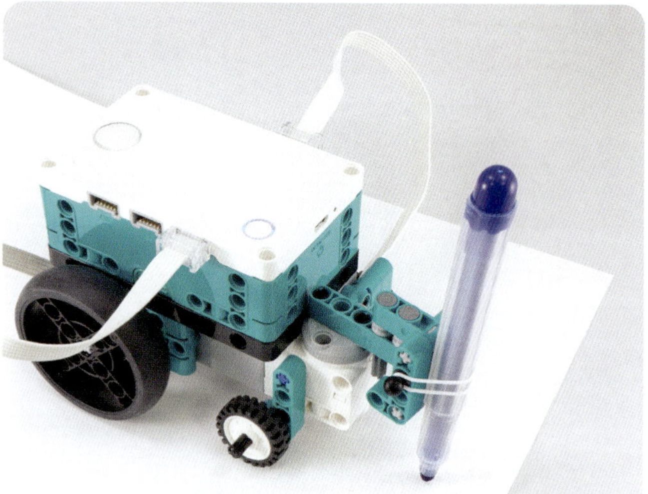

×2 ×2 ×2

×2

×19

7

12

3 ×2

4 ×2

×2

×2

×4

×2

×2

×2

×2

Attach a marker and start the program.

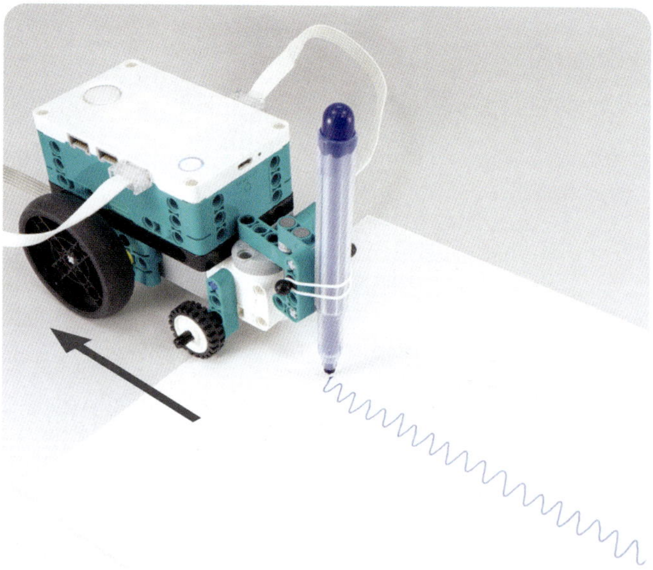

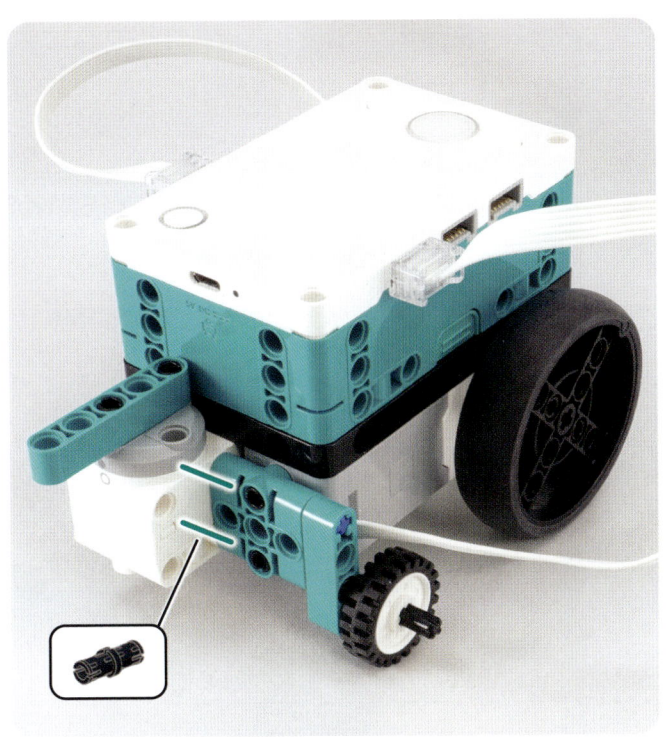

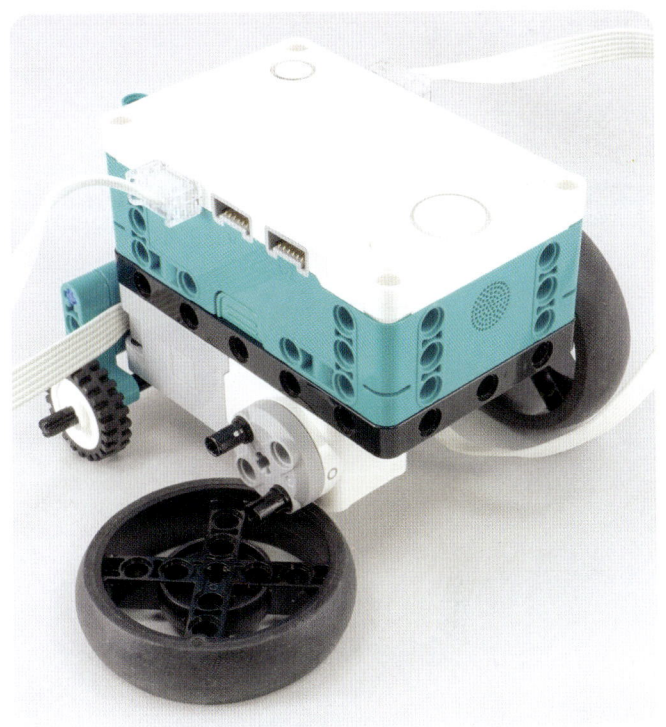

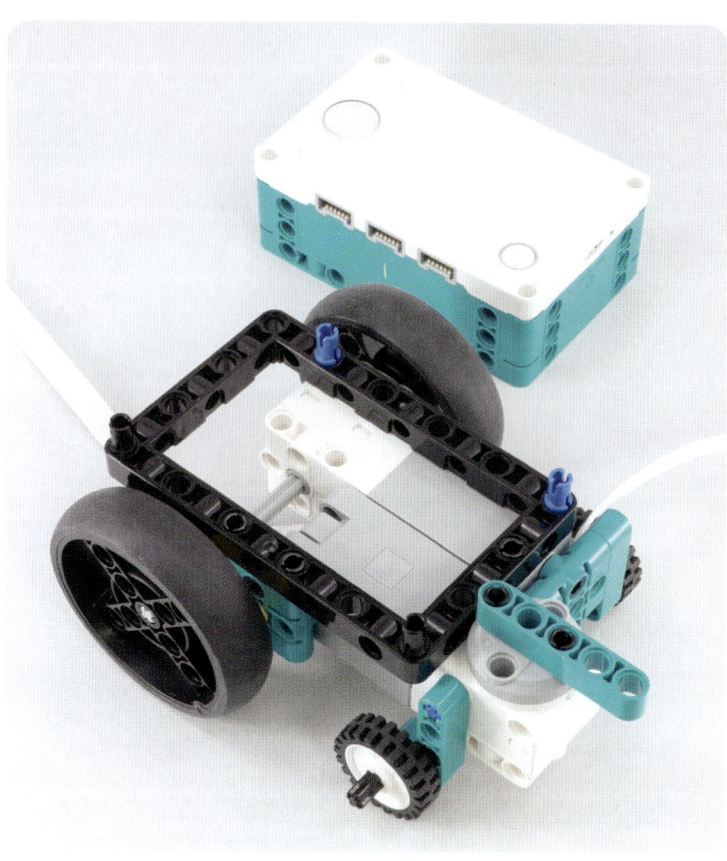

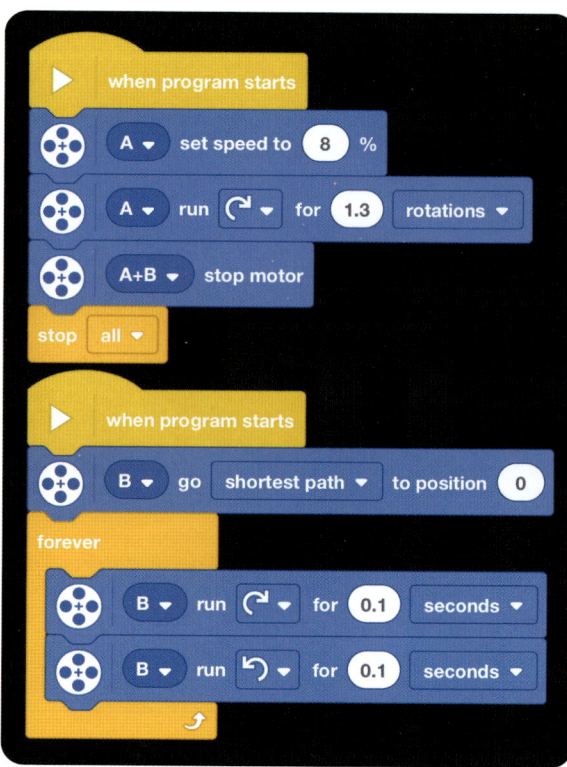

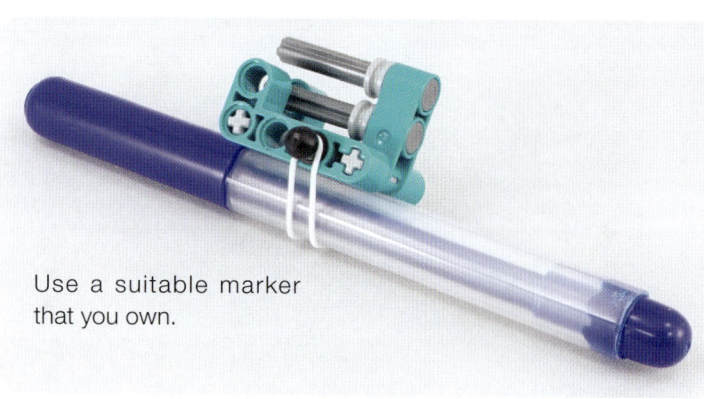

Use a suitable marker that you own.

Adjust the fixed position of the marker so that there is a small gap here when the tip touches the paper.

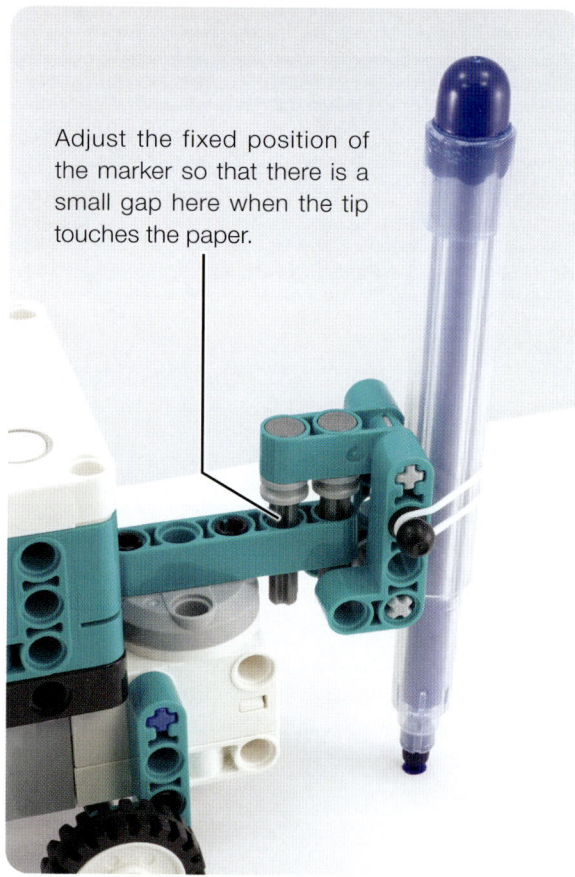

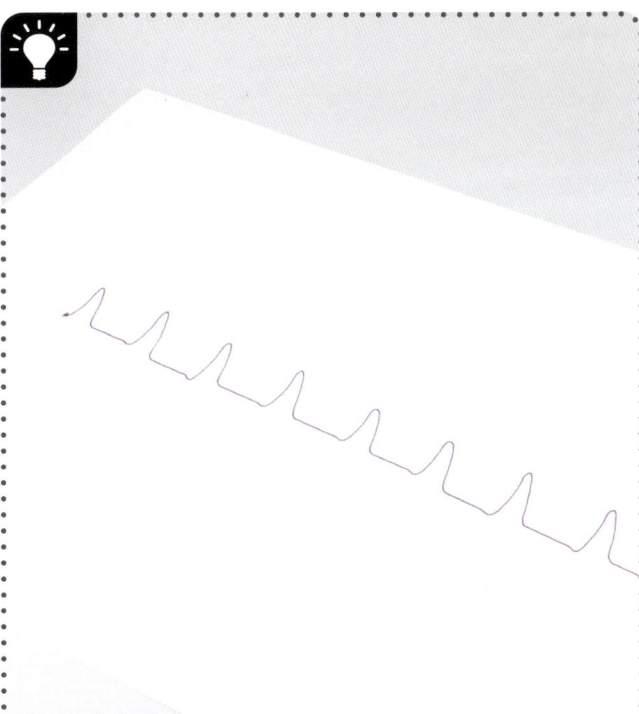

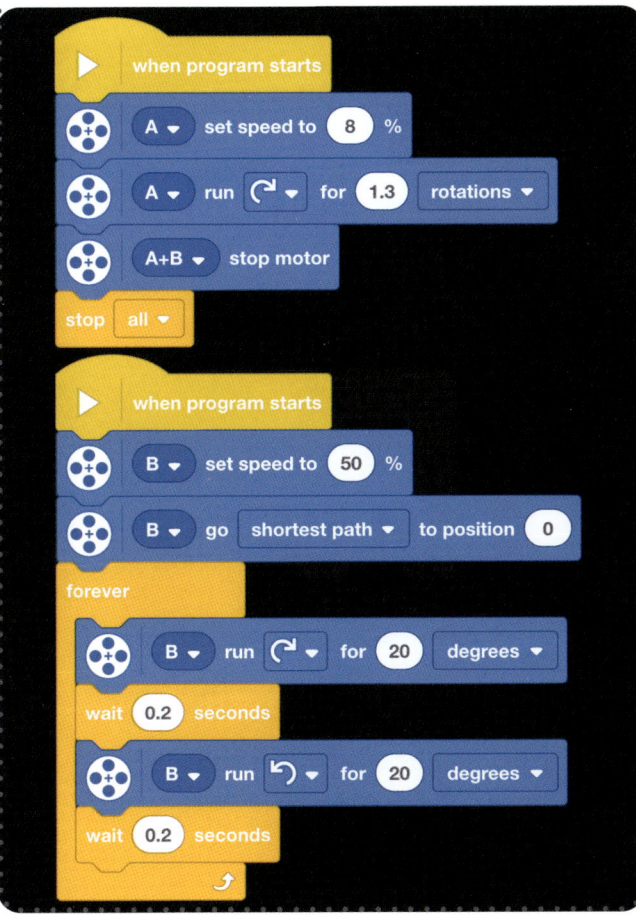

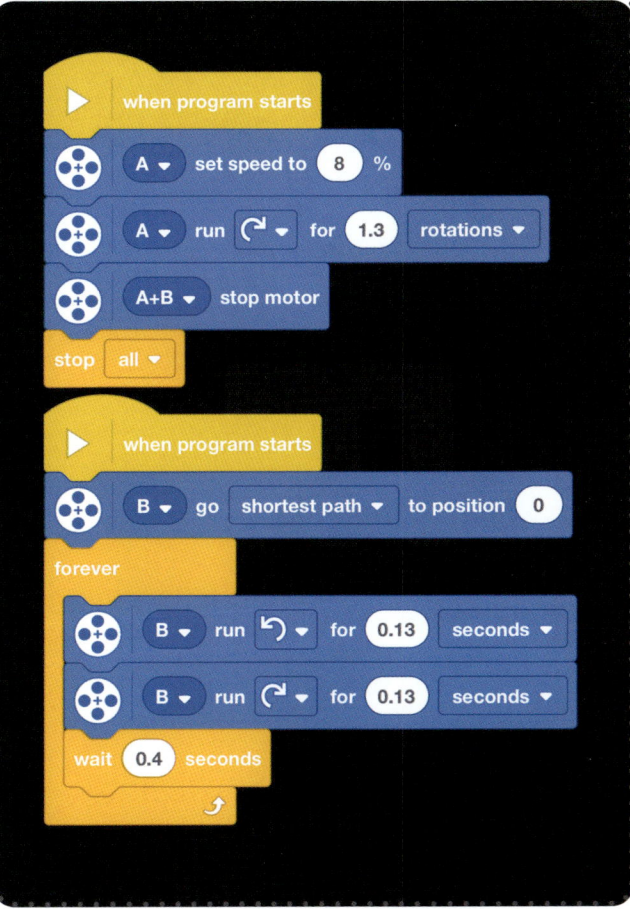

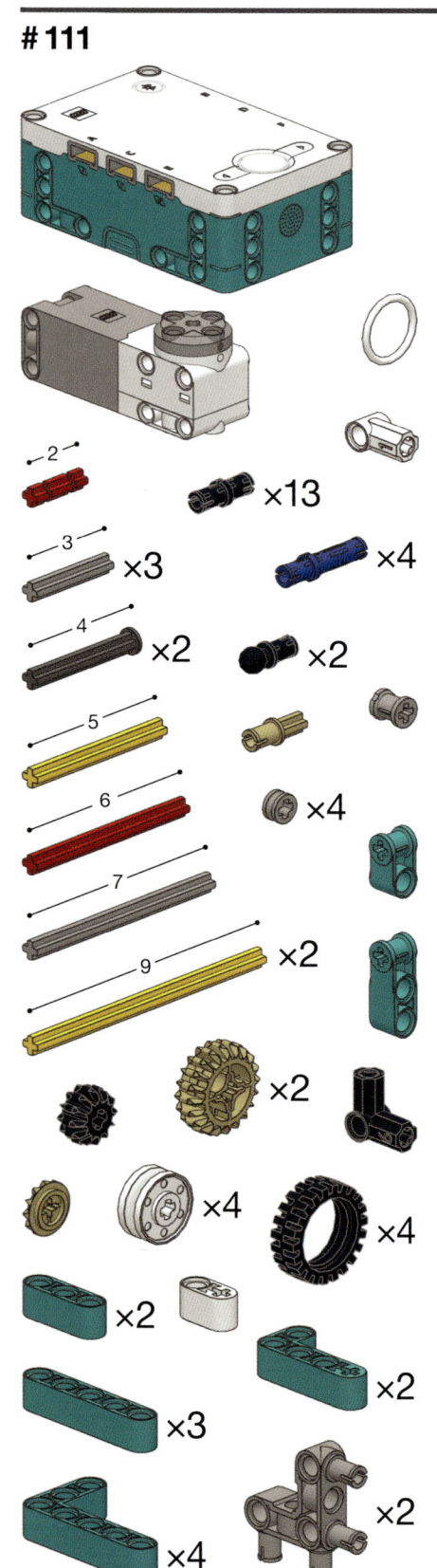

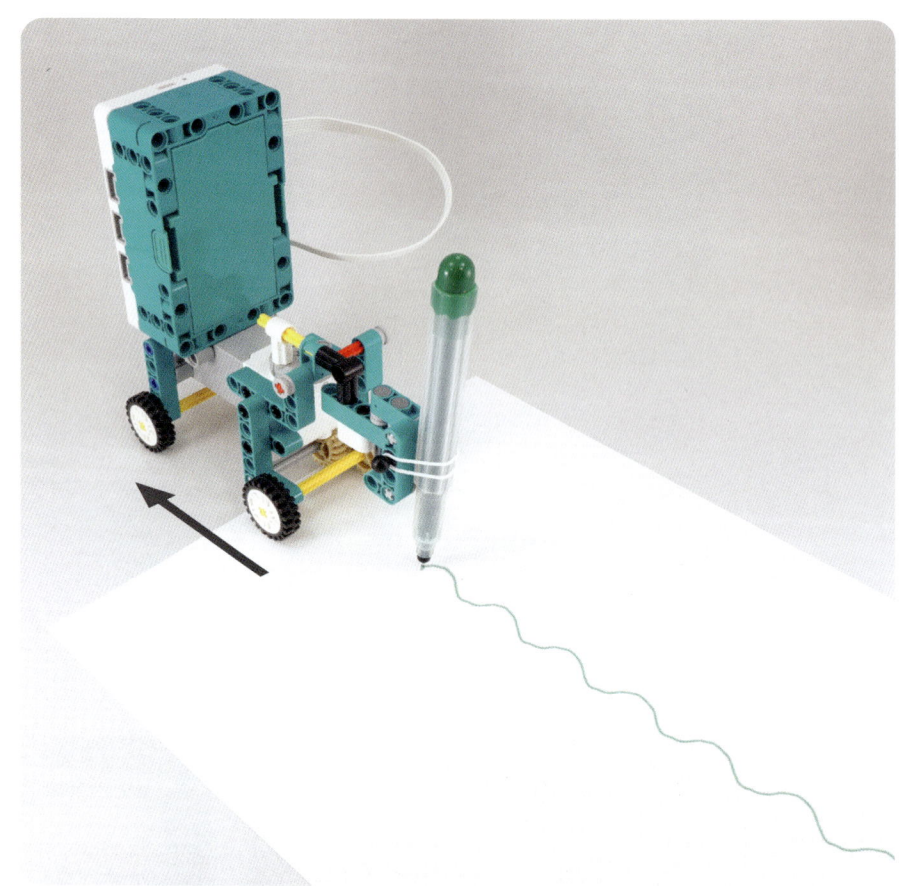

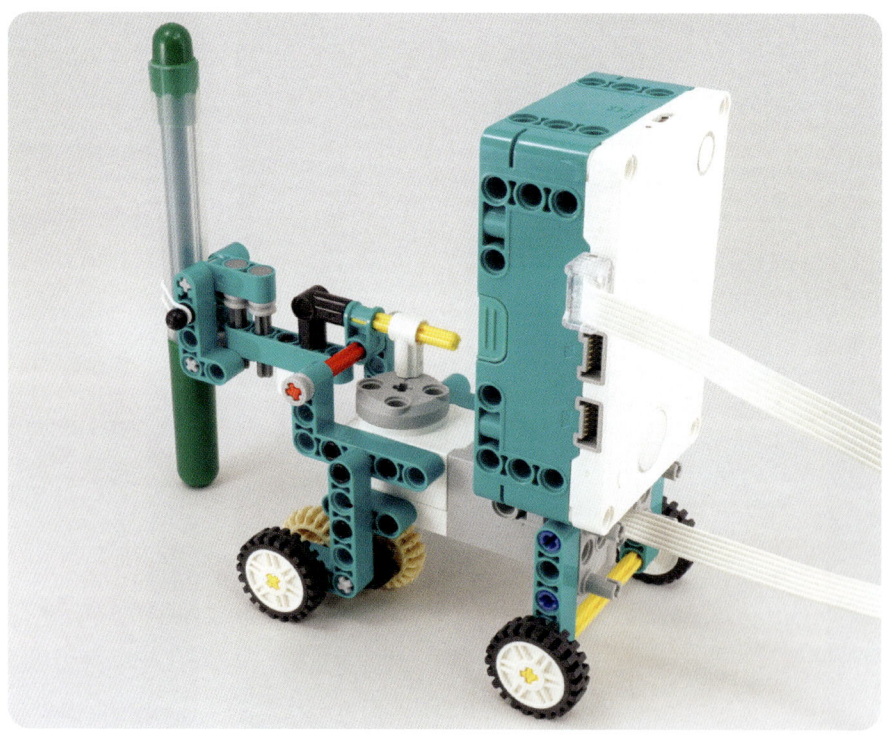

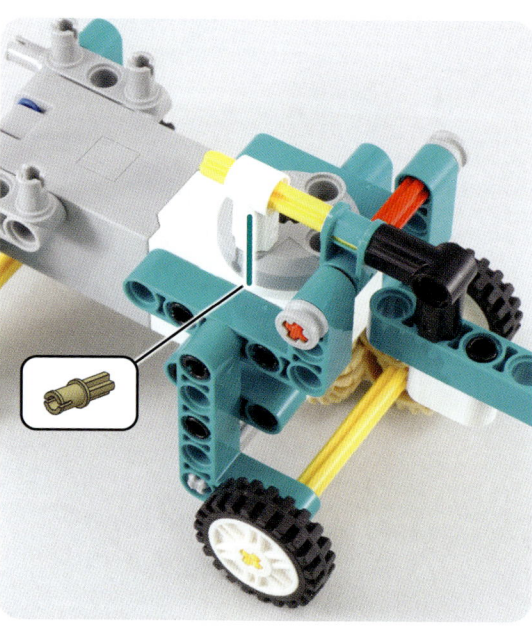

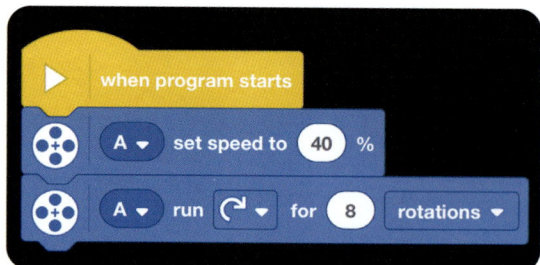

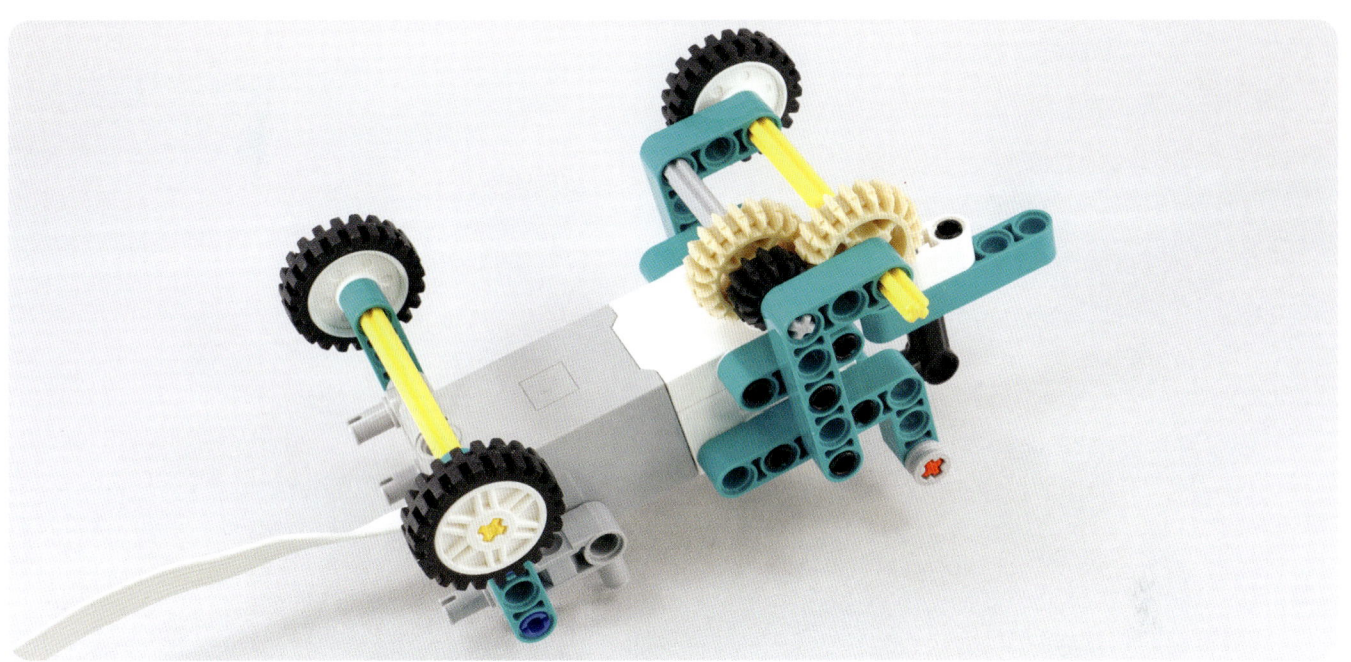

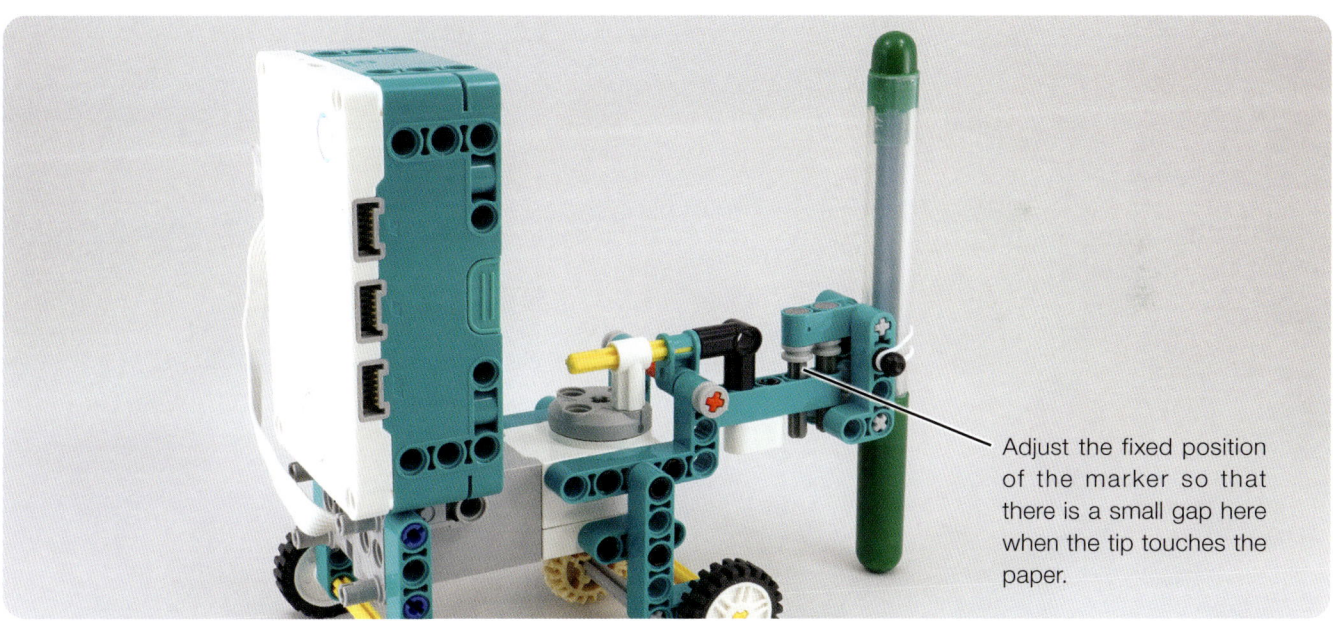

111

Adjust the fixed position of the marker so that there is a small gap here when the tip touches the paper.

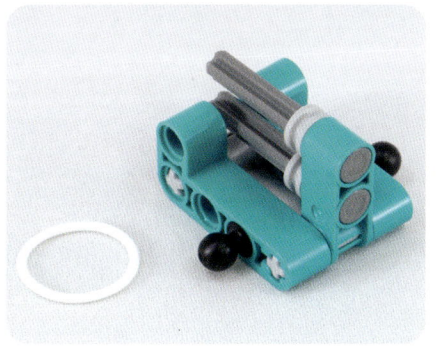

Use a suitable marker that you own.

#112

×2

×3

×9

×4

×2

×2

×2

×2

2

4

3 ×3

4 ×3

5

×3

×2

×2

×2

×2

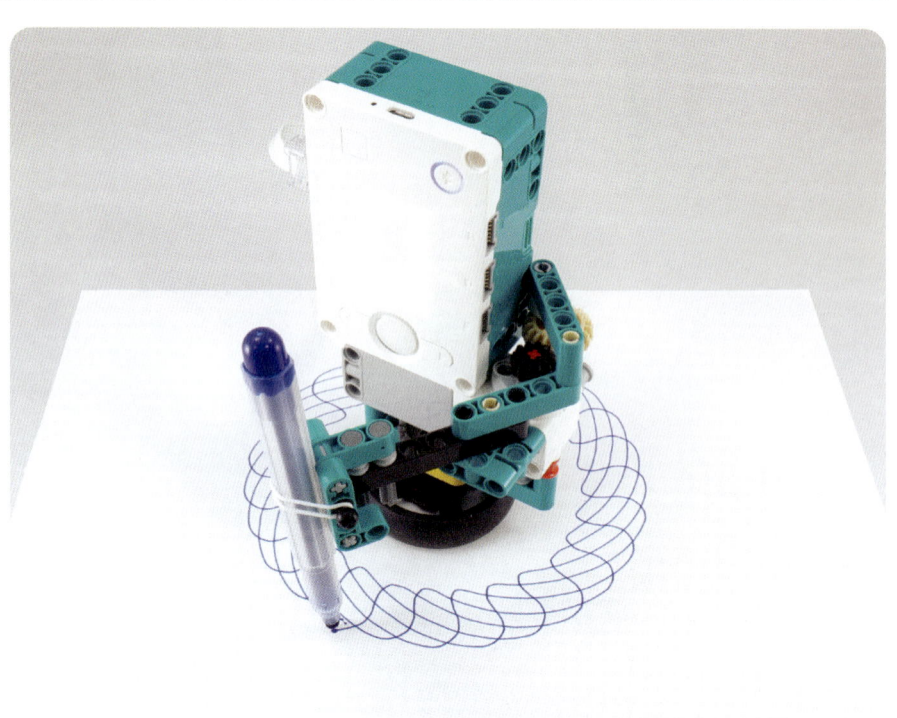

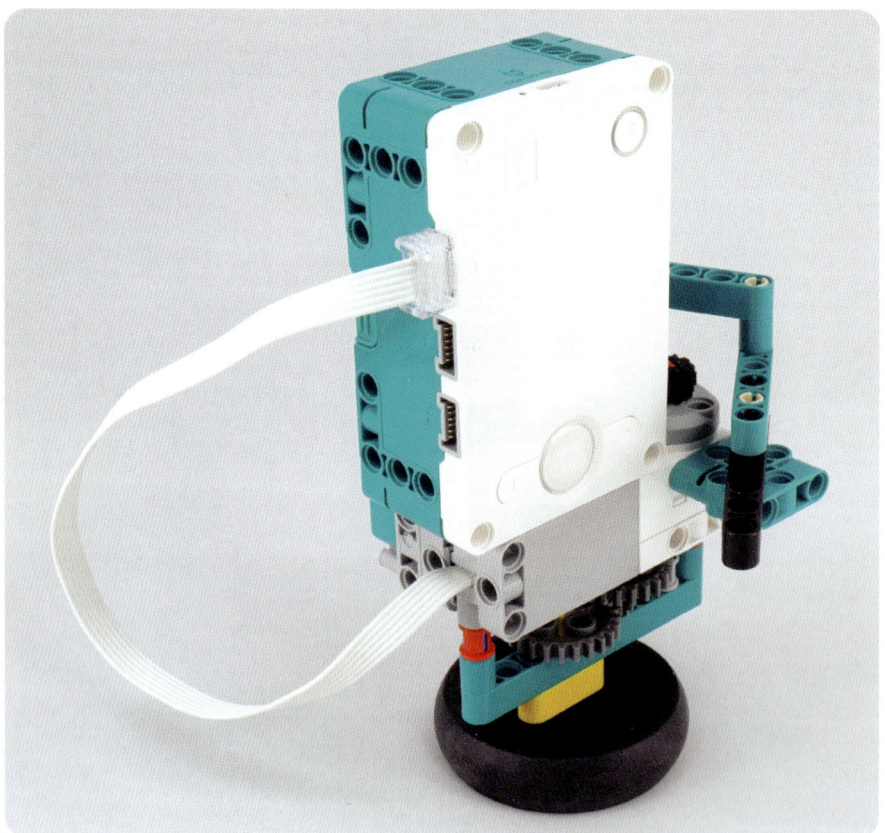

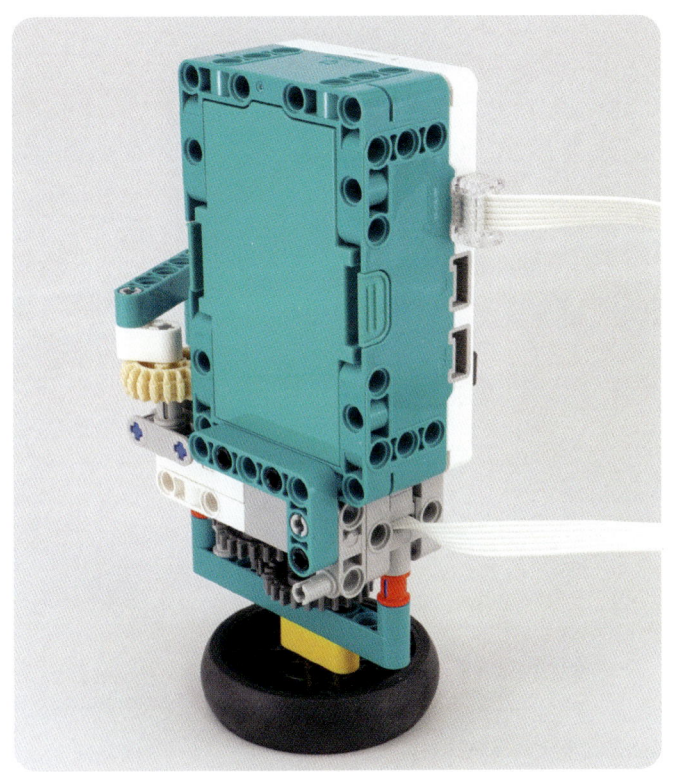

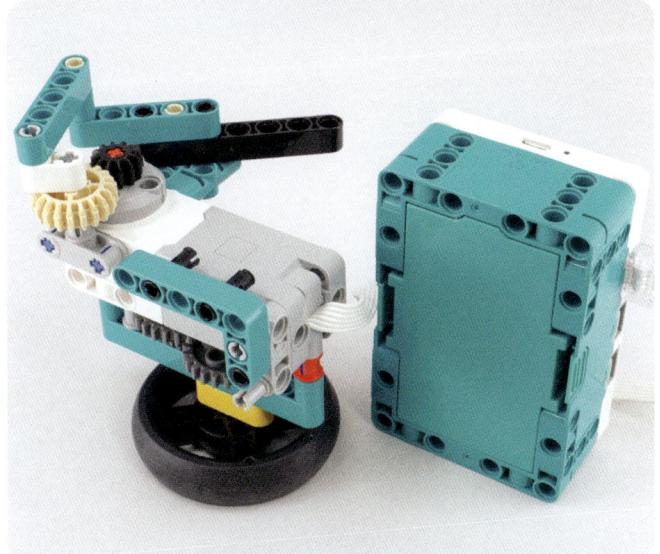

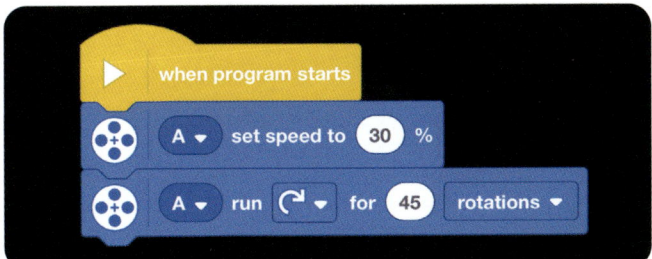

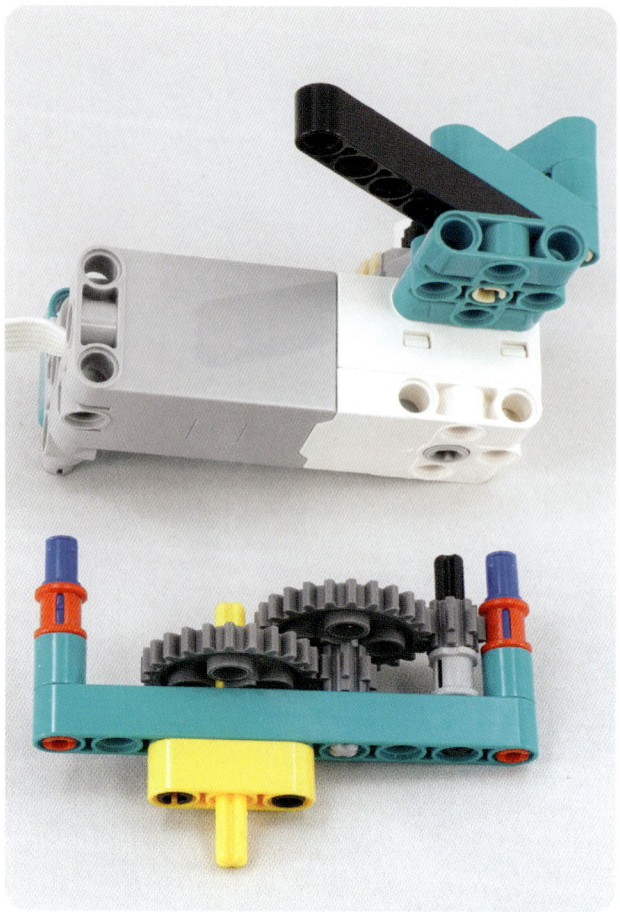

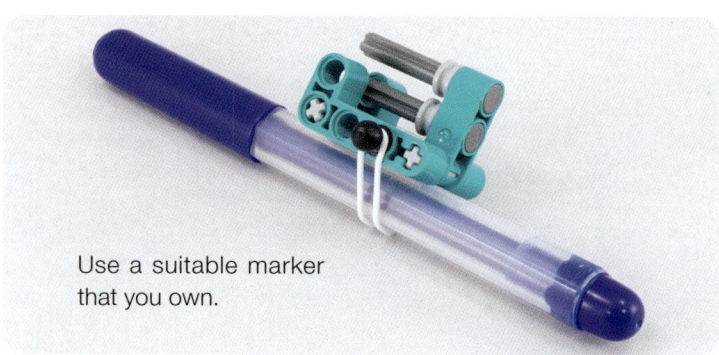

Use a suitable marker that you own.

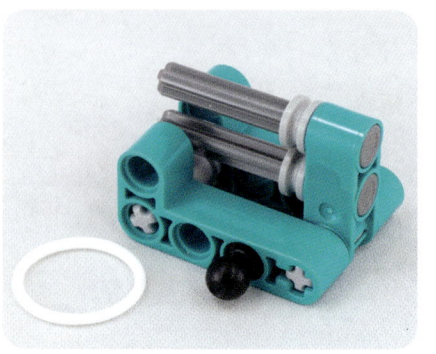

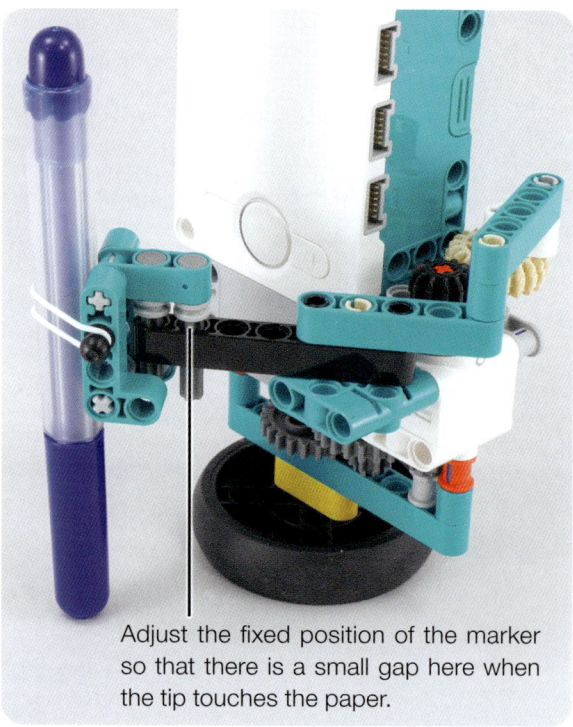

Adjust the fixed position of the marker so that there is a small gap here when the tip touches the paper.

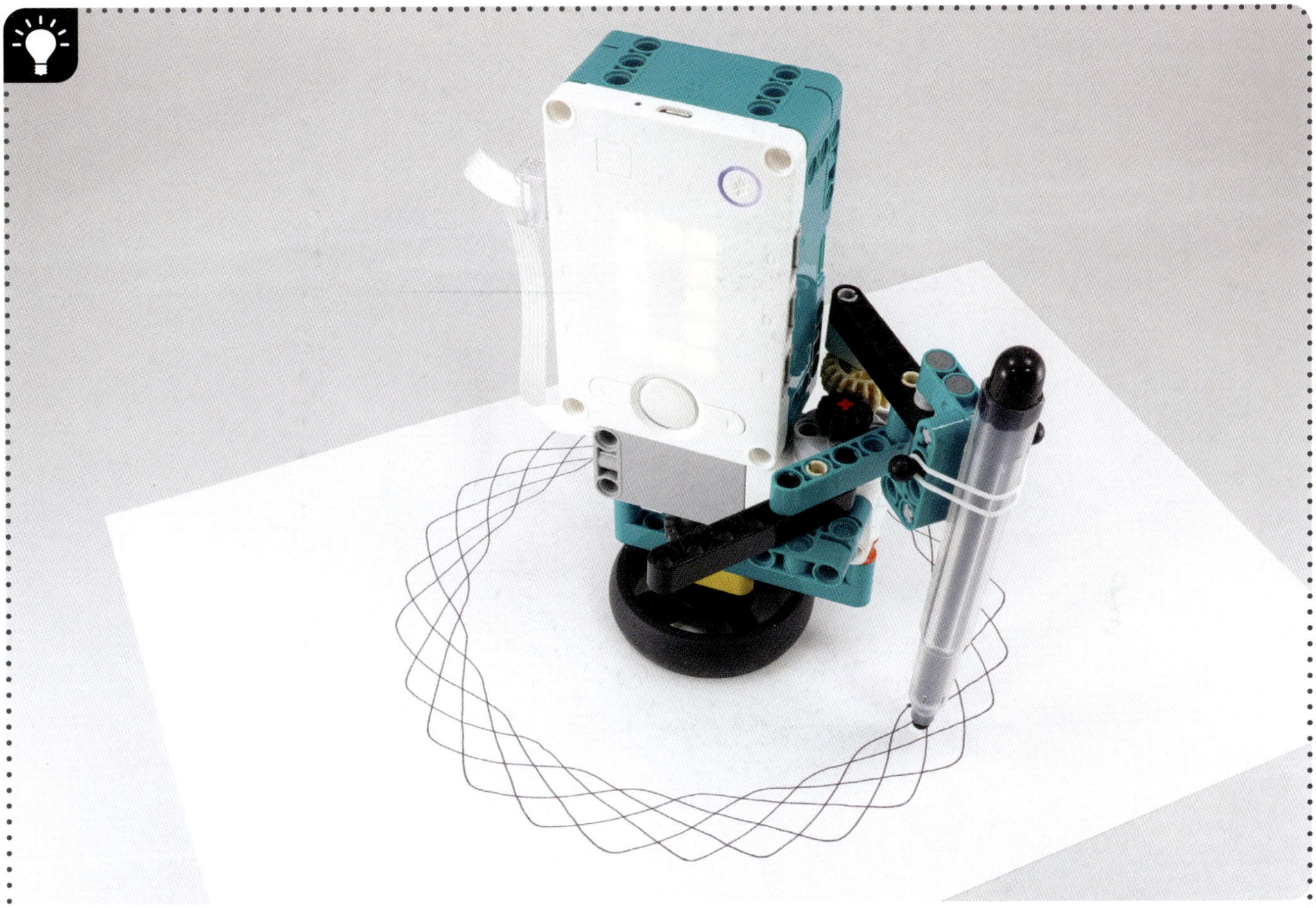

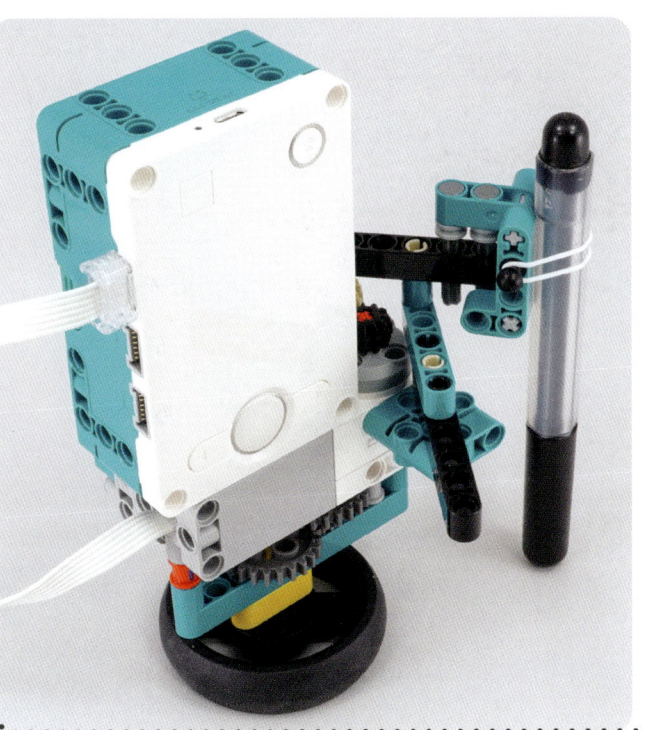

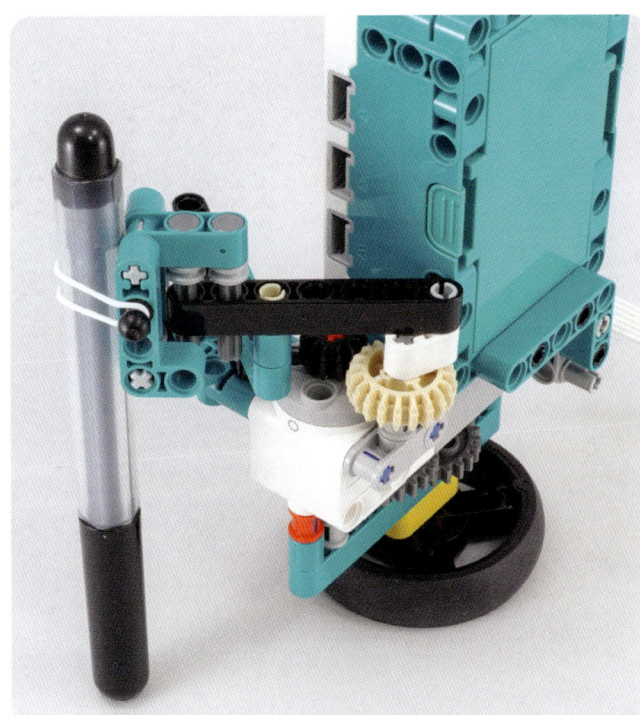

113

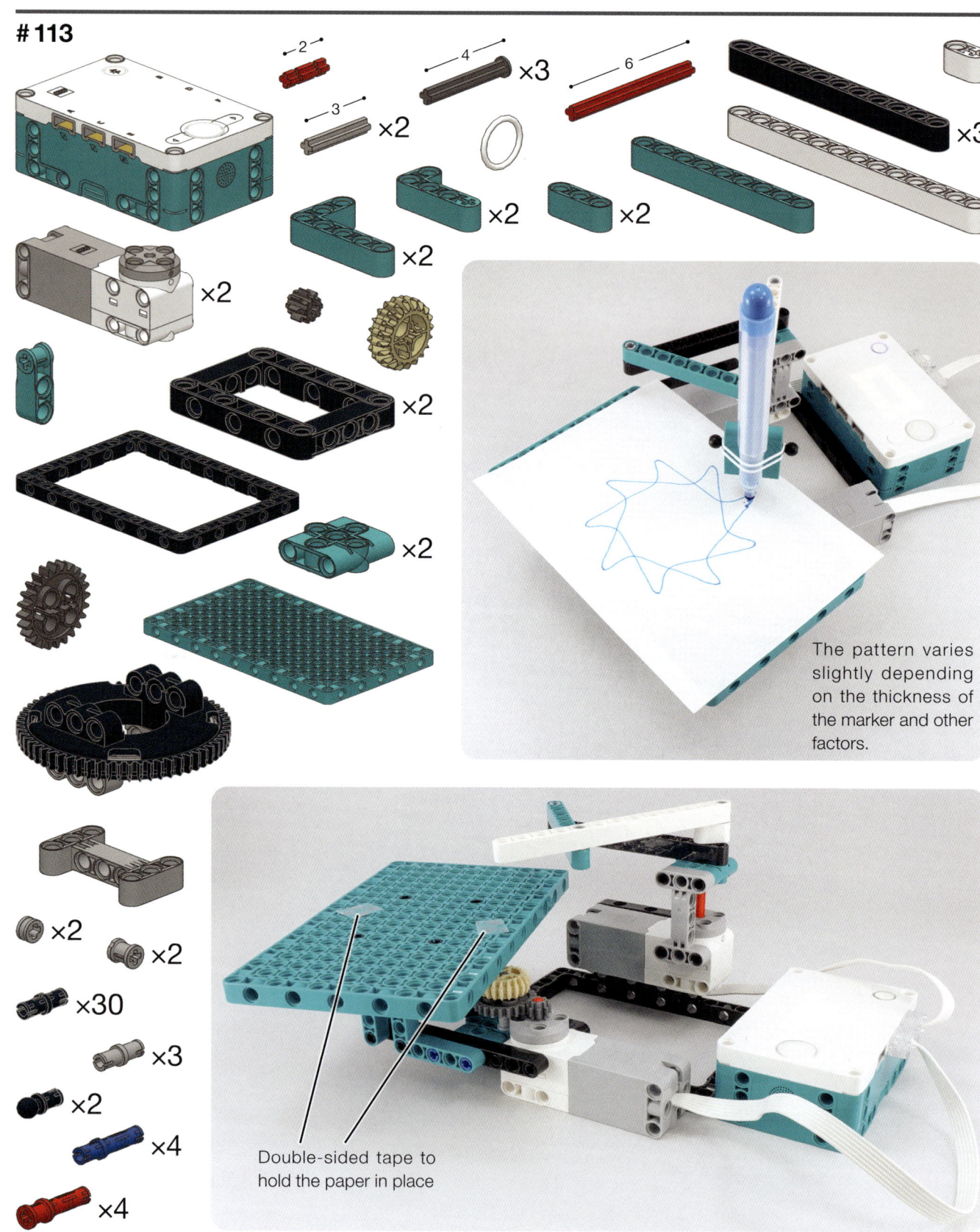

×2

×3

×2

×2

×3

×2

×2

×2

×2

×2

×2

×2

×2

The pattern varies slightly depending on the thickness of the marker and other factors.

×2

×2

×30

×3

×2

×4

×4

Double-sided tape to hold the paper in place

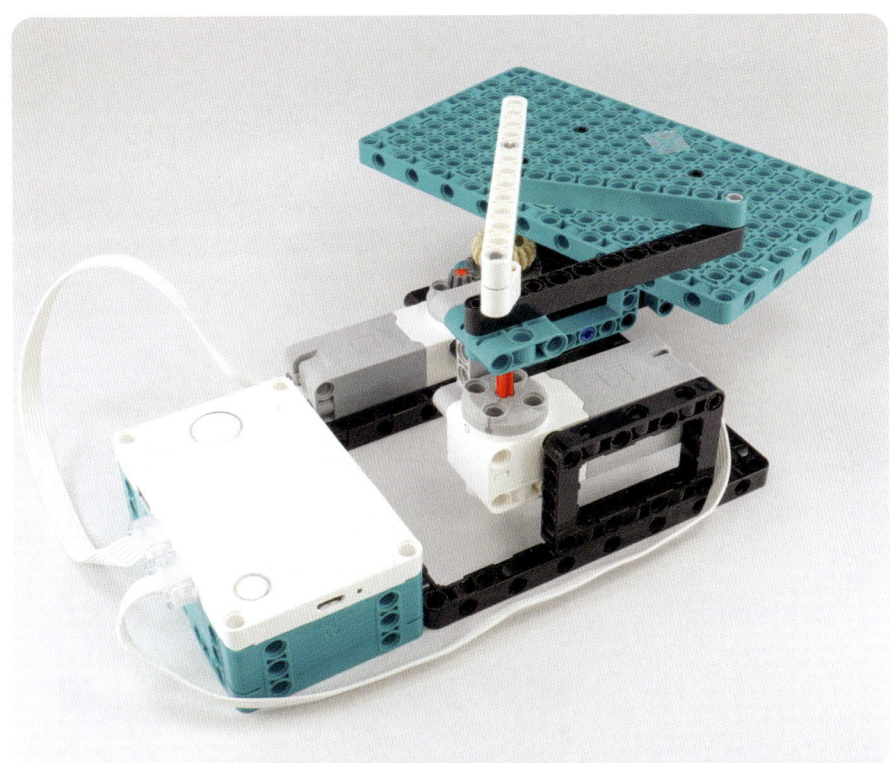

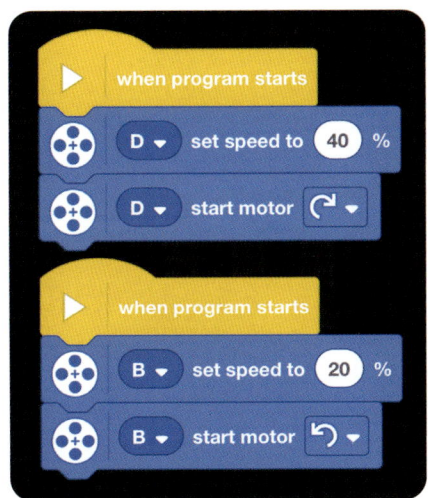

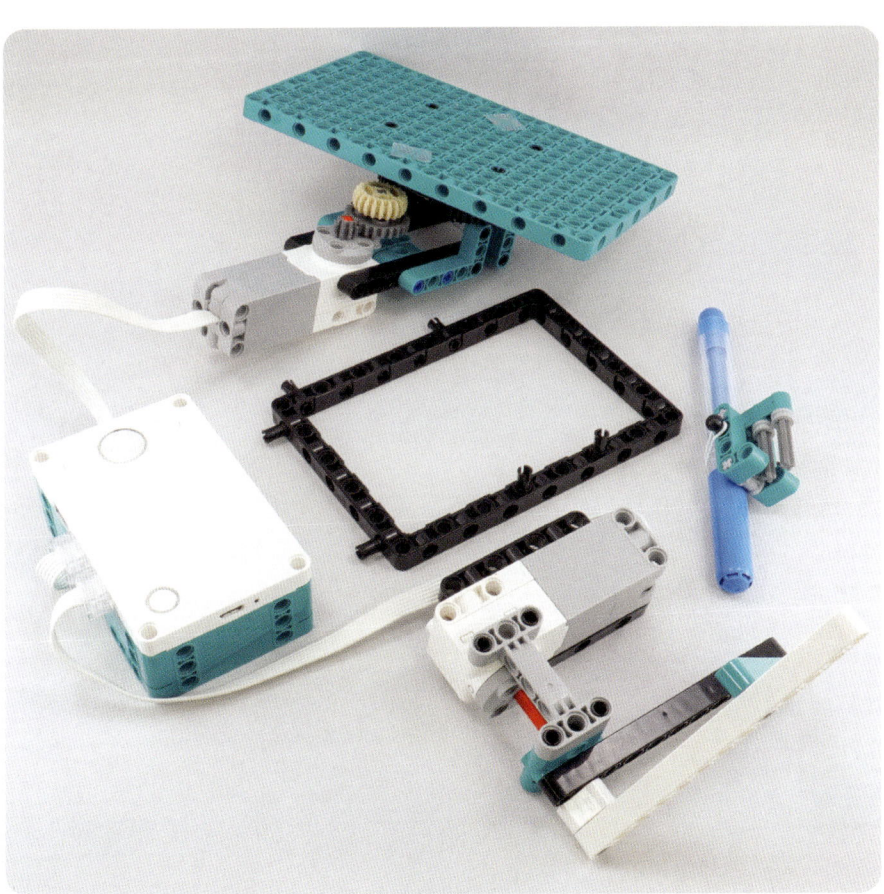

Use a suitable marker that you own.

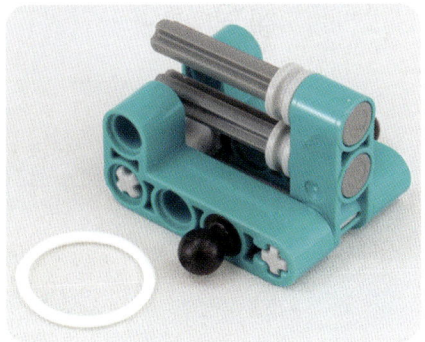

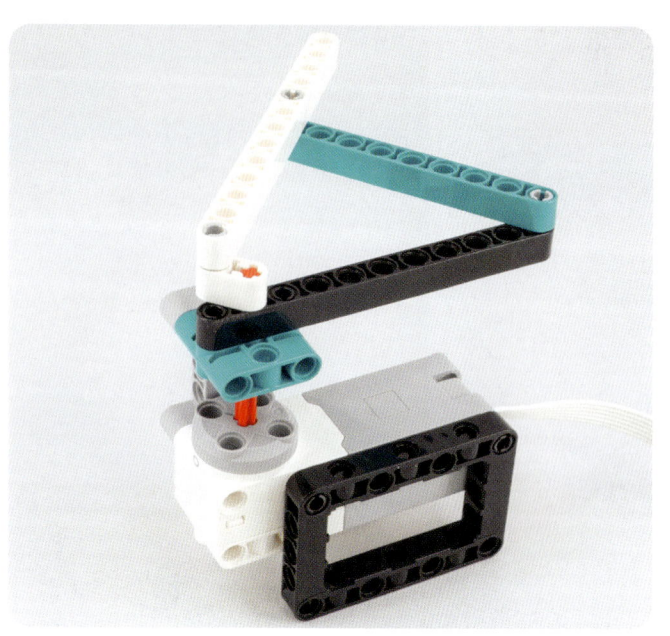

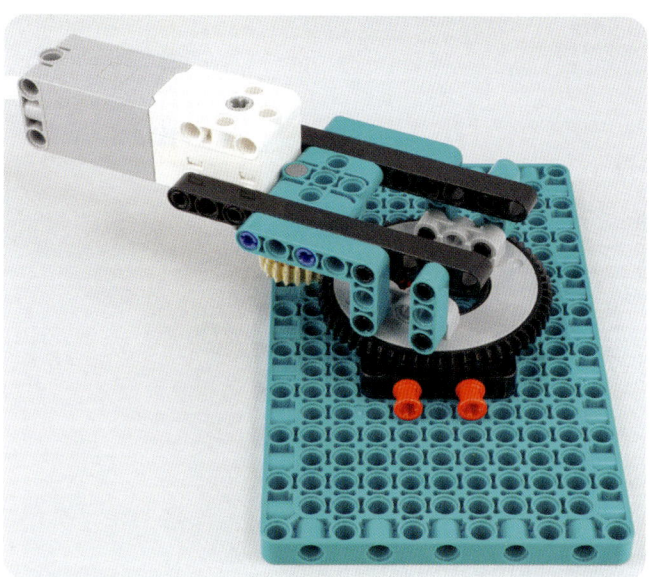

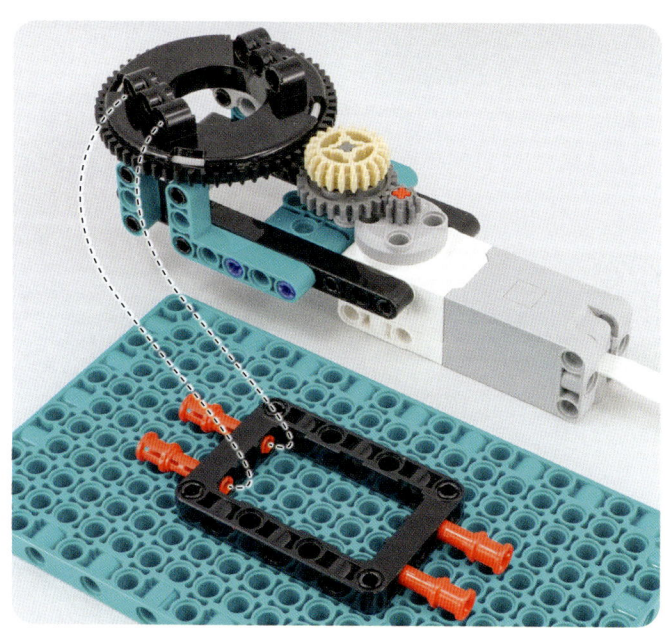

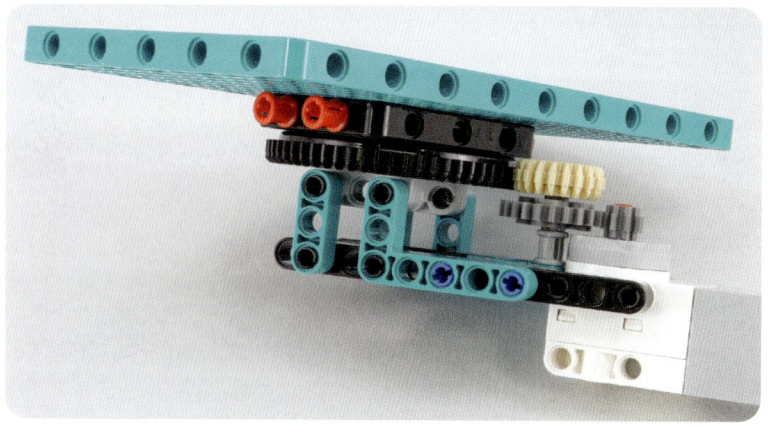

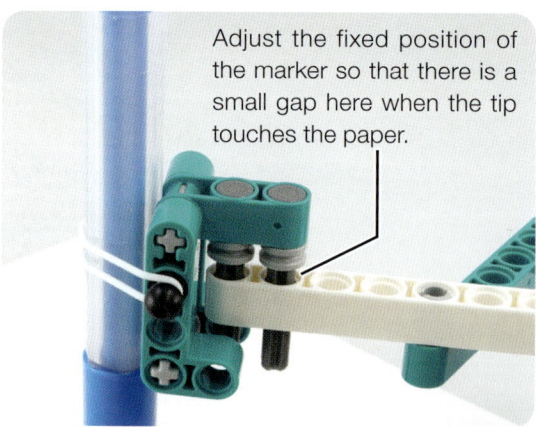

Adjust the fixed position of the marker so that there is a small gap here when the tip touches the paper.

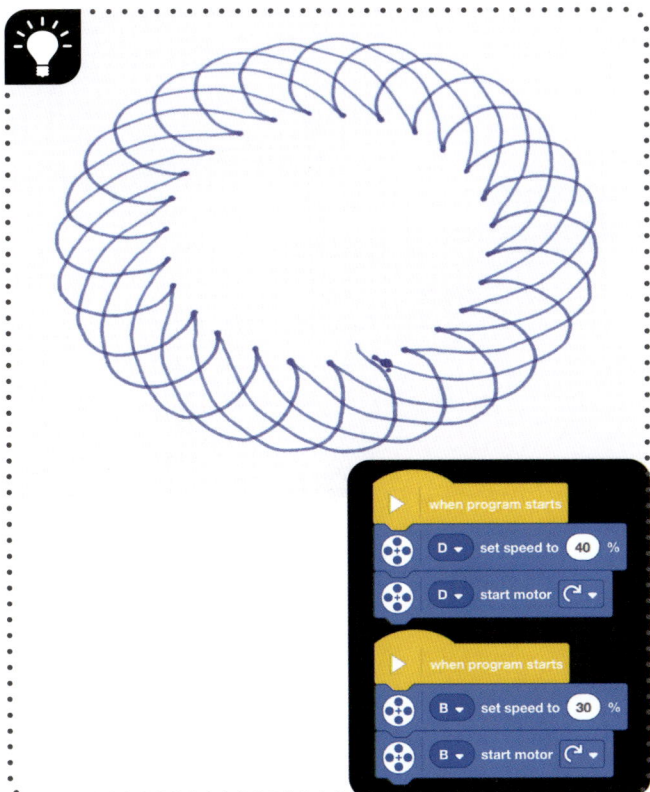

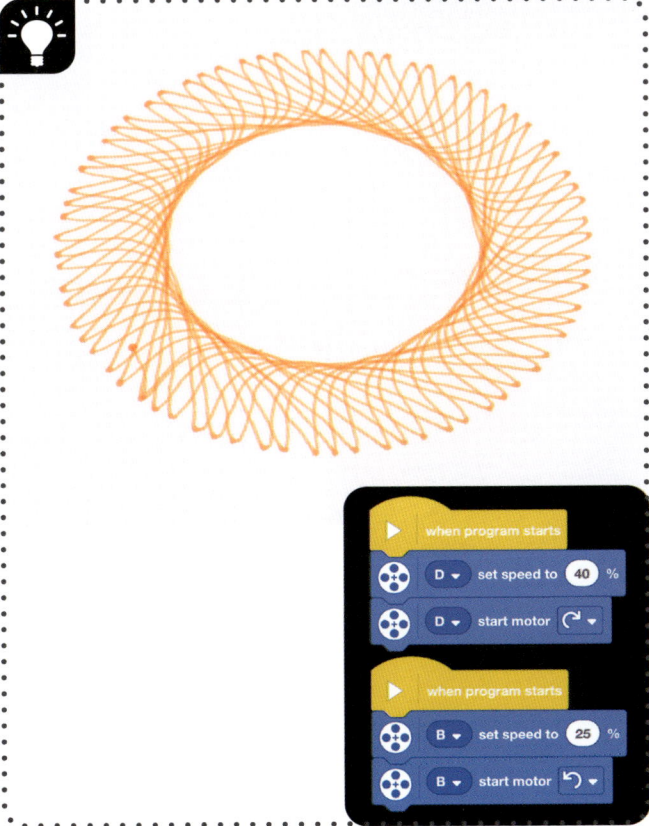

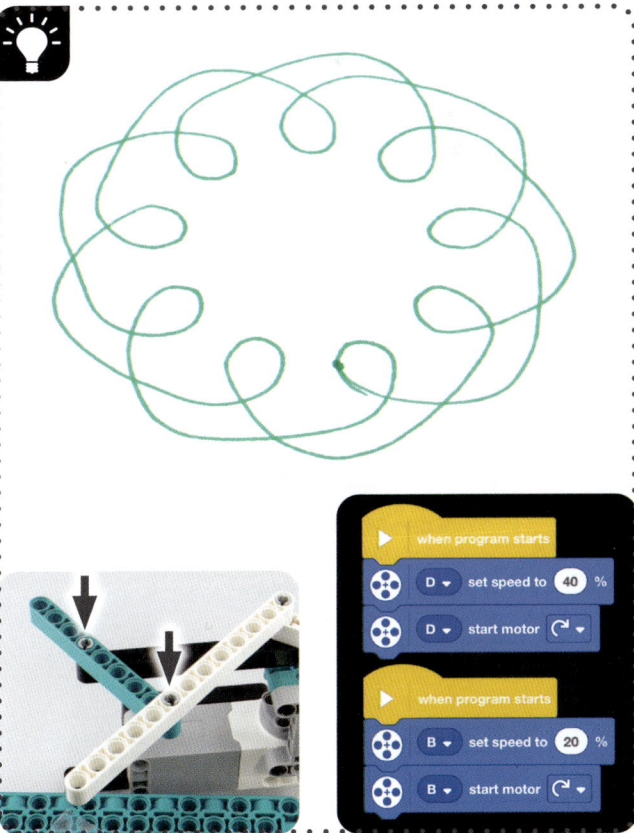

114

×3

×3

—3— ×3

—3—

×27

—4— ×2

×2

×2

×2

×2

×2

×2

×2

×2

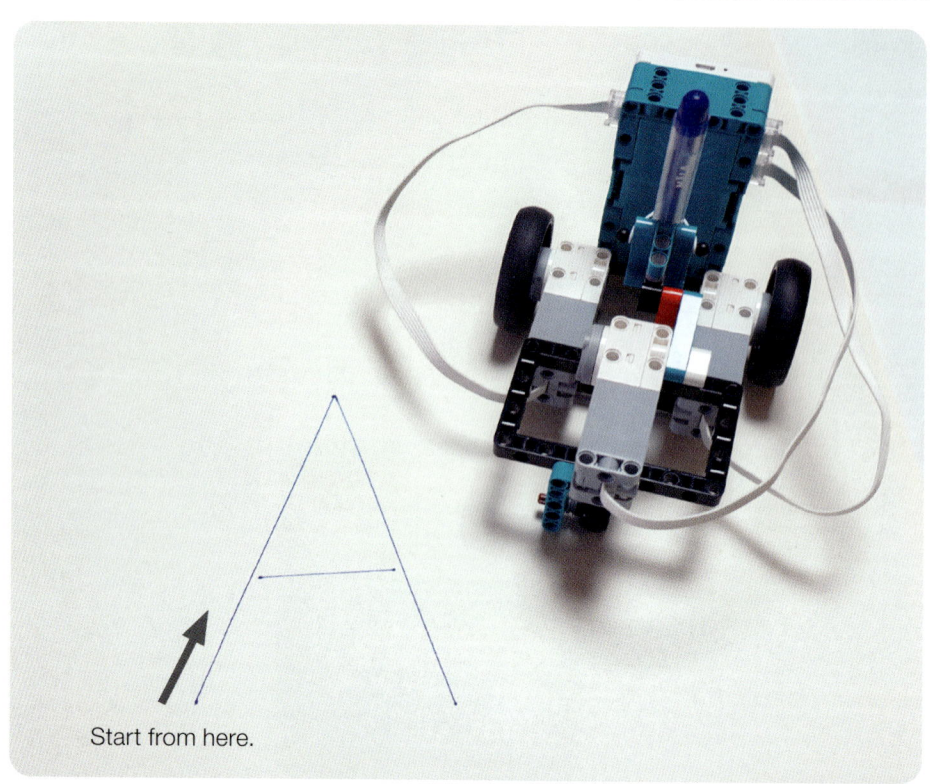

Start from here.

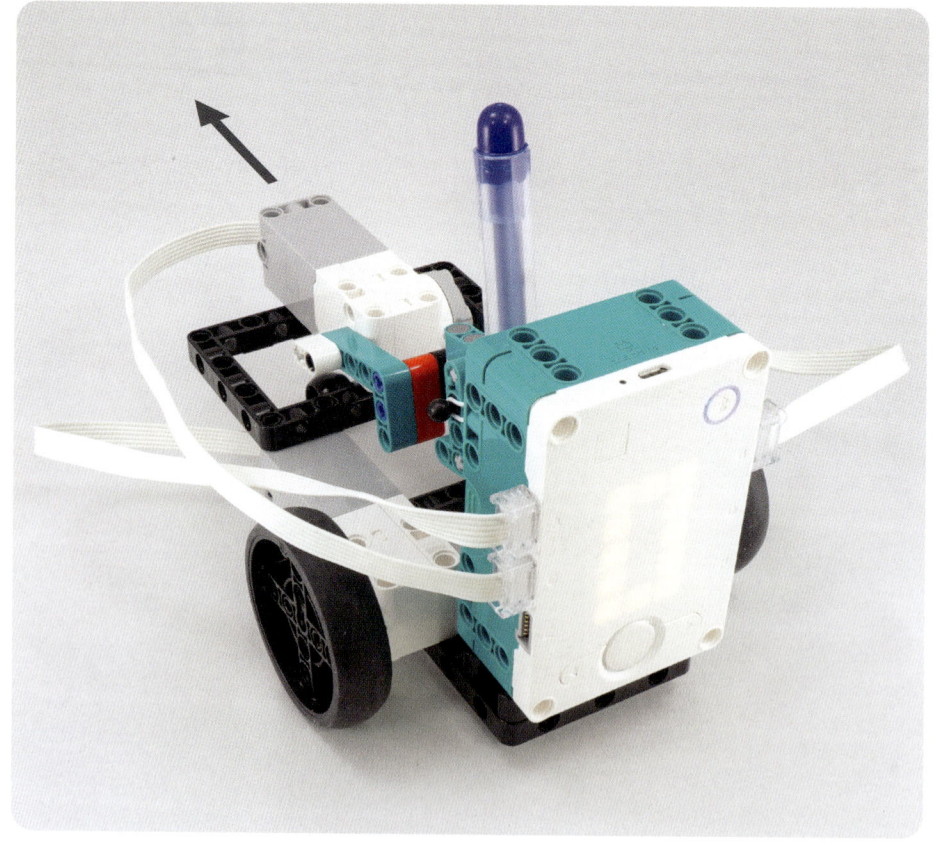

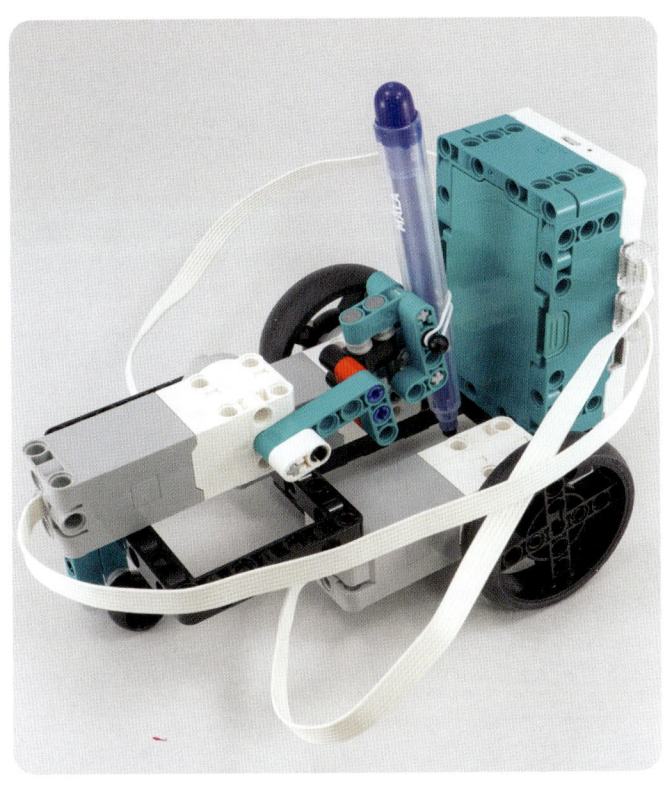

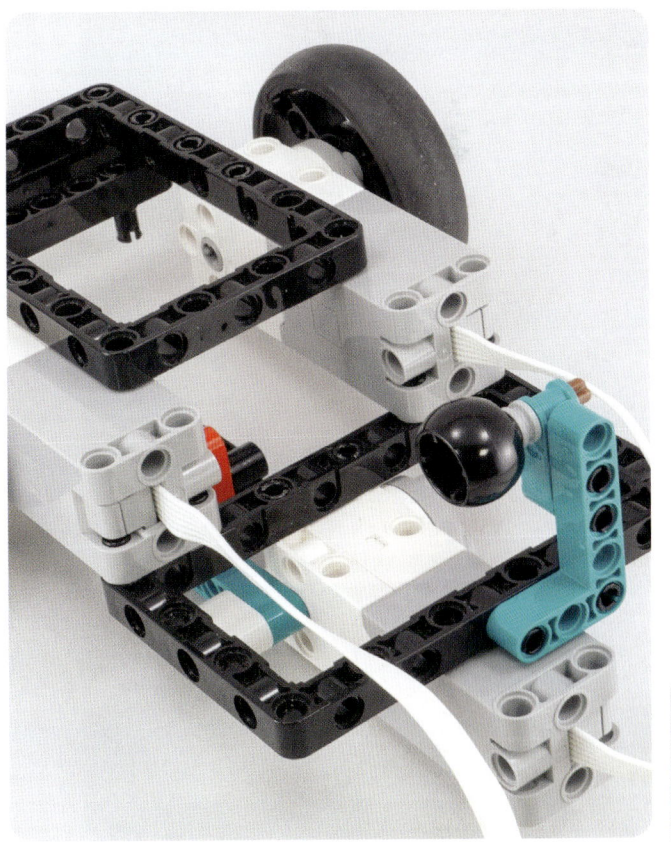

Use a suitable marker that you own.

```
▶ when program starts
  set movement motors to A+B ▼
  set movement speed to 20 %
  C ▼ set speed to 30 %
  C ▼ go shortest path ▼ to position 0
  move ↑ ▼ for 15 cm ▼          = 5.9 inches
  move ↻ ▼ for 14 cm ▼          = 5.5 inches
  move ↑ ▼ for 15 cm ▼          = 5.9 inches
  C ▼ go shortest path ▼ to position 15
  move ↻ ▼ for 14 cm ▼          = 5.5 inches
  move ↑ ▼ for 9 cm ▼           = 3.5 inches
  move ↻ ▼ for 13.7 cm ▼        = 5.4 inches
  C ▼ go shortest path ▼ to position 0
  move ↑ ▼ for 5 cm ▼           = 2 inches
  C ▼ go shortest path ▼ to position 15
```

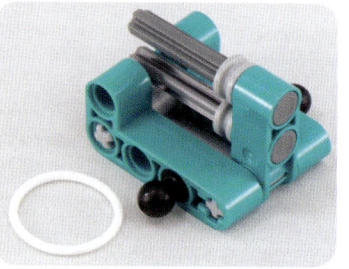

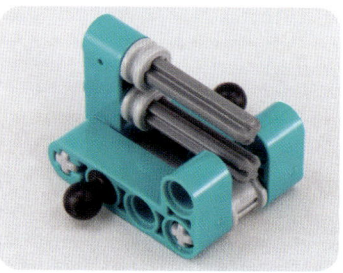

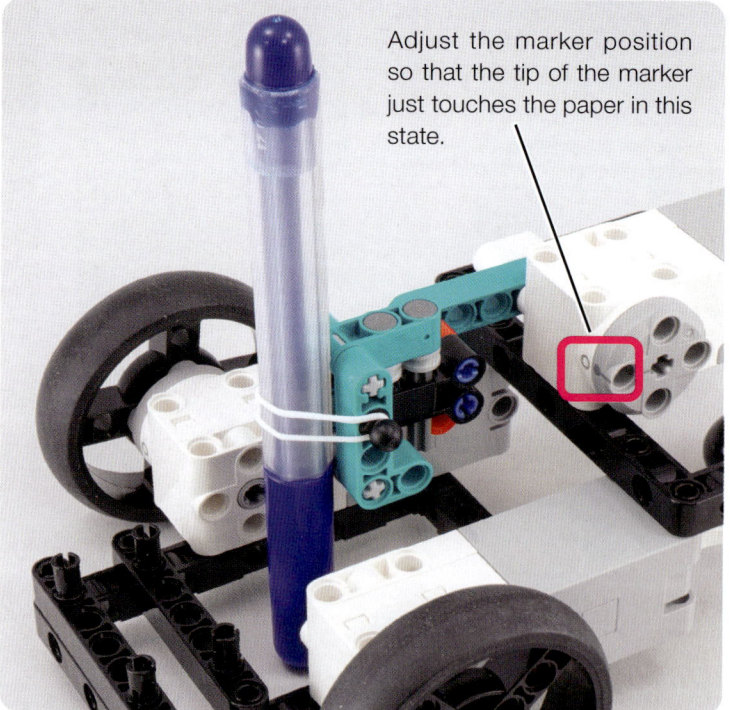

Adjust the marker position so that the tip of the marker just touches the paper in this state.

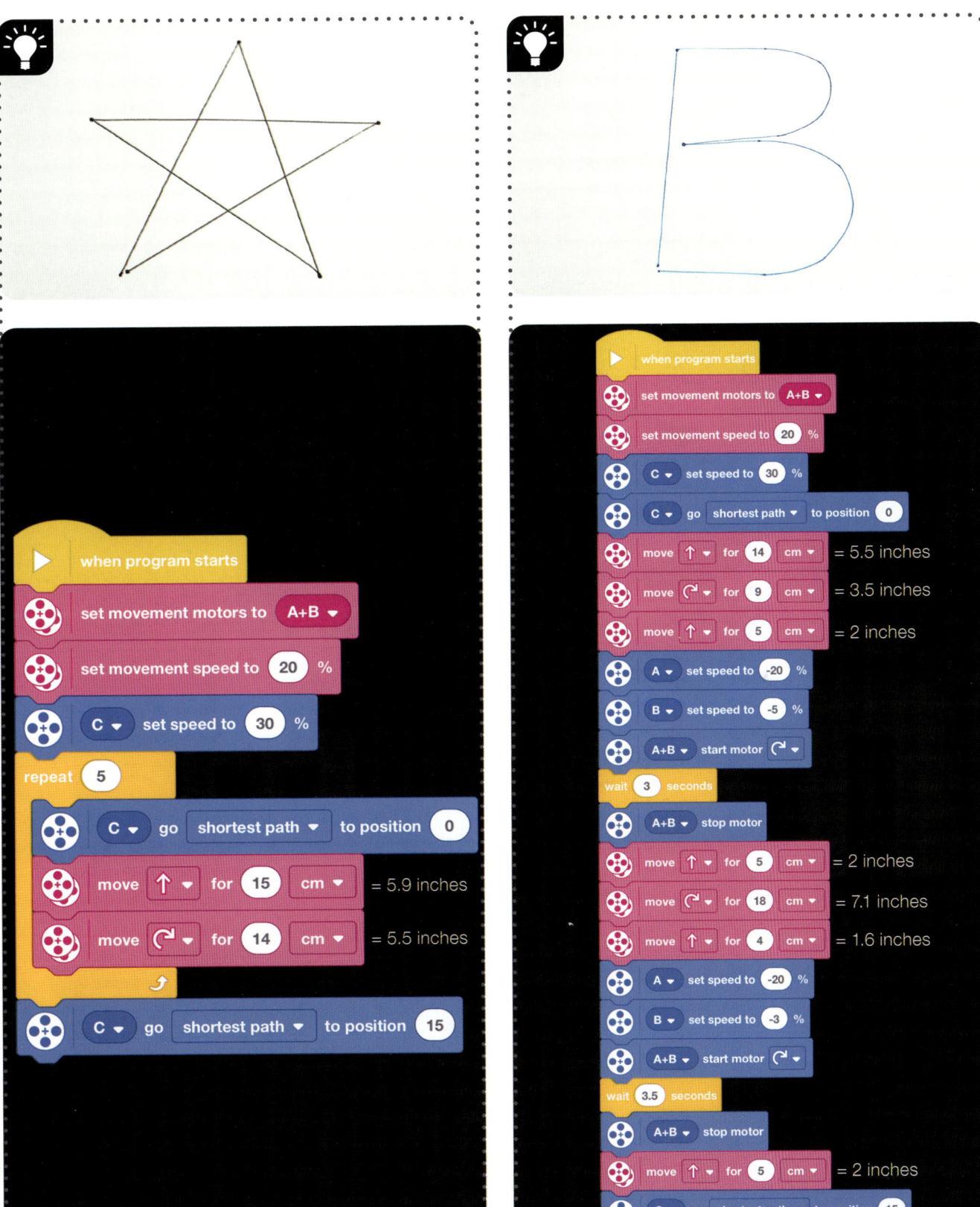

when program starts
set movement motors to A+B ▾
set movement speed to 20 %
C ▾ set speed to 30 %
repeat 5
    C ▾ go shortest path ▾ to position 0
    move ↑ ▾ for 15 cm ▾ = 5.9 inches
    move ↷ ▾ for 14 cm ▾ = 5.5 inches
C ▾ go shortest path ▾ to position 15

when program starts
set movement motors to A+B ▾
set movement speed to 20 %
C ▾ set speed to 30 %
C ▾ go shortest path ▾ to position 0
move ↑ ▾ for 14 cm ▾ = 5.5 inches
move ↷ ▾ for 9 cm ▾ = 3.5 inches
move ↑ ▾ for 5 cm ▾ = 2 inches
A ▾ set speed to -20 %
B ▾ set speed to -5 %
A+B ▾ start motor ↷ ▾
wait 3 seconds
A+B ▾ stop motor
move ↑ ▾ for 5 cm ▾ = 2 inches
move ↷ ▾ for 18 cm ▾ = 7.1 inches
move ↑ ▾ for 4 cm ▾ = 1.6 inches
A ▾ set speed to -20 %
B ▾ set speed to -3 %
A+B ▾ start motor ↷ ▾
wait 3.5 seconds
A+B ▾ stop motor
move ↑ ▾ for 5 cm ▾ = 2 inches
C ▾ go shortest path ▾ to position 15

Automatic doors

115

×9

×2

6

12

×2

×3

×3

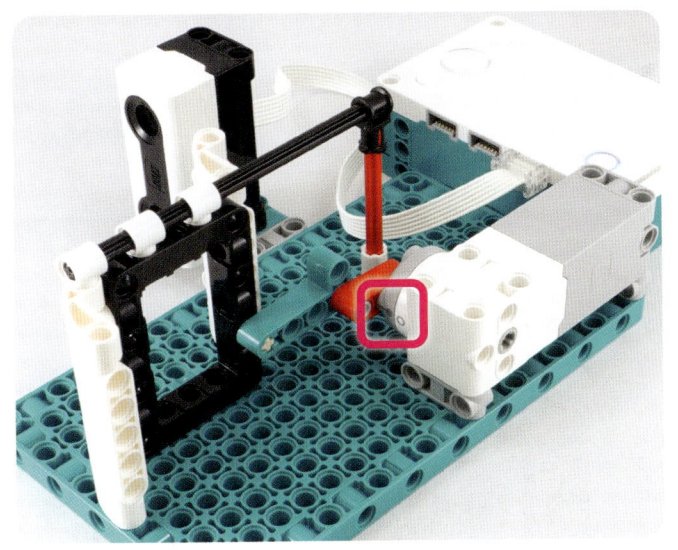

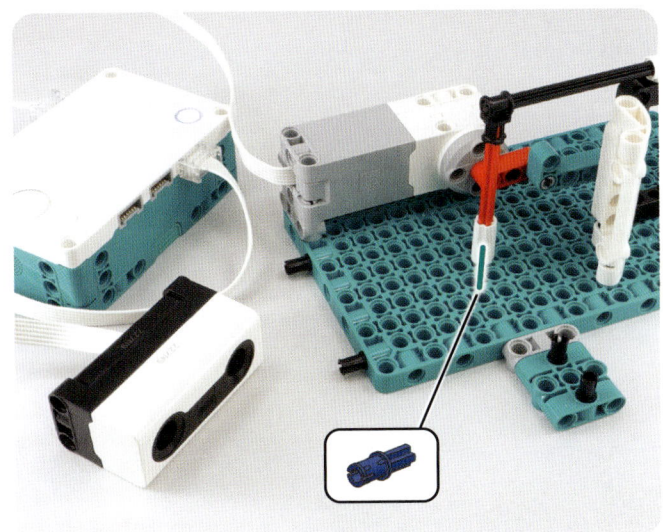

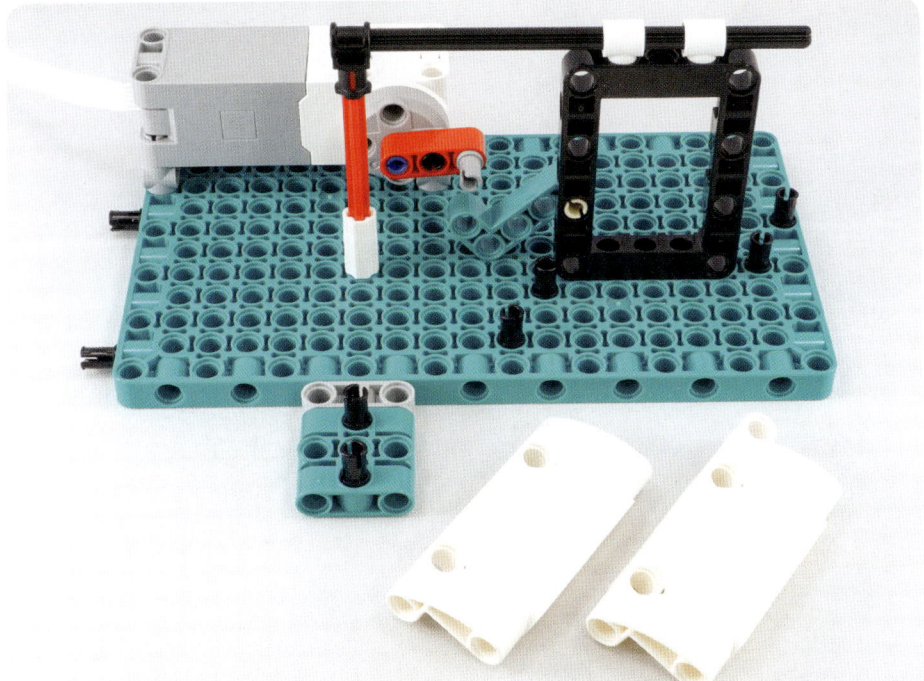

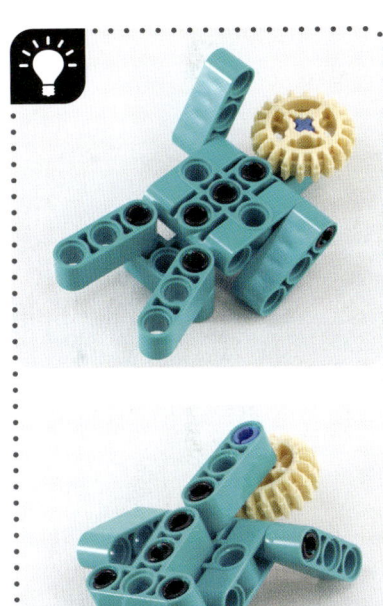

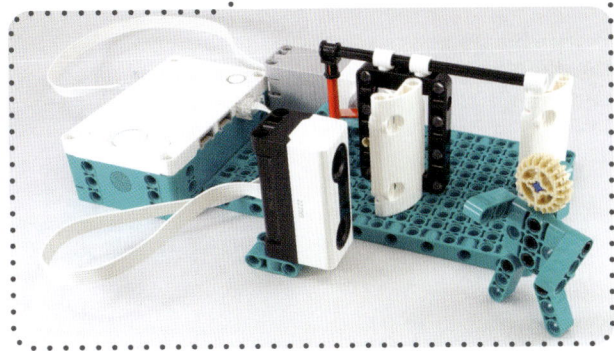

116

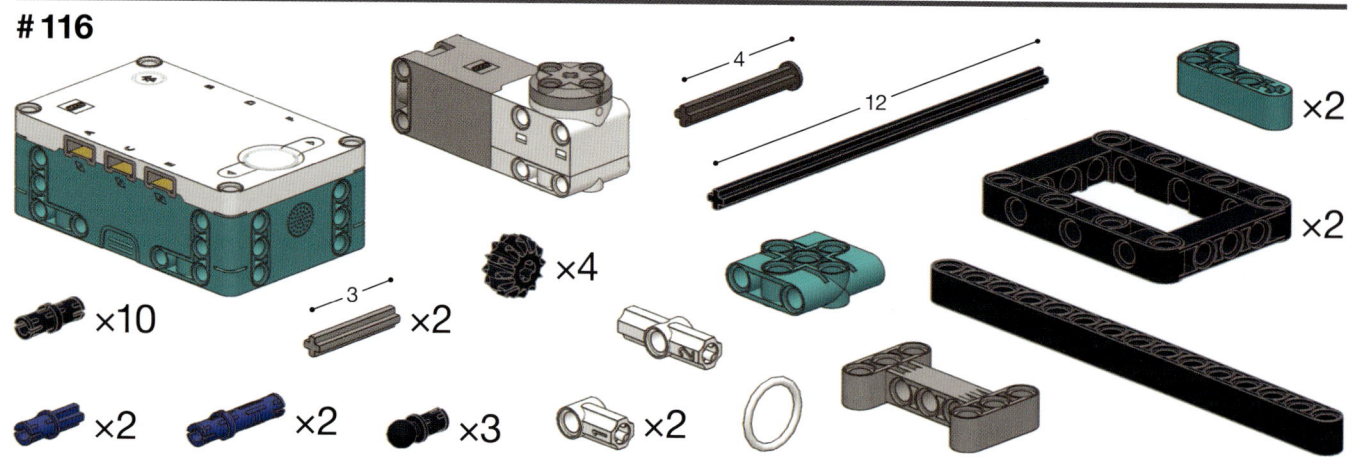

×10 ×2 ×4 ×2 ×3 ×2 ×2 ×2

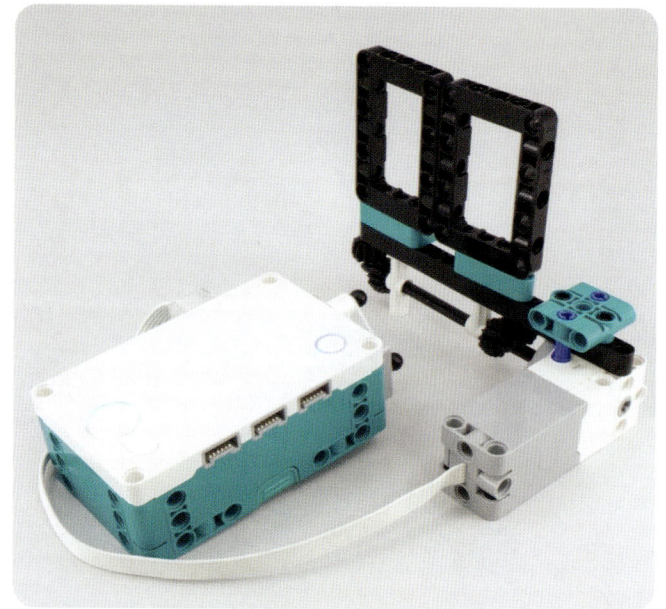

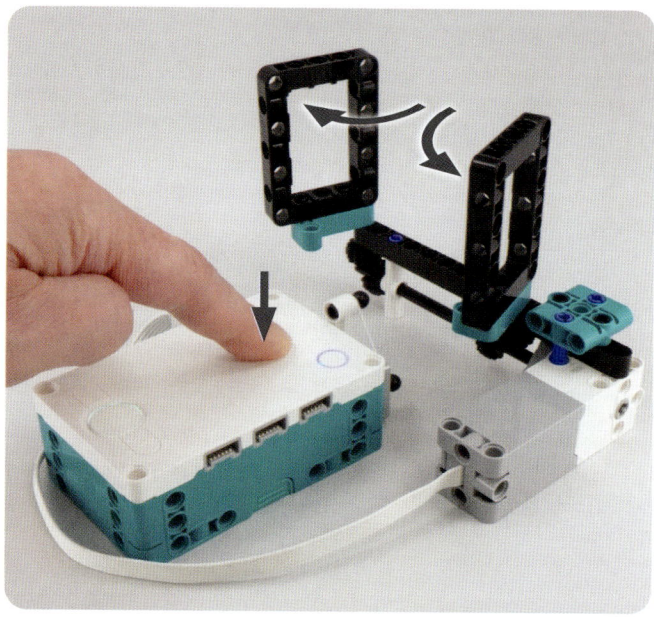

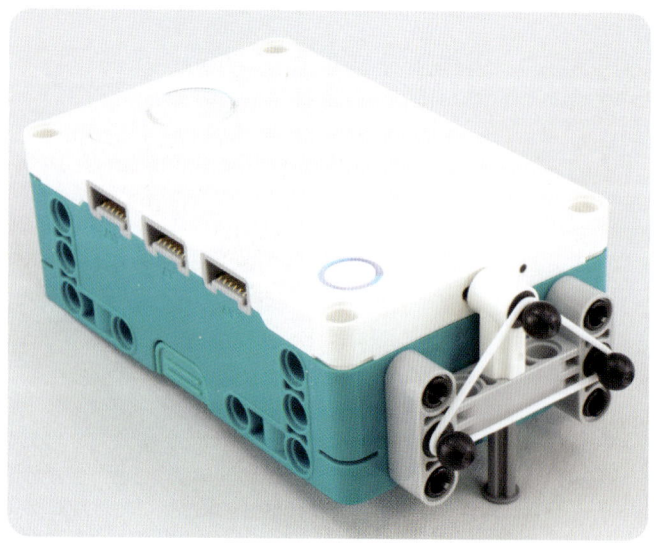

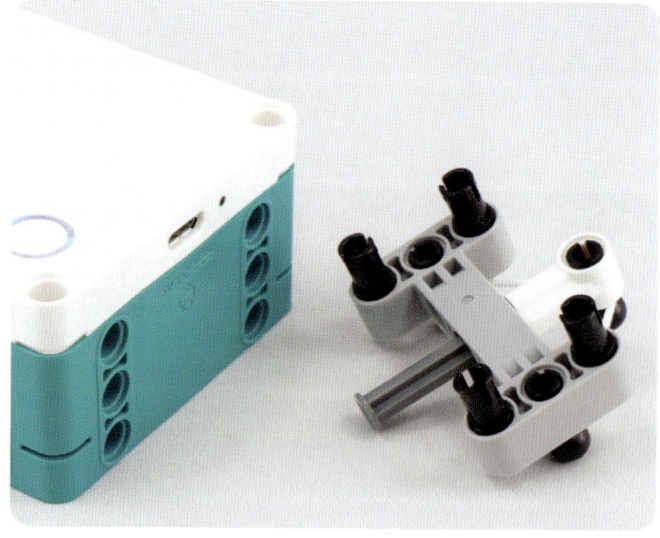

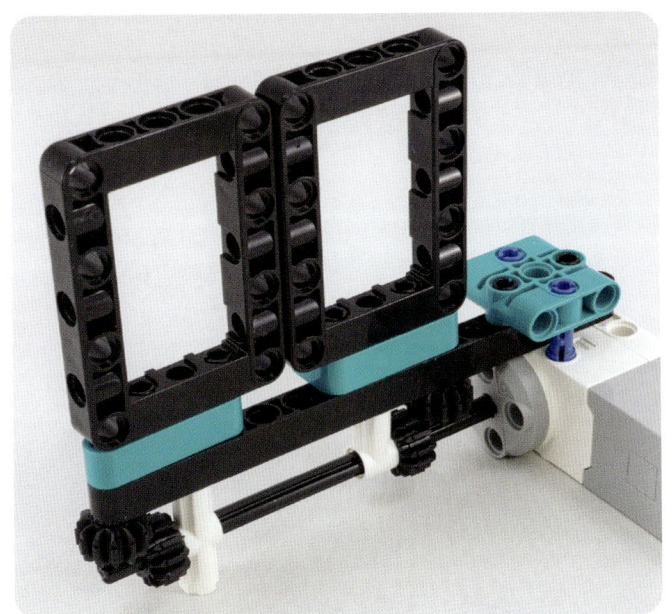

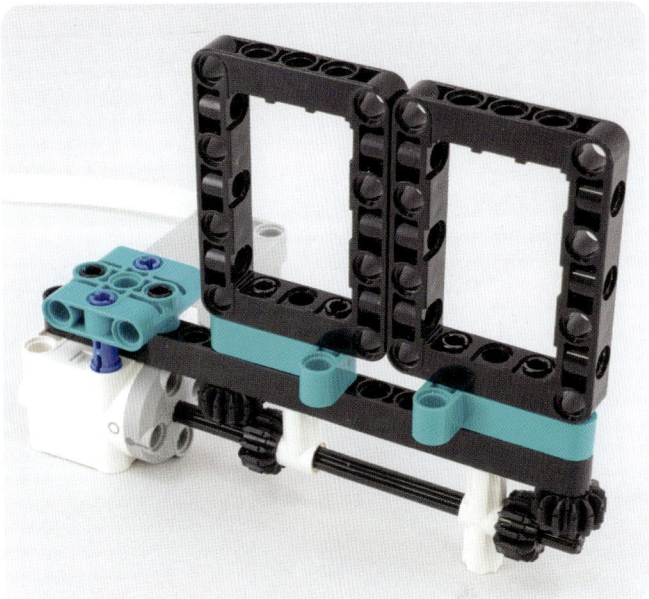

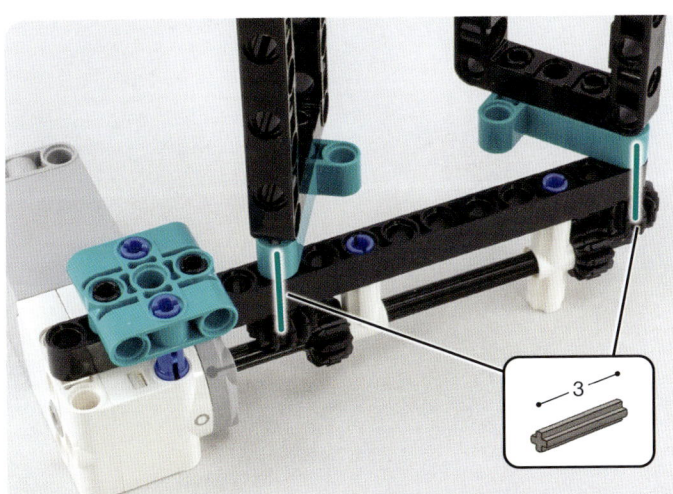

3

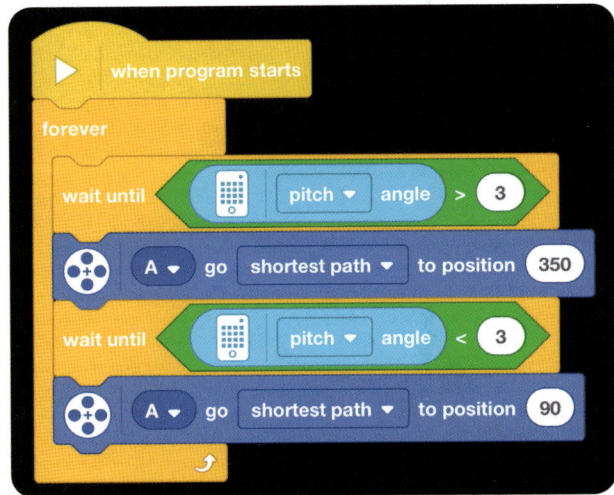

```
▶  when program starts

forever
    wait until    [::]  pitch ▼  angle  >  3
    [⊕]  A ▼  go  shortest path ▼  to position  350
    wait until    [::]  pitch ▼  angle  <  3
    [⊕]  A ▼  go  shortest path ▼  to position  90
```

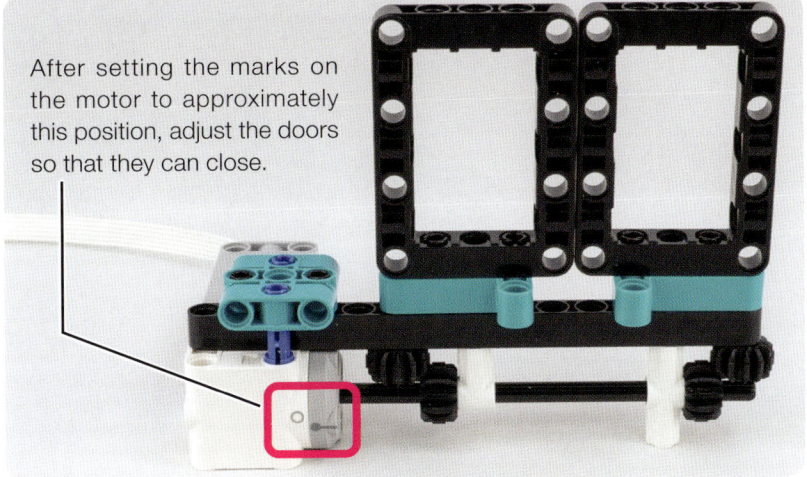

After setting the marks on the motor to approximately this position, adjust the doors so that they can close.

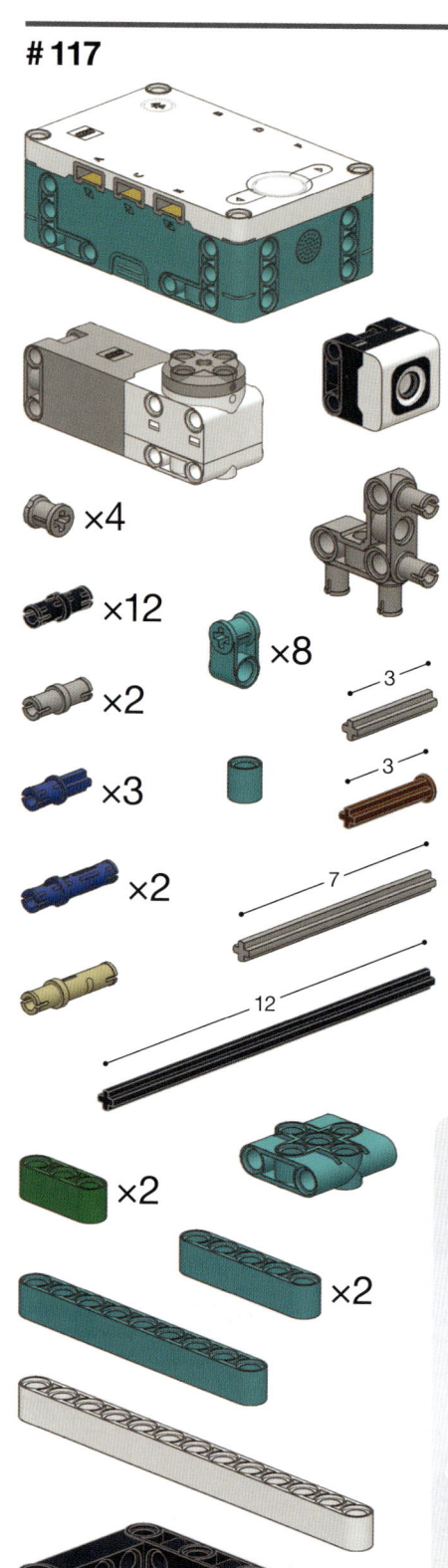

117

×4

×12

×8

×2

3

×3

3

×2

7

12

×2

×2

×2

×2

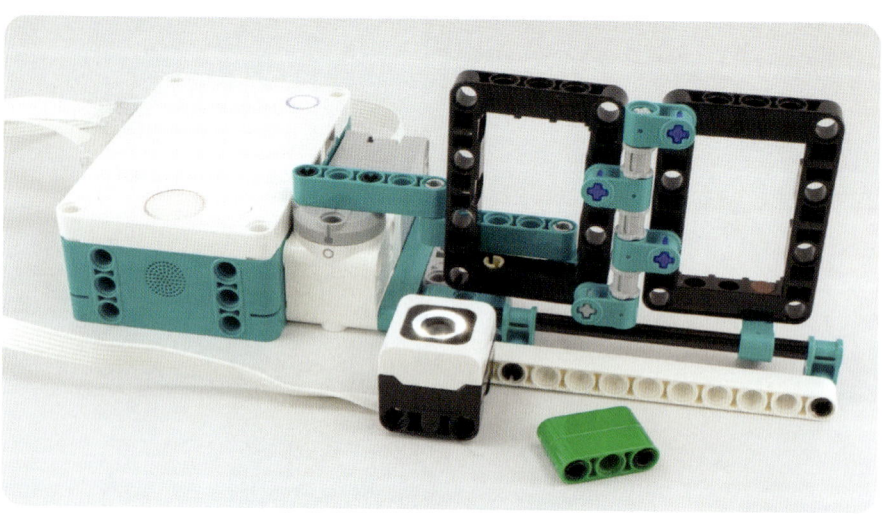

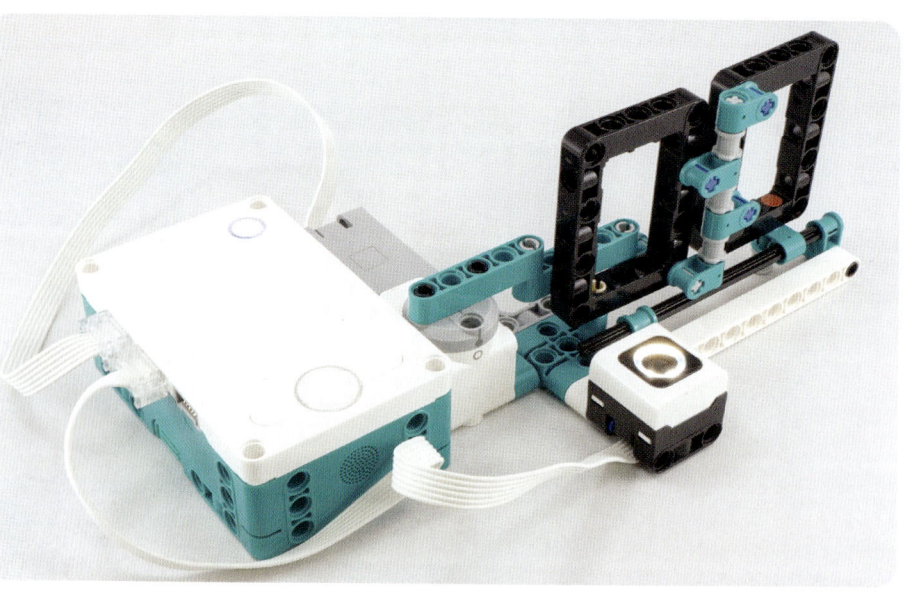

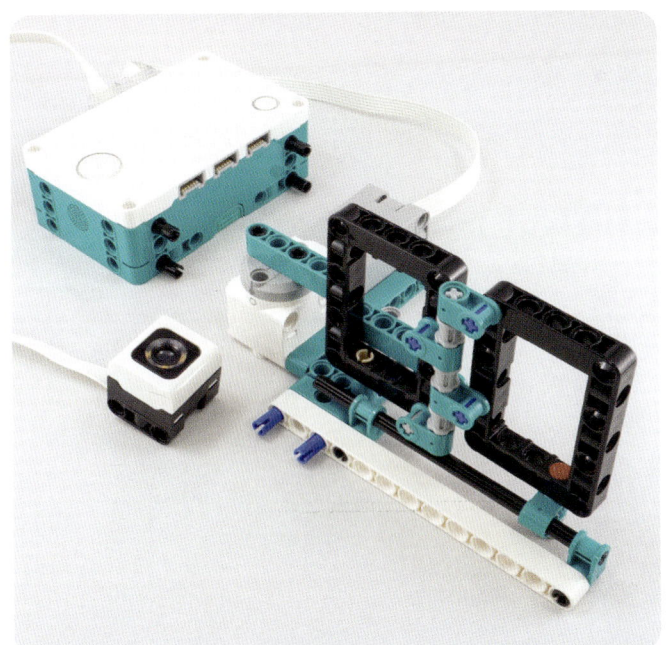

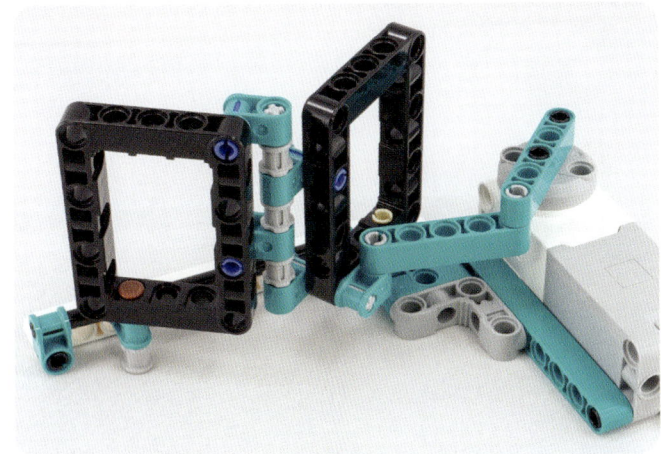

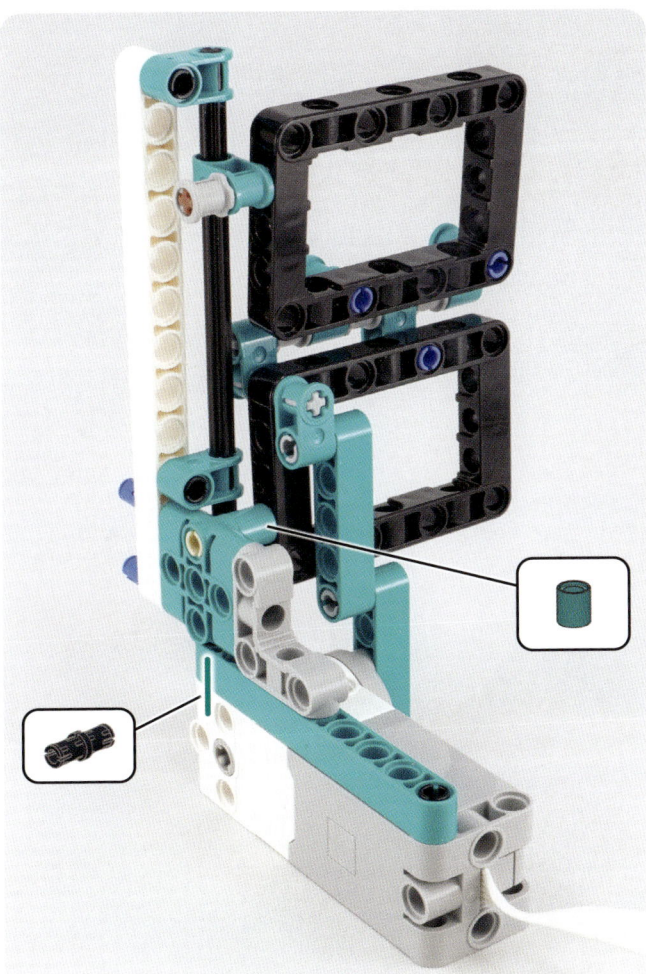

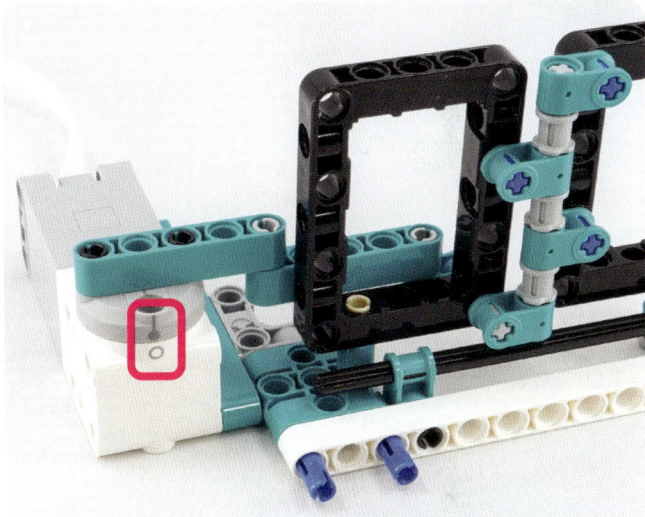

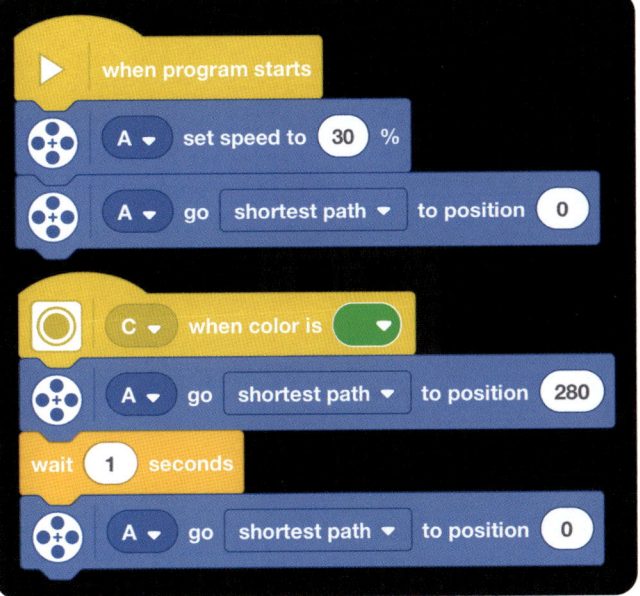

Making fun games and toys

 # 118

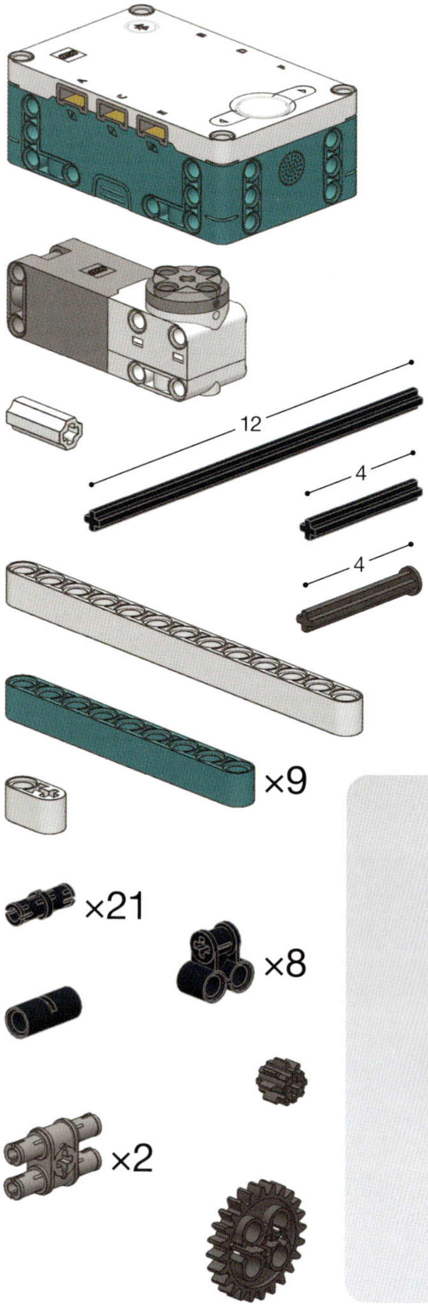

12

4

4

×9

×21

×8

×2

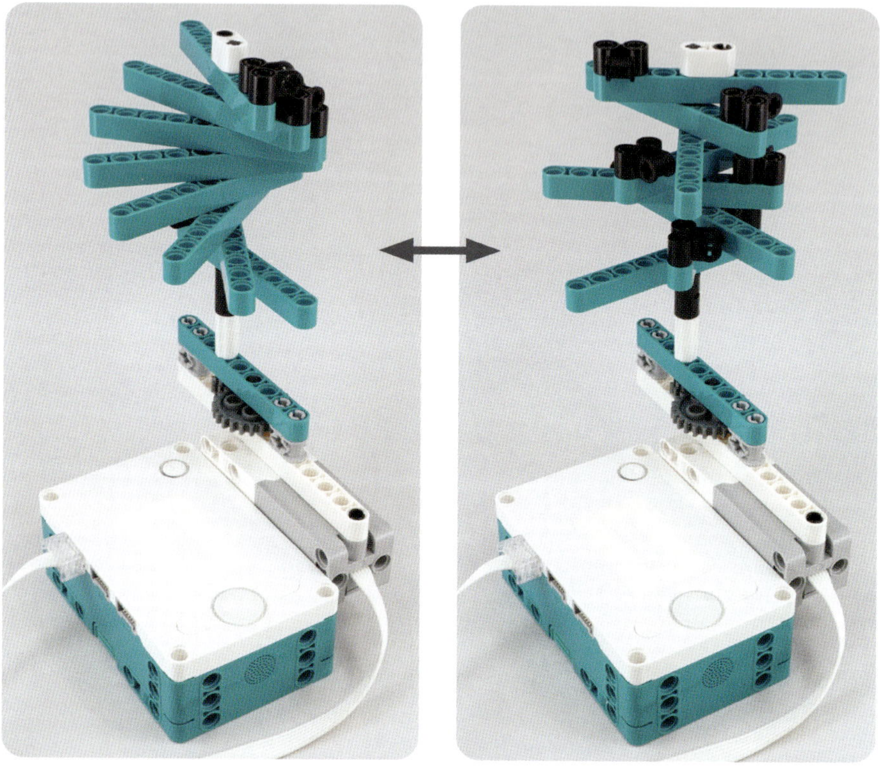

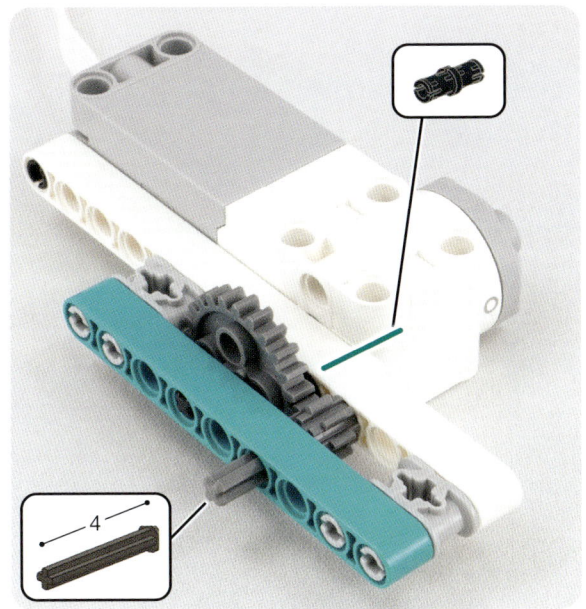

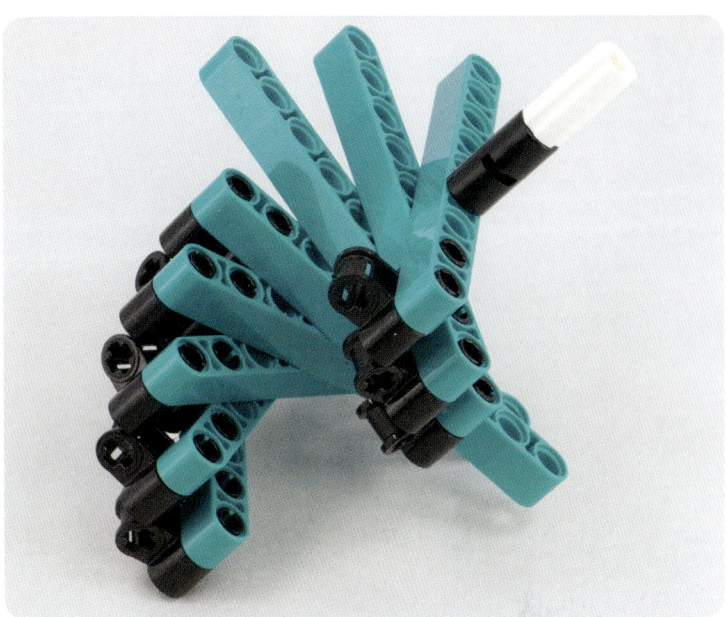

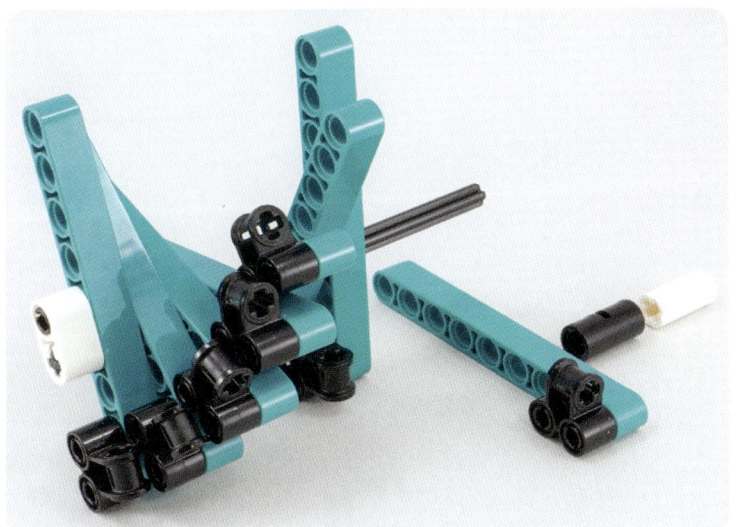

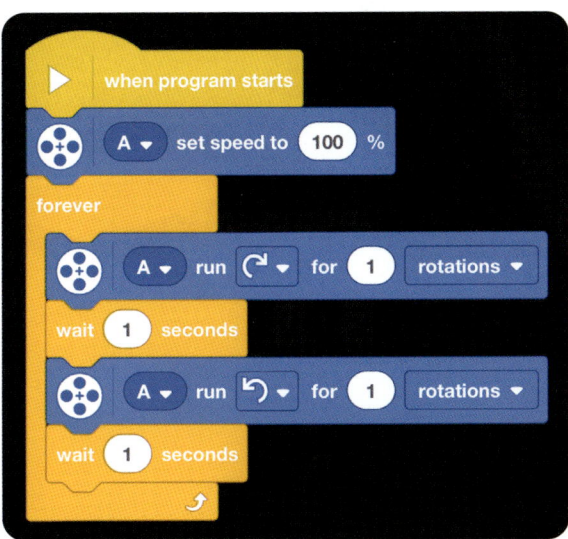

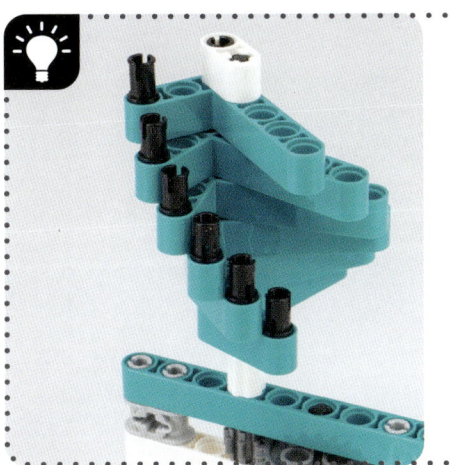

#119

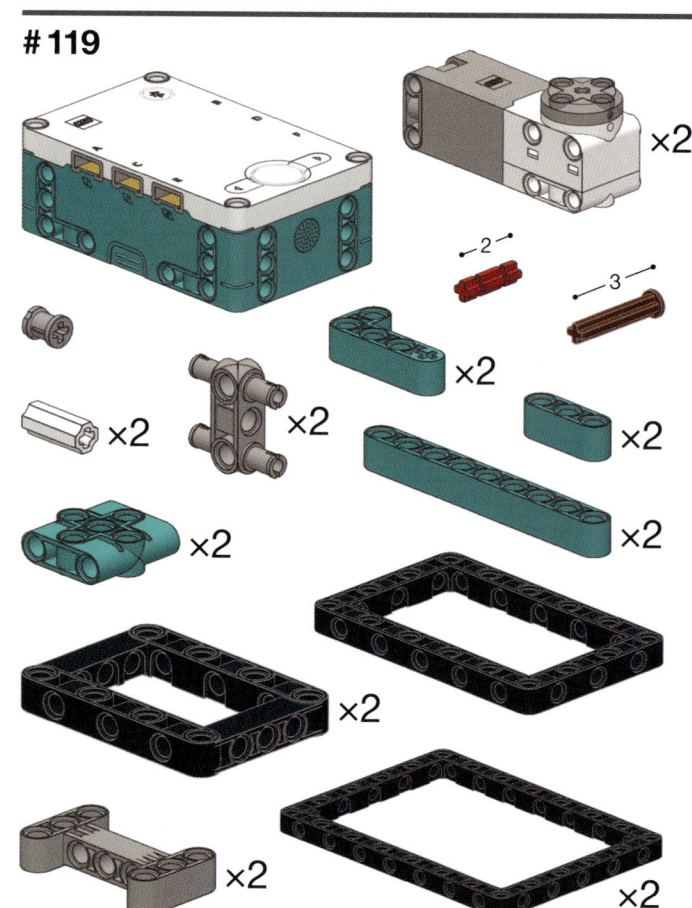

×2

×24

×2

×4

×2

×2

×2

×2

×2

×2

×2

×2

×2

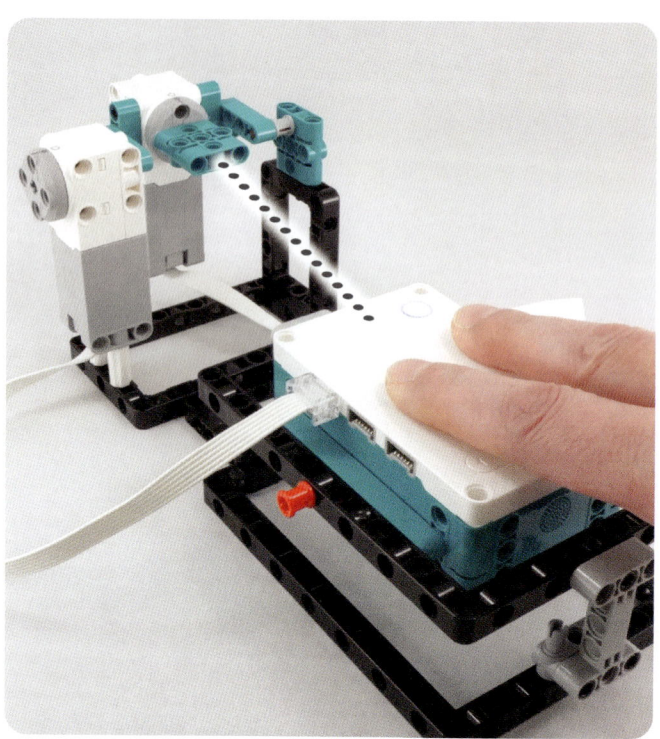

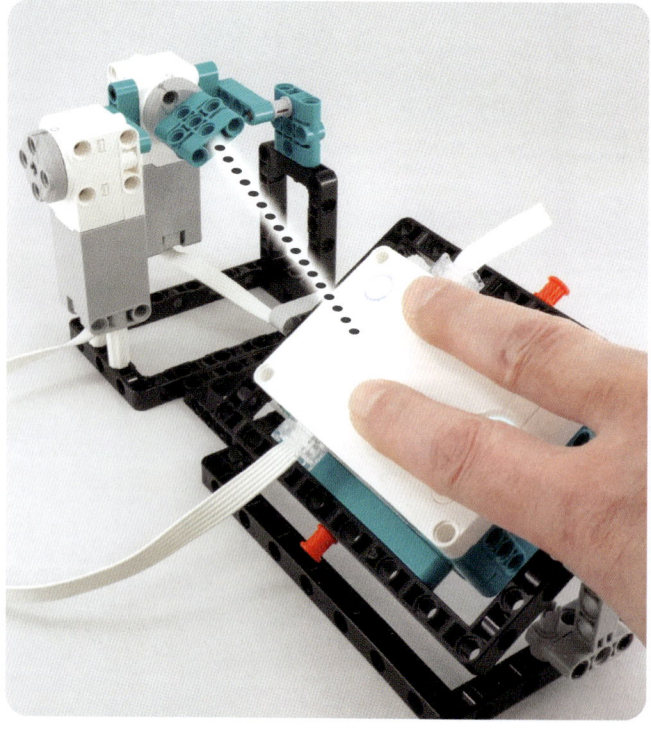

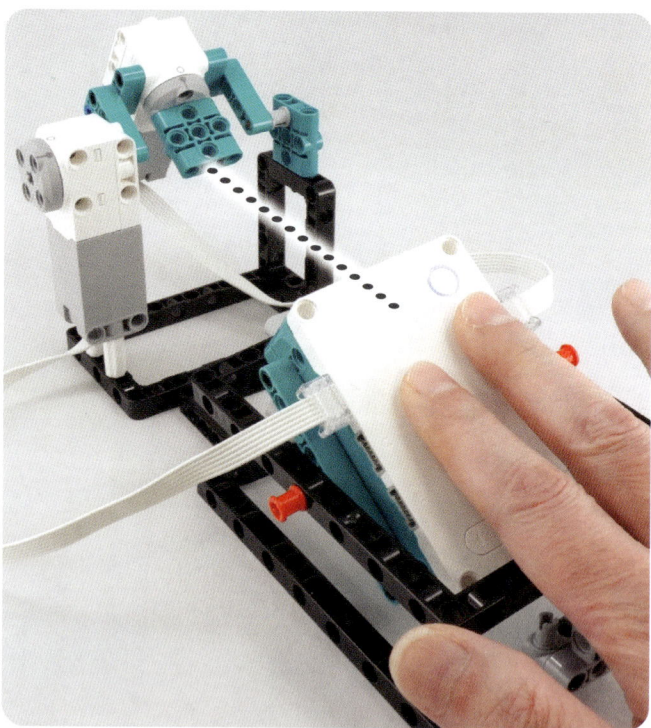

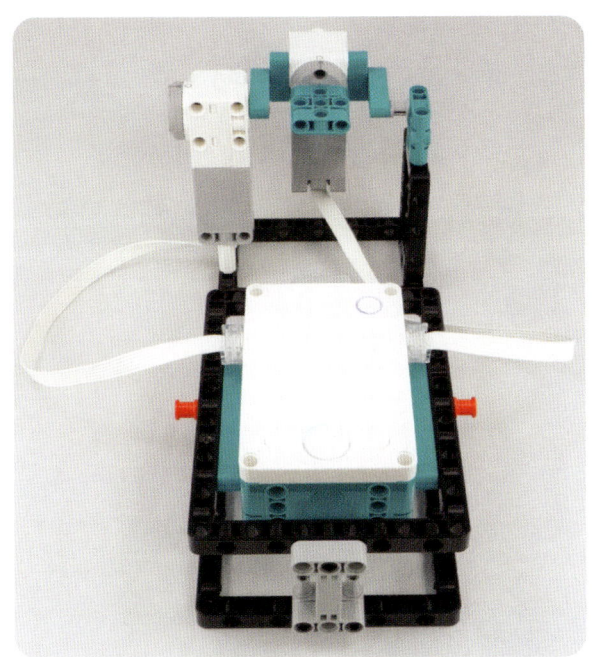

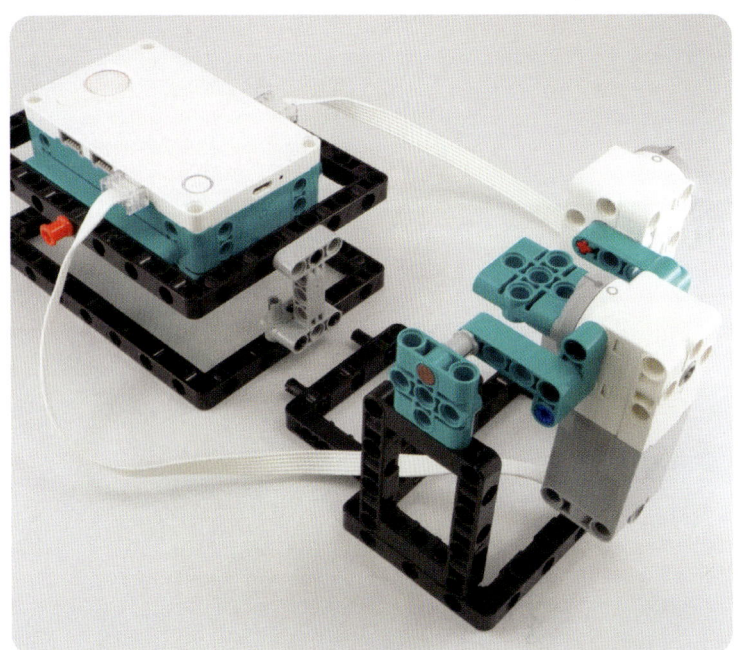

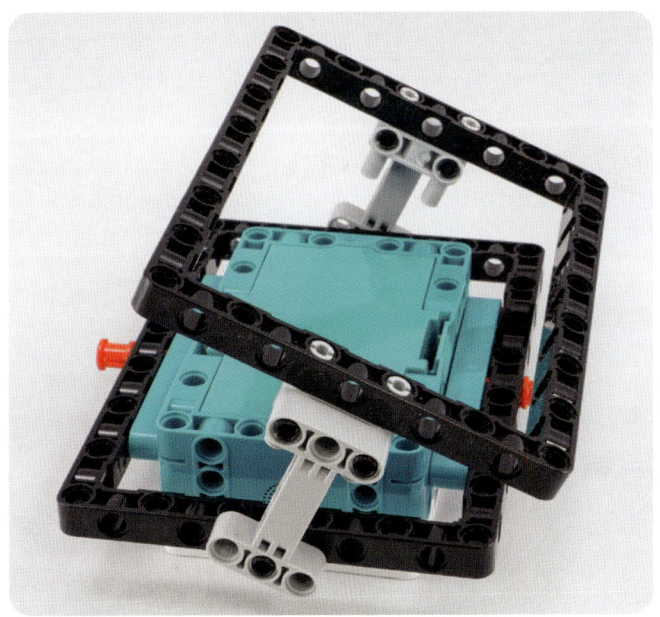

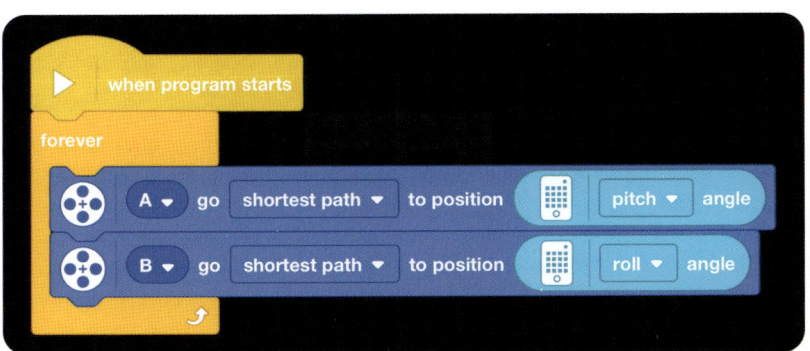

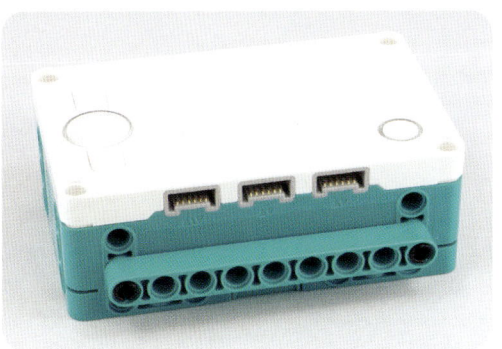

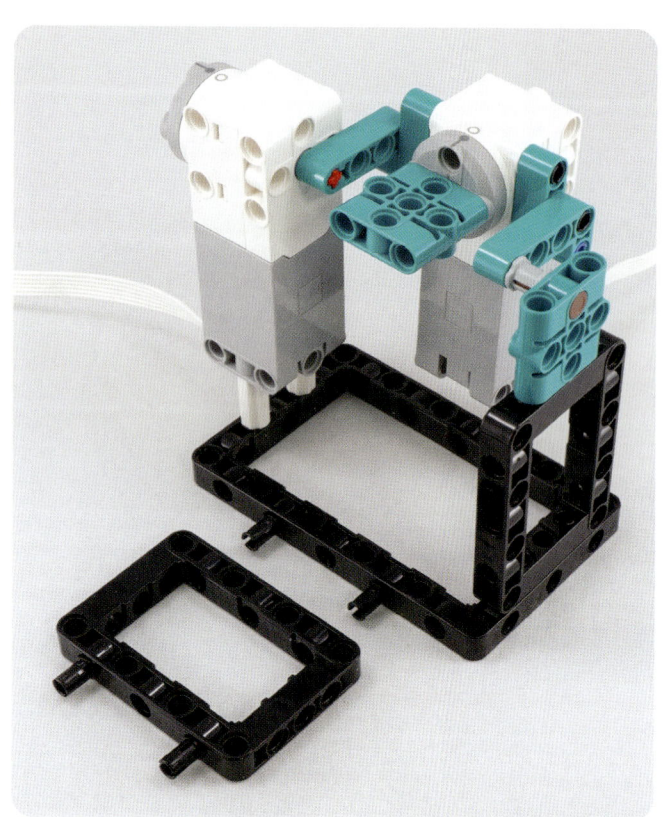

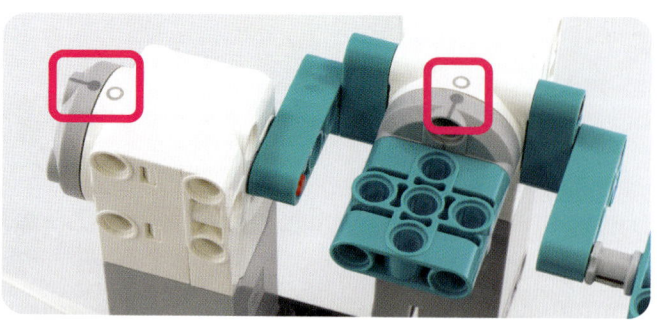

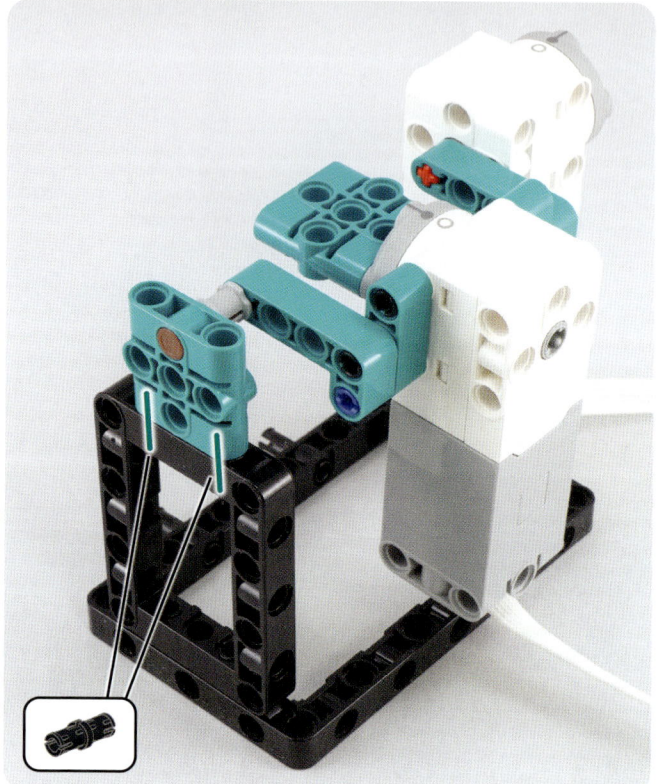

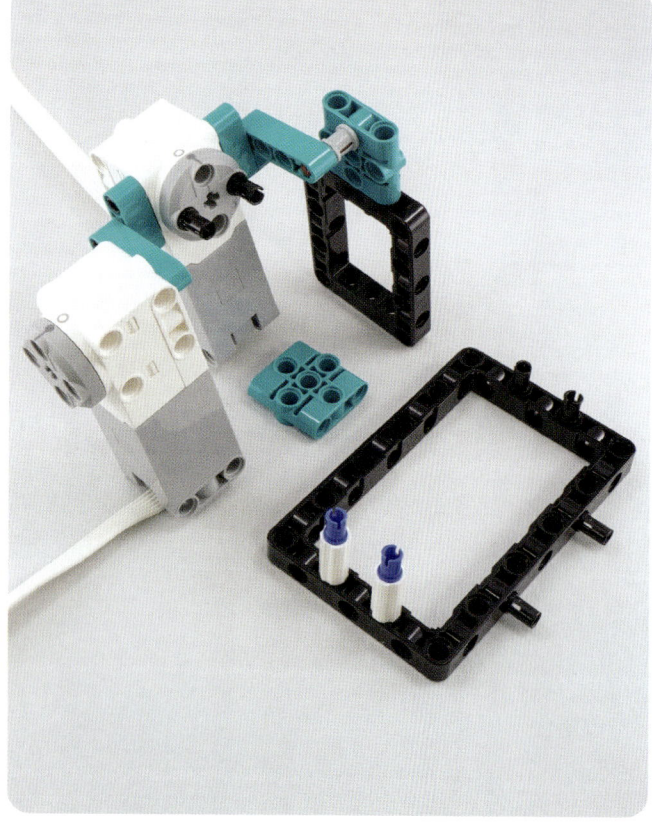

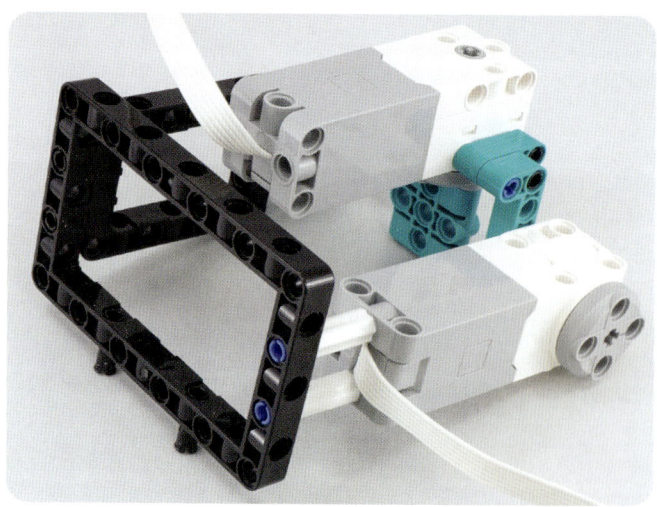

Start

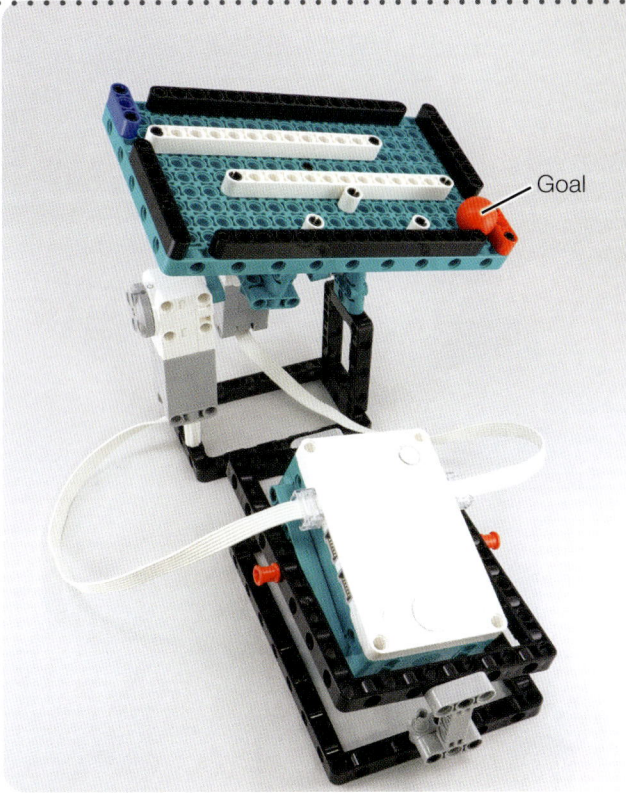

Goal

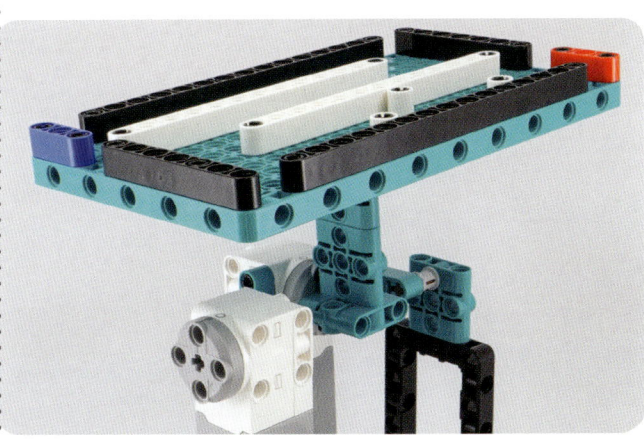

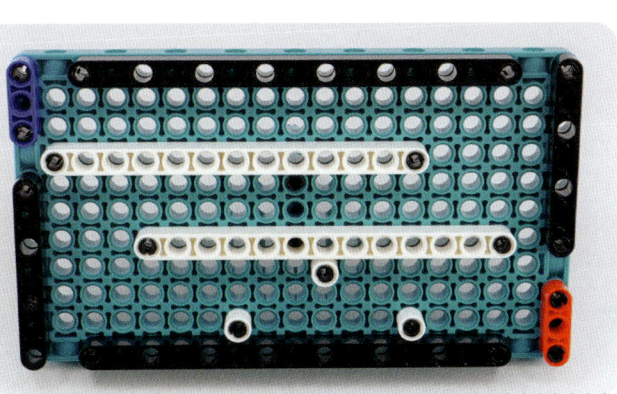

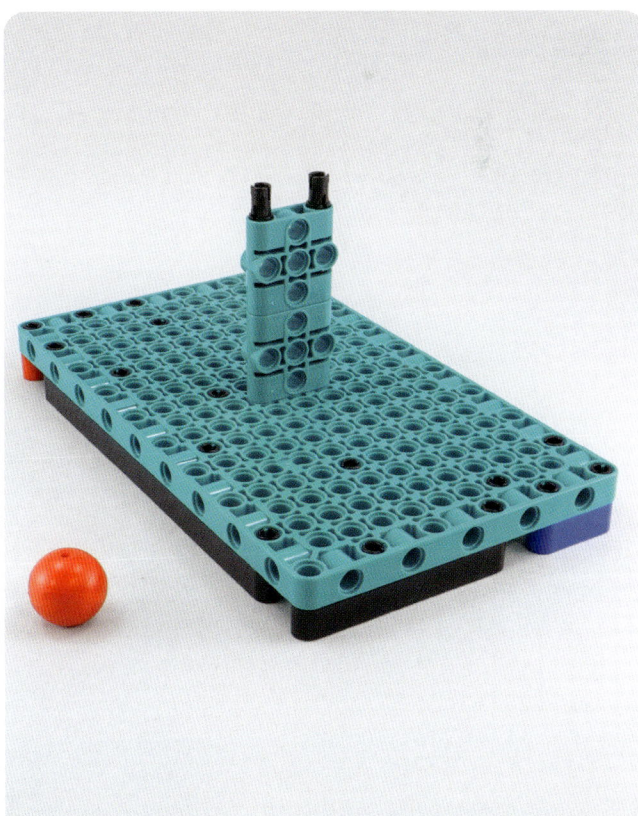

#120

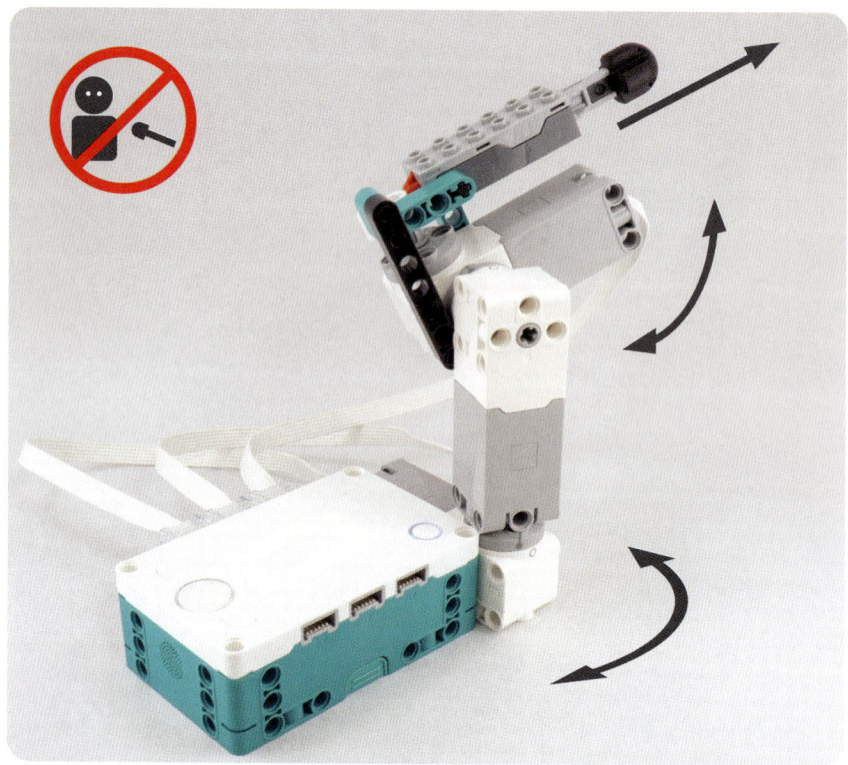

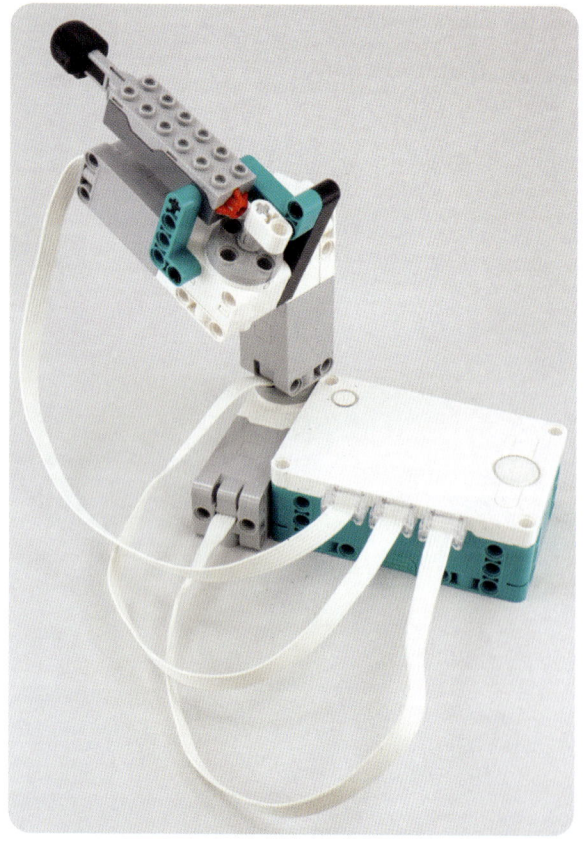

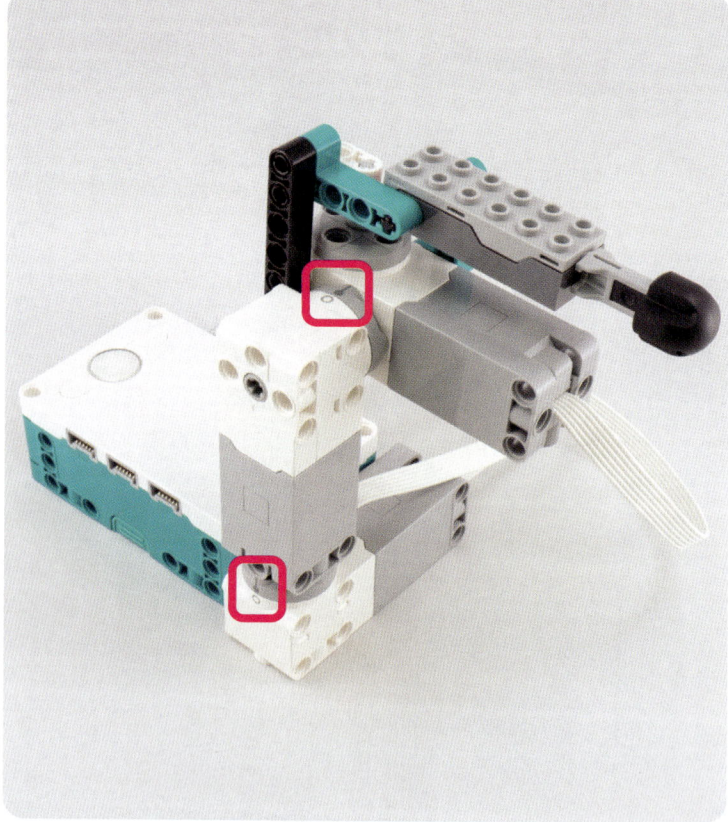

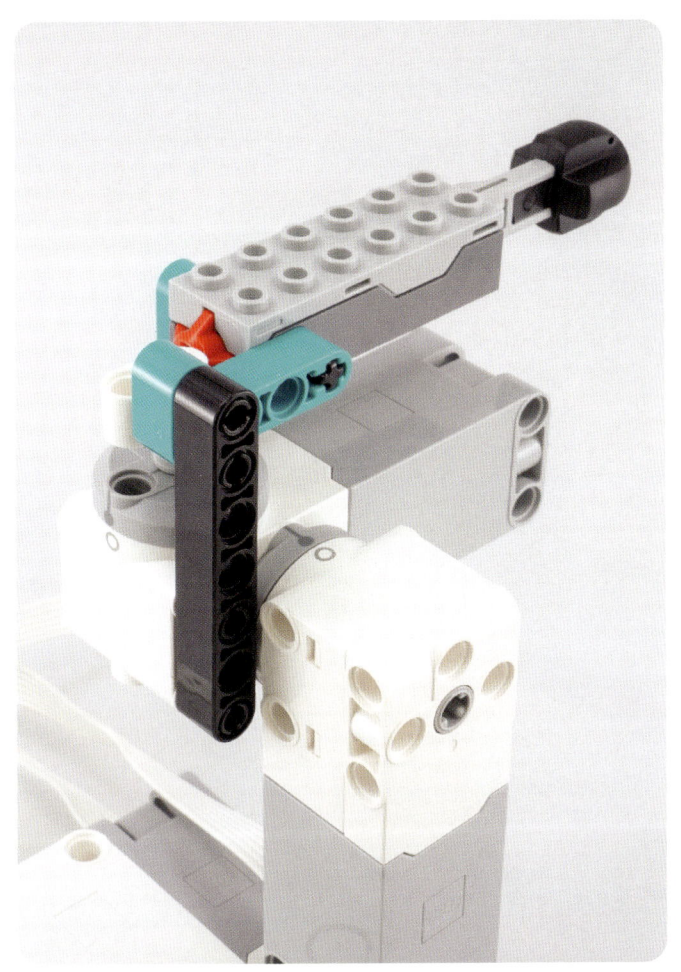

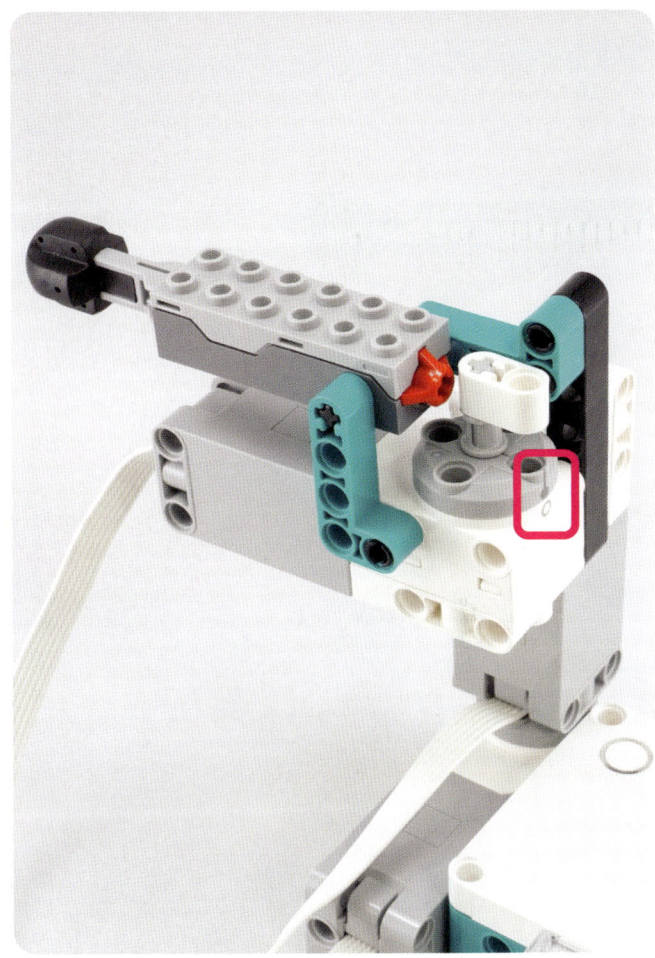

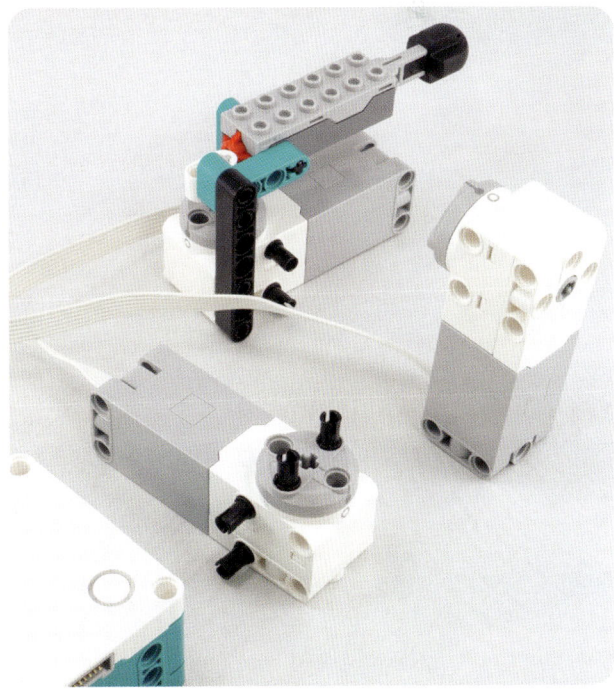

#121

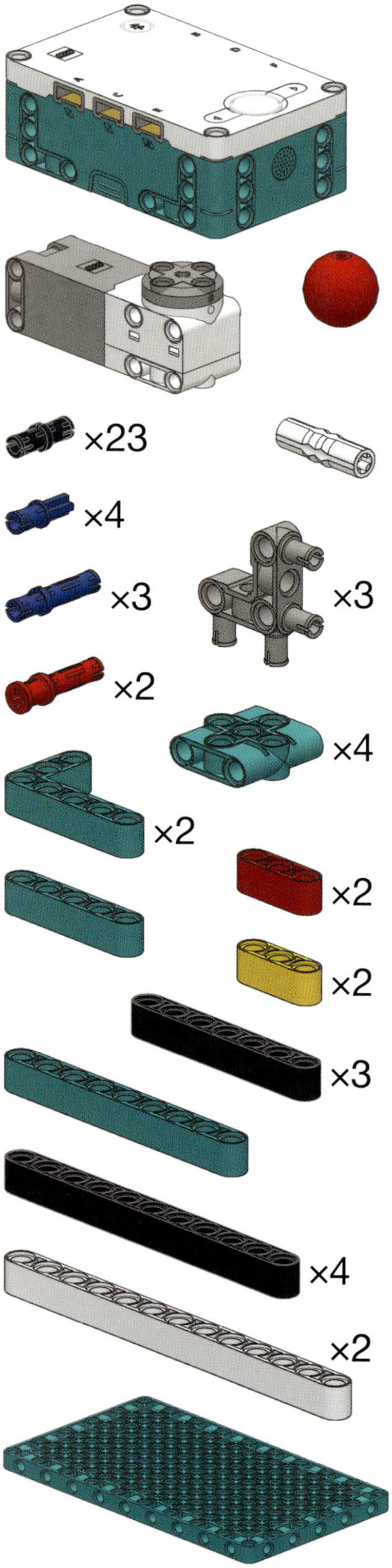

×23
×4
×3   ×3
×2
×4
×2
×2
×2
×3
×4
×2

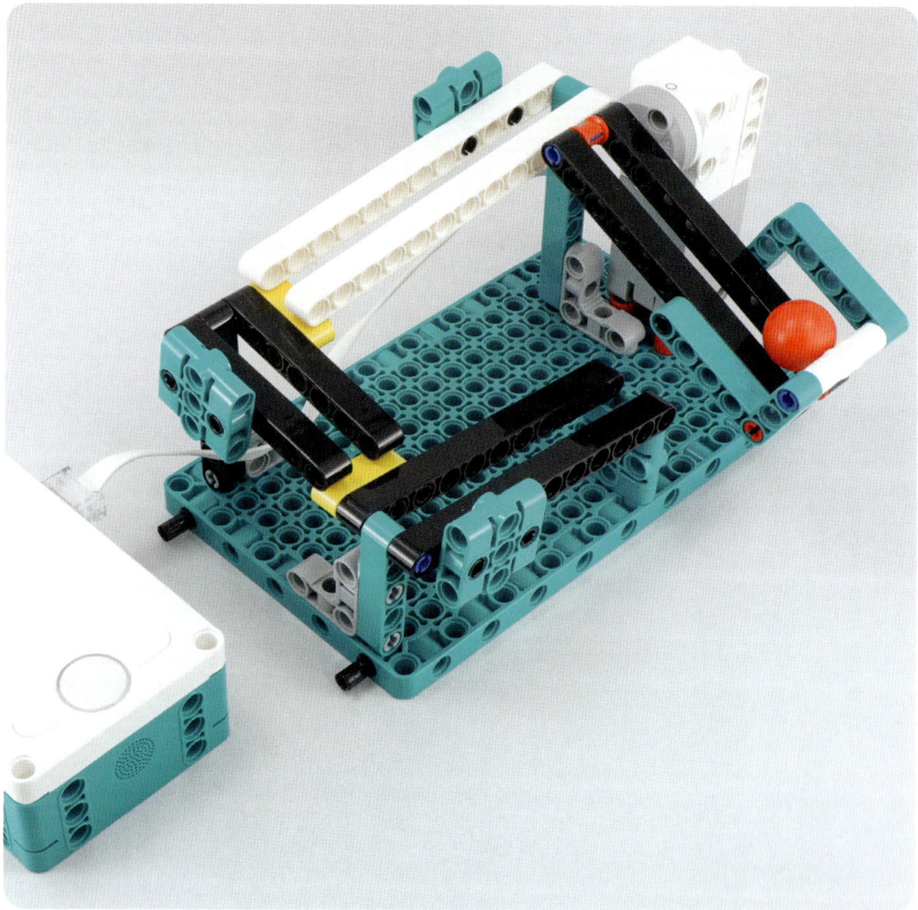

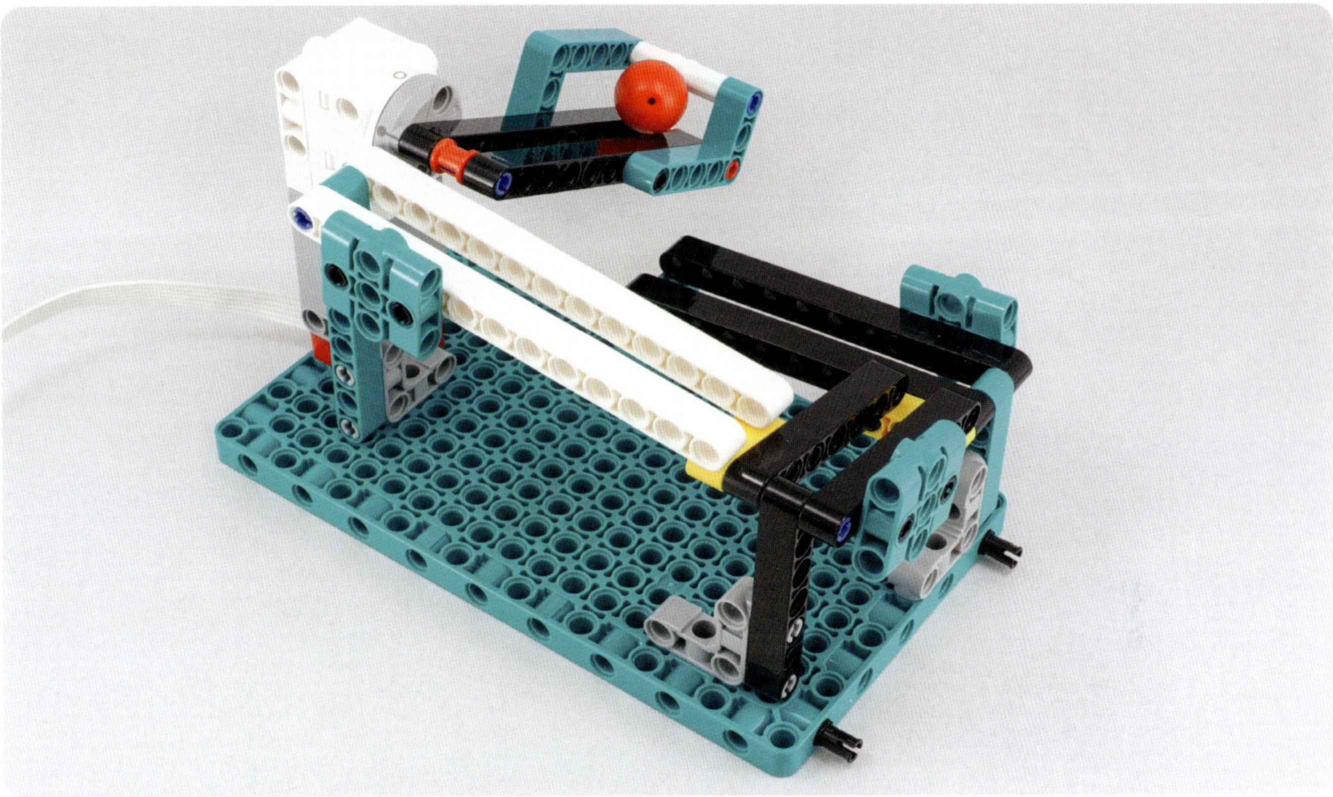

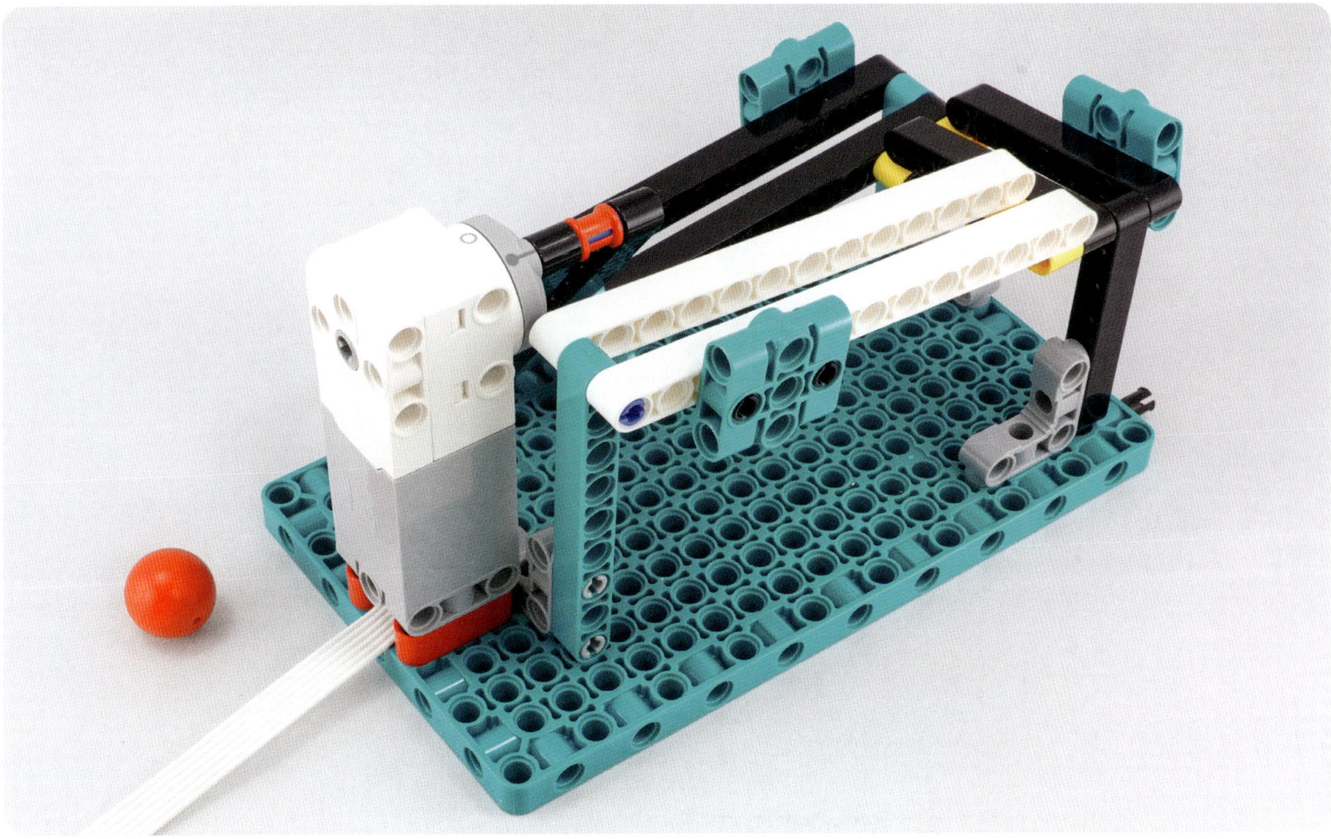

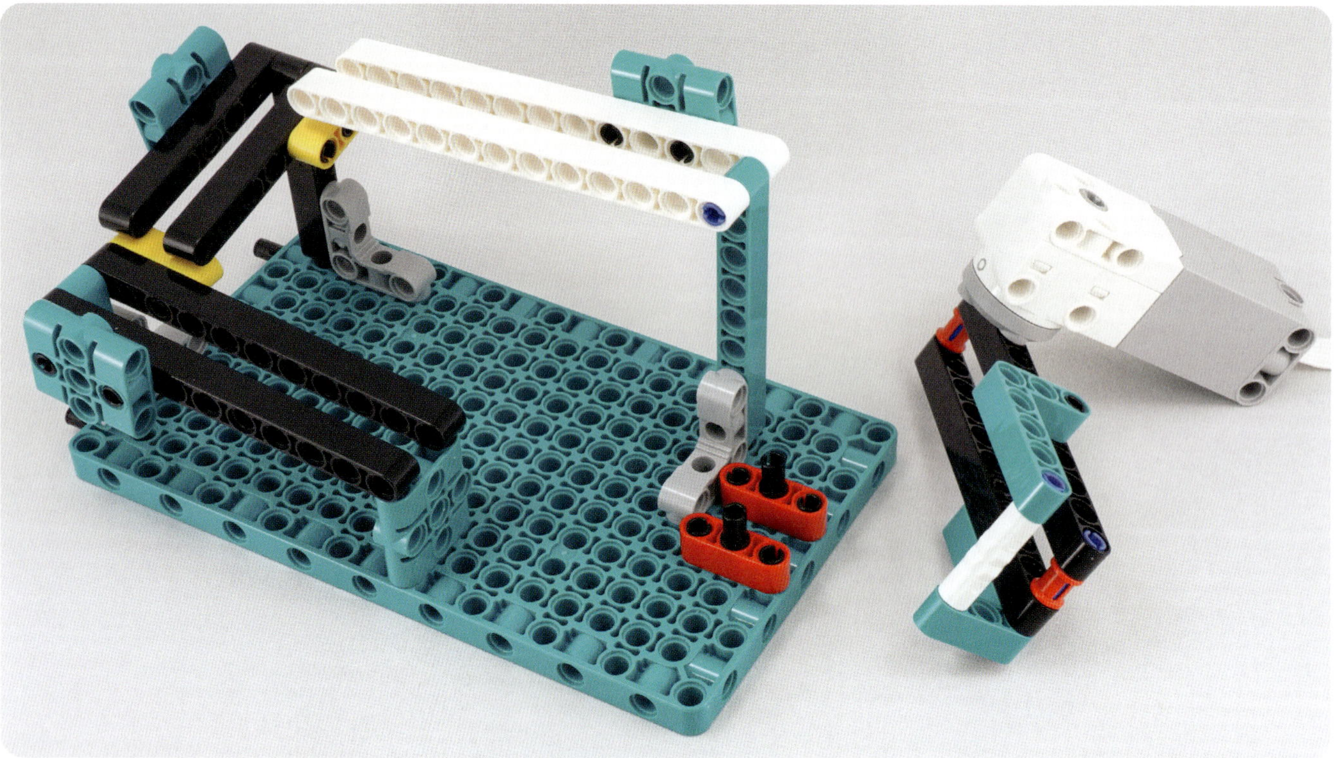

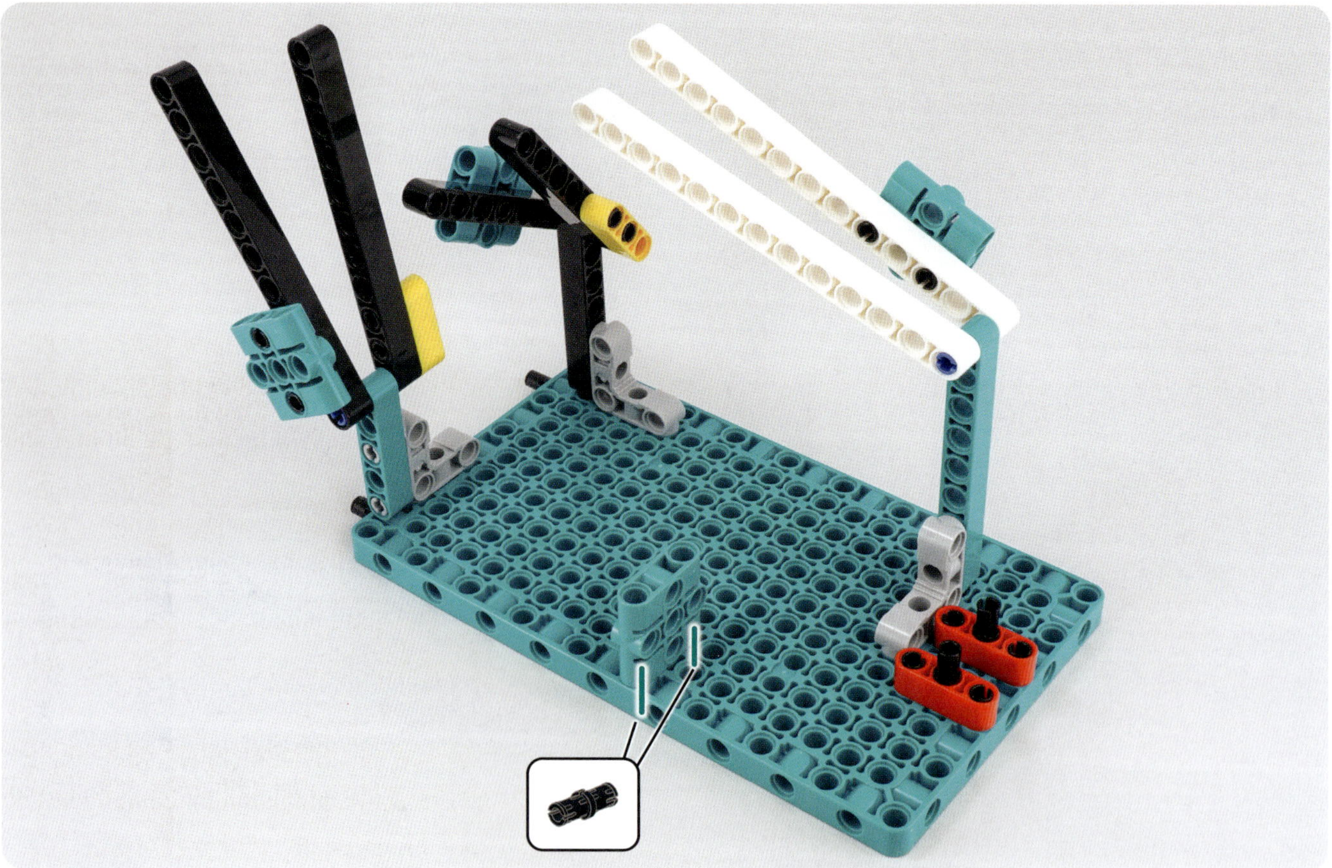

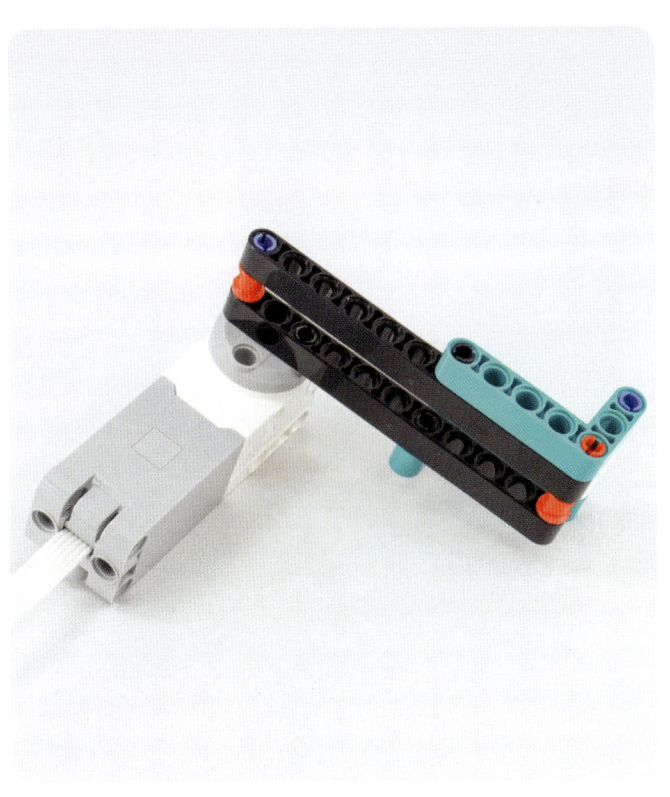

#122

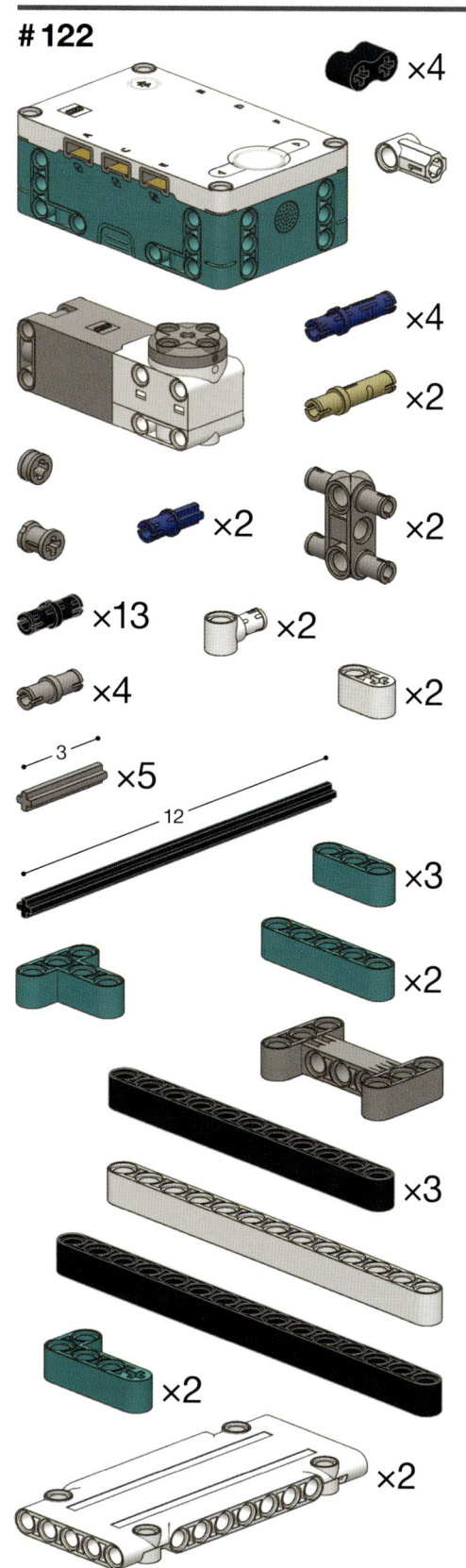

×4
×4
×2
×2
×2
×2
×13
×2
×4
×2
3 ×5
12
×3
×2
×3
×2
×2

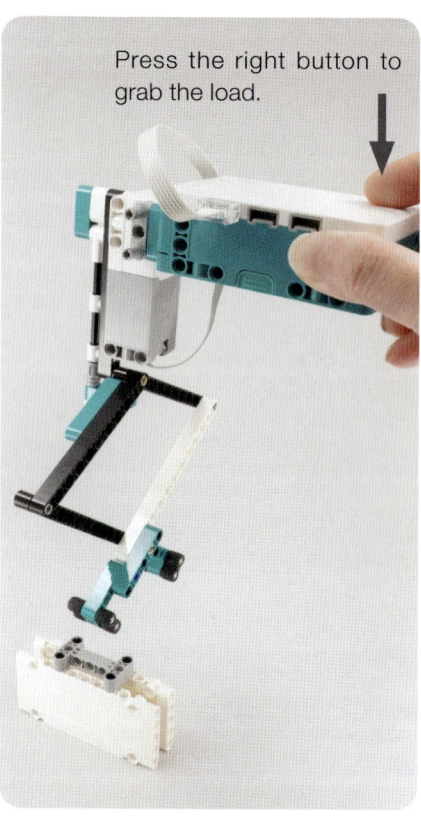

Press the right button to grab the load.

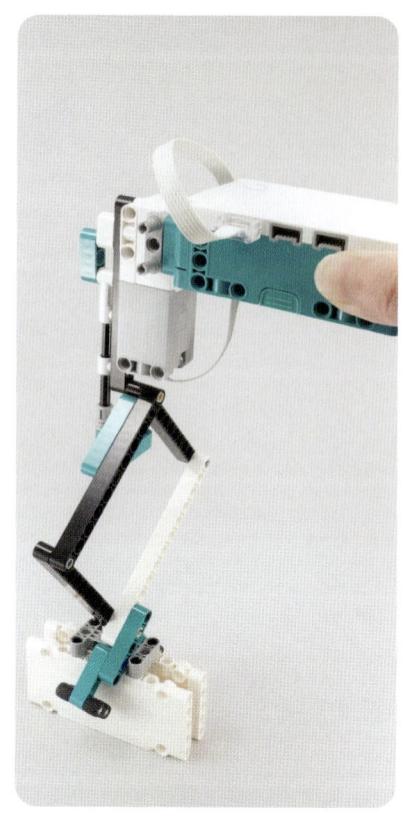

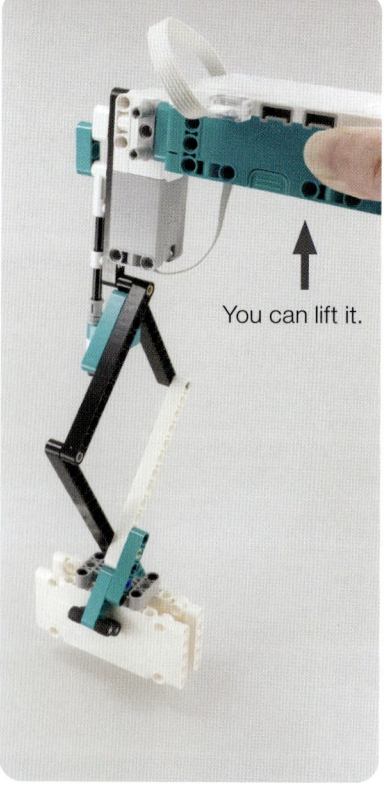

You can lift it.

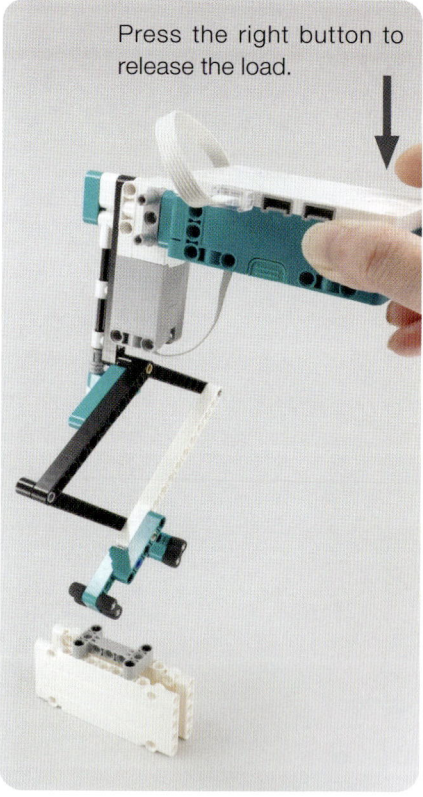

Press the right button to release the load.

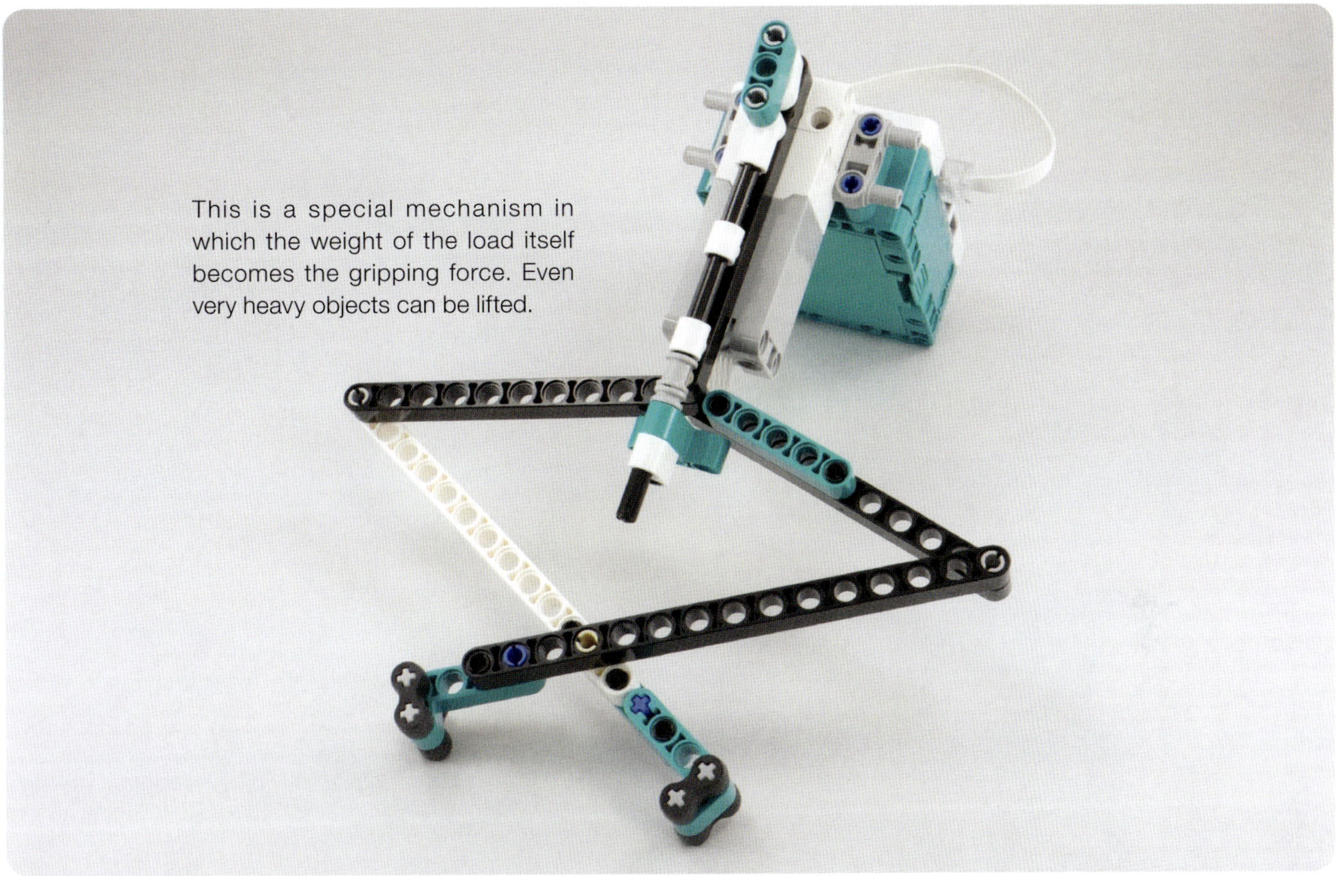

This is a special mechanism in which the weight of the load itself becomes the gripping force. Even very heavy objects can be lifted.

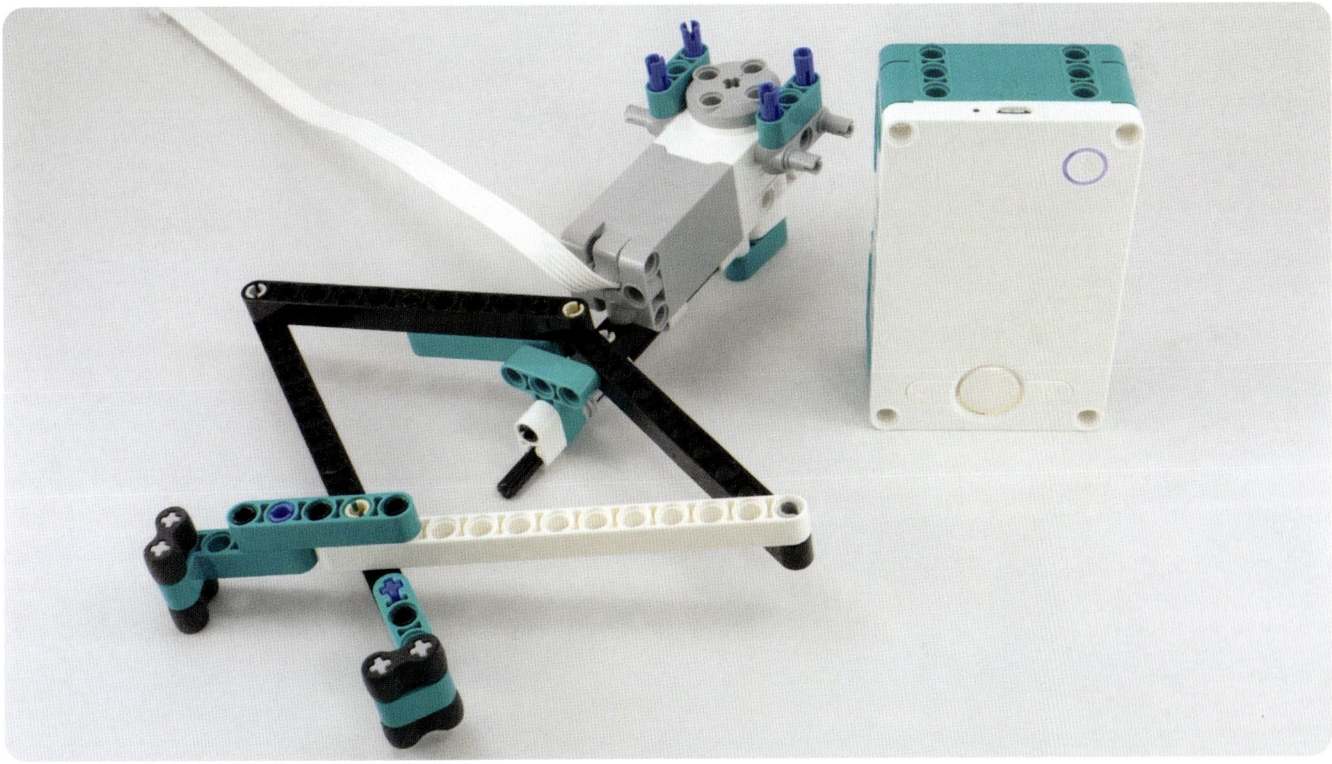

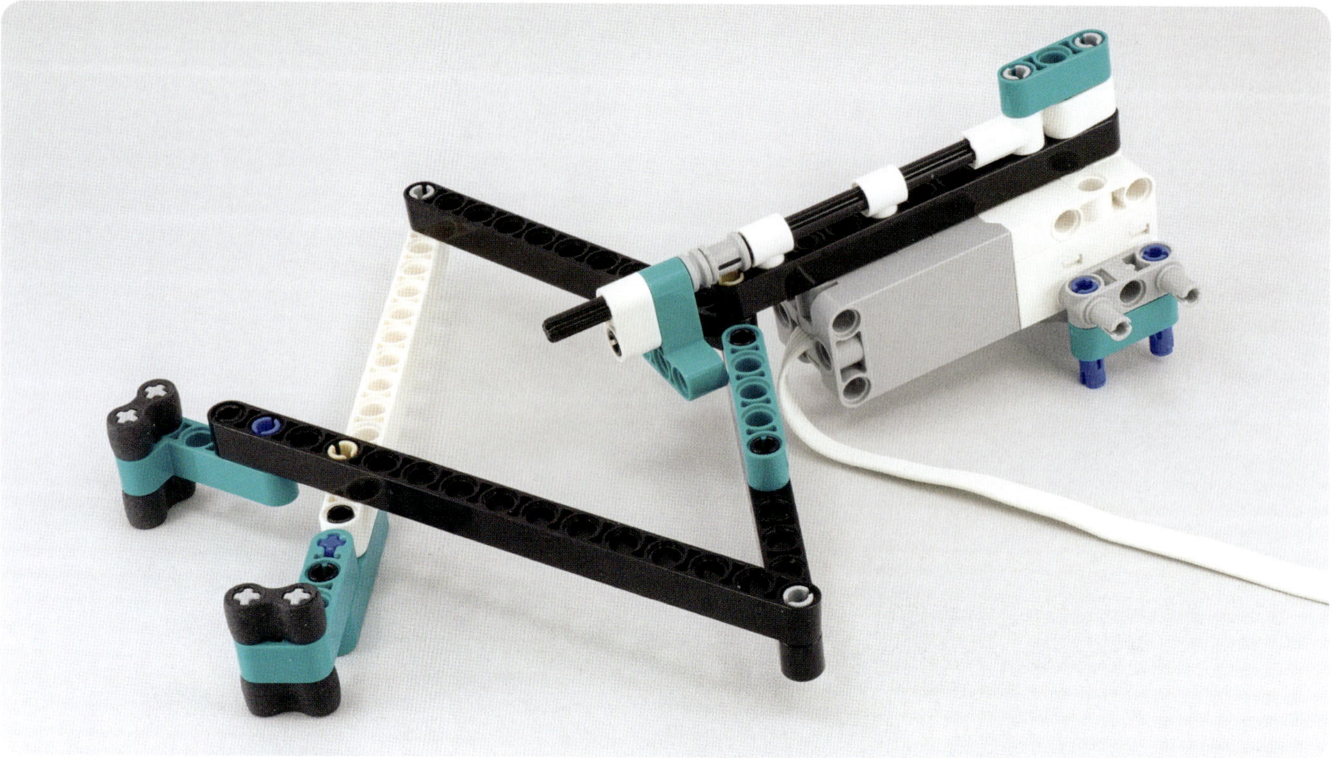

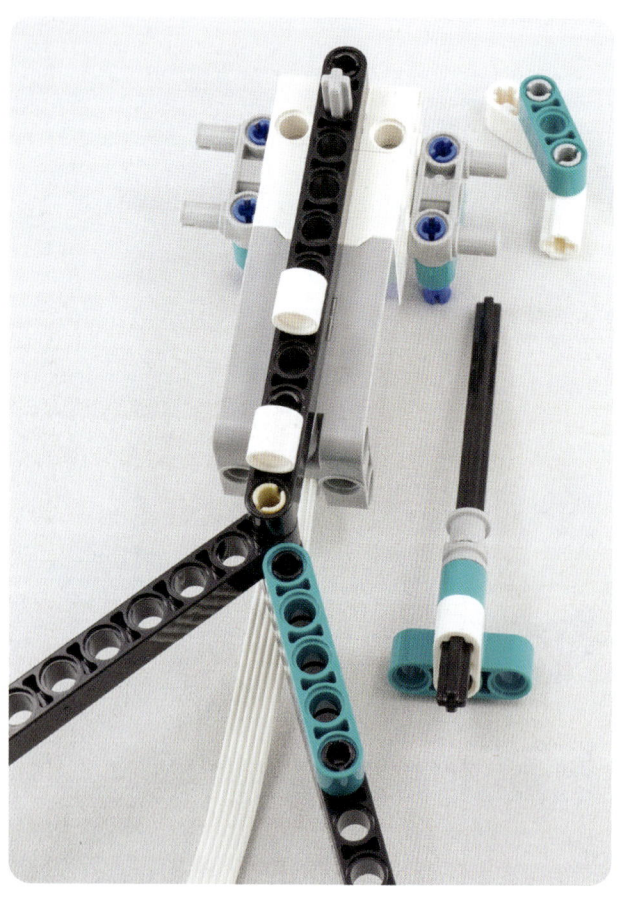

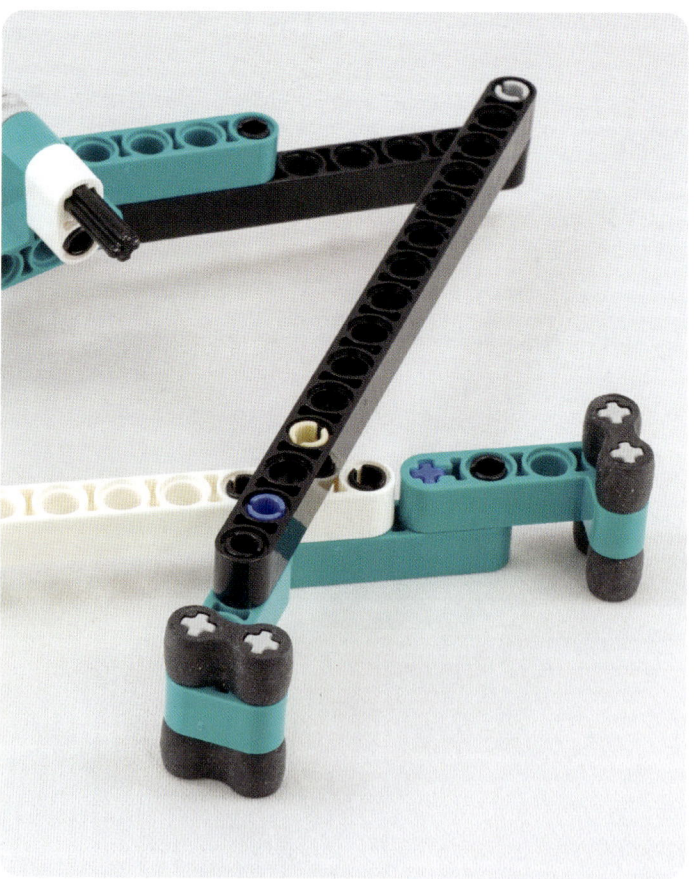

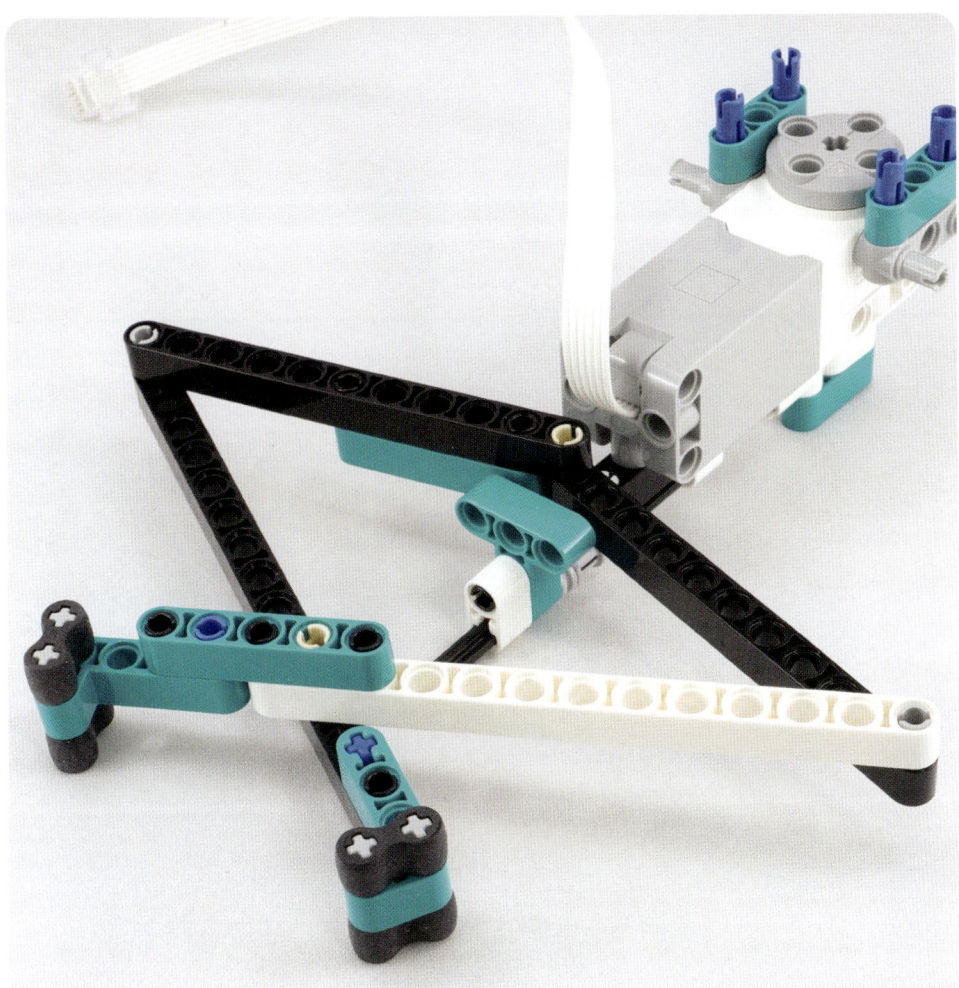

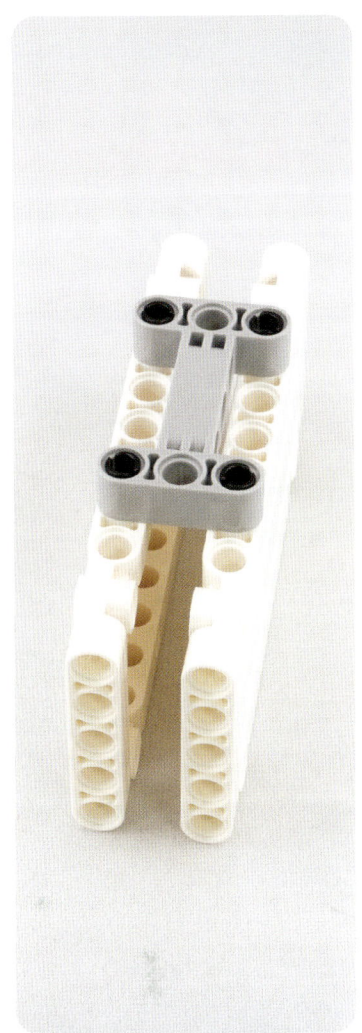

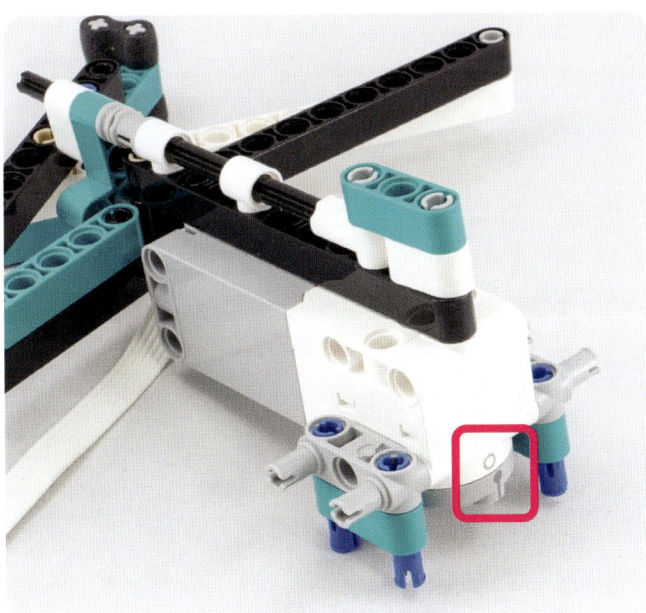

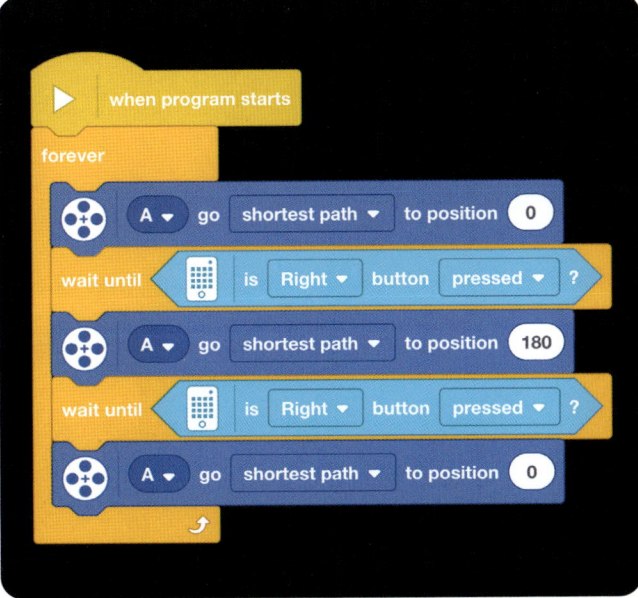

Bonus mechanisms

 # 123

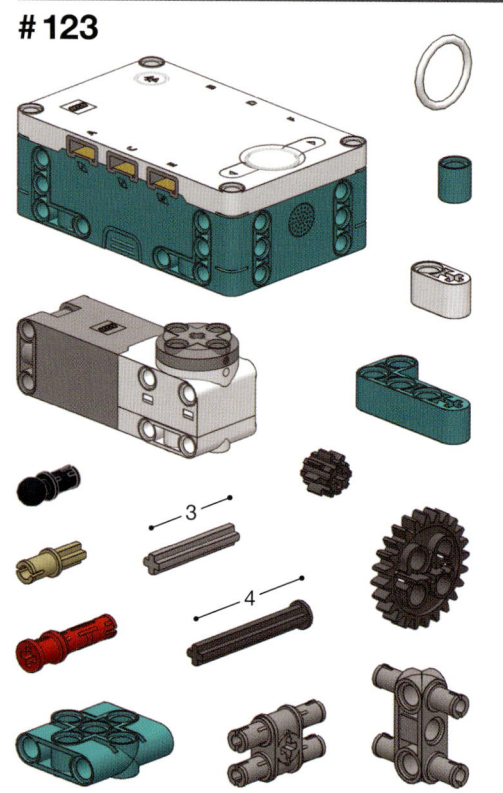

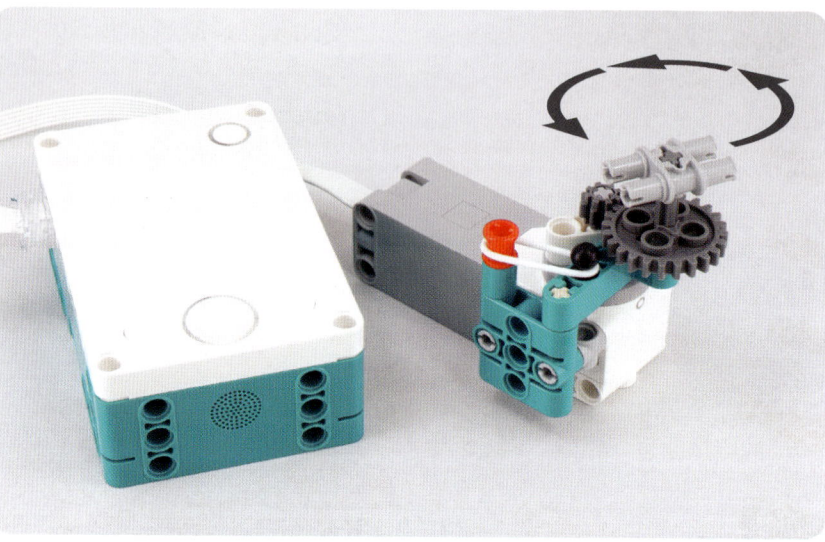

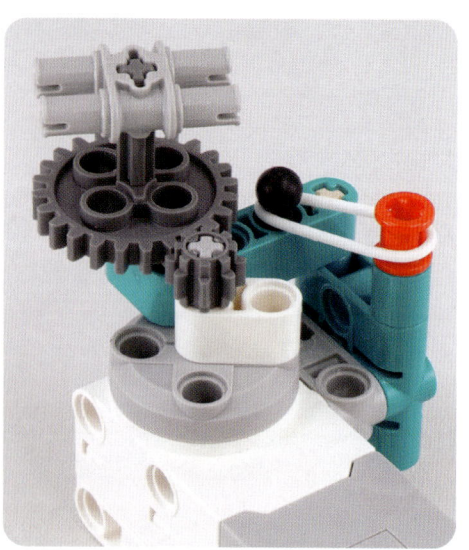

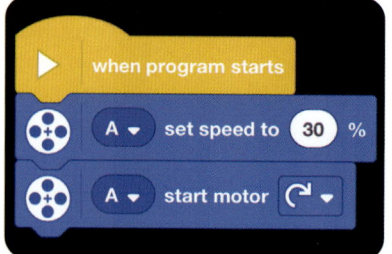

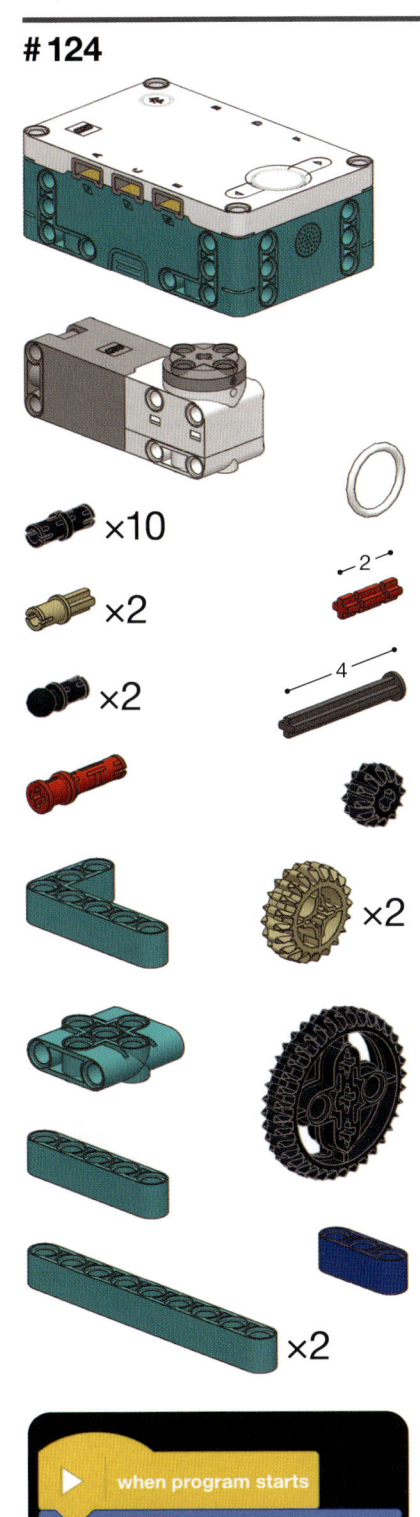

124

×10

×2

×2

2

4

×2

×2

It spins quickly or slowly.

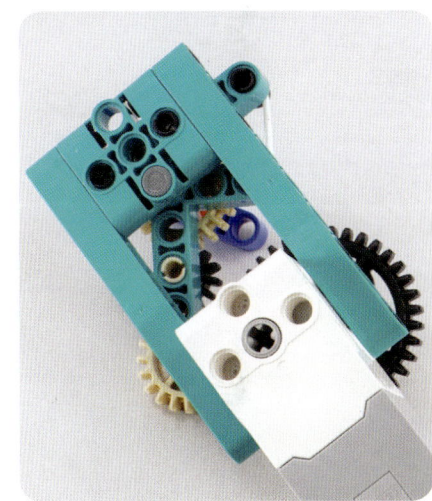

```
when program starts
A ▾  set speed to  30  %
A ▾  start motor  ↻ ▾
```

#125

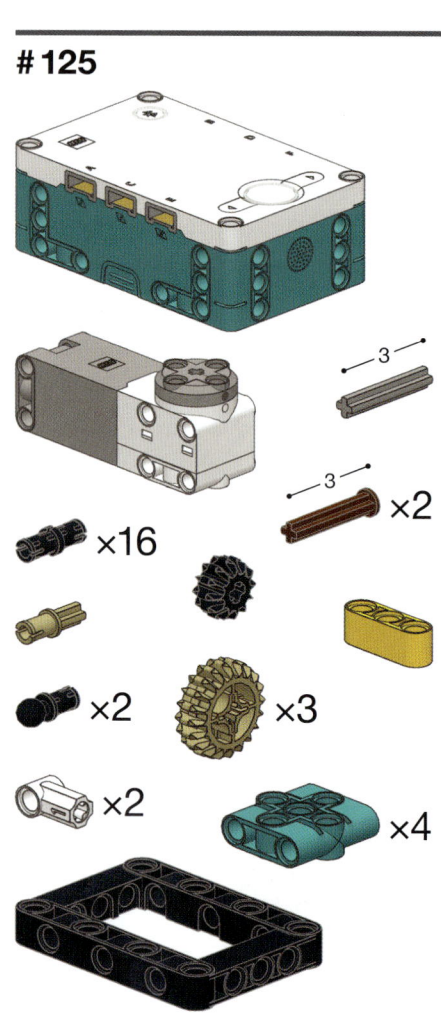

×16

3

×2

×2

×3

×2

×4

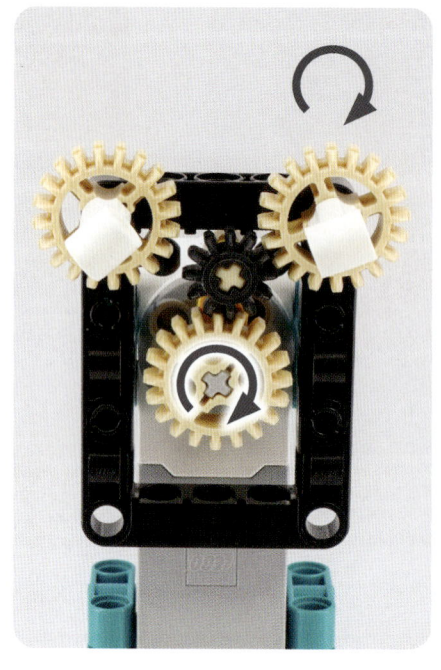

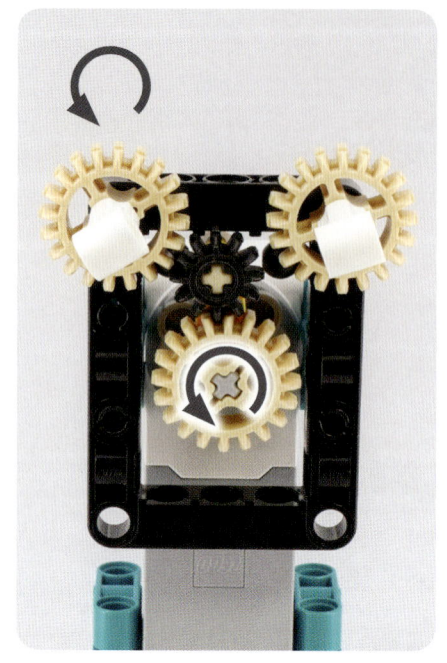

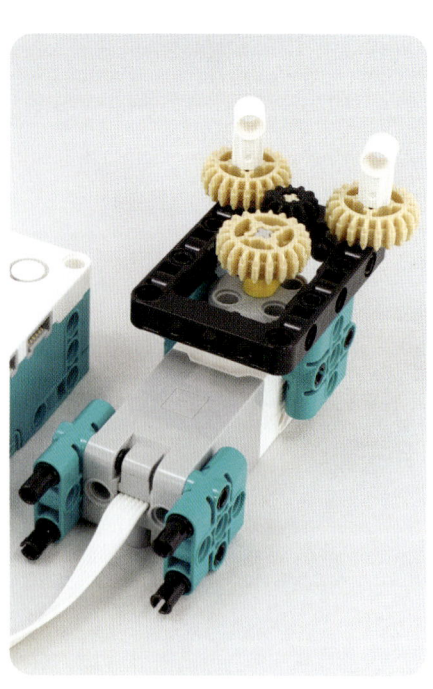

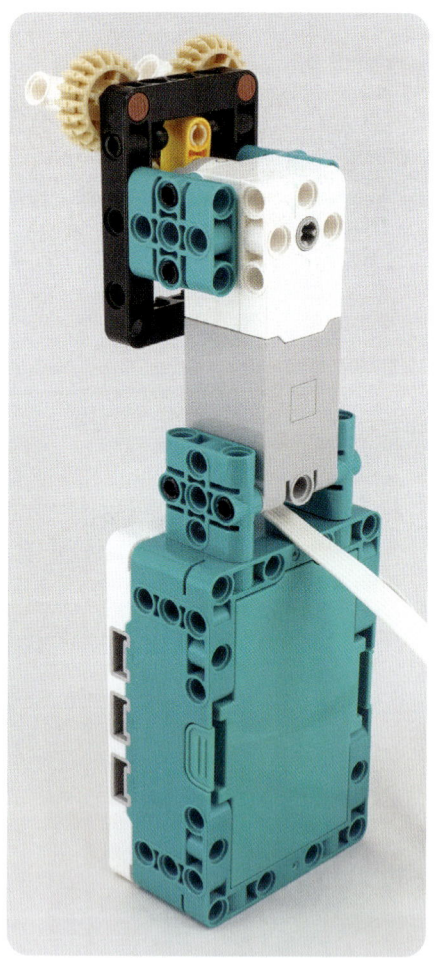

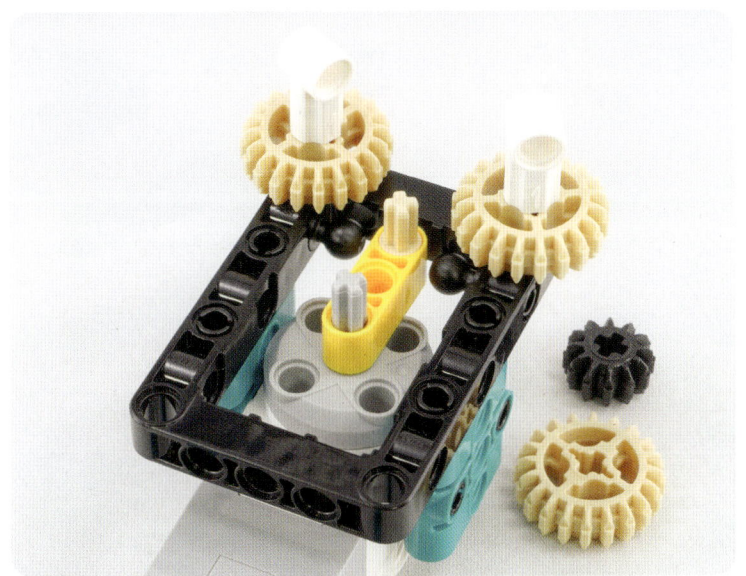

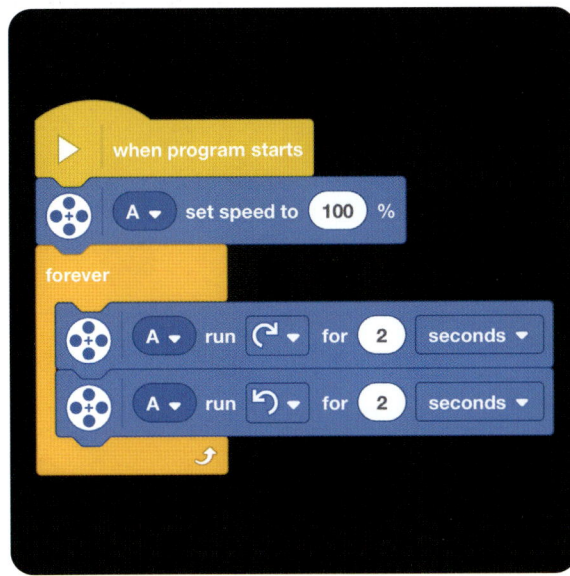

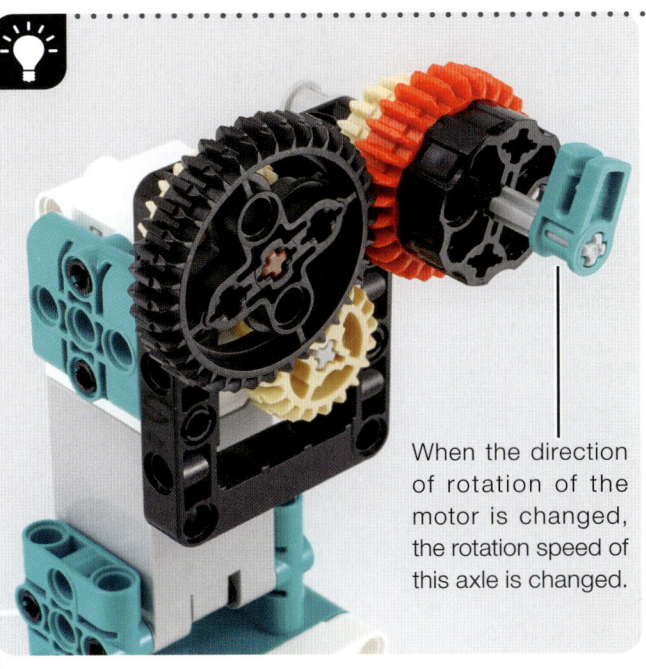

When the direction of rotation of the motor is changed, the rotation speed of this axle is changed.

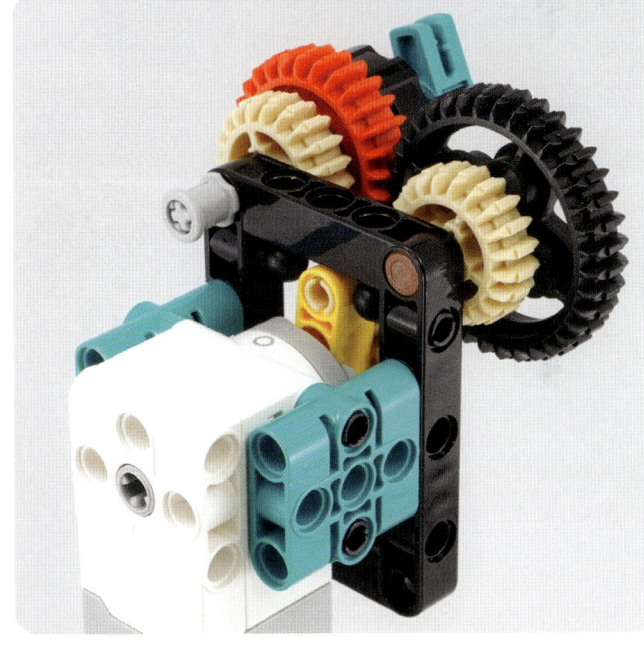

#126

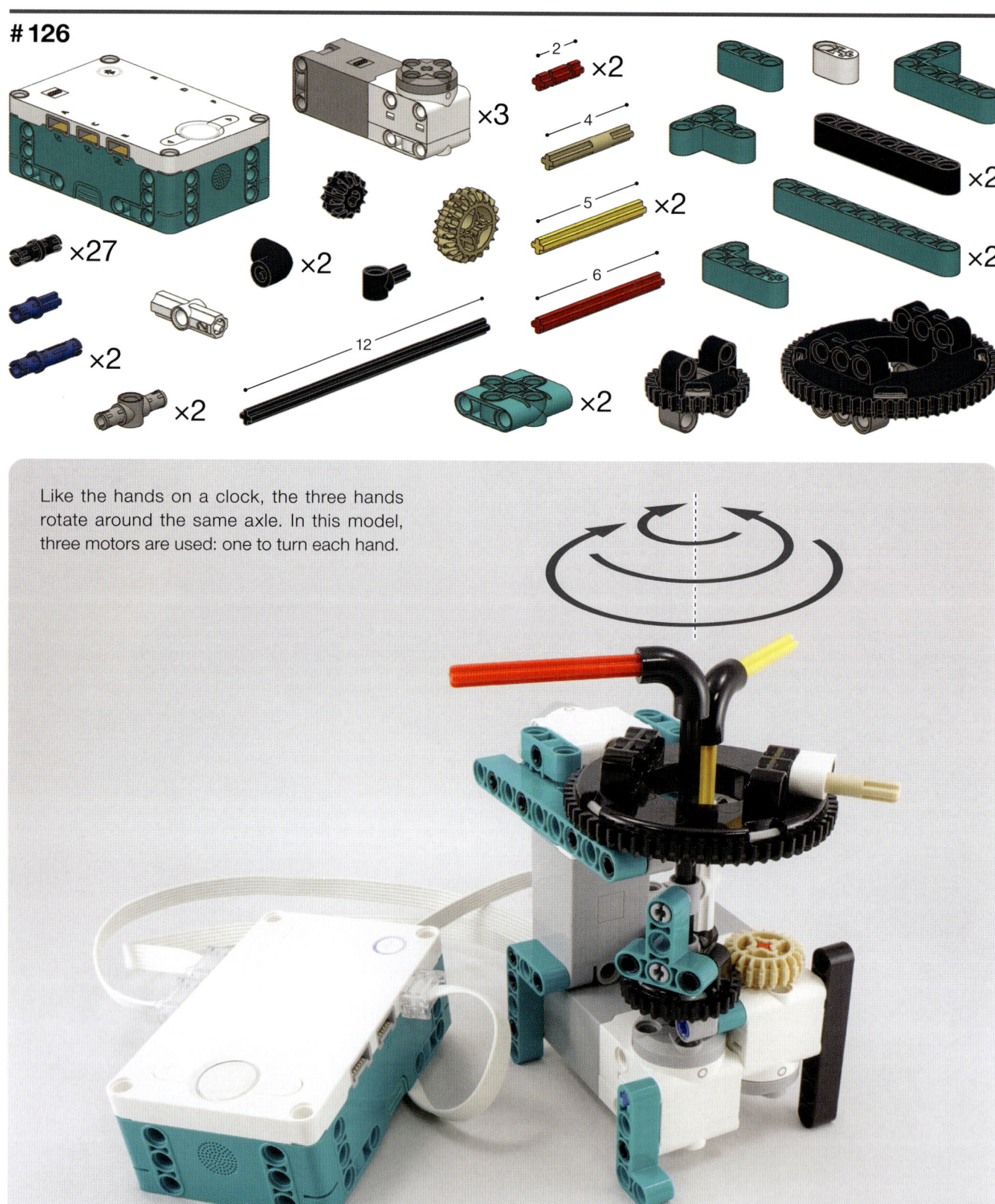

×27
×2
×2
×3
×2
×2
×2
×2
×4
×5 ×2
×6
12
×2
×2
×2
×2

Like the hands on a clock, the three hands rotate around the same axle. In this model, three motors are used: one to turn each hand.

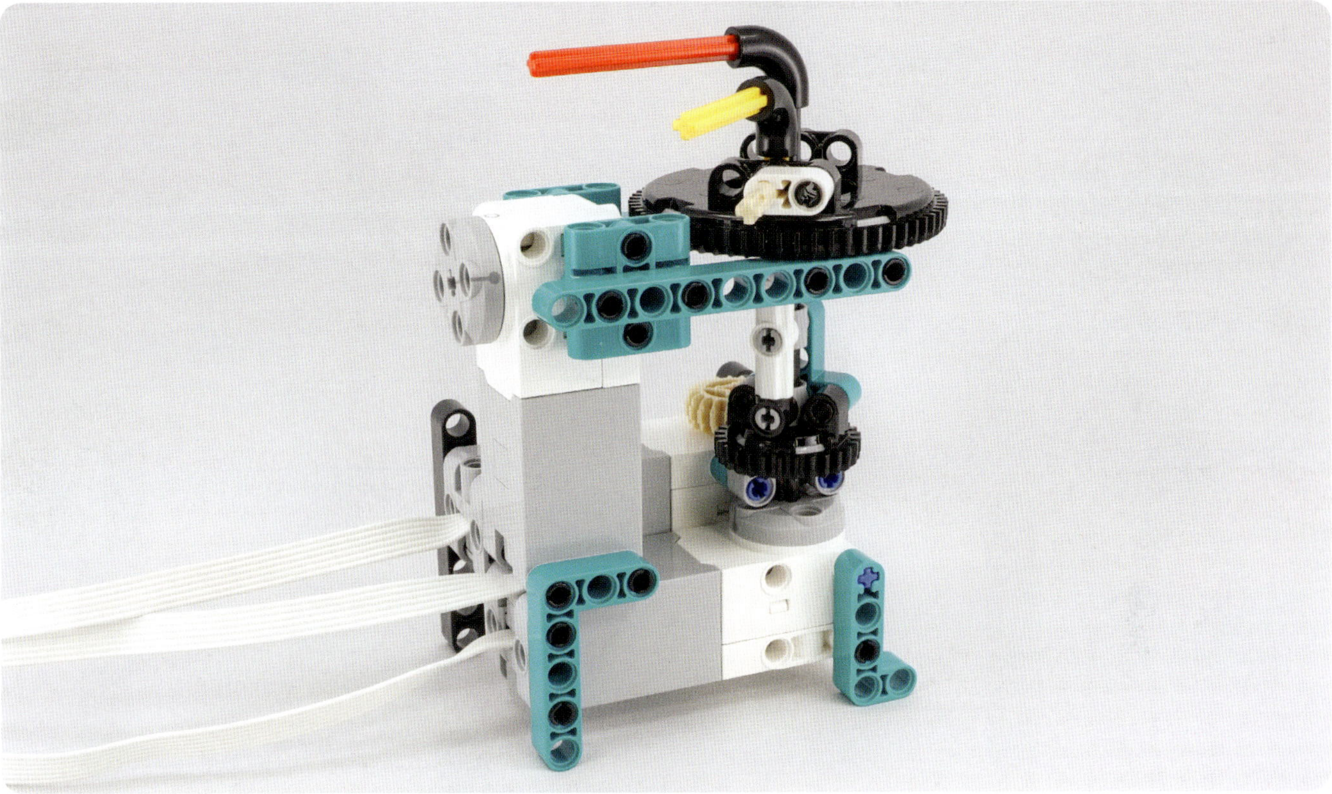

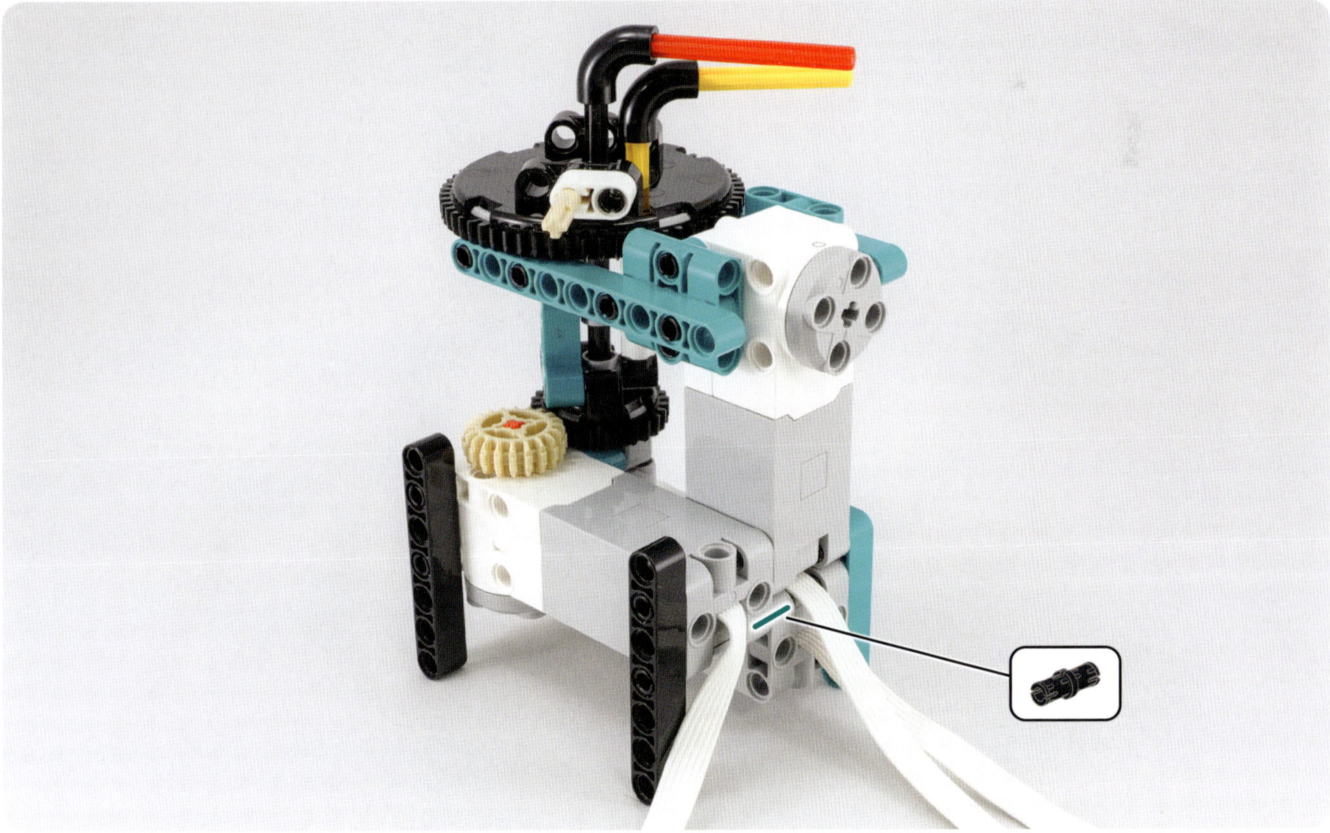

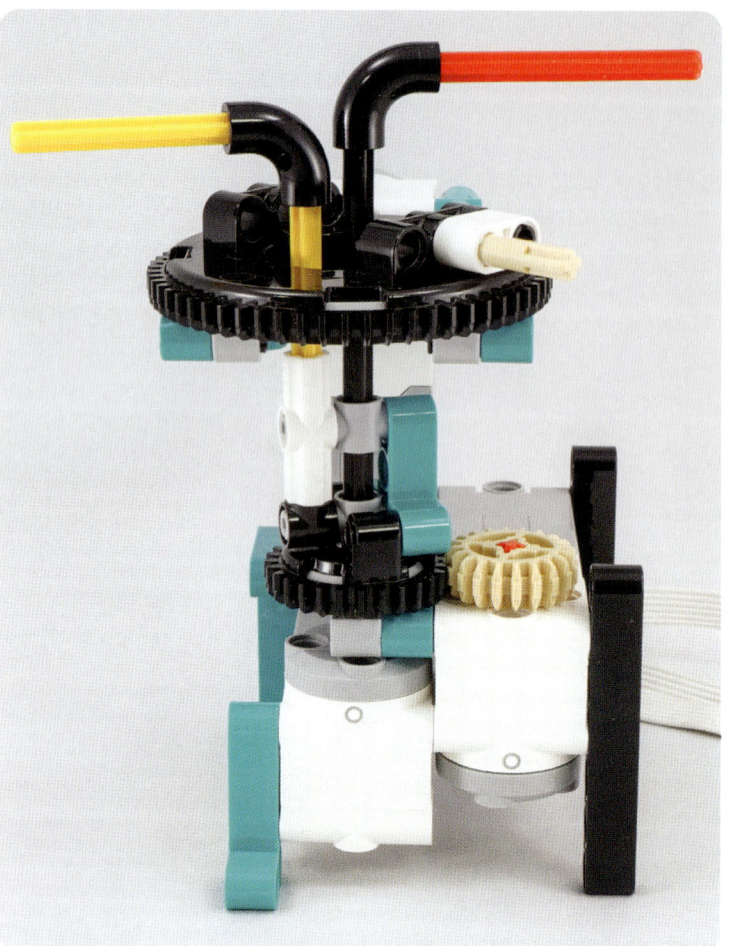

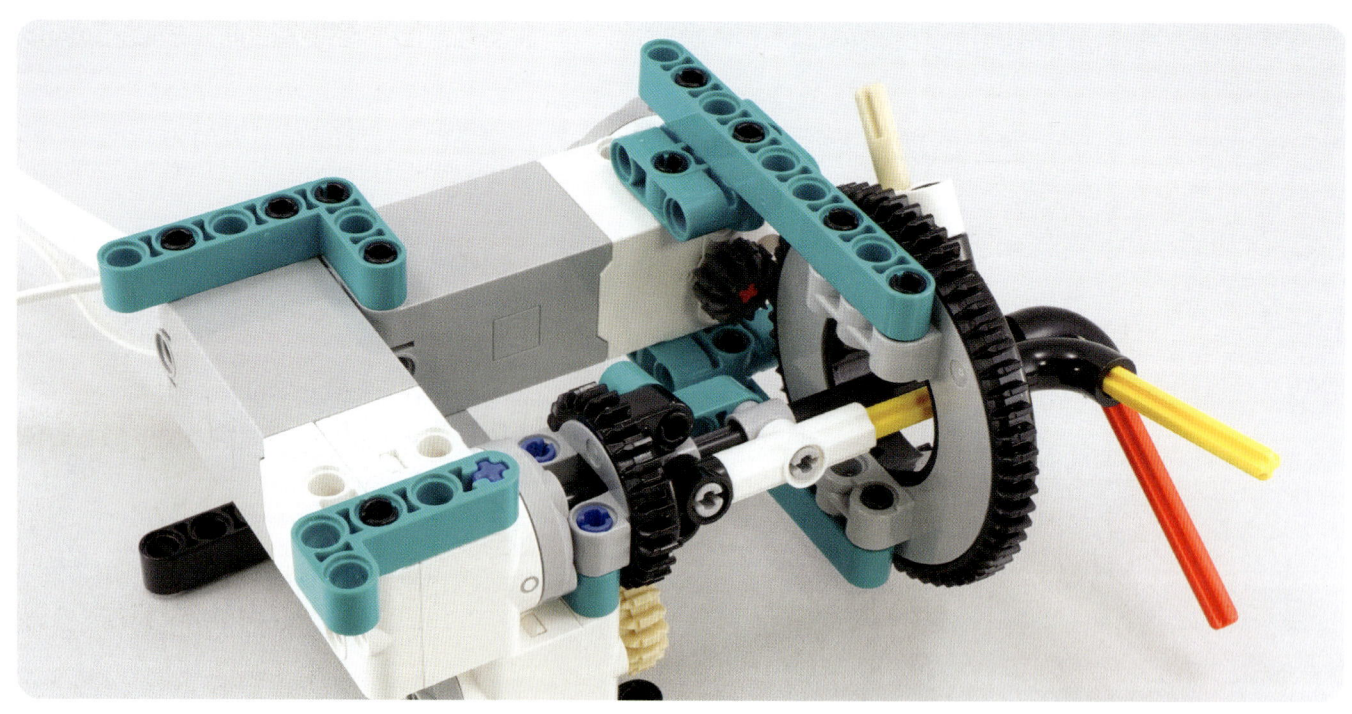

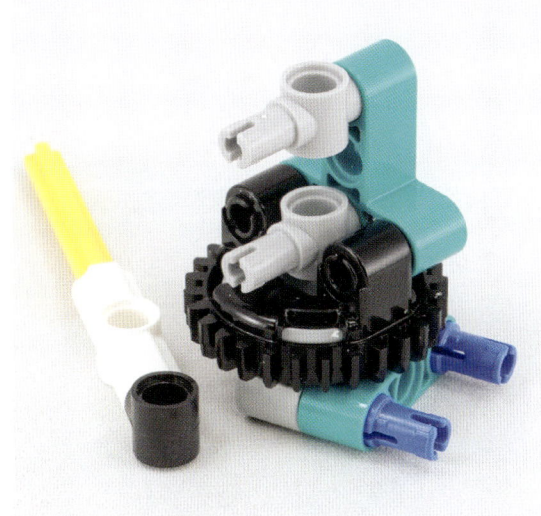

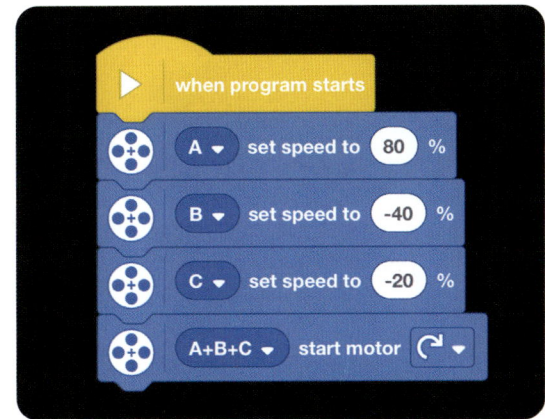

The three hands rotate at approximately the same speed. For details of how to create variables, see page 57.

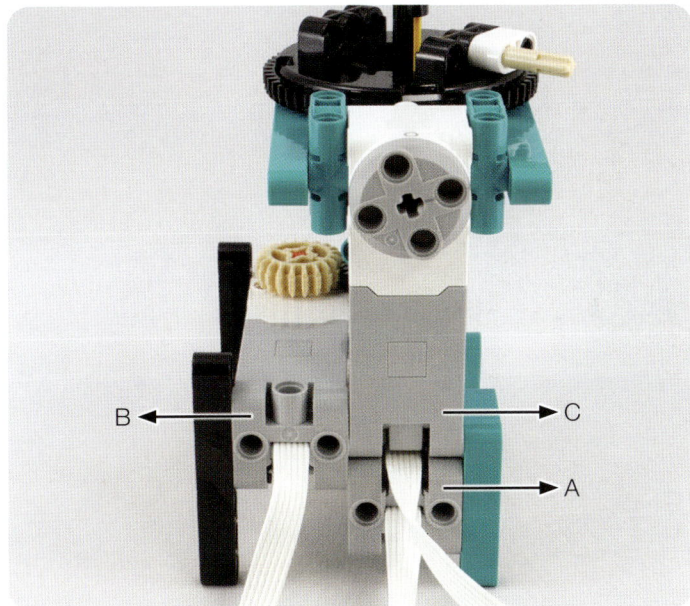

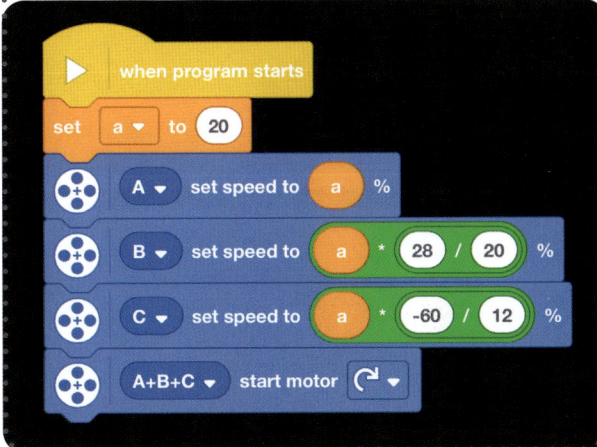

#127

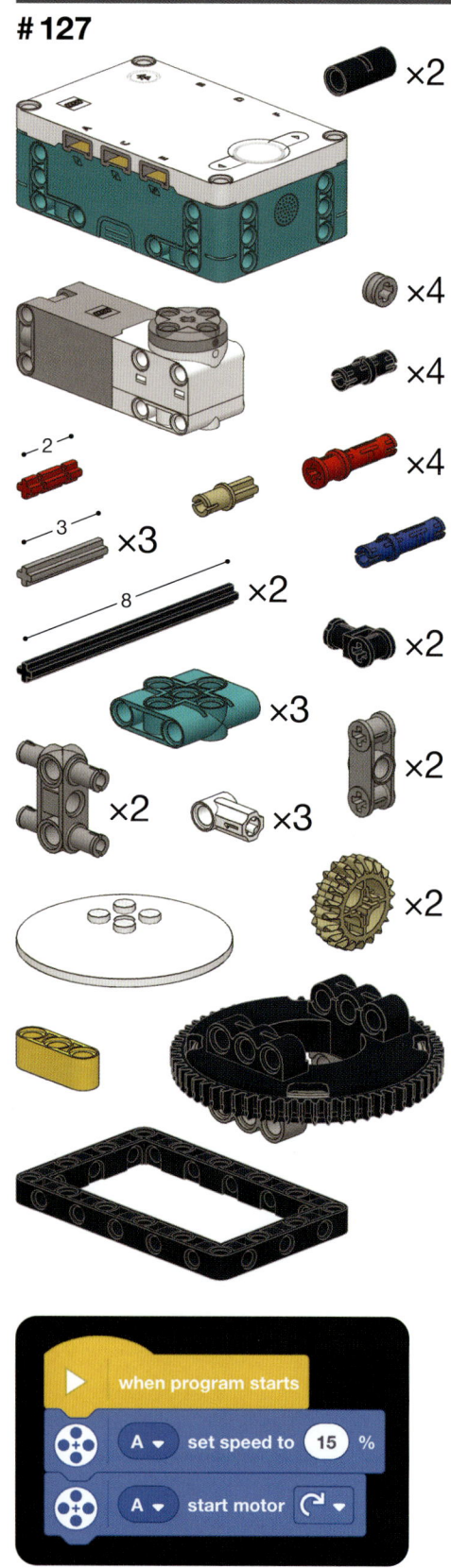

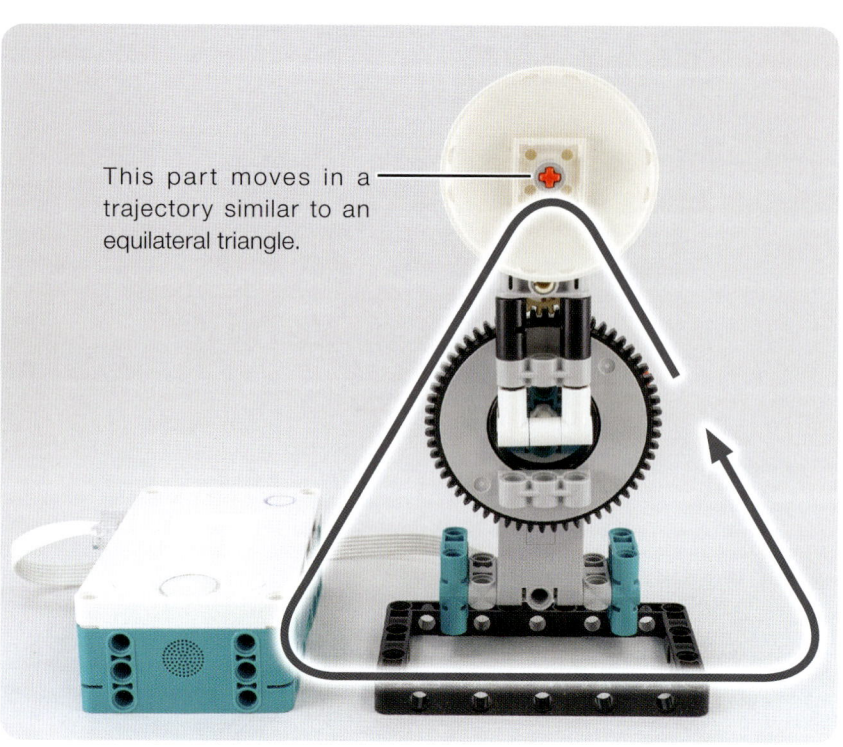

This part moves in a trajectory similar to an equilateral triangle.

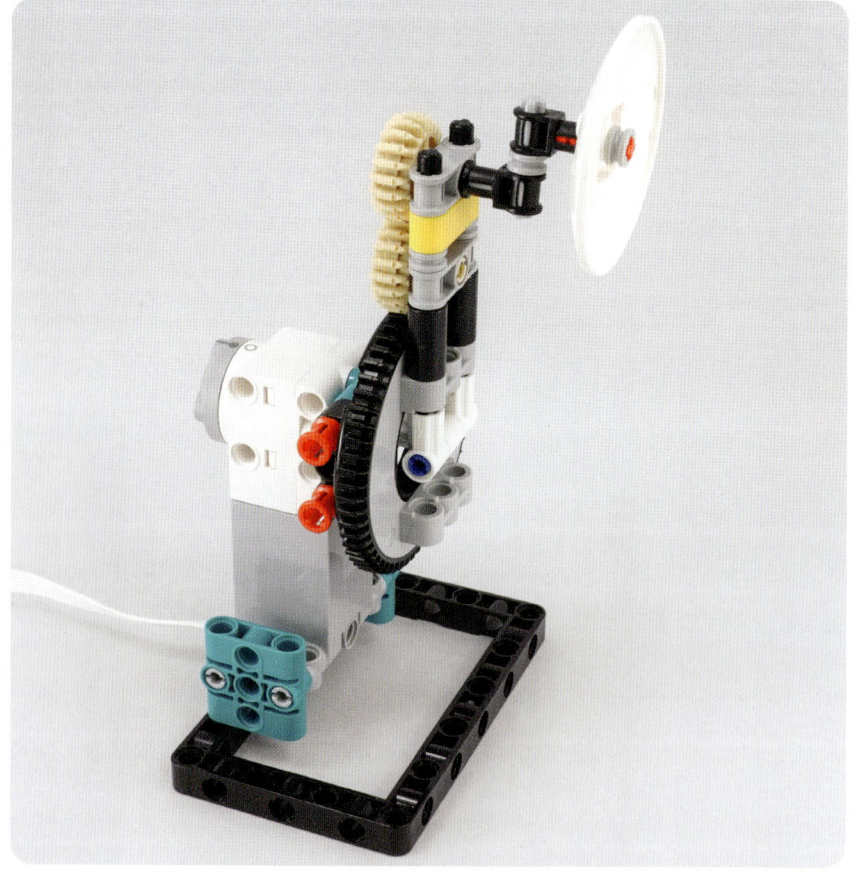

```
when program starts
A ▾ set speed to 15 %
A ▾ start motor ↻ ▾
```

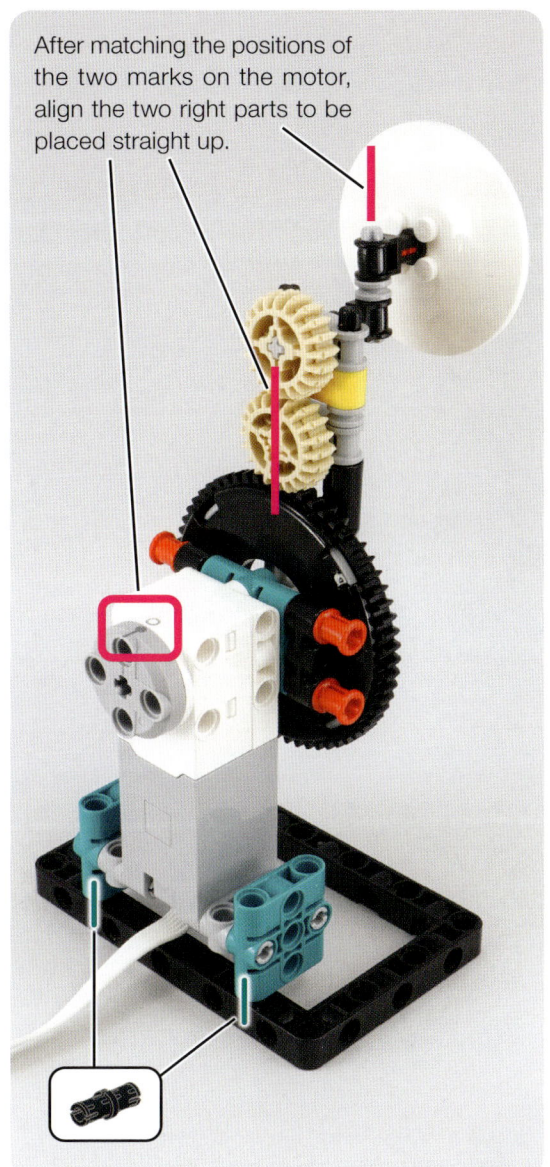

After matching the positions of the two marks on the motor, align the two right parts to be placed straight up.

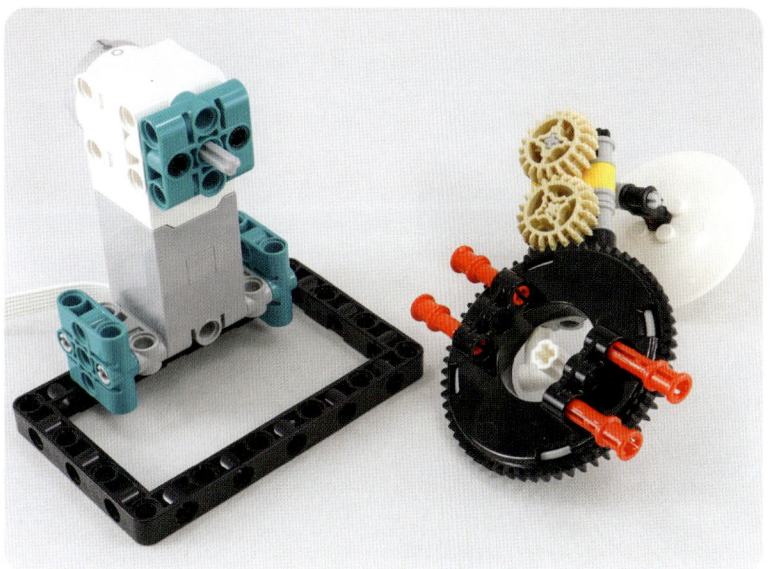

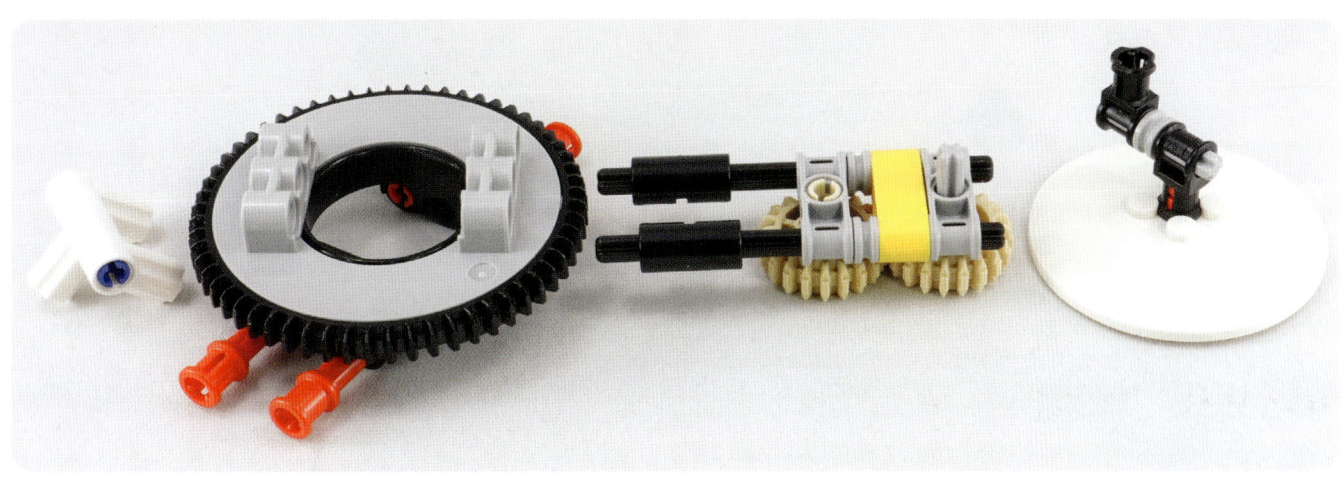

#128

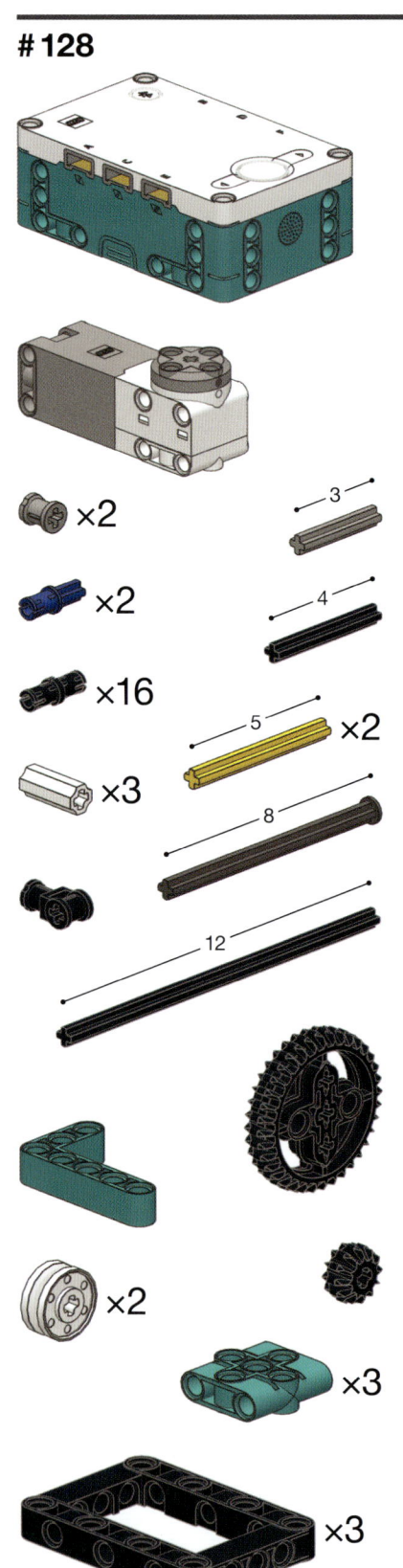

×2

— 3 —

×2

— 4 —

×16

— 5 — ×2

×3

— 8 —

— 12 —

×2

×3

×3

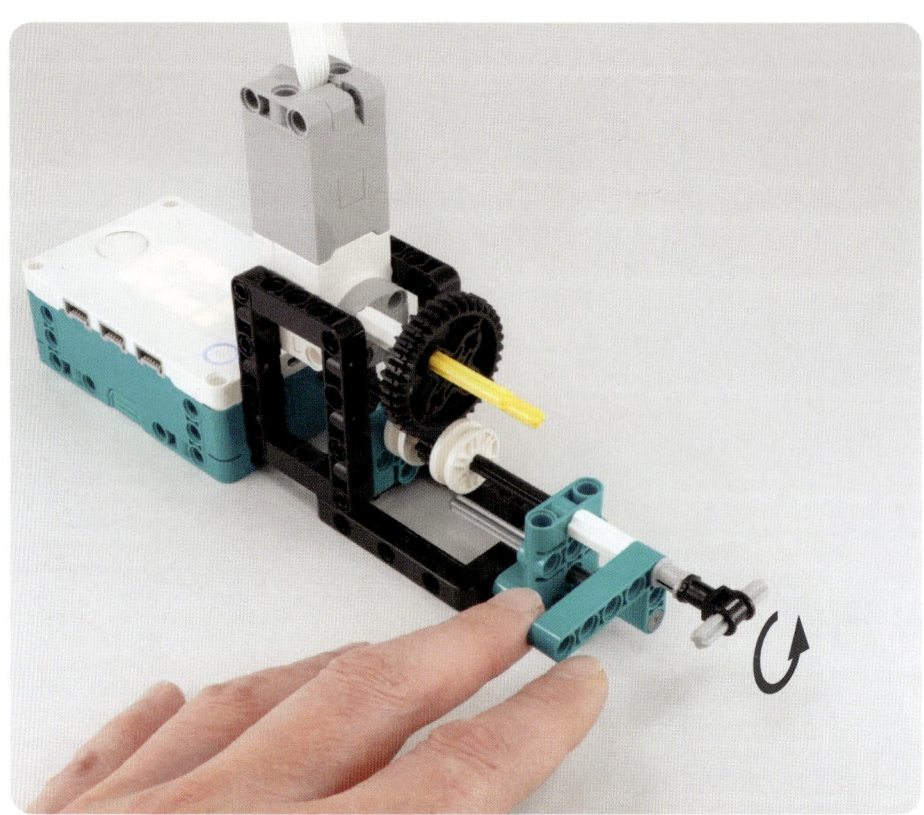

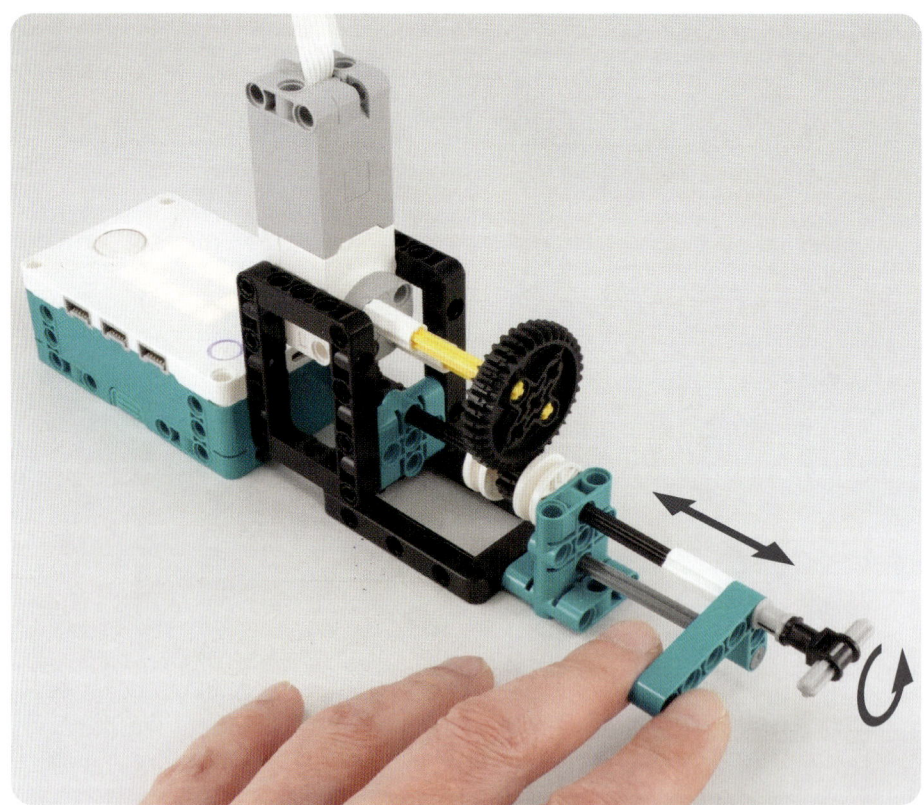

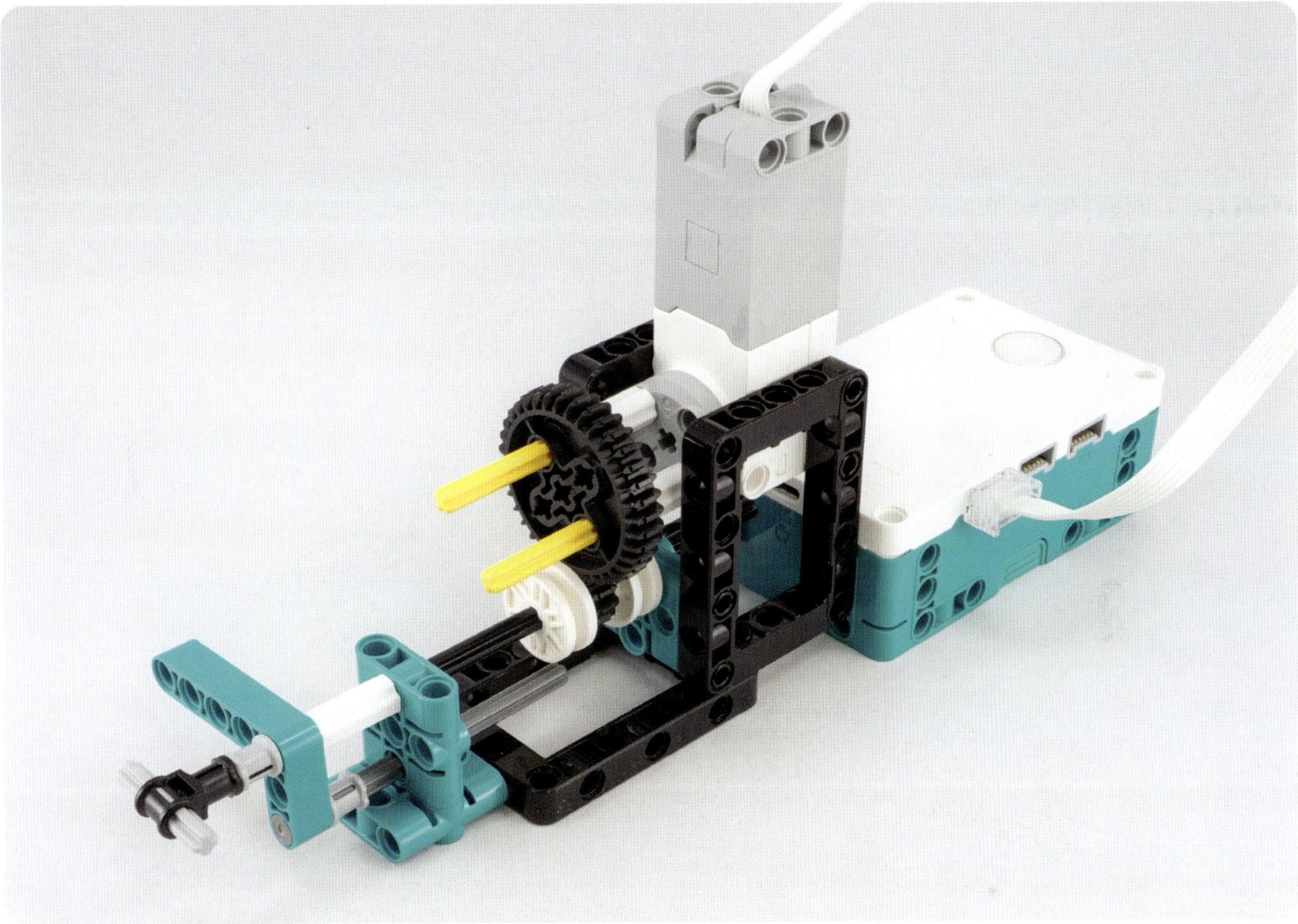

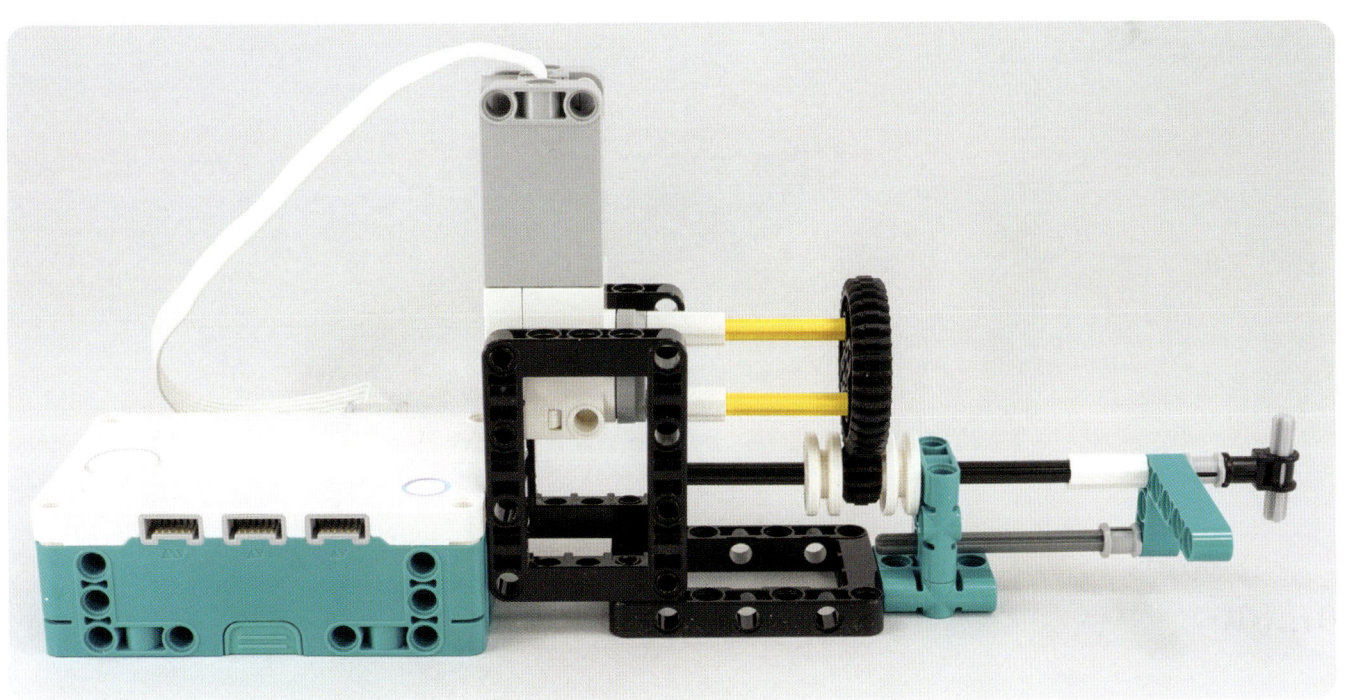

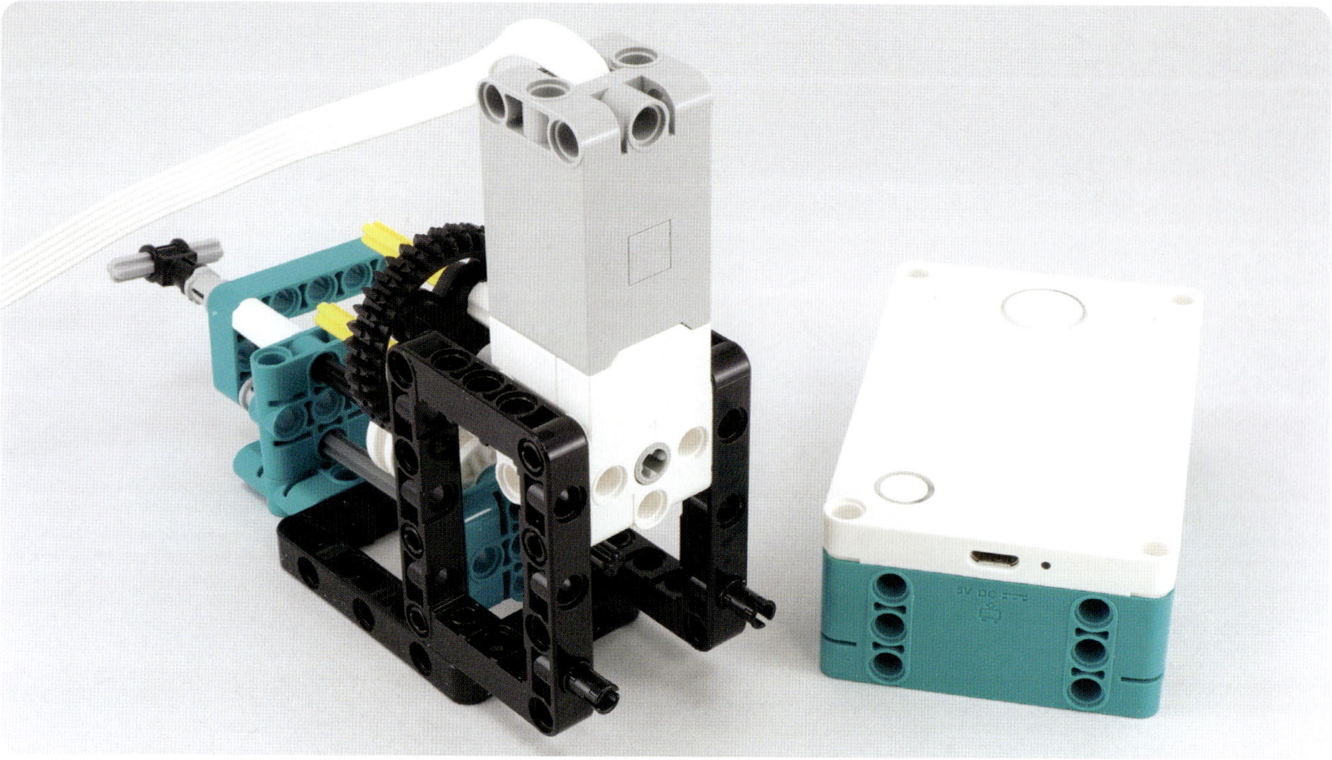

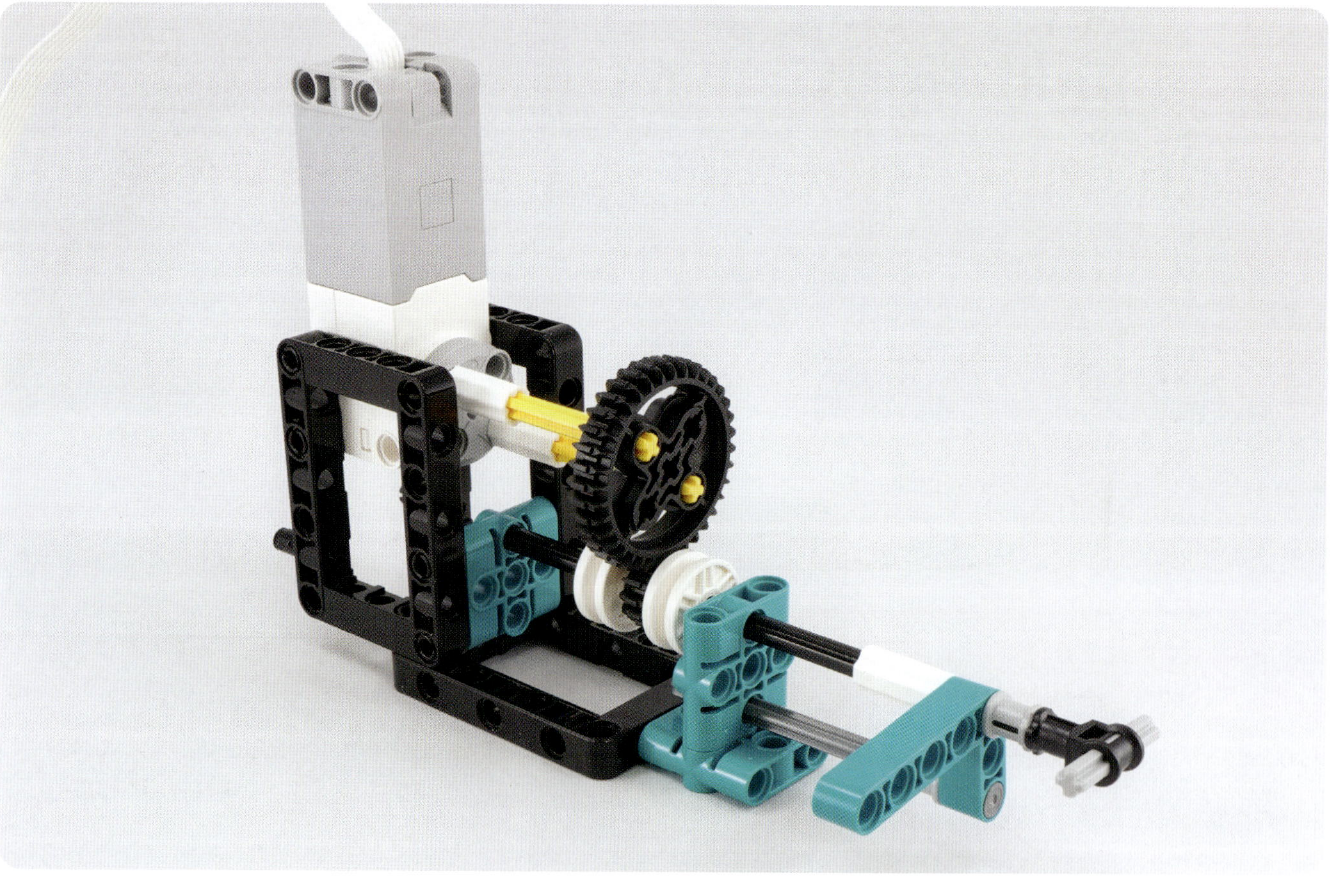

●●●●●

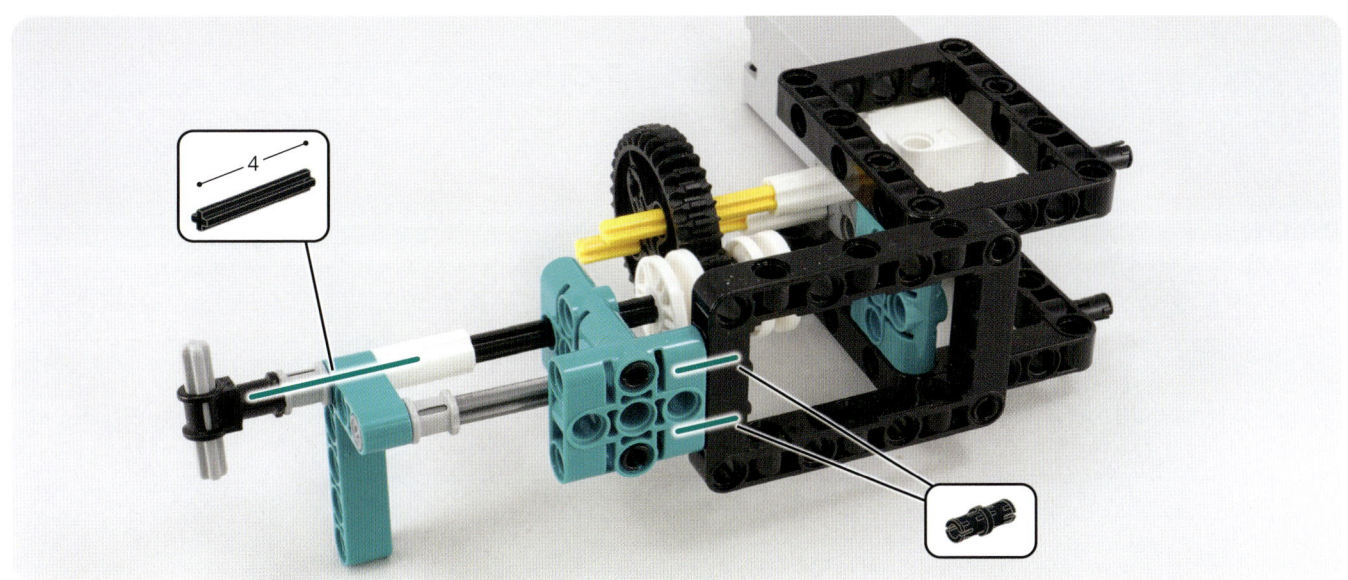

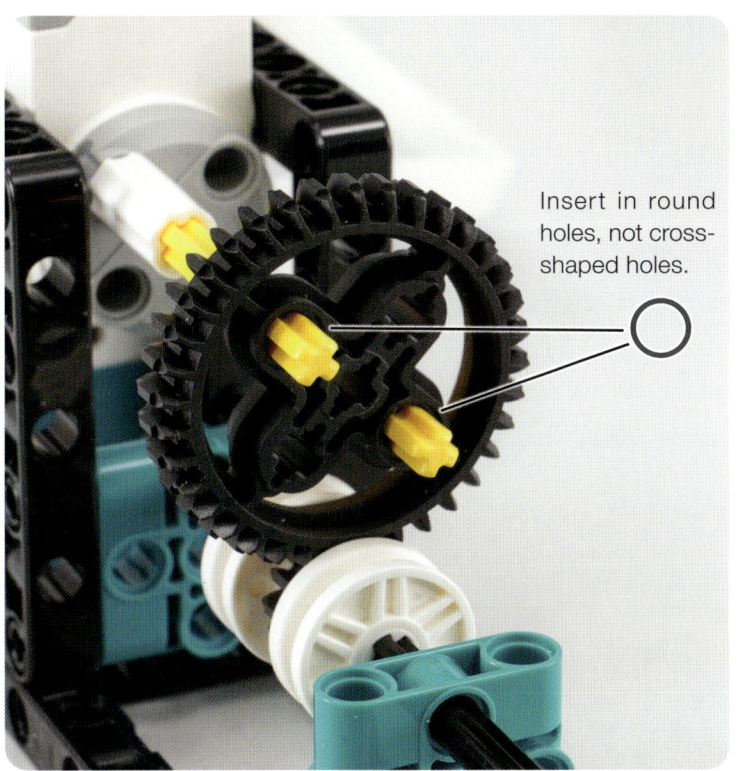

Insert in round holes, not cross-shaped holes.

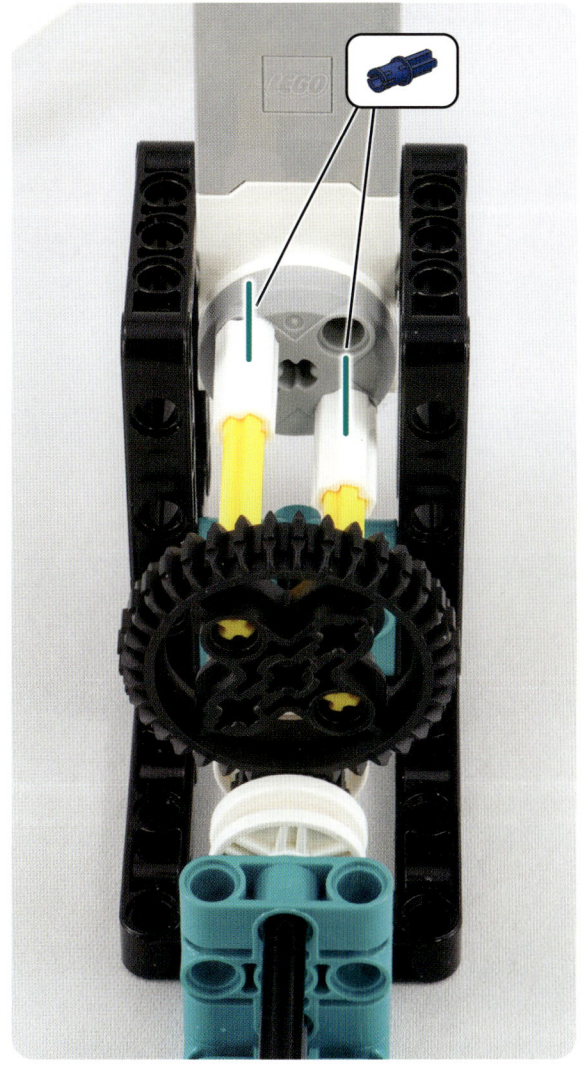

The LEGO® MINDSTORMS® Robot Inventor Idea Book: 128 Simple Machines and Clever Contraptions is set in Helvetica Neue, Avant Garde, and Trade Gothic. The book was printed and bound by Versa Printing in East Peoria, Illinois. The paper is 70# White Coated (Matte), which is certified by the Forest Stewardship Council (FSC).

The book uses a perfect binding with polyurethane reactive (PUR) glue, the most durable bookbinding glue available. Its superior flexibility prevents the spine from cracking when the book is opened wide or pressed down flat.

Resources

Visit *https://nostarch.com/lego-mindstorms-robot-inventor-idea-book/* for errata and more information.

More no-nonsense books from **no starch press**

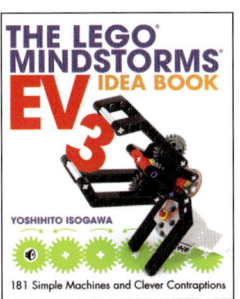

The LEGO MINDSTORMS EV3 Idea Book
181 Simple Machines and Clever Contraptions
by YOSHIHITO ISOGAWA
232 PP., $24.95
ISBN 978-1-59327-600-3

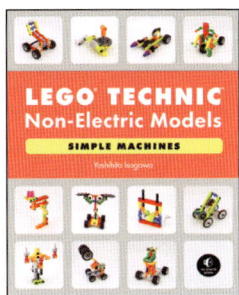

Lego Technic Non-Electric Models
Simple Machines
by YOSHIHITO ISOGAWA
192 PP., $24.99
ISBN 978-1-7185-0120-1

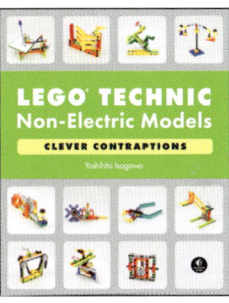

Lego Technic Non-Electric Models
Clever Contraptions
by YOSHIHITO ISOGAWA
192 PP., $24.99
ISBN 978-1-7185-0170-6

The LEGO MINDSTORMS Robot Inventor Activity Book
A Beginner's Guide to Building and Programming LEGO Robots
by DANIELE BENEDETTELLI
248 PP., $34.99
ISBN 978-1-7185-0181-2

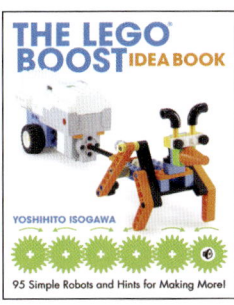

The LEGO BOOST Idea Book
by YOSHIHITO ISOGAWA
264 PP., $24.95
ISBN 978-1-59327-984-4

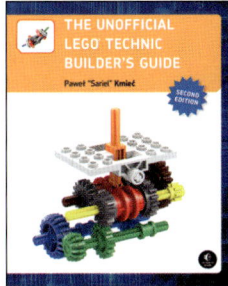

The Unofficial LEGO Technic Builder's Guide, 2nd Edition
by PAWEŁ "SARIEL" KMIEĆ
424 PP., $34.95
ISBN 978-1-59327-760-4

PHONE:
800.420.7240 OR
415.863.9900

EMAIL:
sales@nostarch.com
WEB:
www.nostarch.com